# *Practical Security in Commerce and Industry*

*Eric Oliver and John Wilson*

Fifth edition by John Wilson and Ted Slater

Approved by the International Council of the International Professional Security Association

Gower

First published 1968 by Gower Press Limited
Second edition 1972
Third edition 1978
Fourth edition 1983

Fifth edition published by
Gower Publishing Company Limited,
Gower House,
Croft Road
Aldershot
Hants GU11 3HR,
England

Gower Publishing Company,
Old Post Road,
Brookfield,
Vermont 05036,
USA

**British Library Cataloguing in Publication Data**

Oliver, Eric
Practical security in commerce and industry.
—5th ed.
1. Industry—Security measures 2. Retail trade—Security measures
I. Title II. Wilson, John III. Slater, Ted
658.4′7 HV8290

**Library of Congress Cataloging-in-Publication Data**

Oliver, Eric.
Practical security in commerce and industry / Eric Oliver and John Wilson—5th ed. / by John Wilson and Ted Slater.
p. cm.
1. Industry—Security measures—Great Britain. 2. Retail trade—Great Britain—Security measures. 3. Security systems—Great Britain. I. Wilson, John. II. Slater, Ted. III. Title.
HV8290.042 1988
658.4′7—dc 19 88-1124
CIP

ISBN 0–566–02712–7

Printed in Great Britain at the
University Press, Cambridge

# *Contents*

## APPENDICES

# *Preface*

Two decades have passed since the first edition of this book appeared. During that period the commercial and industrial environment has changed drastically, a different social climate has developed and there is a growing acceptance that wrongdoers may be found in any stratum of society. A burgeoning crime rate has coincided with the deployment of police resources on an unprecedented scale to control mass picketing and crowd disturbances of various kinds – even the expectation of such disturbances can occupy large numbers of police. The training needed to deal with events of this sort further depletes the manpower available for traditional duties, while overtime working is severely limited by financial constraints.

Violence has become a common feature of robbery and other serious crime, and guns are more frequently displayed and used. Knives and boots are used to replace fists in what was once ordinary disorderly or drunken behaviour. Drug abuse has reached such proportions that the CBI has thought fit to warn of its spread among executives; trafficking takes place on an international scale despite universal condemnation and regular seizures of consignments. Fraud has invaded the hallowed precincts of the City and even of Lloyds itself – and at the highest level of management, where it might have been thought that no temptation existed. It is difficult to avoid the conclusion that such crimes have always existed but have been concealed up to now by procedural laxity, fear of publicity or personal relationships. It gives me no satisfaction to have predicted this development in the preface to the previous edition.

Northern Ireland and the Middle East apart, there has been something of a lull in terrorist activities and kidnapping. However, a flare-up could occur at any time and a mere handful of fanatics could be responsible for a large number of incidents. Even the highly principled groups connected with animal welfare contain a fanatical fringe who from time to time publicize their cause by break-ins and violence. Letter bombs sent to government officials by unknown sources became fashionable again for a brief period and presumably led to some tightening of security in the departments concerned.

Thefts from shops have not abated; professional travelling 'teams' still operate but the increasing use of closed-circuit television and video recording is proving invaluable both as deterrent and detector. The system of 'tagging' high-priced items is effective but costly. One disturbing feature is the frequency with which violence is now being offered to store detectives and shop assistants by thieves trying to avoid arrest; in some city centres, knives are increasingly being flourished to intimidate. The larger stores will have to consider providing better alarm systems if they expect their staff to carry out their job with determination. In many towns, shopping precincts have been built which form virtually separate enclaves demanding their own form of security supervision. The problems are controllable provided the managements of these centres acknowledge them and act accordingly.

Computers have continued to be a source of apprehension. Their relatively complex method of operation, coupled with the voluminous paper output, complicates both the establishment and the detection of offences. The media have helped to create an exaggerated fear by sensationalizing the efforts of 'hackers' to gain access to confidential files. The activities of industrial spies are regularly spotlighted but, naturally, are hard to quantify. A successful agent, by definition, does not come to notice, though the results of his or her efforts may sometimes be suspected. In the highly-competitive business world of today it would be foolish not to take precautions.

One consequence of the retrenchment in police activity has been a virtual explosion in that of the contract security firms and alarm installers who, with Home Office approval, continue their unsatisfactory self-regulation. Shopping precincts are nowadays policed almost entirely by contract forces. It is likely that in the near future police will cease to attend alarm calls from business premises and will expect occupiers to arrange for contract security. Police manpower is strictly limited; in deciding which cases to pursue most vigorously, their priorities may well be decided by the gravity or complexity of the case. If a file is presented to them which is complete or virtually so, their co-operation is much more likely.

The standard of service required of both contract and in-company security forces may rise. If this happens training – or perhaps better training – is going to be needed. If investigations into internal fraud or persistent theft are going to be called for, senior personnel should have specialist instruction. In fact, many of those employed by the numerous small security companies which have sprung up during the boom are without any kind of training.

Why has the situation deteriorated so badly over the past few years? It is difficult to avoid the feeling that officialdom has been concerned more with the offender than with the victim. It is true that legislation has been passed to prevent convicted drug traders from retaining the profits of their crime, but that is an isolated measure and may provide another lucrative outlet for lawyers' skills. The Police and Criminal Evidence Act gave the police additional powers of 'stop and search' – and then hedged them about with so many conditions, requirements and disciplinary threats as to discourage their use. The one source of satisfaction is to be found in the field of industrial relations, where a pattern of decided cases has been established by Appeal Court judges who have made it clear that they prefer commonsense to legal technicalities.

A Crown Prosecutions Service has been established which has yet to prove itself. Already there have been suggestions of US style 'plea bargaining', of inadequate staffing and delays, and inexperienced practitioners. It is too early to pass judgement but it is doubtful whether complainants will receive satisfaction.

It is against this background that the fifth edition of *Practical Security in Commerce and Industry* has been prepared. In carrying out the work of updating and expanding the text I have been fortunate enough to secure the co-operation of Ted Slater, a prolific and highly-qualified writer on security matters.

The chapters on arrest and search, and questioning, have been extensively amended in the light of the Police and Criminal Evidence Act. The old Judges Rules have been replaced by virtually identical Codes of Practice which have made them mandatory. Because of the possibility of less involvement by the police in the future, the need to interview and to question may become more frequent, so the legal aspects have been dealt with at length. Similarly, the process of investigating, from outset to reporting, will become a more necessary skill and a complete new chapter is devoted to it.

The increasing number of shopping precincts has made it useful to formulate rules for security in them. Hotel thefts, always rife, have become more professional while all the old problems have remained. Hospitals once relatively quiet and respected, now suffer from most varieties of security problem, including a growing incidence of violent attacks on staff. These three environments call for specialized types of security and are considered in detail in a new chapter.

With progressive privatization of armaments factories and the like, it is not clear what the security implications are, but the Official Secrets Act assumes greater importance for the practitioner who might find himself responsible for such premises, so the legal aspects are outlined in the text. More detail is given about drug abuse. A new Forgery and Counterfeiting Act has consolidated and clarified the law on matters which extend beyond commonplace forgery to credit cards, discs, coins, bank notes, etc. The Public Order Act of 1986 has increased powers to deal with trespass and the text now deals with the whole subject on a wider basis. There is a new section on sexual harassment which, while not strictly a security matter, is worth noting as a potential problem.

The entire text has been updated to take account of developments in the years since the previous edition was prepared. Pertinent case studies have been added and no apologies are offered for picking the brains of colleagues engaged in different projects and industries. Our common aim is to create an environment in which the law-abiding can live and work with minimum fear of violence, theft, fraud and the other illegal and unethical activities that debase the quality of life.

For those who aspire to perfect security, here are two final quotations: 'That a precaution is physically possible does not mean it is reasonably practicable' – an Appeal Judge in a case involving inadequate security measures; and – an ominous one, this – 'The time may come when the police have to abandon some of their traditional duties' – the Chairman of the Association of Chief Police Officers at their conference in June 1987.

I wish it were possible to end on a note of optimism. There is little to indicate any impending action which will have a material effect on the crime rate and in all honesty, therefore, I cannot.

John Wilson

*PART ONE*

# *THE SECURITY FUNCTION AND ITS PERFORMANCE*

# 1

# *Appraising the risk*

'*Security*' – 'that which serves as a guard or guarantee'

The last decade has brought fundamental changes in our society. Long-established industries, faced with falling demand and inability to compete in both home and overseas markets, have been decimated. Many companies have ceased to exist, others have been obliged to drastically retrench and modernize their operations to achieve greater efficiency. Inevitably there has been a revaluation of the cost effectiveness of all functions and procedures from which security has rightly not been excluded. Manpower is a major expense and, where security is equated solely with its original concept of protection of property, its use can be minimized by electronic aids; where it remains necessary for wider reasons, the value for money principle demands that the individuals are adequately trained and motivated to meet the demands made upon them.

To the conventional threats of theft, fire and damage have been added increased fraud and malpractice amongst employees, illegal obtaining of information important to profitability or industrial harmony, and the ever present, albeit remote in many cases, threat of some form of terrorism or extortion. To limit the possibilities of these threats, a clearly defined set of rules and policies is needed, to be strictly adhered to and impartially implemented whether it be directed towards employees *of any grade or connections*, customers, or outsiders. If members of management step out of line, they become vulnerable and unable to give the firm's interests priority in the face of danger of disclosure. Their status should not give them any protection, rather the reverse, or damaging precedents may be created.

In commerce and industry the original conception of security was that of protection of property; but the circumstances of recent years, the increasing professionalism and the demands of cost effectiveness alike have created the need for the safeguarding of assets, personnel, and even the profitability of an organization against theft, fraud, fire, criminal damage, and terrorist acts. To achieve these objectives, formulation and implementation of strict rules and policies by employers are required.

At the senior level there should be responsibility for the evaluation of systems and procedures for making recommendations to minimize temptation and opportunity to cause loss illegally – or unethically by the wrongful use of restricted information; this should be an ongoing process of appraisal of risks coupled with that of potential remedial or preventive action. It is also the responsibility of staff at this level to investigate losses and discrepancies, to identify culprits or causation and prevent a recurrence.

Routine implementation of security duties are control of access, patrol of premises, prevention, detection and fighting of fires, administration of first aid, and enforcement of rules or regulations. Obligations such as weighbridge operation, search of employees and vehicles, traffic control, and supervision of electronic alarm and surveillance equipment are common additions to the basic tasks.

The objective is to furnish a service to the employers which is a positive contribution to the efficient, harmonious, and profitable running of the business. There are many facets to this – as will be seen.

## *THE PROBLEM FOR MANAGEMENT*

Industrial and commercial management of the late 1980s has some unpalatable facts to consider in connection with security matters without being able to predict what their effects will be on the profitability of the company.

The increase in crime has continued unabated and, with a further overall 7 per cent increase in 1986, there are no grounds for any optimism about future trends. There is substantial evidence, in recent findings, that the very highest ranks of management are not all immune to temptation if the inducement is great enough. Sheer greed is a motivating factor too often overlooked. White collar crime as predicted in the United States has become more prevalent with industrial espionage assuming enhanced importance in a highly-competitive environment where research is a major expenditure and knowledge of a competitor's quotations of a value which hardly need be stressed. The real incidence of computer-linked thefts and frauds remains a matter for conjecture – but then that applies to the official figures too; it is an accepted truism that only a fraction of actual crime is reported. Moreover, there is a possibility that many employee offences, such as embezzlement and false accounting, are never discovered, and, in the case of bribery and corruption – a probability.

The preventive and detective roles of the police have been limited by the impact of new commitments, revised priorities, and an overpowering workload which is leading to suggestions that reports of thefts below a certain value should be noted but no action taken. Whilst prepared to tender guidance through Crime Prevention Officers, the view officially expressed by a spokesman for the Association of Chief Police Officers is that 'industry must put its own house in order' and 'responsibility for the safeguarding of property is that of the owner'. It is quite clear, therefore, that the amount of assistance that can be expected has finite limits and management will have to consider what action it can take itself to avoid being the target of internal and external predators.

Courts and prisons alike suffer from overloading, trials can be delayed for months and even years in the more complex frauds, with eventual punishment often being totally inconsistent with the loss to the victim and rarely a deterrent to others who might be tempted to emulate the offender.

The full effect of the Police and Criminal Evidence Act (see p. 135) and the advent of the Crown Prosecution Service (see p. 227) is still to be felt but accentuated delays and difficulty in finding out what is happening to offenders are almost inevitable consequences.

Publicity in the media, and by television presentations or commentaries, has brought to the public eye successful criminal ploys to the extent that carbon copy re-enactments of fictional incidents have subsequently taken place. Inevitably the emphasis laid upon computer-linked frauds and their potential will produce an unwanted reaction in a field where it is often difficult to establish any misbehaviour has taken place, never mind detect it. There is evidence that this process has already begun.

Terrorism continues its attempts to change society by violent action against unpredictable targets and the maliciously minded utilize its threat to disrupt business activity by hoax threats. Militancy amongst a workforce may be fostered by outside malcontents for purely doctrinal purposes seizing on minor disputes which would otherwise be amicably settled.

During periods of recession, competition for orders, and search for means of survival in many cases, creates a breeding ground for business-information gatherers whose ways of acquiring data for clients may offend against ethical and legal standards. In other words, the danger of industrial spying is apt to be on the increase in the present climate – and there is some indication that the level of activity is greater than originally suspected.

Criminal techniques are constantly improving, rapid movement of stolen materials into neighbouring countries and beyond is commonplace, and corruption of employees has extended on many occasions into the ranks of security personnel – those actually employed to protect. Prosecutions in more than one large industry have shown that the canker of corrupt practices affects even the highest echelons of management.

Not a pretty picture and, all in all, the immediate prospect is one of increasing problems with limited cash available to devote to combating them. They will not be solved by being ignored: at the least, serious thought should be given to organizational vulnerability and the protective options that are available.

It is not proposed to mention fire in any detail until later (Chapter 30) though it is more often than not regarded as part of a security remit. Nevertheless, it should be borne in mind that infinitely more firms are put out of business permanently by fire than theft and that 'arson' in 1980 reached an all time peak. Whether the causation is malice or intended concealment of evidence of theft is academic to a firm whose records or productive capacity have been destroyed.

## *RISK APPRAISAL*

Risk appraisal is a process of systematic identification and evaluation prior to making decisions of acceptance, attempted avoidance, mitigation, or transfer of the risk by insurance, contracting out or other means. The most difficult phase is that which precedes the exercise – convincing senior management that a risk does exist and that action should be taken because of it.

Security is easily established as a board level agenda item in the aftermath of a disastrous incident or in the knowledge of something catastrophic which has happened to a local or closely related firm. It is a matter of record that little attention was paid to security of computer installations in one of the largest UK groups until master tapes and discs were stolen by trusted employees in an attempted blackmail – no expense was spared in an immediate clamp-down throughout the group itself and all its subsidiaries. Similarly, the greatest impetus to letter bomb precautions in the industrial world came after the receipt of such a device by a prominent and well-known managing director who had apparently done nothing to merit the honour. It is likely that the sudden spate of letter bombs to senior Civil Servants which occurred in April 1987 will increase vigilance in that sphere.

Inertia is understandable when a dispensation of immunity against unpleasant occurrences appears to have been granted and the only impetus to discussion are the forebodings of a security specialist who, though probably accused of self-interest, is more likely to be feeling like a prophet crying in the wilderness. But there are no divine dispensations and the ultimate responsibility for the safekeeping of a firm's property and the lives and possessions of its employees cannot be disclaimed by the controlling management. Criticism has occasionally followed when the predictable has occurred and stemmed from an absence of precautions – but not as often as it should.

The nature of a business may be such that potential loss is likely to be minimal, or at least acceptable in the context of the turnover. Mitigation may be possible by transferring uninsurable risk to a customer as a price increase, though in this competitive age that policy has decided drawbacks. The addition of 10 per cent on cost to allow for theft and vandalism in the building industry certainly did not appeal to purchasers and has lapsed. However, it is impossible to evaluate alternatives, and particularly economically feasible precautions, without giving positive thought to the risks that exist and have to be catered for. Remedies do not, of necessity, involve high expenditure or use of extra manpower, much can be done by changing procedures and documentation which introduce deterrents into systems without impeding the normal flow of business activity.

### *Sources of advice*

The larger commercial security firms will provide consultants but it must be appreciated that use of their employers' equipment and/or manpower will colour their approach and comments. Independent consultants and advisers are available and it is a matter of regret that there are few criteria by which their efficiency can be measured; the salesmanship of some exceeds more desirable attributes and personal recommendation from a reliable source is highly desirable. Indicative perhaps of what might be encountered – an advert in a UK newspaper by a previously reputable firm – 'Security Consultants required. No previous experience needed'!

The standards of salesmanship vary considerably, for example, guidance notes on countering bomb threats were advertised to prospective customers which contained highly-dangerous advice. Their very youthful and inexperienced originator was subsequently invited to address an august seminar and embarrassed the organizers by the triviality of his presentation which included a series of slides showing likely places for bombs to be left – they included toilet flushes, letter boxes in variety, dustbins, telephone kiosks, etc! A personally applied 'consultant' appellation had been accepted by a body of international security importance at face value without any enquiry.

Police guidance may be forthcoming but public commitments limit their available time and it may take the form of recommendations on physical security of buildings plus a generalized indication of where further thought should be given. No true appraisal can be made by a brief visit; some problems have to be lived with before their presence is appreciated – despite the fact that a well known 'consultant' once made recommendations based on an abbreviated day-visit to a steel complex of 500 acres which employed a workforce of several thousand. Amongst a string of platitudes at least they were told where their fencing was defective!

Physical protection of buildings and sites against outside attack can be the subject of conjoint advice from police, insurers, and electronic alarm installers; the result will be effective and easily costed. Other aspects of appraisal may be more contentious and they deal with the less obvious but potentially more damaging risks which can arise within the organization. In these matters, departmental heads must be fully involved as no one should be more conversant with the loss susceptibilities of their areas. They should be aware of the deterrents that are possible to use without detracting from the efficiency of the department, and be responsible for monitoring that agreed measures are implemented. A practical point – when arranging a discussion on this, see managers away from their own environment where they will not be subject to repeated interference by telephone calls and routine matters which totally disrupt their train of thought.

## *Evaluation of loss*

The cost consequences of a loss may be direct and immediately identifiable, or indirect, arising from it but more difficult to assess; costs of permanent replacement and productivity loss may have to be made and other non-recoverables for which there is lack of insurance cover.

### *Direct loss*

A typical example is money itself, rarely identifiable and at once 100 per cent profitable to the thief, without the usual risks and value depreciation attendant on the disposal of other forms of stolen property, which makes it a priority target for internal and external predators. It can also disappear by internal fraud and increasingly by corruption in contract dealings.

Negotiable pre-signed cheques, cheque books, travellers cheques, credit cards, etc. seem rarely regarded as being in the same risk category as cash, but they are equally disposable and need at least equal care. Firms increasingly deal with their purchase ledgers via a computer, in such instances a pre-signed cheque may be validated up to £10,000 and becomes a potentially very valuable acquisition.

Tools and equipment, indeed property in general, personal or company, can be the subject of theft, malicious damage, or plain vandalism all of which have the same end result – profit loss. In the metal-using environment of one author, life is further complicated by the fact that raw materials, work in progress, finished goods and even waste products are equally attractive targets to some.

Less tangible, but possibly even more important in the long term for a highly technically-orientated firm in a competitive market, is loss of information in the form of research data, process secrets, take-over evaluations, tenders, and the like. It is deplorable that in few countries is there any law directly applicable to what is conventionally termed 'industrial espionage' so that charges of stealing pieces of paper on which the information is printed may be the only ones which can be preferred. Unauthorized access to personnel papers and those relating to industrial and wage negotiations can be both embarrassing and costly to firms, particularly at times of industrial unrest; as often as not this violation of privacy results from the absence of any procedures for safekeeping, or simply lack of managerial care.

Direct interference with production and distribution of finished goods can result from the activities of malcontents in the workforce, or outsiders deliberately intruding to incite disputes, possibly leading to sabotage of equipment. Painful examples of the latter have involved computers where production schedules, customer accounts and wage records have been deleted or altered. Terrorist threats, bomb hoaxes and the like are unpredictable except in the one respect – taking precautionary measures will be costly in terms of productivity.

These of course are only generalizations in the forms of direct loss/causation which may be sustained but they in turn may lead to less immediately obvious consequences.

### *Indirect loss*

Any serious direct loss will have repercussions of some kind, none likely to be of benefit to the firm concerned. Payroll thefts have led to employee walk-outs since wages were not paid on time; a huge advertising campaign was pre-empted by a competitor's prior knowledge of its scheduling; a firm was involved in years of costly litigation after a competitor claimed patent rights in respect of a new process for which the competitor had done no research; large overseas orders and contracts have been lost because of non-delivery caused by thefts at ports of dispatch, and a factory workforce was laid off for two weeks for lack of essential raw material, stolen en route to it. Such examples are numerous. Transit losses may be particularly painful in that the fact that they occurred may not become known for some time during which delivery may be disputed, penalty clauses may be invoked, and goodwill and future orders jeopardized.

Loss certainly does not terminate with the original theft, for example the 'unethical' acquisition of papers relating to former wage negotiations, which should have been destroyed, led to a costly breakdown in industrial relations when the managerial reasoning was noted by the workforce representatives.

### *Factors defining required degree of response*

It has been said that the principles of risk appraisal and control were devised with the objective of selling insurance, but insuring is only one means of risk avoidance and the conditions attached to it may be such as to render it unattractive. Moreover, if insurers require precautions which virtually eliminate risk, these may be put into effect and the expense offset by simply not insuring – as happens increasingly. Obviously if the cost of excluding risks exceeds what is reasonably assumed that can be lost, then the situation might be deemed acceptable as it is. However, how accurate are the assumptions that are made and have some risks been accepted simply because they have been overlooked or not foreseen?

Known losses rarely reflect a true situation and should be expanded by the inclusion of unexplained deficiencies. Departmental heads tend to place theft as a final alternative to consider when looking for causation of shortages; they may even conceal these as reflecting on the efficiency of their departments. This delay factor is distinctly irritating when dealing with discrepancies in deliveries to customers, particularly when contract transport is used; before the probability of theft is accepted, time and correspondence has obliterated all certainty of what has happened. Where systems of stock recording and documentation are inadequate, the first indication that anything is wrong may be a monumental shortfall at an annual stock check. An appraisal recommendation might well be that of more frequent and irregular stock checks and auditing, plus better documentation to throw up queries at a much earlier stage – a deterrent in itself.

Suspicious incidents for which there is no obvious explanation call for precautionary measures. These could be one of the following: restricted information on research and projects becoming known to competitors; industrial unrest springing up for no apparent reason, perhaps accompanied by interference with production machinery and processes; disappearance of documents during or prior to audit. Inadequate enquiries during recruitment and absence of measures to ensure the privacy of the operation are likely to be thrown up as possible causation factors by subsequent investigation.

A number of industries, locations, and products are traditionally known as sources of recurrent and damaging pilferage. Docks, airports and sectors of transport come into this category; industries located in rural or semi-rural areas may be totally without incident whereas in certain city areas it would be the height of irresponsibility not to enforce maximum security in identical factories; photographic and cosmetic products, easily portable and of considerable value are typical of those presenting high temptation to employees. These are matters which must be taken into consideration when assessing probabilities.

On occasions a firm may have no option as to whether it will instigate security measures; this may be made a condition of contract by a client, or an insurer, particularly in connection with governmental tenders, and those where a new product is being developed by the client.

Overlying the other factors is that of the maximum potential damage which can be visualized if certain possible contingencies arise – major theft, uncontrolled pilferage (in the retail field perhaps inhibiting competitive pricing), fire, explosion, flooding, sabotage, information loss, etc. If, for example, a firm's profitability is centred on efficient operation of its computer, any assessment will make this a priority for security attention. If a preventable circumstance could grossly impair a firm's future profitability, it is commonsense to do something about it.

## *Departments for special risk scrutiny*

A full assessment will examine all sectors of a firm's activity. The attitude of the person carrying it out should be that of a prospective and well informed malefactor – after all, it has been said, with much justification, that the best detectives are those who can most closely identify with the thinking of the criminals they pursue. Even the functioning of an existing security department needs inspection to reduce sources of temptation, after all its members have uncontrolled access everywhere when no one else is on the premises. The managing director of a firm of distillers once asked at a seminar for advice on how to stop his chief security officer stealing bonded whisky – as he was sure he was doing!

Not all firms will have a full quota of the departments that are listed and each will have different degrees of priority for them:

*Research and development* may have no control of entry, inadequate facilities for document storage, members who write technical articles for magazines on their work, little checking on new staff recruitment, negligible supervision of paperwork left out on desks, visual observation from outside, etc.

*Sales and marketing* again may have unrestricted access for unauthorized persons to offices containing information of planning, pricing, advertising campaigns, tenders, customer discount lists; flowcharts and other material may be hung on walls where clearly visible; confidential data may be given to salespeople without instructions on its safekeeping. A classic example of the latter was the divulging of a firm's discounting policy to a competitor over drinks in a bar – and producing the confidential papers to prove his point! Incredible but true.

*Computer* – there may be free access for other personnel as well as operators; room may be badly sited and not purpose built – creating fire and safety problems; staff recruitment may not question integrity adequately; no instructions of dos and don'ts may be laid down to minimize fraud or other risks; no alternative breakdown standby may have been arranged. The limitation of outsider access by remote means into data banks (as shown in many TV programmes) may not have been adequately considered and catered for.

*Cashiers and wages* – premises may be inadequately strong and badly positioned; cash handling and procedures may be 'slap-happy'; insufficient importance may be attributed to cheque signing machines, cheques, credit cards, etc; petty cash payments may be made without necessary vouchers or unchecked ones; recording of advances and loans to employees may be limited to the filing of receipts that can be destroyed, and so on.

*Distribution* (transport and warehousing) affords opportunities for driver frauds; collusion between loaders/drivers and customers for over-deliveries under cover of bona fide loads; thefts by contract hauliers and corruption in the manner whereby they are engaged; stealing during customer collection of goods due to lack of supervision; bad documentation covering actual losses sustained; failure to include insurance requirements and binding delivery instructions in contracts.

*Purchasing and materials disposal* is the area where preferential treatment can be given to suppliers or purchasers of unwanted machinery, vehicles and materials. Systems of tendering may be non-existent with unquestioned acceptance of single quotations. The same possibilities exist where the placing of orders is done directly by a department, for example plant services or its equivalent. Orders may be collected direct from suppliers without going through any independent reception point, thereby allowing part to be diverted for personal use without any trace being left; procedures of obtaining confirmatory authorization by a senior manager may be being bypassed. An area where suspicion is readily raised but proof is difficult in the extreme – hence a tight system is needed.

*Stores/goods inwards* affords opportunities for staff pilferage which can continue for some time if there is no periodic stocktaking which throws up shortages. Collusion between staff and delivering drivers can lead to under-deliveries being accepted if the supervisor has no clear view of the reception area. Issues to employees may not be adequately regulated to prevent fraud.

*Production* may not have controls to prevent bonus and overtime frauds which are rarely accorded the financial importance they merit. Reclaimable material may be dumped purely for convenience of supervisory management, a fact which may come to light later as an unexplained shortfall at stock check. Special processes may be accessible to casual or even unauthorized visitors. Laxity in supervision coverage may give opportunities and temptation for stealing.

*Selling areas* in all types of business merit careful scrutiny with retail as a prime example; in wholesale or combined wholesale and retail premises, too much latitude may be allowed to customers carrying out bulk collection. Closed-circuit televisions, mirrors, protective wiring and other devices may be good mitigators of a loss that complacent management has disregarded.

*Personnel and industrial relations departments* contain material the potential nuisance value of which does not always seem to be recognized by those controlling it. Filing cabinets with individual locks, or fitted with padlock and bar seem rarely used in these departments; questioning should be directed at ascertaining what precisely is held and what damage could be done if the information were released, or inadvertently or otherwise destroyed.

*Company secretary's and directors' offices* are focal points for all the really important data and documents of a firm, yet out-of-hours visits may find desks still covered with files, drawers unlocked, in fact everything available for cleaners, security staff or even intruders to look at. No policy of classification or safe handling, storage and disposal of confidential papers may exist, or, if it does, it may not be enforced. Offices may be on communal locks; secretaries may be in the habit of leaving their keys in typewriter tops or drawer. In other words, where security in a company should be at its highest, it is sometimes disregarded to the point of being embarrassing to those responsible when brought to notice.

### *Contracts*

This is most unlikely to be a separate department as such but it merits special mention as a common factor in the work of many which is often totally overlooked in security surveys. Malpractice losses can go unchallenged, and indeed unknown for indefinite periods, because the paperwork will be right where a contract issuer has a free and unquestioned hand. Authorization limits may be avoided by breaking down major contracts into individual ones within those limits; a belated construction enquiry showed several hundred so restricted all going to the same supplier and a mere handful of minor ones to another. The facts and obvious implications would have been noted at the time if a simple system of recording quotations and awarding of contracts had been enforced and periodically inspected, preferably by director level. Most companies, public authorities and departments will have adequate systems but the missing factor will be that of senior level inspection and reluctance to query an authorizer who presumably has status and outward respectability. Frequency of lunchtime social meals with suppliers gives grounds for suspicion and quiet logging of such by a security chief might pay long-term dividends.

### *Action parameters and likely restraints*

In making an assessment, the 'facts of life' in respect of the firm must be recognized and accepted; it is a waste of time making suggestions which are uneconomic in relation to the risk, or those which will cause more than minimum interference with essential business activity and good employee relations. Nevertheless, if measures are recommended which are inadequate, they will be greeted by apathy amongst management and something of the nature of contempt by employees. In other words the assessment should be a true one which also mentions the known difficulties of implementation.

Other factors will exist that might be more awkward to put on paper; the reaction to projected changes in routine may meet apathy, inertia and downright opposition at supervisory levels satisfied with the status quo and not wishing to acknowledge deficiencies exist. Finance will be quoted as a stumbling block since the possible expenditure apparently has no direct bearing on profitability. Custom and practice precedents, with suggestions of tradition perks and privileges and consequential unrest, may be put forward as a contra argument. In many departments, and indeed firms, an attitude of benevolence and of reluctance to mistrust employees – or show mistrust by security measures – may be encountered, though this is crumbling in face of the universal crime escalation.

A danger that always lurks in the future of any assessment lies in a firm's acceptance of the need for a manager to have direct responsibility for security matters – there may be a tendency to give this to an under-utilized manager who has neither interest nor qualifications. If the assessor feels such an appointment is merited, this possibility should be recognized by outlining the expertise required for the post or the necessary training an inexperienced incumbent should undergo.

### *'Defensive' departments*

During an assessment certain departments will emerge as sources of strength in the process of tightening security and their role should be stressed in the presentation to management.

*Personnel* by care in recruiting, especially for posts of risk – and in making positive statements of policy on security matters.

*Internal audit stock control* can bring frauds and discrepancies to early notice.

*Civil engineers/architects* in planning new buildings or alterations can eliminate structural security weaknesses (security codes of practice for building have been operative in several US states for many years, and a like British Standard has now been prepared in the UK but it is too early to appreciate its impact).

*Secretarial/legal* can arrange fidelity bonding of employees and pursue claims for compensation – also probably deal with offsetting risk by insurance if no specific insurance department exists.

*Credit control* can weed out potential customer frauds and provide information where it is believed a false claim of non-delivery is being made.

*An existing security department* should have records and knowledge helpful to an assessment and may have scope for much improved efficiency.

A bona fide assessment exercise cannot be hurried if it is to have real value. Use must be made of the goodwill and internal knowledge of the company from managers, supervisors and those members of the workforce deemed capable of assisting. Where assistance and suggestions from those sources are beneficial, they must be acknowledged – this is part of the process of building-up a sound relationship for the future.

## *OPTIONS*

With a detailed appraisal, a board does not have to rely on mere speculation in deciding its action. Inevitably, even if no comprehensive scheme is implemented, weaknesses will have been spotlighted in a manner which must be beneficial.

1 *Avoidance* – A situation may have been reached, or the risks attendant on a new project may be such, that the option of complete avoidance is desirable. A large supermarket found that it could not control the amount of shoplifting and staff thefts at a particular outlet regardless of strict security measures so it had to be closed down. Similarly, a distribution depot where pilferage at all levels by employees reached an unacceptable level was also closed. The cost of ensuring security during precious metal processing was such that the work had to be transferred elsewhere. Requirements of a contract to create total confidentiality for the work to be done were too stringent to be economically acceptable and the firm pulled out of the contract. Thefts of incoming material and outgoing exports, the latter accentuated by loss of customer goodwill, caused a major industry to transfer its operations to another port with heavy consequential loss to that at which the thefts had become unacceptable. It is noteworthy in this instance that this action did induce a security appraisal and revision of preventive action – but all the trade did not return to that port.

2 *Insurance* – Insurance is certainly one way of catering for risk but conditions may be applied which make it as costly as implementing internal action. If, for instance, cover for a warehouse excludes theft by employees and calls for a full alarm system, it may seem reasonable to put in the alarm as specified and offset its cost by waiving the insurance. This has also occurred in a particular transport field. Cash insurance is almost universally used and will require approved safes and handling procedures. Insurance in itself is no answer to risk. A board cannot therefore make a simple decision to cover potential loss by insurance; a package will be presented which admittedly will mitigate the financial consequences but will also contain sufficient obligatory precautions to minimize the insurer's risk.

3 *Contracting out* – Some risks can be contracted out to others so that it becomes their responsibility, for example wage carrying and transport of goods, and small, high value items may be cheaper to buy-in than to incur the losses when manufacturing in a difficult environment.

4 *No action* – Even in this category, something of value may be found in the most unlikely business. For example, one large road-making contractor was nearly put out of business by frauds in the supply of hardcore of all things – a case of quantity rather than quality of theft!

Whatever option is taken, the basics of a security policy will have been developed during the appraisal. If it is decided to nominate or appoint someone to have responsibility for security in general within the organization, or to create a security department, the effectiveness of this will be very dependent on the attitude of all grades of management.

### *Management's role in security*

If there is indifference at board level, it will be reflected throughout the management structure, any policy will only be partially efficient and the whole exercise and expenditure becomes pointless. The continuing interest and goodwill of the board is most desirable; all matters of internal importance should be notified as soon as possible to appropriate directors who can tell colleagues if they so desire. Things of lesser import which for any reason are likely to be brought to their notice by another source, where they may be embarrassed by lack of prior knowledge, should also be notified. Information about major security incidents affecting competitors, or in their own specialist field, should be passed on, especially if consideration of internal protective measures is likely to be necessary. Recommendations and reports should be clear and terse and, where expenditure is involved, show alternatives and be costed adequately. Much of the directors' attitudes to a security force will be based on the clarity of reporting to them, but having set the parameters within which security effort and principles will be applied, they will not want to be bothered with trivia.

### *Departmental heads*

Managers are primarily interested in their own discipline; their career advancement is dependent upon success there, and their initial reaction to any suggestion that they have a responsibility for security may well be irritation. This has got to be overcome in some way, or only lip service will be paid in their own areas. As much thought has to go into obtaining the desired degree of co-operation as into any planning of physical and procedural measures. It must be remembered that many individuals have a psychological objection to exercising anything savouring of a disciplinary role and an understandable reluctance to risk being unpopular with their staff. In addition, there are many different outlooks nowadays on dishonesty, and these have to be taken into account in deciding the best approach. Typical examples are:

1 'Stealing is a fact of life and we cannot stop it' (a view that justifies a careful look at the manager's own potential activities!).
2 'Security is not necessary, my staff are not thieves' (a naive but sometimes genuinely held belief).
3 'The interference with normal procedures is unacceptable' (production oriented, or just does not want to be disturbed out of a personally comfortable routine – needs to be convinced by examples and consulted in detail on every proposed measure).
4 'We must do something or be criticized' (reluctant acceptance of suggestions, but regular monitoring thereafter is needed to ensure that they are being carried out).
5 'We must stop all avenues regardless of cost' (usually follows a substantial loss, and the manager has in effect to be held back or he may want extreme measures, over costly or liable to cause industrial trouble).
6 'The cost is greater than our losses' (the accountant's attitude and an awkward one to refute – potential loss, unexplained discrepancies, examples of occurrences in comparable firms, all need to be stressed).
7 'We are at risk, what can we reasonably do about it?' (the ideal to aim at).

Nevertheless, most departments have risks over and above those of petty theft and pilferage. The best people to define these, if they are correctly interviewed and their interest aroused, are managers. The best people to evaluate probable adequacy, acceptability and effectiveness of agreed countermeasures, if expertly guided, are managers. The best people to enforce and monitor procedures, who are on the spot, suitably indoctrinated and consulted, are managers. They play a vital role in security and in compiling reports or recommendations to board level, and any assistance or helpful suggestions from them should be duly acknowledged. Liaison must be maintained by keeping managers informed on everything of a security nature bearing on their department: new legislation, losses of competitors and like departments, ploys and gimmicks that have been encountered, etc. While doing this, a discreet eye can of course be kept on whether the agreed procedures are indeed being adhered to!

Whilst managers should perforce be regarded as colleagues in the security effort, there should not be an automatic belief in their individual reliability and honesty because of status. Though the disgrace which would follow discovery of dishonesty is a more potent deterrent at their level, the same temptations apply to them and they have less scrutiny of their actions plus greater opportunity. Without labouring the point, practically all computer frauds and cases of corruption are perpetrated by persons of managerial status.

## *ROLE OF SUPERVISORY STAFF*

The goodwill and co-operation of senior management is fundamental in establishing sound security principles and procedures, but day-to-day efficient security practice is more closely dependent on the acceptance and understanding of the principles and procedures amongst the workforce. Shopfloor and office supervisors have implicit responsibility for the protection of their employers' property against loss. There may be an understandable reluctance for them to take action against persons under their charge who may have offended but nevertheless they and not security personnel are the most likely to see or learn of dishonest actions by employees. It is important that they are guided to a realization that they have this responsibility to the firm and that inaction will be looked upon as impugning their integrity and as indicative of an inability to lead.

Disciplinary action has been taken for 'turning a blind eye' and it was held in *Sybron Corporation* v *Rochem Ltd* 1983 IRLR 253, a Court of Appeal decision, that there was a duty to report the fraudulent activities of subordinates where these constituted serious misconduct, but there was no general duty to report on an employee's conduct and whether there is a duty in a particular case depends upon the circumstances.

### *Training of supervisory grades*

Instruction on security responsibilities and practice should be incorporated in the training of management and supervisory grades. An outline of such a course is given in Appendix 1.

*Checklist for supervisors on curbing losses*

A good supervisor can do more to stop stealing than the most effective security force or management crash programme. Here are nine suggested methods to discourage amateur thieves.

1 Don't let them get started. Stealing is contagious. Once supervisors let workers get away with 'borrowing' they have shown them that it is acceptable. They will not be one-time thieves. The stealing virus will spread fast until the whole department is nibbling away at inventories. Probe all losses however small and make a fuss about them.
2 Remove temptation. Supervisors should not encourage pilfering by leaving broken cases or cartons of material, small tools, hardware, typewriter ribbons, or other items people find useful, lying around unguarded. They should keep unauthorized employees and people from other departments out of sensitive areas by courteously challenging strangers. They should discourage visiting.
3 Control the inventory. In departments where useful, easily concealed parts are used, or in the final assembly and packaging area, a rigid inventory control system is the supervisor's most effective theft preventive. Nothing discourages an industrial thief more than an early detection system.
4 Curb the borrowing. If a company does not permit employees to borrow certain equipment without permission, supervisors should be sure to enforce the rule. This goes for workers, other supervisors, and themselves.
5 Check the hiding places. There are probably many good hiding places around the shop – between bales or boxes, in ventilators, and electric panel boxes. Many a thief stashes his loot away to pick up later when the coast is clear. Except for emergencies, supervisors should not allow workers to visit parked cars during working hours. They should not hesitate to question suspicious packages or bundles leaving the department.
6 Make tools and supplies hard to get. One of the best ways to impress workers with the value of tools or the small common supplies is to set up a system that makes them hard to get. If workers have to sign a chit every time they want a screwdriver, and if a hard-and-fast quota is placed on the perishable items, people will be careful about returning tools and taking supplies for their own use.
7 If possible, limit the avenues of leaving the department, strictly control and record access to locked storerooms.

8 Keep a special eye on employees whose behaviour gives grounds for suspicion (see p. 23).
9 Be completely aware of the company policy for dealing with theft and disciplinary matters so as to be able to act promptly and confidently.

## *CO-OPERATION OF WORKFORCE*

This is much more difficult to achieve; a high percentage of shopfloor workers would not contemplate stealing from their employers, but, on the other hand, they will not want to expose themselves to accusations of informing on their fellows. If there is genuine acceptance that what is being done in security terms is intended for the benefit of all, and there is respect for those in charge, there is a greater probability that the existence of dishonest behaviour will be brought to notice. If anyone, of any status, asks a security officer to treat what is said as confidential, this request must be regarded as binding – unless what is said is ill-founded and malicious. Failure to do so will virtually destroy future possibilities of similar communication.

In several long-established firms with good industrial relations and a relatively static workforce, security committees, akin to safety committees, have been established. Not only have these produced valuable suggestions but their presence sometimes acts as a spur to early action by management where expenditure is concerned. In all cases where an idea originates from a works council, or its equivalent, it should be seriously considered and answered. The informal posing to shop stewards of an apparently irresolvable problem affecting their members and wage frauds, was productive of a system that management never dreamed would be acceptable – and it worked! Co-operation may not be forthcoming, but if the effort to obtain it is not made, one will never know. Personal approach and the personality of the security incumbent are all-important. Too much weight should not be attached to the pessimistic comments of a few senior security personnel who possibly operate in hostile environments, attempt a policy of splendid isolation and ignore potential assistance from fellow employees.

### *Responsibilities to employees*

Many security job descriptions at supervisory level mention obligations to fellow employees whilst at work. These should be strictly observed; no less, or even more, attention should be given to investigating their losses than the firm's, and their interests safeguarded as much as possible. Assistance or helpful advice should always be given when sought and no opportunity missed to create goodwill. Again, where assistance has been given in dealing with dishonest outsiders, this must be acknowledged in reports to management – unless the person wishes otherwise.

Publicity should be given in works magazines and the like, or at works councils, to offences detected or prevented. This occasionally dispels unwanted rumours. Where information or advice bearing upon the safety and well-being of employees can be given, this should be regarded as a duty.

Promises made to employees to 'look into' something or to do something for them, such as passing messages etc., must be kept, and any personal confidences entrusted to a security officer must be observed.

## *THEFT BY EMPLOYEES*

Many studies have been undertaken on the causes of stealing and from the erudite verbiage produced, some positive conclusions for security practitioners emerge. Dishonesty can be found in any stratum of society or status in the business world; previous integrity is not conclusive in predicting future behaviour, which makes it impossible to recognize all potential theft sources. Extreme personal pressures can change a trustworthy employee overnight; debts, hire purchase commitments, family illnesses and crises, drug addiction, marital problems alike – matters often unknown to colleagues – can create an intending criminal where least expected. Opportunity may be the deciding factor, and if that is coupled with little likelihood of detection, and good remuneration for such risk as there is, the normally potent deterrent, fear of disgrace that might ensue, fades in importance.

Good security practices will therefore aim at evolving systems to reduce those opportunities, to make the risk of being caught unacceptable, and provide means of bringing losses to early notice. Any business which relies solely on an annual stock-check to show what is happening deserves all that happens to it.

Top management must lead by example. If an executive arranges for work to be done at his home by employees he controls, using the firm's materials and during their normal working hours, he is both naive and under a delusion if he thinks that it will not become common knowledge. Whilst this may not lead to any action at the time, thereafter he is faced with an impossible situation if he wishes to discipline someone who has that information. A director who instructs a subordinate manager to have such work done, puts himself in the power of that manager; abuse of power at that level will have repercussions throughout the structure. This can operate down to the lowest level of supervision until supervisors who have been involved in malpractices are obligated to turn a blind eye to bonus frauds, clocking offences and petty stealing.

Dual standards are unacceptable in any organization which wants an efficient security system with co-operation from all personnel involved. Employment protection legislation (see Chapter 4) has emphasized this by rightly punishing instances where the disciplinary norm has been mitigated without good reason – and good reason would be found lacking where action had been influenced by managerial status, personal relationships, or any other source for preferential treatment. Indeed, industrial courts have laid down that the more senior the post, the greater integrity is to be expected. If an ordinary clerk is to be sacked for dishonesty, it is quite essential that a factory manager should be similarly treated – and equally dispassionately.

### *Grounds for suspicion*

When there are grounds for believing petty pilfering is taking place, or that an actual theft under investigation must be the work of an employee, in the absence of direct evidence as to who is responsible the automatic reaction is to narrow the field as far as possible and as quickly as possible. It has been said in police circles that a good informant to point you in the right direction is of far more value than the best forensic laboratory whose work often convicts, but rarely detects. In other words, if an enquiry can be concentrated on a particular individual(s), it has a much better chance of success.

On the premise that anyone might be induced to steal it is necessary to systematically ask questions, listen to theories and keep one's eyes open. Known gamblers and betting men are likely to have sudden needs for sums greater than their salaries, as are those paying alimony or supporting illegitimate children. Gossip will pinpoint people who are showing signs of living above what could normally be expected on their income – evident perhaps by a superior new car or an extravagant holiday. A good indicator is that a person has been showing signs of strain or acting out of character: for example a bookmaker's clerk was physically sick for a week before disappearing for good with some £40,000! Behaviour 'out of character' is understandably pertinent to first offenders who may even panic to the extent of seeking excuses to leave work early after committing the theft or feigning sickness afterwards to keep out of the way. One symptom a culprit finds hard to conceal when an enquiry is being carried out in an open area is that of continually looking up to see where and what the investigator is doing! – an indicator only perhaps and far from infallible but it might be useful.

Employees seen in areas where they have no reason to be, or at times when they should be elsewhere, may cause comment and, where outsider involvement is suspected in conjunction with an employee, that too may have been noted. The archetypal cashier or accountant who converts the firm's money to his own account by 'rigging the books', is never off sick, never takes holidays, is always first in and never leaves work until everyone else in his department has gone. Many such people have come to grief by the accident of a sudden illness which has given their records into the hands of an inquisitive assistant – at that moment of revelation of course everyone will claim to have suspected something was wrong! The difficulty is getting those suspicions conveyed beforehand.

The unexpected and unusual reaction should raise mental queries and it might take the form of an antagonistic stance towards security without apparent cause. If this occurs at managerial level a closer scrutiny of what is going on in that department is merited.

Comments on theft by employees might most adequately be summarized by the findings of a very thorough US survey. These are not necessarily the same results that would be shown in the UK or elsewhere in the world, nor are they in the sequence of importance that might be anticipated:

1 Those involved in stealing would be likely to be 'anti-employer' in their normal working attitudes (there is an opinion here that at shopfloor level there is a nexus between absenteeism and dishonesty).
2 The highest level of theft lies amongst those who, because of the special nature of their jobs, have greatest access to items that can be stolen (warehouse staff, storekeepers, electricians, engineers, plumbers and the like – we would not like to be dogmatic about this for the UK).
3 Younger, unmarried employees without personal obligations are more likely to be involved.
4 Employees over-concerned with status, financial remuneration, and career advancement are liable to be vulnerable to temptation.
5 Regular association with other employees outside working hours may lead to greater possibilities of co-operation in stealing from their employers.
6 Lack of job satisfaction, coupled with grievances against supervision and a sense of unfair treatment makes some prone to stealing.

7 The higher the chance of detection, the less the incidence of theft in an organization.
8 Organizations which have a clear anti-theft policy have a lower level of theft by the employees.

In the UK the 'batting order' in sequence of importance is almost certainly the reverse.

# 2

# *Formulating a security policy*

During the winter of 1980/81 a long report on 'Theft in metal-using industries' was submitted to the British Non Ferrous Metals Federation (BNFMF). This summarized the experience of the federation's Security Liaison Committee from its formation in 1965 and therefore represented a pooling of the expertise of some 100 top ranking security practitioners, augmented by senior police officers and delegates from areas such as insurance, alarms and closed-circuit television, brewing, chemicals, docks, and transport industries. The validity of the contents has been endorsed by its circulation within several 'multinationals'.

An appendix dealt with 'Sources of security weakness' and is reproduced here because of the importance of a sound security policy. The report relates to the metals industry, but it is not difficult to see how its principles could be applied in other industries. In a cosmetics company, for example, returned defective or sub-standard items were checked for quantity on receipt and then disregarded; two things happened, the percentage of returns immediately grew out of all proportion, and customers complained that a trade in cut-price items had developed in the city where the factory was sited. Employees were raiding the 'reject' stock and substituting what they took for goods awaiting dispatch so that no shortage was recorded – and of course the flow of returns increased! Alternatively, their takings were disposed of to small back-street shops prepared to sell sub-standard goods.

## *SOURCES OF SECURITY WEAKNESS*

(a) Any analysis of loss-incidents to minimize opportunities of recurrence inevitably includes broad discussion of causation and prevention. So far as employees are concerned, a strong impression has been gained that *those industrial groups who have clearly defined and publicized policies for dealing with dishonesty, coupled with consistent implementation irrespective of status, have fewer problems than those who deal with each incident individually*. This latter option risks the danger of creating undesirable precedents, and may be creative of doubt as to consequences tantamount to a form of temptation.

(b) Certain adverse factors have been observed over the years and attention given to those undermentioned could be calculated to result in a progressive fall in loss of profit due to theft and fraud:

1 Management failure to fully investigate discrepancies; to instigate enquiries at a sufficiently early stage; or to accept that theft is a possible causation before all other alternatives have been exhausted.
2 Failing to take note, and appropriate precautionary measures, when costly incidents involving substantial loss have occurred in like industries, and in circumstances which could be duplicated internally.
3 Inadequate internal liaison on matters which may give rise to risk of loss through dishonesty or other criminal action, i.e. new procedures, new products or processes, new building or structural alterations; failure to communicate information on losses that have become known.
4 Frequent failure by management to appreciate the potential contribution to profitability from reclaimable metal waste correctly handled, and disposed of to best advantage; allowing accumulation to a point that sheer quantity provides opportunity and temptation, or conceals loss.
5 Accepting transit losses as primarily insurance matters and indulging in prolonged correspondence before making positive enquiries. Failure to check that contract carriers have adequate insurance cover.
6 Inadequate care in driver recruitment and other posts where integrity is of importance; not taking up references when recruitment is that of supervisory staff status.
7 Adhering to a belief that managerial status in itself is an endorse-

ment of integrity in the holder, i.e. failure to accept that problems may derive from a manager rather from other possible alternatives.

8 Absence of adequate inclusion of security-based requirements in 'Conditions of Contract' applied to contractors working on sites to ensure adequate supervision of their employees and transport.

9 Allocating localized security responsibility to a disinterested manager; recruiting manpower for security purposes and failing to adequately train, instruct and utilize.

## *TARGETS OF SECURITY POLICY*

The appraisal of risks to which an organization is exposed (Chapter 1) will have made it clear that there are security considerations in practically all departmental activity; certainly of varying importance and often verging on being disciplinary in nature or in the good housekeeping category. The checklist in Appendix 2 shows policies with which security is mainly apt to be concerned, but it is not suggested that it is comprehensive. There is scope for interest in the documentation of goods dispatch, in stores receivals, and other routine matters. This chapter deals mainly with personnel matters but subsequent chapters will spotlight those shown in the appendix or not specifically mentioned.

## *COMPANY RULES*

Conditions of employment have to be agreed by all employees before being engaged; contained amongst these have to be, by law, details of disciplinary rules which will be applicable to the employee. Frequently, this is done by referring to acceptance of company or works rules which are made available separately in booklet form, or other means of easy reference, and expanded to include extra information over and above the purely disciplinary.

The amount of space accorded to security matters could reflect the importance attached to them. Conventionally, policies in connection with theft (and possibly prosecution), right of search, taking of materials off site without permission, and 'clocking' offences would be likely inclusions in rules; additionally there could be a disclaimer of responsibility for employees property and vehicles whilst on site. 'Right of search' is a most beneficial section which provides a positive deterrent to employee stealing but it is extremely difficult to get it added to existing rules (see Chapter 9).

Dismissal and prosecution are not considered 'double jeopardy' by industrial relations courts. Indeed a case of *Saeed* v *Greater London Council* 1986 QBD, declared:

> An employer is not precluded from taking disciplinary proceedings against an employee in respect of an identical offence for which he has been acquitted in criminal proceedings. Double jeopardy cannot apply as between criminal and civil proceedings as there is a different burden of proof.

*Company rule for theft and dishonesty*

All types of serious misconduct meriting dismissal should be listed by a firm amongst its rules. That relating to theft is of particular importance and should be quite explicit to the extent of taking care of any legal niceties which might permit an unfair dismissal appeal. For this reason an extension of the ruling previously advocated is now recommended in the light of the ever changing legislation and decided cases.

> If any act of theft, or attempted theft, or other form of dishonesty, or criminal damage, is admitted, or on reasonable grounds found to have been committed by an employee in connection with Company property, or the property of another employee, then the employee responsible will be dismissed and, where Company property is involved, the Company will exercise its discretion in whether the matter is to be reported to the Police.

Of course, this does not cover all the criminally actionable offences that may be committed – for example, those of a sexual nature or appertaining to company customers – but there is no doubt that other rules which merit summary dismissal will cover these abnormalities.

Other circumstances can involve security. These would include clocking offences, being under the influence of drink or drugs during working hours, endangering the health, safety, or welfare of other employees, sleeping on the job, etc. When rules are being compiled, the person in charge of security must accept the obligations and ask to be given access to the draft to scrutinize and recommend such additions as he feels are required.

*'Prosecution without dismissal'*

A practice of not dismissing for theft is followed in some industries – notably those associated with building and construction – where it is not uncommon for employees to be charged with theft from their employer and to continue working almost as if nothing had happened, unless a prison sentence ensues.

This does not have wide appeal, and it is conventionally accepted that dismissal is automatic but prosecution thereafter discretionary. There are cogent arguments in favour of dismissal:

1 It is an incisive, clear-cut policy which cannot be manipulated for personal reasons or because of status or relationships.
2 There is no semblance of leniency which might encourage others to steal by giving the impression that a firm does not regard theft of its property as being of particular importance.
3 There is no question of retaining a person who may steal again thereby occasioning recriminations from all and sundry.
4 An employee will not be retained who will feel he is regarded as suspect for every incident which occurs thereafter.
5 Fellow employees will not have in their midst an individual of known dishonesty whose actions will be a source of mistrust.
6 If any employee's property has been concerned, strained personal relationships will inevitably ensue.
7 If it has been decided to prosecute, the retention of the individual's services will weaken credence in the evidence and possibly cause adverse comment by a court.

These factors weigh against the idea that the firm is punishing itself twice by losing the services of a trained worker, an argument that will be used by a manager who gives priority to personal convenience or has a personal friendship with, or obligation to, the individual. If this is pressed to a degree that seems unreasonable, enquiries might prove productive. Normally, managers would go along with the rules to avoid controversy or precedents; if they act otherwise, it might be a sign of apprehension of something else coming to light.

## *PROSECUTION AFTER DISMISSAL*

The end product of a prosecution may be a trivial penalty of little detrimental value to either offenders or those who may be tempted to emulate them. Where the circumstances indicate that this is likely, dismissal is a far more punitive measure, particularly where the individual is of long service or holds a post of importance or responsibility. The police, as may be anticipated, would prefer all detected offences to be prosecuted so that the culprit is recorded for any future occasions; but from the point of view of the complainant this may be a timewasting exercise with an outcome insignificant in comparison to the effort and cost expended. Successful prosecution does clarify any issues of unfair dismissal; there is much to be said for automatic prosecution where a person, obviously responsible, obstinately denies his guilt. One point is quite certain, however: the police/Crown prosecuting service will not be sympathetic to a case being belatedly referred to them in anticipation of complications arising from a disputed dismissal being appealed to a tribunal. The current tendency is towards automatic prosecution in all cases, on the grounds that the firm is the sufferer and should not expose itself to the inconvenience of an appeal simply by acting leniently.

Successful prosecution is by no means essential to dismiss for theft, nor for that matter is there any need to wait for the police even to make up their minds as to whether they will take action upon the facts reported to them (Chapter 10).

There are good reasons however for using prosecution as a deterrent to further theft where there is a constant and obvious temptation in front of employees – as occurs in all phases of food handling, in wholesale and retail trading, in the carriage of goods, or in working with high-value material (metal, cloth, etc.). Indication of a lax attitude where there is gross temptation is exceptionally dangerous, which many retail concerns have found to their cost and will continue to do so.

The bogey of trade union sympathy towards dishonesty when members are involved is often overstated as an excuse for lack of action. In such matters, not only are trade unions as responsible in their outlook as management, but perhaps they are even more dogmatic about the action which should be taken. After all, their members have to work with the individual concerned. If they do display this sensible attitude, there is no danger whatsoever of management trying to apply different standards for different levels of employee. It should be noted that where an employee's personal property is concerned, or where he has sustained injury in any way, his agreement must be obtained before any action is taken involving the police. Nevertheless it is essential that any ruling in connection with dismissal must be adhered to; otherwise the injured employee may request that no disciplinary action be taken because of pressures put upon him.

There are factors that militate against prosecution of an employee in addition to that purely of inconvenience, such as:

1 Will the firm receive adverse publicity by:

   (a) A lax system putting temptation in the way of the employee?
   (b) Having put a lowly-paid person in a position where he is handling property of a value out of all proportion to his salary?

(c) Permitting the ventilation of a grievance in open court which would reflect adversely on the firm?
(d) Compulsorily disclosing matters in court which the firm would desire should not be known?

2 Will the prosecution spotlight a weakness in the security system which could be exploited by others and which may be difficult or impractical to remove?
3 Is the end product of prosecution worth the effort and inconvenience; or considering the age, length of service, and good repute of an employee, would dismissal be a greater penalty than any the court might impose?

It must be remembered that malpractices, actual or alleged, by an employer have greater news value than routine theft by an employee.

*Authorization for prosecution of employees*

In many instances of detected internal theft there will be no immediate necessity for an 'on the spot' decision to prosecute; offenders will be known, property will be recovered, there will be no danger of them absconding, and generally speaking there will be ample time to consider how the firm's interests can best be served. A policy for such circumstances should be laid down – for example, that the facts and a recommendation should be given to a specified director for consideration and decree. This would ensure a uniform approach throughout the organization and have the advantage of stopping personal or departmental relationships from having any bearing.

Conversely, the employees may run away, there may be a strong suspicion that they have other stolen material at home, the offence may be a serious one of gross theft or causing grave injury, and they may turn violent on being challenged. These factors would merit an immediate call for police assistance. Clear definitions of the discretion allowed to security personnel and of the procedure that they should follow ought to be laid down. For the sake of industrial relations the sequence of action must include informing the individual's union representative at a reasonably early stage. This should be done via the offender's personnel officer or line manager, not directly by security unless this is specifically agreed.

*Dismissal without prosecution*

To reiterate: there is nothing wrong in operating this option and there is ample confirmation by industrial tribunals of the legality of doing so. A precaution should be taken to pre-empt the possibility of a claim for unfair dismissal which includes a denial of things previously admitted. This usually follows a realization on the part of offenders that they are not going to be prosecuted and as they have nothing to lose might therefore just as well take a chance. It may sound ridiculous to some, but firms have been known to make a compensatory payment to have a claim withdrawn rather than have the cost and inconvenience of contesting it, even when the facts are undeniable.

The precaution is a simple one, obtain a signed admission or have it made under circumstances which brook no denial – in the presence of witnesses and/or the offender's own representative. If this is not forthcoming, it seems only reasonable to say 'be it on his/her own head, the firm's responsibilities are to the employees and shareholders, not to an offender', pass the matter to the police, and exercise the firm's undoubtedly legitimate right to dismiss forthwith. That will remove an unfair dismissal talking point that the firm had not informed the police because they had not complete confidence in the evidence, a suggestion that has frequently been made.

### *Convictions not connected with employee or company property*

These are dealt with in detail as matters of industrial relations about which there is considerable precedent in cases already decided (Chapter 4), but it is worth stressing here that precipitate action should be avoided until all facts are known and considered. The criterion is whether the offence makes the person unacceptable either for his type of work or to his fellow employees, or, in exceptional circumstances, to customers with whom he must come in contact to do his job.

### *Non-reporting of thefts and offenders – the law*

When the Criminal Law Act 1967 came into being it gave respectability to a technically illegal practice which for many years had been widespread but tactfully ignored – that of not reporting offences or offenders when the complainant does not wish to do so. Prior to the Act, an employer might have committed 'misprision of felony' by not reporting a theft to the police, or 'compounding a felony' by agreeing with a thief not to prosecute if the property, or its value, were returned. Categories of crime were however changed and practically all felonies became 'arrestable offences' (see Chapter 9). Misprision and compounding were not reaffirmed and Section 5(1) legalized a conditional agreement not to prosecute:

> Where a person has committed an arrestable offence, any other person who, knowing or believing that the offence or some other arrestable offence has been committed, and that he has information which might be of material assistance in securing a prosecution or conviction of an offender for it, accepts or agrees to accept for not disclosing that information any consideration *other than the making good of loss or injury caused by the offence, or the making of reasonable compensation for that loss or injury*, shall be liable to conviction on indictment to imprisonment for not more than two years.

The interpretation which has been given to the wording is that the 'consideration' should not exceed the actual loss entailed in the offence itself. Therefore, if a wages clerk steals £25 from a float, his employer cannot say that what he has done has resulted in three days of enquiries by the chief security officer and chief auditor, therefore the loss to be compensated is £25 + £200 worth of the employer's time. No case appears to have been reported where this interpretation has been tested but it is most certainly *not* suggested anyone should do other than accept it!

Arrestable offences are dealt with and defined in a later chapter but it may be well to note now that they include all those of a serious nature – theft, criminal damage, inflicting serious injuries in particular.

Closely allied to Section 5 is the preceding Section 4 which might cause unfounded misgivings. This restates the principles involved in the earlier offence of being 'an accessory after the fact' using the terminology of the new Act:

> Where a person has committed an arrestable offence, any other person who, knowing or believing him to be guilty of the offence or some other arrestable offence, *does without lawful authority or reasonable excuse any act* with intent to impede his apprehension or prosecution shall be guilty of an offence.

Section 5 provides the 'lawful authority of reasonable excuse' if a loser makes an agreement with the thief; and Section 4 specifically refers to 'act', not 'act or omission', so that simple failure to inform the police is not a contravention.

It would be most unwise for management to take advantage of this section to justify, without good cause, failure to notify the police where a substantial loss is involved. Information nowadays is readily leaked, shareholders could pose questions, and the investigative Press bring unwanted publicity – as a nationalized industry attempting to 'bury' a £50,000 fraud once found.

It is of interest that no proceedings under these Sections 4 and 5 offences may be taken except by, or with the consent of, the Director of Public Prosecutions – which would seem to imply an intention that they should not be misused.

### *Rewards for return of stolen property*

In the case of a high value insured loss, the insurers themselves may offer a reward for the recovery of the property on a pro-rata basis (they will be found most reluctant to make a payment of any substance for recovery before the offer). A firm may however decide to adverttise a reward itself as perhaps it would if a briefcase of modest value was stolen containing vital papers; care must then be taken with the wording or an offence may unwittingly be incurred (Theft Act 1968, s.23, see Chapter 11). Promises of 'no questions asked' or 'will not be prosecuted' must not be used – the publisher should pick this up, he too could be prosecuted!

If good work or loyalty by an employee results in property being recovered and there is a wish to recognize this action, it is suggested, in the knowledge that this advice might be criticised, that if the offender is an outsider, remuneration should be done openly, if a fellow employee – privately.

## *INFORMING THE POLICE*

There should be a clear understanding as to who decides the police should be informed of an incident. Often, as in bomb threats, fires, explosions, wage attacks and the like, their attendance is both automatic and urgently required; it is in lesser matters that an option can be exercised. Security heads may have complete discretion subject to informing, say, the personnel director as soon as possible; provided they are experienced and competent, this saves time and makes sense. However, down to the lowest security guard everyone should know when they can act upon their own initiative.

Often an approach will come from the police themselves in the tracing of stolen property back to a firm, or after information has been received which links the firm in one way or another to a crime committed or intended. In such cases management would be well advised to give every co-operation, and leave the investigation, prosecution decision and control of what happens entirely to the police. Any semblence of reluctance could create suspicions of managerial condonement, or perhaps even involvement, and impair future relations.

If the matter originates internally, however, the firm's interests can be accorded higher priority. When the police are brought in they are unfamiliar with procedures, persons, and places, all of which have to be explained to them. Their presence automatically attracts attention from workers, with adverse effect on output. Individuals may have to be taken from productive work to be questioned. Loss, from the point of view of production, may be out of all proportion to the original theft. The senior members of security should be capable of carrying out most internal enquiries, and if they have accepted status and ability it is likely that they will receive co-operation from all employee levels, albeit as an alternative to civil police.

In some circumstances the police should be involved from the outset. It is suggested that in all instances where outsiders are caught or identified in cases of theft or matters of like gravity, the police should automatically be informed. If there is a loss of consequence due to intruders or non-employees, all sources of enquiry will be off company property and the police probably represent the only avenue of detection. This is equally true of employee theft where identifiable property has obviously been taken outside the premises in quantity. Generally the factors to be taken into account are value, the possibility of identification and recovery, and the inconvenience that enquiries might occasion. Commonsense coupled with the local police 'track record' is also a consideration. There is a gross overload of crime upon them which militates against the giving of time to investigation of trivia. All that might transpire after such a report has been made could be a brief visit by a uniformed officer to note the details and record the crime in police files. This may be regrettable, but it is inevitable in the conditions now appertaining. Pressures vary between police areas – hence the reference to 'track record'. Resident Beat Officers are allocated to, and identified with, particular areas; firms would be well advised to make contact with the particular officer who is responsible for their premises. This system was common practice before the fetish of mobility eliminated local knowledge and contact – advantages which weigh in industry in favour of own security force as opposed to the services of contract security.

In any court proceedings subsequent to police action, if it is possible for a competent security officer to give the evidence, he ought to do so, lengthy periods can be spent unprofitably in courts and the time of departmental heads should be saved. If security cannot do this a suitable junior from the department concerned should be used. It should be appreciated that the attendance of witnesses at court is not always required; provided the evidence is not controversial, the provisions of the Criminal Justice Act 1967 allow the introduction of the evidence in written form. Before signing the statement prepared for presentation to the court, care must be taken to check that it is complete and totally accurate – circumstances might arise where the presence of the witness might be required for the evidence to be given on oath.

## *RECRUITMENT OF EMPLOYEES*

In the pages that follow, reference is frequently made to the need to be careful in recruitment for sensitive posts, this was a security weakness noted in the BNFMF survey earlier. It cannot be overemphasized that firms often create their own problems by sheer laxity in this field – it is easier to acquire than dismiss. A transport manager whose firm deal in high-value materials interviewed and took on two drivers at a Job Centre without checking references – one had 20 previous convictions, including a jail sentence for lorry theft, the other had been dismissed twice for misbehaviour with customers, the latest of which was an assault! An offhand personnel manager took on without enquiry a militant shopsteward from a local firm; his activities had nearly closed that firm before he and many others were made redundant. After one day he reported sick and went off on a 'right to work' march! On return he resumed his antics in his new environment and the manager's competence and advancement were both undermined.

This is not a weakness solely applicable to junior employees; checks are equally, if not more, necessary for managerial posts – 'he's a good man, we'll lose him if we go through all that' is the usual cry in defence of inaction. Once a rapport has been established between local firms and local personnel officers, verbal reference checks can be made before a job offer thereby cutting out delay. A genuine and intelligent applicant will respect a firm the more for not being taken on 'blind' and accept that such enquiries are good business practice, even if delay is the consequence.

An Act of Parliament has a bearing on the importance that can be attached to the criminal convictions of prospective and indeed present employees. This is a convenient point at which to discuss it.

## *REHABILITATION OF OFFENDERS ACT 1974*

This Act, once termed 'an enactment of social philosophy', has caused infinitely more apprehension amongst personnel officers and those responsible for workforce recruitment than is ever likely to be justified. It is regrettable that the only case of consequence which has been reported, upon which the provisions of the Act have had direct bearing, concerns security personnel and resulted in a dismissal decision being ruled 'unfair' (*Property Guards* v *Taylor and Kershaw* 1982 IRLR 175, see p. 64).

However, note must be taken of its contents, if only to answer the queries that may be raised when it comes to notice that employees or prospective employees have erred.

*Rehabilitation*

A person becomes 'rehabilitated' when he has been of good behaviour for a specified period of time since conviction. The duration is dependent upon the sentence, not the offence for which it was imposed. When this time has elapsed the conviction shall be treated as 'spent', that is, regarded for all legal purposes as if it had never happened. No questions may be asked in judicial proceedings concerning it, no evidence given, nor any of the facts associated with it (s. 4(1)(a) and (b) ).

Away from the courts the person will in no way be prejudiced by failure to disclose spent convictions in job applications, and they must be totally disregarded for all purposes in so far as they affect promotion, exclusion from any office, profession, occupation, or employment, or as a ground for dismissal (s. 4(2) and (3) ). Section 4(4), however, gives the Secretary of State power to exclude certain occupations from the provisions of these two subsections, and this has been done for a surprising range of occupations which nearly stultifies the purpose of the provisions. Unfortunately there is no indication of intent to include security work among the exemptions.

Certain convictions cannot become spent. Briefly these are the sentences of life imprisonment, imprisonment or corrective training for more than 2½ years, preventive detention, and 'detention during Her Majesty's pleasure'.

The periods of rehabilitation vary: for example, for a sentence of over 6 months but not exceeding 2½ years' imprisonment – 10 years are required; for below 6 months – 7 years; for fines 5 years. HM Forces are provided for with 10 years for discharge with ignominy, cashiering, etc. and 7 years for dismissal. Rehabilitation begins from day of sentence, and the periods are halved for persons under 17 at that date. Provisions for young offenders are specified: for Borstal training – 7 years; for detention between 6 months and 30 months – 5 years; for periods in a detention centre – 3 years.

It is not proposed to treat the Act in more detail here. Its main security implications are that job applicants can lie about any spent convictions that they have and that nothing can be done about it subsequently. Also, if an employee is known to have spent convictions they must be totally disregarded in every way, unless in each case the job is an exempted one. Finally, if spent convictions are known they should not be discussed or the person may be enabled to sue for defamation.

*The Rehabilitation of Offenders Act (Exceptions) Order 1975*

This Order became effective on the same day as the parent Act. It lists:

1 Jobs where a spent conviction may still be a lawful reason for dismissing or refusing to employ.
2 Professions where spent convictions may be a bar to admission and be mentioned in any disciplinary proceedings.
3 Licences, permits, and certificates that may be denied because of spent convictions.

In the first provision applicants are not protected by the Act if they fail to disclose a spent conviction, provided that they are warned at the time that the job is one to which the Order relates and they must disclose. No legal penalties accrue from lying, but employment may be refused or, if the spent conviction is learned of after employing, dismissal or demotion may legitimately follow. There are no fewer than 22 such jobs listed in the original Order alone, typical ones being the police, clerks to magistrates' courts, prison-service officers, traffic wardens, probation officers, teachers, and various employments involving access to persons under 18. Despite many parliamentary and pressure group representations there are no indications that security officers will be added to the list in the foreseeable future.

Similarly, in the second provision spent convictions may be considered in evaluating an individual's suitability for certain professions. Ten were originally listed, and they include: barristers, solicitors and advocates; dentists, doctors, midwives, opticians, and chemists; chartered accountants; and veterinary surgeons.

Eight categories are listed under the third provision where again the necessity arises to warn the applicant of exemption from the Act. Examples include: licences, permits, and registrations applicable to firearms and explosives; those granted by the Gaming Board; and those associated with abortion clinics, nursing homes, or homes for the disabled or mentally handicapped.

Though the case for security personnel to be added to the list of exemptions seems formidable and more deserving than many of the present inclusions, there is little doubt that opposition, particularly that of the Home Office, is based on the difficulty of defining what constitutes 'security' and the danger of other bodies claiming comparable or prior responsibilities.

## *FIDELITY BONDING*

Insurance cover may be obtained against the risk of loss from the actions of a dishonest employee who takes either money or goods. Not all insurers will have such policies and they may be restricted by conditions and limitations. Single persons, or departments of employees may be covered, or, alternatively, the posts they hold. The amount of the cover can be agreed for each name/post or a blanket cover may be obtained for all with a total sum specified which would cater for theft by a group of employees. Transport drivers are frequently subject to such bonding as are in-company and contract security.

## *PERSONAL REFERENCES*

References have been repeatedly mentioned as a necessary check on the integrity of prospective employees. The need for care in preserving confidentiality if an adverse reference is received – and if an adverse reference is given – is spotlighted by *Lawton* v *B.O.C. Transhield Ltd* 1987 IRLR 404, a High Court QB Division hearing.

The plaintiff found out that an adverse reference given by a previous employer cost him his job – he had nominated that person as a referee. Though he lost his case, it was stated clearly that the giver of a reference had a duty of care to a previous employee to ensure the accuracy of what was given. It might be thought that the recipients of the reference also had a duty to the sender to ensure that the document was kept confidential . . . there are virtues in the practice of some firms in making verbal enquiries before confirming a job offer.

# 3

# *Responsibility for security*

The ultimate responsibility for every activity in an organization is vested in the managing director, who is answerable to the shareholders. Hence if something catastrophic in a security sense happens which affects the viability of the concern, or if lives are lost through gross neglect, managing directors cannot 'pass the buck' to someone above them. If there are such risks, they should therefore designate responsibility for the function to a director or executive in charge of a department to which, if security staff are employed, they can be attached. In practice there is no uniformity in the departments chosen. Personnel is the most usual, but there are drawbacks to this in that incumbents may be influenced in the decisions that they may have to make by a wish not to do anything to jeopardize industrial relations; they could find themselves in something of the role of prosecutor, judge, and defender all at once. For some reason, production or engineering is also sometimes selected; it is difficult to see any virtue in this. On the other hand, there is merit in attachment to a 'neutral' department such as the company secretariat or audit if this is of adequate size and importance, as in large groups it might be.

On large sites a 'services department' may have been formed which combines such trades as joiners, electricians, plumbers, builders, cleaners, etc. whose main tasks are maintenance of the premises. Security, as a service function, may be dogmatically, and indeed logically, attached to this with the services manager as its titular head. It is doubtful whether this is a good thing or not since the members of the composite department are those with the greatest opportunity to steal and managers may be confronted by situations which test their impartiality.

The basic fact is that the nature of the unit to which security is allocated is of much less importance than the integrity, interest and forcefulness of the person in charge.

Irrespective of whether there are persons employed in a full-time capacity, departmental heads cannot divest themselves of obligations for the protection of the company property entrusted to their care nor for that of the personnel they control – also for the health, safety, and welfare of those persons. Incidents showing a failure to meet these obligations will reflect on the competence of the managers to manage, regardless of their qualifications in their own discipline. It is a misconception that the security manager or security department is solely responsible for security; they are there to assist management of all grades to carry out that aspect of their job.

The status of the individual given special responsibility for security will depend on the size, importance, and nature of the firm. It can range from board level director, as in many American multinational groups which frequently have a designated 'director of security', down through the management structure. At a group engaged on substantial UK government contracts it is likely that a director would be responsible, with full-time security officers reporting to him from different units. According to the need, a manager might be allocated a part-time responsibility either with no staff or with persons whose skills are only at the basic level required for routine duties; conversely a person of supervisory status, with training and experience sufficient to advise on the policy and legal aspects of the work, could be appointed. The choice will depend on the multiplicity of risk factors mentioned in Chapter 1.

The danger with the part-time function lies in the fact that it will be subordinate to the main job and will become progressively more neglected, unless the incumbent is an unusual person and becomes really interested in the subject. If, however, his is a largely administrative or decision-making role with an experienced person in operational charge and reporting to him, the arrangements may well prove effective, and this of course is the most usual structure in UK industry at present.

Security has to be flexible to cope with new risks due to changing trading circumstances and variations in the techniques of thieves and other offenders. Past experience on the part of the practitioner is not enough; he must keep up to date with what is happening around him.

The spate of wage attacks which began in the late 1970s found far too many firms with totally inadequate precautions for both money and employees, a state of affairs which also reflected upon their insurers for permitting it. In many areas there was a lack of urgency to take any steps to tighten up procedures until impetus was given by an incident in the near vicinity; indeed the Home Office decided to circulate an advisory leaflet throughout industry with recommendations.

It is a sad reflection on security attitudes that nearly a decade later the incidents still continue unabated. The person in charge of security at a firm cannot divest himself of responsibility either in allowing a dangerous risk situation to exist initially or for any delay in taking action when external circumstances clearly dictated it is necessary. The initiative in advising management on such matters must come from security without waiting for an outside agency to do so.

Apart from an ongoing study of internal matters, the practitioner must develop liaisons locally with the police and industrial and commercial firms with like problems and, on a wider basis, with competitors in the same line of business. Several excellent associations have been built up with the common objective of theft prevention. Those of the British Non Ferrous Metals Federation, which now is linked to the engineering and British Steel Corporation groups, and the brewers are excellent examples of what can be done through mutual co-operation and exchange of information. As one security chief said: 'I either weep with, or laugh at, our competitors when they get "done", but I make darned sure it doesn't happen to us – and I know they do the same when we suffer.'

A number of other organizations exist for liaison purposes. These are not quite on the same lines as those mentioned but have central offices with a full-time staff. These have done excellent work in their specialized spheres.

| | |
|---|---|
| Construction industry: | Construction Security Advisory Service, (CONSEC)<br>82 New Cavendish Street,<br>London W1M 8AD |
| Tobacco industry: | Tobacco Advisory Council, Security Liaison Office,<br>Glen House,<br>Stag Place,<br>London SW1E 5AG |
| Wines and spirits: | Director,<br>Wines & Spirits Association of GB,<br>356 Kennington Road,<br>London SE11 4LD |

Road haulage: Director,
Atlas Express Group Ltd,
Canon Beck Road,
Rotherhithe, London SE16 1DG

Manpower is one of the great costs in any operation, no less in security than in any other. Electronic or other measures could provide an alternative to an increase in departmental strength; it follows that exhibitions of equipment should be attended to ascertain what is being developed in burglar alarm, fire detection, radio, closed-circuit television, and other directions which may prove helpful. The professional journals contain details and assessments of these developments together with reports and comments on incidents and legislation; these should be purchased and passed down to staff to sustain their interest and add to their knowledge. Conferences and seminars likewise should be considered as part of routine training; all else apart, the liaisons formed there may prove invaluable. Security heads who regard themselves as self-sufficient in all ways delude themselves and are not observing their responsibilities to their employers.

## *APPOINTMENT OF A SECURITY SPECIALIST*

Eventually, a firm with problems has to decide whether it is going to employ a person with special skills to devote his main effort to solving or mitigating them. The police will assist with separate incidents, and their crime prevention officers will give advice. But their manpower is insufficient for the public calls upon them, and they have neither mandate nor wish to provide a regular presence at any business concern. Their advice very soon will be to employ someone who knows what he is doing on a full-time basis.

Analysis of the intended content of the job should be the first step as this will influence the kind of experience required by applicants. Ideally, departmental or divisional heads should be consulted to ascertain where the services of the proposed incumbent would be of assistance to them and what qualities or qualifications would be beneficial. Such consultation could extend the terms of reference beyond that originally intended. Unfortunately this rarely happens in practice. Some series of incidents or external persuasion usually causes a high-level decision to make an appointment. A personnel manager will then probably enquire from friendly firms or from his colleagues in the Institute of Personnel Management, get a job description applicable to an existing post, and amend it to the requirements of his own firm.

He may or may not indulge in internal discussion before finalizing it and may even amend it after gleaning new ideas from the interviews.

Selection will be influenced by the main purpose of the job. If the predominant risk is fire, detailed training or past experience in a fire brigade will carry greater weight than, say, police service. On the other hand, if crime is rampant it is easier to graft on fire training than to build-up desirable legal knowledge coupled with instinct for dealing with the dishonest; the same would be applicable to a combined proposed safety/security post. It may come down to a matter of priorities and judgment as to whether an applicant has the capability to be trained in unfamiliar skills. Management in creating the post should have decided what general training should be given to the proposed incumbent; budgetary preparation, management techniques, company policies, familiarity with departmental functions and personalities, etc. are all matters for inclusion in a programme the applicant should be deemed able to assimilate.

Under *no* circumstances should an internal appointment to a supervisory security position be made purely for welfare, convenience, or personal reasons, as in the case of redundancy. The whole organization will know that the position is a sinecure, respect will be nil from both subordinates and colleagues, and if ever one of these moves has been successful it is, in Scots legal parlance, 'outwith' the knowledge of the authors.

## *JOB DESCRIPTION*

The preparation of a job description is no different from that for any other supervisory post, and the manner in which it will be compiled will be in accordance with the company practice. Amongst desirable inclusions are discretion to vary working hours in accordance with security requirements and to make out-of-hours visits. The limits of the security chief's disciplinary authority outside his own department should be defined. The general rule seems that powers of suspension are vested in an offender's supervising line management, but occasionally this is written into the job description of a chief security officer (CSO).

A form of job description used by one group is shown in Appendix 3. However, standard headings of general application are now more conventional. Those with particular security overtones are as follows:

*Main purpose*

This has to be a brief statement to be expanded upon later. It could, for example, be 'to establish efficient security procedures and supervise the work of 24 security officers to prevent loss to the company from unlawful action'.

*Contacts*

1 Internal – this should list those with whom the job holder has necessary and regular contact, exclusive of his own department. In the case of a security head this could well be 'all levels and departments of the workforce, supervision, and management'.
2 External – this is self-explanatory and would include police, fire brigade, firms in like business and in the immediate locality, commercial security firms, electronic alarm suppliers, and professional security associations.

*Dimensions*

1 Personnel – this simply lists the persons and status of those over whom the job holder has full authority – that is, the security staff and any other ancillary personnel who may have been acquired by additions to the basic job (cleaners, firemen, weighbridge operators, etc.).
2 Annual operating budget – again self-explanatory, being an estimate of the annual expenditure to be incurred by the job holder and his subordinates. Conformance with this budget could be a factor in the appraisal of the efficiency of the security head.
3 Other – this will be determined by specialized requirements of the specific job.

*Extent of authority*

1 Personnel – including:

   (a) Appointment – the degree of discretion that the incumbent is allowed in the selection of staff.
   (b) Discipline – the extent of powers of action without reference to superiors – for example, oral and written warnings or dismissal.
   (c) Overtime – the degree to which overtime may be authorized. Usually this is left to discretion unless it becomes excessive or a strict economy drive is in force.

2 Expenditure – the level to which the incumbent in his own right can authorize expenditure and any restraints upon his action in doing so. This will be applicable normally to the operation of his own department; he will be recommending to other departments that *they* incur expenditure.
3 Other – this would normally specify any other limitations upon his authority. Probably power to suspend in certain circumstances would be given under this heading, if allowed.

*Experience and qualifications*

Self-explanatory, this would list what the employer thinks are essential, desirable, or merely advantageous qualifications for the job. These could take the form of the prior holding of an army or police rank, supervisory experience in security work, the holding of an institution qualification, etc.

*Controls and checks*

These would list the parameters under which the incumbent carries out his work – for example, periodic meetings, reports, statistics of losses, etc.

*Responsibilities*

1 Planning and budgeting – this would show to what extent the incumbent has discretion in planning his work and that of his team, or working to prearranged schedules, and how far ahead he would be expected to plan; the procedures he has to carry out to obtain the materials used by him in his work; the degree to which he would be required to submit budgets; and the discretion he has to change systems and methods of working in his department.
2 Operation and co-ordination – this would be the detailed list of day-to-day activities and responsibilites. Only the most essential should be described. In the case of a security head they should be so framed as to allow maximum latitude in interpretation since he will be frequently involved with unpredictable situations.
3 Personnel – refers to training, welfare, motivation, and merit appraisal responsibilities in respect of his staff.
4 Efficiency – this would specify the criteria whereby the work of the job holder would be judged. It is difficult to define in the case of security work and might be expressed in terms of achievement of objectives and lack of incidents and complaints.

*Conditions*

These would mainly list special conditions calculated to affect the value of the job in so far as hardship and danger were concerned. In the case of an ordinary security officer it is not unusual to find that 'his job may involve the acceptance of a degree of personal danger in its performance'. Necessity to vary hours of working and to pay out-of-hours supervisory visits might be mentioned here.

## *STATUS AND SALARY*

The remuneration and position in the management structure for the security chief must reflect the importance attached to the post. He should be equated with other management of like level for pay scales, merit and incremental awards, and any other concessions or fringe benefits that they enjoy. If during the interviews an applicant with experience who has an occupational pension is asked questions about it, he would be well advised either to walk out of the interview or to tell the interviewer that it is a personal matter with no bearing on his qualifications or application. Petty-minded firms do exist that hope to get a cut-price employee who wants the job because he lives conveniently and thinks he will not be called upon to accept much responsibility. The fact that such a question is posed is indicative of the lack of respect with which the appointment is regarded. As in any occupation, the employer gets what he pays for.

Before accepting the job, the proposed incumbent would be well advised to clarify both his starting salary and the scale of increases he may expect, the wages of his subordinates will be of interest and may have bearing on his acceptance. They will have shifts and inconvenience allowances which, not infrequently, create the ludicrous situation where the employee with no responsibility has a higher wage than the person in charge of him who has all the decision-making and supervisory onus plus the inconvenience of being on permanent call. Many applicants who have not encountered that state of affairs may find it unacceptable.

Status is important: though the security chief may report to perhaps the personnel director, unlike practically all other staff he will come into contact with all levels of management and workforce throughout the organization, and he must be allowed flexibility in the protocol of approaching and being approached. It would be an invidious state of affairs if, for example, a transport manager had to ask the personnel director for the CSO's services to investigate a theft from a lorry or if the CSO, before making an enquiry from a departmental head, had to refer through his own director to the director in charge of the person he wanted to approach. Impossible state of affairs? It has been encountered, and no credit attaches to the firms concerned.

This flexibility should be extended to many of the reports that have to be submitted in the course of security work. Unless the personnel director wishes to see every recommendation or incident report, there is no reason why they should not be sent directly to the individual requiring them, with a copy to the director on a 'need to know' basis. This is a good factor in improving personal liaisons and creating a security interest in the recipient.

Obviously if the status is equivalent to that of manager, this would faciliate dealings with managers, but the responsibilities might not justify anything more than that of senior supervisor. It would be unusual to encounter any 'pulling of rank' and obstructiveness to the carrying out of security duties, but the security chief should have the personality politely to embarrass any obstructionist and to carry out what he should do without having to refer to his own superior for backing, though he should not hesitate to do this if necessary.

## *CATEGORIES OF SUPERVISORY SECURITY OFFICERS*

The terminology used to denote status varies, and firms have their own foibles in this, but generally the following are used:

1. *Chief security officer (CSO)* is in charge of staff employed primarily on security duties. He may have additional responsibilities and be named accordingly '*chief security and safety officer*'. Fire is often an accepted inclusion, but occasionally '*chief fire and security officer*' appears.
2. *Security officer* may be used with the same meaning as CSO where a firm calls its patrol and gate personnel '*security guards*'. The term would have the same permutations with the safety and fire additions.
3. *Company security officer* has a co-ordinating and advisory role where the firm operates a number of sites each with CSOs and staff accountable locally. He will undoubtedly deal with major or sensitive enquiries and those involving senior staff.
4. *Divisional security officer* is used as 'company security officer', but the 'company' forms a division of a larger group.
5. *Group security officer* – The term is only used by very large groups with several divisions. The incumbent usually has an advisory and monitoring role, only becoming operationally involved in the most important matters, and has managerial or executive status.

6 *Security adviser* is very similar to 'group security officer' but has little operational involvement. In some cases it may be a part-time appointment involving supplying advice against specific risks and presenting recommendations. This is also a title used by employees of commercial security firms. If their services are to be used other than for specialist guidance on alarm systems, the customer should satisfy himself as best possible that the 'adviser' has the necessary experience to justify his adoption of the title. There are no legal means of restricting its use or that of its alternative, '*security consultant*'.
7 *Security manager* – In practice the title is not used as much as it should be. It is applied to the head of a large security force, usually on one site, and the incumbent is responsible for all administration, budgeting, etc., as in any other department.

## *ACCOMMODATION*

The incumbent needs a separate office for privacy in his own work, to segregate himself from his staff, and for undisturbed interviews with callers and employees who want to discuss confidential matters. If the extent of his responsibilities does not justify a personal secretary or even a share of one with perhaps the safety officer, he must have recognized access to a source of typing which can be trusted to respect 'confidential' or 'secret' correspondence.

Usual office furniture is required, plus adequate filing cabinets, one at least of which should be fitted with a locking bar if confidential papers are to be held. The normal cabinet is given little security by its lock and is easily forced; duplicate keys can be freely obtained if the number has been left on the lock surface. To improve this, U-shaped pieces of steel are welded or riveted to the top and bottom of the outer frame, directly in line with the handles and of a size to take a rectangular bar of adequate length to protrude through the bottom piece with a head to rest on the upper piece, with a hole immediately beneath the top handle to take a strong combination padlock.

For communications purposes he should have an internal telephone and preferably, in a large or high risk organization, an external line which does not go through the switchboard. As he will have a 24-hour and seven-day-week responsibility he should be on the telephone at home at the cost of the firm whose interests dictate the need. The general practice is that the rental is paid plus the cost of all business calls listed and claimed.

## *ANCILLARY DUTIES*

A CSO can easily have other responsibilities added to that of security. Fire, as mentioned before, is in many cases accepted as part of his job, which is logical since his men must be fire trained and must provide the continuous cover which is rarely given by a safety officer. Safety is a feasible addition, but the pressures of the Health and Safety at Work Act strictly limit the size of unit where both jobs can be combined without loss of efficiency. As a rule of thumb, a factory with over 750 personnel, or with complex or varied processes, really needs a safety officer with a full time commitment.

Civil defence rarely achieves a mention, but a CSO may find himself holding the files together with the responsibility, even if it is purely nominal. The weighbridge could come under his jurisdiction as might cleaners in premises of limited size. Where a firm utilizes hire transport rather than its own fleet for dispatching its products, the booking of this sometimes finds its way under a CSO's wing. In short, an active and intelligent man could find himself developing into a general factotum, deserving the title of 'services manager' which several now have.

Though not a separate duty, a CSO could also find himself as a member of a risk management or total-loss control team. In multinational firms, a total loss controller might well find himself responsible for executive protection – the protection of senior management and their families. In some countries this is deemed increasingly important – the US in particular if the number of 'consultants' is any criteria. A controller would be most unwise if he relied entirely on his own judgment and facilities without seeking guidance and assistance from a genuine specialized company (see Chapter 33).

## *TOTAL LOSS CONTROL*

This is the description given to the co-ordination of security, accident prevention, safety, fire prevention, damage control, and insurance functions in commercial and industrial concerns, to prevent and reduce losses. In larger organizations the various loss prevention responsibilities are co-ordinated under a full-time total-loss controller to whom those responsible for the separate functions described report. The concept was an American one which has found favour with a number of firms in the UK but has not spread to anything like the extent for which its protagonists hoped – perhaps on the supposition that, if the departments concerned are individually efficient and liaising as they should, a combined body is unnecessary.

Another way of securing the same result but without the appointment of such a person is to have periodic meetings of those officials under either a permanent chairman or one drawn from the members by rotation.

The permanent inclusion of departmental managers or individual ones as required when a matter affecting their responsibilities is to be discussed, has advantages. Where an internal audit department exists, the person in charge should be invited to the meetings to collaborate in the assessment of risks and in introducing measures and controls towards the common purpose.

# 4

# *Security practice and industrial relations*

The operational ambit of a security department inevitably involves it in incidents of misbehaviour by employees which fall into the category of 'gross misconduct' – theft is an obvious example. In all commercial and industrial undertakings these are seen as serious disciplinary offences which merit the potential penalty of dismissal. Such punishment may become an emotive issue to other members of the workforce and result in action which disrupts routine functioning of the organization. This is much less likely to happen if the facts are clear and the managerial decision patently reasonable. However, should management have been misled by an inadequate, inaccurate, or incoherent version of the occurrence, its decision is likely to be justifiably and successfully disputed – the effect on the reputation of the security department, and the incumbent in charge, will then be both adverse and long-lasting.

Whilst the importance of clear, concise and unbiased reporting is self-evident, security heads should be thoroughly conversant with their employers' disciplinary procedures and be sufficiently aware of precedents to prevent mistakes which might provide avenues whereby just retribution might be avoided or mitigated. A high percentage of all appeals against dismissal are for security-based offences which emphasizes the necessity for such knowledge. Moreover, senior managers and personnel officers, when adjudicating, may be doing so under distracting pressures of other commitments – a modicum of expertise on the part of the security head might then be invaluable and react favourably for the personal respect in which he was held. As a Hong Kong security officer expressed himself on a similar matter 'he could gain much face with his employer'!

## *LEGISLATION*

The Acts upon which industrial relations law relative to dismissals is based have been in effect now for sufficient time for a sound framework of judicial precedents to have been established. Any changes in Government are deemed unlikely to affect this form of existing legislation. Those decisions bearing on matters likely to arise during security practice are fully covered hereunder with detail to enable them to be quoted in discussion on policy or action. The thought and principles behind the judgments may be of value when similar incidents crop up outside the UK.

The principal UK statutes concerned with unfair dismissals are the Trade Union and Labour Relations Act 1974 and the Employment Protection Act 1975. These have been supplemented by subsequent measures found necessary for clarification and extension of some of the provisions. The Employment Protection (Consolidation) Act 1978 and the Employment Act 1980 are particularly noteworthy in their confirmation of previous tribunal rulings.

The statutes create an appeal sequence that may be followed by an aggrieved employee when representations against dismissal have been rejected at all levels of the employer's own internal appeal procedure. In the first instance recourse is to an industrial tribunal, thereafter, if still dissatisfied, to an employment appeal tribunal (EAT). A limited number may go further to a Court of Appeal and a handful of these, deemed issues of public importance, may go on to the House of Lords. If a tribunal finding is against the employer, he has the same right of appeal to the superior courts. The constitution of these bodies is in accord with their progressively weighty authority of judgment. An industrial tribunal had a legally-qualified chairman (usually a solicitor) and two lay members, one of whom is experienced in the managerial aspects of business and the other versed in trade union matters; an EAT has a High Court Judge presiding, assisted by two similarly qualified assessors; a Court of Appeal comprises three High Court Judges; the House of Lords is represented by a panel of 'Law Lords' – Judges who are members of the House.

Both the grounds for dismissal may be challenged before a tribunal and the manner in which it is carried out. If either is found to be unfair the employer may be ordered to reinstate and/or pay compensation within laid down limits. A successful appeal can be a very costly business for an employer and one which may attract undesirable publicity.

It is noteworthy that the fairness of decisions to dismiss is judged upon the information that was available at that time and upon which the decision was made, not to be justified by anything which might come to light later, though that could have bearing on any compensation award – *W. Devis & Sons Ltd* v *Atkins* 1977 IRLR 314 and confirmed by many subsequent cases.

Disciplinary rules apply equally to a security department and an efficient security head will make sure he knows both the principles of such rules and the manner in which they are applied in his firm.

### *Disciplinary rules*

Employers have to provide written details of their disciplinary rules and procedures and must adhere to them. The essential ingredients are that rules should:

1. be speedy in operating;
2. be in written form;
3. clearly show to whom they apply;
4. outline the punitive options;
5. specify who has power to do what – disciplinarily;
6. provide that the accused is informed of the complaint and given an opportunity to explain and the right to be represented and advised;
7. give a right of appeal, means to acquaint the accused of this, and the reasons for the disciplinary decision;
8. provide that there will be no dismissal for a first offence, *other than for serious misconduct*;
9. provide that no dismissal will take place without adequate investigation and, if necessary, for suspension with pay to apply while this is being done.

### *Wording of rules*

Even the phraseology used in framing rules has been found grounds for rendering a dismissal unfair. That most frequently challenged was 'liable to dismissal' and it was contended that an employee might reasonably take the view with such wording that to be caught once might not lead to instant dismissal. Surprisingly, most of these instances seemed to be associated with 'clocking offences'. Eventually a more realistic view, not subsequently challenged, was taken in *Elliott*

*Brothers (London) Ltd* v *Colverd* 1979 IRLR 92, when the EAT held that 'there is no rule of law that a warning must indicate inevitable dismissal'. However, the moral is clear and 'will be dismissed' or 'will normally be dismissed' has been advocated as preferable.

*Breach of rules by employer*

Similarly, a dogmatic approach was at first applied to instances where an employer did not strictly conform to laid down procedures in dealing with a disciplinary matter. However, a more flexibile approach has been applied here too with a case of Court of Appeal authority, *Hollister* v *National Farmers Union* 1979 IRLR 238, where the ruling was that 'one has to look at all the circumstances of the case and whether what the employer did was fair and reasonable in the circumstances prior to the dismissal'. In other words, any defects in procedure have to be weighed against the other relevant factors. Again, the message is clear, keep to agreed and laid down procedures and that form of controversy will be avoided.

*Stricter enforcement of rules and lack of consistency*

Where rules have been applied leniently or not enforced at all, any tightening should be done with care and due notice of intention and reason. Sudden implementation of a dormant search clause, incorporated in conditions of employment but not put into practical effect, would be grossly resented by employees and could lead to refusals, disciplinary action for which, if carried to the point of dismissal, would be unlikely to be sustained before an industrial tribunal.

Similarly, where various 'perks' or malpractices have been tolerated by management, to dismiss for 'gross misconduct', on the grounds that they really constitute 'theft', could court both an adverse tribunal decision and the unwanted publicity of what may be colloquially referred to as 'washing dirty linen in public'. In *Cambria Mobel* v *Cridland* (11 October 1976) driver Cridland was dismissed for booking hours he had not worked, amongst other reasons. The firm had previously allowed recording of hours to be manipulated to its own advantage – the EAT confirmed a finding of unfair dismissal and said in effect the company had only themselves to blame.

While courts accept some flexibility in the punishments that are handed out for like offences, outstanding and unexplained inconsistencies are not accepted. Several cases have arisen where employees, dismissed for sleeping when they should have been working, have been reinstated because others previously have been suspended or even reprimanded when found doing the same. If there is to be such a change of heart by management, a clear advance notice must be given of the intention and preferably the reasons for it. The pre-eminent in a long line of cases stressing the importance of consistency is a Court of Appeal case relating to assault – *Post Office* v *Fennell* (1981) IRLR 221.

### *Theft as grounds for dismissal*

A common principle has been expressed in all recent decisions in cases concerned with theft – that dishonesty constitutes a breach of trust by an employee and merits dismissal, irrespective of the sums involved, or degree of damage to the employer's reputation. Simple theft from an employer was regarded as automatically justifying a dismissal in *Trust House Forte Hotels Ltd* v *Murphy* 1977 IRLR 186, and similar attitudes have since been confirmed for 'fiddling' expenses, falsely claiming sick benefit, defrauding customers, clocking offences, etc.

### *Necessity for adequate investigation*

For its importance to security action in instances likely to lead to the dismissal of an employee, *British Home Stores Ltd* v *Burchell* 1978 IRLR 379 is of importance comparable to *Christie* v *Leachinsky* in criminal law. Though concerned with a relatively minor matter, the kernel of its ruling has been repeatedly referred to and confirmed by courts since. It merits quoting in full:

> In a case where an employee is dismissed because the employer suspects or believes that he or she has committed an act of misconduct, in determining whether that dismissal is unfair an Industrial Tribunal has to decide whether the employer who discharged the employee on the ground of the misconduct in question entertained a reasonable suspicion amounting to a belief in the guilt of the employee of that misconduct at that time.
>
> This involves three elements. First, there must be established by the employer the fact of that belief; that the employer did believe it. Second, it must be shown that the employer had in his mind reasonable grounds upon which to sustain that belief. And, third, the employer at the stage at which he formed that belief on those grounds, must have carried out as much investigation into the matter as was reasonable in all the circumstances of the case. An employer who discharges the onus of demonstrating these three matters must not be examined further. It is not necessary that the Industrial Tribunal itself would have shared the same view in those circumstances.

Apart from these principles being confirmed by other judgments, an important Court of Appeal case, *W. Weddel & Co. Ltd* v *Tepper* 1980 IRLR 96, extended them to all cases of dismissal – the *Burchell* case being an issue of suspected dishonesty:

> The correct legal test for determining whether an employee dismissed on grounds of alleged misconduct has been fairly dismissed is the three stage test set out by the EAT in the 1978 case of *British Home Stores Ltd* v *Burchell.*

Further confirmation was given of requirements that have to be met and these are very clearly spelt out – their importance to any security officer carrying out an investigation is obvious (except in most unusual circumstances – see *Pritchett and Dyjasek* v *J. McIntyre Ltd* (see p. 60)):

> The effect of this is that where an employee is suspected of having committed a dismissible offence, an employer needs to show: (1) that the dismissal was bona fide for that reason and not for a pretext; (2) that the belief that the employee committed the offence was based on reasonable grounds – i.e. that on the evidence before him, the employer was entitled to say that it was more probable that the employee did, in fact, commit the offence than that he did not; (3) that that belief was based on a reasonable investigation in the circumstances – i.e. that the employer's investigation took place before the employee was dismissed and included an opportunity for the employee to offer an explanation.

### *Opportunity for explanation by an offender*

By disciplinary rules, court judgments, commonsense and fairness alike, an offender or suspected offender must be given a chance to explain his actions. What he says may be an admission, so absurd as to be unbelievable to the extent of self-convicting, or may be a verifiable explanation that negates guilt.

Circumstances, however, may exist which make the simple request for an explanation a much more complex matter – the offender may be in police custody, or may be advised by a legal representative not to answer. This is another area of differing judgments – at least in details, but there is a decisive ruling by the Court of Appeal in *Harris and Shepherd (appellants)* v *Courage Eastern Ltd* 1982 IRLR 509. The gist of the ruling in this instance, in which the two men were awaiting trial, is – if employees were given the opportunity to give an explanation and were told that dismissal was being considered, and they chose not to make a statement, a reasonable employer was entitled to consider whether the material before him was sufficiently indicative of guilt as to justify dismissing without waiting for the criminal trial.

There is further indication of flexibility of approach in *Gibbard* v *C. W. Wantage Ltd* EAT 556/79 where Gibbard, when challenged with stealing diesel, but not asked specifically for an explanation, said nothing. The tribunal ruling was that had the correct procedure been followed, would there have been a different outcome? They found that even if Gibbard had been given a proper opportunity to state his case, it would have made no difference. In like vein, *Scottish Special Housing Association* v *Linnen* 1979 IRLR 265 where Linnen was arrested by the police with property admittedly stolen from his employer who was told the circumstances and dismissed Linnen without further enquiry. This was held to constitute reasonable grounds on the basis of information from the police. Conversely, in *Scottish Special Housing Association* v *Cooke and others*, 1979 IRLR 264, the alleged theft was of articles from the house where the men were working, not from the employer. The offence was denied but the men were dismissed, without any enquiry by the employer, as soon as it was known that the police had charged them. In making an 'unfair' finding, the EAT said 'the mere fact of a charge of theft being preferred, standing by itself and without any further information being available to the employer, is not sufficient to constitute reasonable grounds'.

From a security-action point of view, avoiding pitfalls is simply the carrying out of good security practice – ask for an explanation even when an offender is caught redhanded and probably thinks you are joking – and put that reply on record.

A final case that shows what might happen – *Qualcast (Wolverhampton) Ltd* v *Ross* 1979 IRLR 98. Ross, a gateman, was found asleep on duty by the managing director of his firm; next morning he was dismissed without being asked for an explanation – he had one which could have influenced the disciplinary decision – the EAT confirmed a finding against the firm.

Requesting explanations inevitably involves a degree of questioning and security officers, particularly those with a police background, may be concerned about the limitations imposed by the Codes of Practice of the Police and Criminal Evidence Act 1984 (see Chapter 13) on the asking of questions – the old Judges Rules which are encapsulated in the Act.

Fortunately there has been a definitive EAT on this point – *Morley's of Brixton* v *Minott* 1982 IRLR 220. In this a sales assistant was seen on three occasions not to ring sales on her till. She was interviewed by internal and external security but denied all knowledge; when left with the managing director however she admitted the third incident after a comment by him claimed to be of the type 'I would like to put this behind us' or 'I would like to close the incident'. This, she said, she took to imply she would not be dismissed – as she in fact was. In the first instant, the industrial tribunal held the admission had been induced and was therefore to be disregarded as being in contravention of the Judges Rules. The EAT however ruled positively otherwise and amongst the points made were: no cases could be traced where the Judges Rules had been applied to any branch of civil litigation; the Rules were intended to protect against criminal conviction whereas an employer was concerned with reasonable grounds for belief in guilt only; finally that circumstances might affect the weight attached to an admission but would not automatically exclude it. The comment was also made that encouragement to 'come clean' in disciplinary matters was not improper and that to import the technicalities of the Judges Rules into industrial matters would make the life of employers impossible.

This attitude of industrial courts has been reinforced by cases subsequent to the introduction of the Police and Criminal Evidence Act 1984. In the Court of Appeal, *R* v *Hampshire County Council ex parte Ellerton* TLR (3 January 1985), it was declared 'the burden of proof in disciplinary proceedings is a civil rather than a criminal standard of proof'.

Perhaps the ultimate contrast between criminal and industrial courts is provided by *Pritchett and Dyjasek* v *J. McIntyre Ltd*, CA 1987 IRLR 18, a unanimous judgment of three Law Lords which confirmed the principles of *British Home Stores Ltd* v *Burchell* (see p. 57). This appeal against dismissal was concerned with theft of metal, over a long period, from a firm where there had been a previous prosecution in which, by the express wish of the police, the identities of their informants had been withheld. In this instance, information was forthcoming from fellow employees whom it was necessary to protect by anonymity; this in itself was not sufficient to frame specific charges, and eventually to cut losses the two were discharged without being asked for explanations or told the allegations against them. The appeal was based on the impropriety of the action but the Judges found that the employer had acted in genuine belief of guilt 'after as much investigation as could be reasonably expected in the circumstances'. It was said to be impossible for the employer to put the allegations to the employees because of the assurance of confidentiality to the informers. It would have been meaningless to put generalizations to the accused, such interview would not have affected the employer's decision, hence the appeal was dismissed – this goes beyond the *Burchell* case but only because of special circumstances. A criminal prosecution on the facts given would never have been started, which emphasizes the independence of the industrial courts from legal technicalities.

*Dismissal in cases of court action and police involvement*

At the time an arrest is made or a prosecution process commences, it is nowadays quite impossible to forecast when the hearing will take place. This is due to work pressures on the police which may delay the preparation of the file, and similar overloading of the solicitors prosecuting; it can then be further aggravated by adjournments at defence request and complicated by fitting into an overflowing court calendar. If the case is one for trial before a Crown Court, a biased observer might think any time delay factor has minimal priority after the committal proceedings have taken place.

The consequences for a firm that has a policy of suspension with pay, prior to a charge of dishonesty being proven before a court, can be alarming. In mid 1981, five employees of a large firm pleaded guilty to a series of charges in connection with numerous thefts from their employer; it was 2½ years since they had been suspended on pay after conclusive evidence had been found. An estimated figure of the cost of this was £50,000 and the entire group policy was changed when this was brought to notice. This is perhaps a negligible sum in the context of a huge turnover, but nevertheless the money would come directly out of profits, and possibly be only one of a series of amounts which could make the difference between deficit and profit for a company. In one instance, intervention by head of security, immediately a benevolent personnel officer proposed suspension with pay for four employees, saved his employer nearly £10,000 – they had been caught in the middle of the night by security and management stealing a van load of metal. It was 10 months before the case was tried before the local Magistrates Court – and that with a certain plea of guilty from the outset!

Suspension without pay is tantamount to dismissal if the power is not shown in the contract of employment (*D & G McKensie Ltd* v *Smith*, Court of Session 1976) but numerous authorities exist confirming the right of management to act without delay based on their own reasonable belief as per *British Home Stores Ltd* v *Burchell* and *W. Weddel & Co. Ltd* v *Tepper* mentioned earlier. If time is required to complete an investigation and evaluate explanations, the process should be as speedy as possible and the offender suspended on pay for that period.

As mentioned later, there is no requirement that an employer should notify the police of a suspected or detected offence of dishonesty. If he does so, whether the offender is convicted or otherwise, the court decision has limited relevance to the validity of a dismissal. A criminal court has to have a matter proved beyond any reasonable doubt; a tribunal however has only to decide whether the employer was acting reasonably in forming an opinion that the employee had done what was alleged – the burdens of proof are different. The EAT in *Harris (Ipswich) Ltd* v *Harrison* 1978 IRLR 382 made a further point of importance – that dishonesty may be associated with a breach of rules sufficient in itself to justify dismissal.

> Thus, where an employee is charged with a criminal offence alleged to have been committed in the course of employment, and consequently dismissed, it does not follow that because he is later acquitted, the dismissal was unfair. It will not always be wrong to dismiss the employee before his guilt has been established. For, quite apart from guilt, involvement in the alleged criminal offence often involves a serious breach of duty or discipline. For example, the cashier charged with a till offence, guilty or not, is often in breach of company rules in the way in which the till has been operated. The employee who removes goods from the premises without express permission, guilty or not, is often in breach of company rules in taking his employer's goods from the premises without express permission; and it is irrelevant to that matter that a jury may be in doubt whether he intended to steal them.

Thus, a factory manager, who used his position of authority to cause goods to be sent out for his own purposes, without any records or documentation, and without payment, was dismissed on the dual grounds of breach of works rules (aggravated by his status) and gross misconduct (dishonesty) so that what the police did after the matter was reported to them was largely a source of interest rather than concern to the employer.

Whilst the power to immediately dismiss is not affected, an industrial tribunal is empowered to postpone unfair dismissal proceedings when High Court action is pending (*Carter* v *Credit Exchange Co. Ltd* 1979 IRLR 309). This they can do if, in their opinion, it is in the interests of justice. It is also a further argument, if any were needed, against policies which involve prolonged suspension with pay. This ruling was subsequently confirmed by a Court of Appeal decision in *R* v *BBC ex parte Lavelle* 1982 IRLR 404.

The fact that the police take no action against an offender reported to them does not affect the fairness of that person's dismissal – as in *Patterson* v *Mecca Bookmakers* 1976 in which a Procurator-fiscal decided not to press charges in a Scottish misappropriation of money case.

If there is clear indication of guilt in an offence against an employer, for him to continue to employ the offender pending trial, and then dismiss after conviction, creates a positive risk of a successful appeal against that dismissal. After all the trial would add little to the knowledge already available which had not been acted upon (*Donson & Judd* v *Conoco* (1973) IRLR 258). If, however, an employee receives a prison sentence for a matter not necessarily connected with his work, this may make it impossible for him to perform his part of the contract of employment and creates a yet more complex problem for the employer. The Court of Appeal gave authoritative guidance on this in *F. C. Shepherd & Co. Ltd* v *Jerrom* 1986 IRLR 358, in which an employee was given six months to two years Borstal, saying clearly that the imposition of a custodial sentence on an employee is capable in law of frustrating a contract of employment. The employer therefore has an option and the criteria as usual will be whether the dismissal is reasonable in all the circumstances. Duration of sentence will be a prime factor and careful consideration will be needed if it is measured in weeks rather than, say, months.

A Court of Appeal ruling in *British Leyland* v *Swift* 1981 IRLR 91 is worth bearing in mind to bring to the notice of those who think it unfair to punish by both disciplinary and court action. In this the court commented adversely on a tribunal reference to the offence as a relatively minor one and the implication that the criminal penalty itself was adequate. Per Lord Denning 'That was the wrong approach. If a man is convicted and fined, it is a ground for dismissing him, not for keeping him on.'

### *Criminal offences outside employment*

The conviction of an employee for dishonesty unconnected with his normal work will inevitably raise a doubt in the mind of an employer about his integrity and reliability. There may be a temptation to use the conviction for what might be termed a precautionary dismissal, which is a most unwise procedure in most circumstances. It has been laid down that 'the main considerations should be whether the offence is one which makes the individual unsuitable for his or her type of work or unacceptable to other employees'. Duration of penalty has already been mentioned as 'frustration of contract' and that might be a solution, but a remand in custody should not be used as the sole reason for dismissal. Alternatively, the nature of the offence might well have a bearing upon the capabilities of the individual to carry out his job.

A cashier convicted of stealing funds from a club where he was honorary treasurer could no longer be trusted; a delivery driver who stole from customers; a store supervisor who stole from an adjoining retail store – in each instance what they did directly reflected upon their acceptability to do their own job. Each case has to be considered upon its own merits.

### *Convictions prior to employment and false information by job applicants*

Not infrequently, management find that a person recruited for a specialist job is useless, having obtained it by claiming qualifications he did not possess; less often, that someone in a position where integrity is essential has been convicted of dishonesty. There is no difficulty in dismissing the former who, in addition to contravening any 'satisfactory references' or correctness of information (supplied conditions on his application or contract of employment), almost certainly will have commited an offence against his new employer by obtaining a pecuniary advantage by deception (Theft Act 1968 s. 16 – see p. 165), that is, the job and its financial remuneration would not have been given but for the false statements.

Failure to disclose convictions for dishonesty is complicated by the Rehabilitation of Offenders Act (see p. 38) which makes it illegal to take any action in respect of offences which have been cancelled, under the provisions of the Act, by the passage of time since the conviction. The only instance which has come to notice in official reporting where the provisions of the Act have been applied as the decisive factor in an unfair dismissal finding is that of *Property Guards Ltd* v *Taylor and Kershaw* 1982 IRLR 175. This coincidentally concerns a contract security company which accepted two applicants for employment who signed statements to the effect that neither they nor any members of their families had ever been convicted of any criminal offence. Subsequently it was found that both had been so convicted, but lapse of time had caused their convictions to become 'spent' and therefore to be disregarded. They were dismissed but successfully appealed to a tribunal whose findings were upheld by the EAT. It is obvious that such 'spent' convictions have to be treated as if they had never happened, unless the employment is an exempted one (see p. 39).

*Torr* v *British Railways Board* 1977 IRLR 184 is the most frequently quoted authority. In it the EAT found it was not unfair to dismiss an employee who had failed to disclose an unspent conviction, even though he had held his job 16 months before the fact came to light. It is necessary that the question of having previous convictions is specifically raised and that the answer should have a bearing on the decision to employ. It is strongly recommended that a special section should be filled in on a signed application form; where an interviewer has written in himself 'no previous convictions' on the side of a standard form, it has been contested that the question was never asked. It is good practice to include that section on all forms where integrity is an essential ingredient of the job, for example security, transport drivers, cash handlers, etc. There is no obligation to qualify the question by wording such as 'any unspent conviction under the conditions of the Rehabilitation of Offenders Act – otherwise an interviewer had better familiarize himself with the intricacies of the Act and be prepared to waste explanatory time with each applicant!

Tribunals have taken a similar strict view where there was failure to disclose an adverse medical history, for example *O'Brien* v *The Prudential Assurance Co. Ltd* 1979 IRLR 140, which is worthy of note since specific answers should be required from security guard applicants about their health record. However, where the misleading information relates to previous employments, there is sufficient ambiguity in judgments to suggest each instance should be dealt with on its own merits and commonsense applied.

### *Miscellaneous*

A number of other rulings are of importance to a security head:

1 A security guard was fairly dismissed for refusing to sign a witness statement in connection with an employer's legal action against secondary pickets. It was held that his employer was entitled to expect that co-operation from him in their intended litigation (*MacKenzie* v *York Trailer Co. Ltd*).
2 An employee, in certain circumstances, can be dismissed for negligence resulting in the loss of his employer's property – milkroundsman lost his cash float for a second time after receiving a warning for carelessness (*Jackson* v *Home Counties Dairies Ltd* (IT Brighton, 10 October 1977)).
3 A commercial traveller was fairly dismissed for refusing to comply with an express term of his contract that the company's merchandise should be taken out of his car at night for safety. He was off sick and claimed this suspended his contract (*Marshall* v *Alexander Sloan & Co. Ltd* 1981 IRLR 264).
4 If an employer considers a clocking offence, which is admitted, and is known to be serious, is one which justifies instant dismissal it is not for a tribunal to usurp the function of the employer – EAT ruling (*Taylor* v *British Sisalkraft* 1981).
5 An employee admitted involvement in stealing company property and was given the option of 'resign or police will be called'. He resigned and later claimed unfair dismissal. The tribunal said 'fair',

it was not a case of resign or be sacked (*Thomas* v *Everest Frozen Foods Birmingham* (23 October 1978) ). This was later confirmed by a Court of Appeal decision in *Martin* v *MBS Fastenings Distribution Ltd* 1983 IRLR 198. A drunken employee damaged a firm's vehicle, was breathalyzed with a positive result and was told the enquiry would probably end in dismissal. He was allowed to resign and he then appealed on the grounds that this was tantamount to a dismissal – ruling, not a dismissal.

6 Inordinate delay in confronting an employee suspected of dishonesty may render dismissal unfair – nine days elapsed between discrepancy queried and confrontation when option shown in 5 above was given; EAT criticized delay rather than option given (*Allders International Ltd* v *Parkins* (1981) IRLR 68).
7 The giving of a police-style caution to an employee was held to have effectively stopped him from giving an explanation and dismissal was therefore unfair – EAT (*Ladbroke Racing Ltd* v *Mason* 1978).
8 A police request not to correspond with a dismissed employee is not in itself a good reason to refuse a request for written reasons for dismissal (*Daynecourt Insurance Brokers* v *Iles* EAT (21 June 1978)).
9 Breaches of trust by persons in positions of authority are to be more seriously regarded because of their status (*Mansard Precision Engineering Co. Ltd* v *Taylor and another* EAT (15 December 1976)).

## *GENERAL COMMENT*

The foregoing constitute only a part, albeit a major one, of the instances in which specialized knowledge on the part of the security head may be of assistance to his employer in dismissals likely to be appealed to industrial tribunals. Security staff may become involved in complaints of sexual assault or indecency (see p. 179), assault (see p. 177) or the use of obscene language. Suffice it to say that court rulings have been sufficiently diverse on sexual matters for the comment to be made that it was safer for an employee to press ahead with an indecent assault than talk about doing it! In the case of fighting and bad language it is obvious that particular circumstances rather than binding rules are the criteria. The former generally accepted blanket rule of automatic dismissal for those fighting may now be unfair and it is essential to establish who was the aggressor and who the victim; in verbal violence, what may be obscenely offensive to some listeners may be everyday parlance to others.

Important though security involvement with dismissals may be, and indeed it is that in which the department comes most forcibly to the notice of the rest of the workforce, there are other aspects. The maintenance of disciplined behaviour by employees is very largely dependent upon their voluntary co-operation in the observance of works rules. This can be thoroughly jeopardized by thoughtless or aggressive performance of duties by security staff. There is no quicker or surer way of initiating an industrial dispute than by trespassing on the dignity or rights of a fellow employee without justification. Any tendency by a member of his department to act with unnecessary officiousness should be reprimanded by the security head before a confrontation situation develops.

Seeking co-operation does not mean currying favour by inaction to breaches of discipline; any apparent friendship that is shown after such an incident is quite spurious and the circumstances are apt to be brought to notice when an attempt at enforcement is subsequently made. The noting of precedents is not confined to court proceedings! Impartiality is essential, smartness is conducive to respect, humour mitigates resentment, ability to give informed answers to queries breeds better communication and engenders confidence, restraint and keeping of temper in the face of deliberate provocation reduces opportunity for diversionary malicious complaints, helpfulness induces trust and co-operation – security personnel of all levels should note and deport themselves accordingly.

Relationship with fellow employees inevitably leads to consideration of security membership of unions and participation in their activities. There are ample forebodings, but little evidence, that this acts adversely to an employer's interests.

### *Trade union membership*

In the interests of absolute impartiality, it would perhaps be better for security officers not to be members of a trade union, though this would deprive them of the advantages of various benefits and representation when personal interests were threatened, as would arise in connection with compensation after accidents and unjustified complaints. The nature of a security officer's work gives a clear priority to the interests of his employer and the danger is that union membership, especially where a man became actively involved, might induce him, or perhaps expose him to pressures, to act other than he should. He has almost unrestricted access to offices during quiet periods and may acquire, accidentally or otherwise, classified information that management would not wish to be divulged – there must be no clash of loyalties in those circumstances.

By far the greatest number of in-company security personnel are now of staff status – which is commonsense in view of their relationship *vis-à-vis* the managerial function. It follows that membership of a staff union is preferable. Clearly linked with such membership connotations are the duties that are required during strikes and disputes.

## *DUTIES DURING STRIKES*

Where security staff have not been union members, no difficulties have arisen in respect of their non-participation at times of industrial action. It has been accepted that they will continue their normal duties but do no additional ones. This is the attitude of responsible unions when the people are their members, though the exclusion is almost always a verbal agreement at official union level and not necessarily endorsed by local shop stewards at the times of unofficial disputes.

Several acrimonious clashes have followed a return to work when security personnel have not participated in strike action. Their position has been somewhat clarified by paras. 21, 22 and 23 of the Code of Practice as endorsed by the Trade Union and Labour Relations Act 1974. These state that some employees have special obligations arising from membership of a profession. While they should respect their union obligations, they should not be called upon by the union to take action which would conflict with the standards of work or conduct laid down for their profession if that action would endanger:

1 public health or safety;
2 the health of an individual needing medical or other treatment;
3 the well-being of an individual needing care through the personal social services.

Professional associations, employers and trade unions are asked to co-operate in resolving any conflicts that might ensue.

Security officers, by and large, now have a first-aid responsibility in addition to others relating to safety of premises and individuals, and the term 'professional' can nowadays be applied to their functions. Therefore it seems probable that non-participation in strike action is even less likely to be condemned than formerly.

It is as well to emphasize here that security responsibilities stop at the perimeter of the employer's premises. On no account must security officers become involved in any dispute over the behaviour of pickets towards employees, transport, or others beyond that limit.

## *PICKETING*

Enforcement of the law relating to picketing has been the subject of acrimonious bickering inside and outside Parliament. It is one of the most difficult and thankless tasks that the police have to cope with and there is an increasing tendency for employers to seek court injunctions when strikers' behaviour goes beyond reasonable grounds. In major confrontations, such as the year-long miners strike or that of the print workers at Wapping, there is likely to be a massive police presence and they will exercise complete on-the-spot control. Nevertheless, the person in charge of security should have a working knowledge of the legalities of a situation which can be as sensitive and animosity-provoking as any with which his department will have to deal. A Code of Practice has been approved by Parliament and has been quoted in court actions arising from the strikes.

Picketing as a form of industrial action has become an almost automatic corollary of the withdrawal of labour in trade disputes. In the majority of instances, which naturally attract little publicity, it is carried out without violence or acrimony, but on occasions insufficient knowledge of what is acceptable behaviour, legally and in commonsense terms, coupled with lack of tact or insufficient liaison between the two sides, leads to a flare-up which can produce lasting ill-feeling and mistrust. A considerable responsibility therefore rests on those still at work who are in contact with the strikers at the entrances; obviously these will be security staff, where such are employed.

Recriminations are bound to fall on the head of anyone who sparks off an avoidable incident and it behoves whoever is entrusted with overall security responsibility to be conversant with the law relating to picketing; and the practicalities of such situations, so as to be an informed source of reference to both sides if needed. It goes without saying that he should be aware of the employers' policy and instructions so that he can adequately brief his own staff, and apprise them of any contingency plans or arrangements that have been made.

Traditionally, politics have caused reluctance to create effective powers to deter the mass picketing likely to lead to disorder and violence or the employer-resented secondary picketing. The police attempts to maintain a low profile have proved impossible in the strikes previously mentioned and it is a matter for speculation whether they will now be tempted to provide a presence earlier and more frequently – a matter for local judgment. The Code of Practice approved in December 1980 gave clear guidelines but expediency has continued to be the deciding factor whenever a large number of people are involved. Fortunately the legal requirement to ballot those who would be called out on strike seems now to be producing a reluctance to do so if any avenue of negotiation remains.

The Code imposes no legal sanctions and non-conformity does not render anyone liable to proceedings. However, its provisions are admissible in evidence and to be taken into account, if considered relevant, in any proceedings before any court, industrial tribunal or Central Arbitration Committee (s. 3(8), Employment Protection Act 1980). It does define the existing law conveniently but its interpretation will be a matter for the courts and tribunals.

*Code of Practice (main points)*

> The Code is intended to provide practical guidance on picketing in trade disputes for those who may be contemplating, organizing, or taking part in a picket and for those who as employers or workers or members of the general public may be affected by it.

Though there is no legal right to picket, it has long been accepted that to peacefully picket is lawful, but certain limits are imposed on how and where this can be done to protect those who wish to work as usual or are otherwise affected by it.

Normally, to persuade a person to break his contract of employment or secure the breaking of a commercial contract, is actionable through a Civil Court but dispensation is given to those 'acting in contemplation or furtherance of a trade dispute including pickets, providing they are picketing only at their own place of work'. Some forms of picketing are permissible elsewhere, but there is no latitude allowed for offences which contravene the criminal law. Indeed, pickets who so offend may forfeit their immunity against civil action.

*Basic rules*

The Trade Union and Labour Relations Act 1974, s. 15 (as amended by the Employment Act 1980) lays down the basic rules for lawful industrial picketing:

1 It may only be undertaken in contemplation of furtherance of a trade dispute.
2 It may only be carried out by a person attending at or near his own place of work. A trade union official may also attend elsewhere providing he is accompanying a member (legitimately picketing) whom he represents in the normal course of his union duties.
3 Its only purpose must be that of peacefully obtaining or communicating information or peacefully persuading a person to work or not to work.

Pickets may endeavour to persuade orally, by leaflets, or banners and placards but they have no power to require people to stop or compel them to listen or do what they have asked them to do. A person who decides to cross a picket line must be allowed to do so.

It is lawful for a worker, unemployed for reasons connected with the dispute, to picket his former employer but he cannot do so if he has obtained new employment.

*'At or near his own place of work'*

This is not strictly defined but, in general, lawful picketing involves attendance at an entrance or exit from the factory, site or office at which the picket works. He cannot do this at other premises, even if those working there have the same employer, or are covered by the same collective bargaining arrangements. There is no protection against civil action for trespass for those who picket on or inside premises which are private property.

*Secondary picketing*

Employees may picket their own place of work in support of a dispute elsewhere but they can only have as their target the supply of goods between their employer and the firm in dispute.

1 In the case of customers and suppliers of the employer in dispute, on the business being carried out during the dispute between the customer or supplier and the employer in dispute.
2 In the case of an associated employer (part of same group), on work which has been transferred from the employer in dispute because of the dispute.

There is no immunity for interfering with commercial contracts by indiscriminate picketing at customers and suppliers or associated employers of the employer in dispute.

*Essential supplies and services*

Pickets are asked to take very great care to ensure their activities do not cause distress, hardship or inconvenience to members of the public not involved in the dispute; also that the movement of essential materials, the carrying out of essential maintenance of plant and equipment and provision of services essential to the life of the community should not be impeded. Similar advice was given in a TUC guide issued in February 1979.

*Picketing and the criminal law*

The Code defines what offences may be committed, but does not create any additional ones; they are not offences confined to picketing but worded in that respect, the examples given by the Code are:

1 To use threatening or abusive language or behaviour directed against any person, whether a worker seeking to cross a picket line, an employer, an ordinary member of the public or the police.
2 To use or threaten violence to a person or to his family.
3 To intimidate a person by threatening words or behaviour which cause him to fear, harm or damage if he fails to comply with the pickets' demands.
4 To obstruct the highway or the entrance to premises or to seek to physically bar the passage of vehicles or persons by lying down in the road, linking arms across or circling in the road, or jostling or physically restraining those entering or leaving the premises.
5 To be in possession of an offensive weapon.
6 Intentionally or recklessly to damage property.
7 To engage in violent, disorderly or unruly behaviour or to take any action which is likely to lead to a breach of the peace.
8 To obstruct a police officer in the execution of his duty.

A picket has no legal right to require a vehicle to stop or be stopped and may not physically obstruct a vehicle if the driver decides to drive on. The driver himself must take due care and attention to avoid accidents in the circumstances.

*Role of police*

The police have to impartially uphold the law; they have no responsibility to enforce the civil law and will not identify pickets against whom an employer wishes to take action, nor enforce the terms of an Order. In the event of the latter being made by a court, the police may assist the officers of the court, if a breach of the peace is anticipated.

The police may limit picket numbers where they fear disorder; there is no specific number fixed and it is a matter for police discretion. If a picket does not leave when so requested he can be arrested for one of the offences of obstruction or being likely to cause a breach of the peace.

### *Picket organizers*

The Code advises that an experienced person, preferably a union official, should be in charge on the picket line which should not exceed six in number. Amongst other things, he should brief the pickets, be available to advise them, hand out badges and armbands, and keep outsiders off his line.

### *'Lock-outs' and 'sit-ins'*

These are self explanatory forms of action during industrial disputes that no one, least of all the legislators, wants to give advice about.

In a lock-out, an employer in effect suspends his employees and denies them access to the workplace. The reason is usually that he considers the continuing of the business operation is being rendered financially unrealistic because of a 'work to rule', 'go-slow', 'blacking' of plant or some other restrictive action – or to pre-empt a sit-in. Fortunately such situations are few and far between and the 'locking-out' is verbal rather than physical in that the workforce is simply sent home on most occasions.

'Sit-ins' are the occupying of premises by strikers to prevent production, or the removal of goods or machinery. It most frequently now seems associated with protest against closure, threatened or actual, of the business – other occasions are extremely rare. Court orders have been sought by employers against the occupying workers who are to all intents and purposes trespassers – this is an extreme measure when all forms of mediation have failed.

Security's role will depend upon too many different factors to allow the giving of any general guidance – they may be locked-out or sitting-in themselves! They are likely to be excluded by the strikers during the latter, but may carry out their own basic protective duties during the former. Low profile, commonsense, and a realization that when all is over continued good relationships are essential are the guiding principles.

For the person in charge of security however, cautionary words – if your files contain items which should not become generally known, take those papers home or destroy them if a sit-in seems imminent. Strikers take care to avoid damage, but the sanctity of information is an entirely different matter. This would be especially true in academic surroundings where students would not hesitate to bring to notice material which they considered should not have been put on record or showed duplicity by those in authority – as a senior CSO, who should have known better, found to his cost. The same advice could with advantage be passed on to those members of management who hold comparable documents – at least as a reminder – it should not be necessary, but some managers ignore the obvious, and the higher up the management tree, the more likely they are to be negligent in such matters.

### *UNFAIR DISMISSAL: HOUSE OF LORDS RULING*

A case said to be the most important of its kind in the 1980s merits late inclusion and by its origin is a binding authority – *Polkey* v *A.E. Dayton Services Ltd (formerly E.W. Holdings Ltd)* 1987 IRLR 503.

The circumstances are immaterial but the implications of the judgment should be noted by anyone responsible for a sacking on security-linked reasons who wants the person to stay sacked. The main judgments supplement each other:

'If the employer could reasonably have concluded in the light of the circumstances known to him at the time of the dismissal that consultation or warning would be utterly useless he might well act reasonably even if he did not observe the provisions of the Code' and 'An employer having prima facie grounds to dismiss will in the greater number of cases not act reasonably in treating the reason as sufficient reason for dismissal, unless he has taken the steps conveniently classified in most of the authorities as procedural which are necessary in the circumstances of the case to justify that action'.

In other words, however strong the evidence, stick to the laid down procedures (both security and the responsibility of management); if the system has to be short circuited, have a sound and acceptable reason for doing so or an appeal may lead to an unfair dismissal finding by an Industrial Tribunal.

Another shrewd ruling was made, namely that the issue in proceedings was what the employer did, not what he might have done.

Much of what has been said earlier is summed up in these rulings which approve the unusual *Pritchett and Dyjasek* v *J. McIntyre Ltd* outlined on page 60.

# 5

# *Employment of security staff*

Amongst the many changes that have taken place during recent years in attitudes to security work, none is more important than the realizations that incapacitated employees are totally unsuitable and that it is no longer a welfare-oriented dumping ground for those unfortunates. The increased violence in society dictates a need for able-bodied people. The balance of probability is that they will never be individually attacked, but such chances as there are will be considerably reduced if it is apparent that they are capable of defending themselves – physique and smartness command respect in several ways.

Those in charge of the department must have in mind the age range of their staff when vacancies occur. The lower the average the better, but there should be sufficient spread to ensure that at some future time there is not going to be a simultaneous exodus of half the staff. Maturity is an essential quality, and domestic responsibilities are conducive to stability in the job, patience in dealing with people, and resistance to corruptive influences.

Health is most important because, as security is a non-productive unit, manning is kept to a minimum and a higher standard of attendance is required than in other sections of the workforce. With this in mind every effort must be made to avoid acquiring anyone who has a slipped disc or back trouble which may be recurrent and could be aggravated by lengthy spells of standing or walking. Similarly, an asthma or bronchitis sufferer is going to be an increasingly frequent absentee as he grows older and at the time when numerical strength is most needed, that is, in the winter months. Impaired vision or hearing, apart from making a person something of a menace in an environment of machinery, could lead to embarrassment in relations with the workforce. Embarrassment was hardly an adequate word in one other instance where an excellently qualified applicant, who was accepted for a post with an essential first-aid content, was found at his medical examination to have a most persistent form of dermatitis on his hands; he had not thought fit to mention it. Matters of this kind should be specifically included in a special application form, so that they do not slip the memory of the interviewer and there is no excuse for not disclosing them. A form to meet these requirements is shown in Appendix 4.

## *RECRUITMENT*

Surprisingly, there is difficulty in obtaining suitable recruits for the security staffs of both firms and the service-supplying companies, despite the facts that the work has become increasingly professionalized, offers steady employment, and no longer has first place in the queue when staff reduction is being considered. There may be several reasons for this, perhaps predominantly the 'unsocial' hours of working; in addition, the wearing of a uniform does not appeal to everyone, and there may be an antipathy to exercising what is considered to be a disciplinary role.

With these reservations in mind it is advisable, when interviewing prospective recruits, to ensure they know exactly what is involved; otherwise there will be an unnecessary turnover of labour and a wastage of training. These are matters which must be of particular concern to the commercial firms where this turnover may easily exceed half their manpower per annum in small units.

A series of points which an interviewer should ensure that an applicant fully comprehends and considers is:

1 Security demands continuity – has he decided that he now wants to settle down into a regular job routine where opportunities of internal promotion are bound to be limited?
2 If not used to shift and weekend working, has he discussed this with his wife and family to find their reaction to the prospect? Has he considered the effect that the change in routine may have on any physical weaknesses, for example stomach troubles or insomnia?
3 Does he appreciate that overtime and unexpected shift changes must be expected in emergencies, will not be voluntary, and may cause him social inconvenience?

4 Does he accept that, with a limited number of colleagues, days off for minor sickness will throw extra work upon them and that a higher degree of attendance is normally expected in security work than in most other jobs?
5 What is his reaction to the prospect of working alone, and at night, in circumstances where there may be an element of personal danger which he cannot avoid simply by running?
6 If not used to it, does he think that he will acclimatize to wearing a uniform which makes him stand out and attract questions to which he will be expected to know the answers?
7 Does he realize that occasions will occur when he will have to act upon his own initiative and perhaps make decisions affecting life, property, and industrial relations?
8 If his previous status has been supervisory, does he now appreciate that he must be prepared to accept and carry out orders?
9 Does he appreciate that there will be instances of extreme provocation, though these will be very rare, and that he will in all circumstances have to control his temper?

All these are matters divorced from the qualifications of experience, desirable skills, integrity, and acceptability by age or physique upon which interviews usually concentrate. Nevetheless, getting rid of an unsuitable employee is made so difficult by the Employment Protection Act that time spent in ensuring that the situation does not arise is more than repaid both financially and in a security force's harmonious working.

*Sources of recruitment*

It is conventional to advertise vacancies internally and also in the local press. It is unlikely that at the level of guard any advantage would be obtained by doing so in the specialized security press, though this can be helpful for supervisory status.

The main sources of recruits are:

1 *Armed forces* – These provide fit and disciplined men who may have had the added advantage of pre-release training in security subjects. The obvious snag is that they have no industrial experience and may have operated in an environment where contact with the working public has been limited. There may be also a risk that NCOs – the usual applicants – may find it hard to adjust to not giving orders.

2 *Police* – The majority of police officers leaving the service before pensionable time elect to do so in the earlier years of their service; if they have not settled down in the police there is a distinct question

as to whether they will do so in security. Those leaving later, again without a pension, should be closely questioned as to their reasons for doing so and asked to produce the conduct certificate that they will have been given.

An experienced officer who decides to take up a second career as soon as he reaches pensionable service may have 19 or 20 years to offer his new employer, plus many of the other security virtues which are required, provided that he has remained fit and agile. Again, there may be the difficulty of adjustment to an environment where he will not be able to exercise his previous powers. The limiting factor on this group will be the extreme reluctance of those most suitable to undertake further shiftwork after many years of performance. A further judgment which an interviewer has to make is whether the applicant may regard the job as one in which he simply has to 'go through the motions' and do as little as is necessary.

With the substantial increase in salaries and pensions this source has virtually dried up for other than supervisory posts, though some ranking officers do use their status and experience to enter the security field as 'consultants' or 'advisers' setting up their own firms.

3 *Fire brigade* – This is the obvious source for specialized recruitment where fire is an outstanding risk and there is likely to be the additional advantage of a first-aid qualification, but for some reason there seem to be few applications. The same limitations as apply to the police may also be found here, and there will be no counterbalancing experience, local knowledge of criminals, and practice in establishing amicable relationships with all types of individuals.

4 *Internal recruitment* – In principle, this is undesirable and not in the interests of management, the workforce, or the applicant. Long association with other employees could interfere with the essential impartiality; there is the risk that applicants might regard undesirable practices as acceptable because they too have participated in them; their freedom of action might well be curtailed because of others' knowledge of such past participation. On the credit side, they will be familiar with the premises, processes, and procedures of the firm, and it will be relatively easy to establish their previous attendance records, job performance, etc. There will be exceptions, where persons have first-class qualifications for the work and are known to be the type who would loyally carry out responsibilities, but on balance external recruitment is preferable to internal.

5 *Commercial security firms* – The physical requirements of in-house security are apt to be higher than those of service-supplying firms, and the experience gained in that employment may be of an entirely different nature to that which is demanded – for example, concentration on cash carrying or spot visits to premises. Any training

undergone would be an advantage, but there is a tendency for this to be sketchy. Provided that there is a good record of continual employment, this can be a good and fairly numerous source of recruits who are already familiar with the inconvenience considerations.

6 *Other external recruitment* – A large number of applications are made purely because those doing so do not realize that the current requirement is something more than a watchman or gatekeeper. A high percentage of these will be eliminated without interview. Nevertheless, the bulk of candidates will come from people without directly relevant experience, and the interviewer will have to evaluate the chances of their assimilating training and developing the desirable attitudes and a questioning state of mind – in effect, to back his own judgment of likely 'winners'.

Sources which should be treated with extreme care are those industries where there is a history of continual acrimony, strikes, go-slows, and known addiction to absenteeism – and this is said from sad experience!

These are the major groupings. Further possibilities with some training qualifications are the forces Special Investigation Branches, which obviously merit serious consideration, the prison service, and ambulance service.

### *Selection*

Desirable qualifications include a current first-aid certificate, fire training, driving capability, clerical competence, and any specialized requirement of the job. An effort to familiarize himself with the prospective employer's business would show commendable interest by the applicant.

A method of emphasizing the points which should be looked for by the interviewer, other than those previously established, is perhaps to indicate what the candidate should do and expect at his interview.

1. Remember that first impressions are very important. Cleanliness and smartness in appearance are essential for security officers; so dress and behave accordingly at the interview.
2. If you have references, service documents, licences, etc., bring them to the interview.
3. Sit-up, look interested, and do not overstate your case.
4. Expect direct and personal questions aimed at testing your integrity.
5. Expect your previous employment record and reasons for leaving to be closely examined.

6 Expect not only questions on your medical record and attendance but also a medical examination.
7 Expect to be questioned on any special qualifications that you claim, to establish whether you are exaggerating.
8 Expect test incidents to be outlined to you to gauge your reaction, how you would deal with them, and your general initiative.
9 Expect references to be taken up and that your subsequent employment will be conditional upon the replies being to the standard you have claimed.
10 Expect the interview to be more lengthy and detailed than for the normal job, as aptitude and other routine tests may also be included.
11 Ask questions of the interviewers about any points that occur to you – in connection with duties, fringe benefits, remuneration, or anything else that may affect your decision to accept the job.

The latter point is highly important to interviewers. They should ensure that the applicant fully understands what is expected of him and what he will be paid. If the applicant is being seriously considered, it may be advisable to give him a copy of the standing orders or job description to think about at a break in the interview.

One of the last things a security department would wish to acquire is a militant barrackroom lawyer type and if replies to questions hint at this possibility they should be pursued to the interviewer's satisfaction.

*References*

Risks must not be taken in the employment of security staff, as they have outstanding opportunities to steal from their employers in the absence of other personnel on the premises and are rarely under continual observation to ensure they are carrying out their duties. No offer of employment should be made until references from previous employers have been checked. It is desirable that a special application form should be used which will give the prospective employer better grounds for dismissal if he has been deliberately misled. The Employment Appeal Tribunal ruling in the case of *Torr* v *British Railways Board* (1977 IRLR 184) may have an important bearing if criminal convictions are not disclosed. Disqualifying features should include a history of frequent changes of occupation without reasonable excuse, unaccounted breaks in employment, reluctance to expand upon any points of query, indications of aggression or arrogance, and plain apathy about the content of the job.

Unfortunately, instances have occurred of security officers being charged with criminal offences involving the property of their employers, who were subsequently grossly embarrassed when proceedings showed that the officers had previous convictions for crime – even served periods of imprisonment. These are likely to get headlines in newspapers which bring discredit on both the security profession as a whole and upon the employer for having failed to exercise the care that he should have done in selection. In the commercial security field it would be a nice legal point as to whether a firm was negligent if, in offering a service to a client, one of its staff stole from the client and then was found to have undisclosed convictions which should have disqualified him from the position. The fine print on agreements might not afford complete protection.

Any hope of a registration and licensing system for security personnel and commercial service-supplying security firms would appear to have faded for at least a long period of time. Proposals for legislation to ensure the past integrity of prospective employees and the bona fides of the service firms were turned down by the Home Office in December 1980 after a prolonged investigation into the opinions of interested parties. In doing so, it is suspected a minority view was accepted and at least one Private Member's Bill has since been introduced with the object of making it an offence for a person with convictions for dishonesty to undertake work of a security nature – the position may therefore change at short notice and a much needed control be imposed.

The main difficulty is that of definition – as is easily envisaged when watching uniformed 'security' personnel, badges and all, opening doors to customers, collecting trolleys, sweeping up, and carrying empty boxes in supermarkets.

## *UNIFORMS*

One of the primary questions which has to be decided when ordering uniforms for the security staff is what colour are they to be. Navy blue is the most popular because of its serviceability and because of its similarity with the colour worn by the forces of the law, the civil police. By this it is hoped that security staff will attract similar respect as symbolizing law and order within the areas of their responsibilities.

However, the Police Act 1964, s. 52(2) says:

> Any person not being a constable wearing any article of police uniform in circumstances where it gives him an appearance so nearly resembling that of a member of the police force as to be calculated

to deceive shall be guilty of an offence and liable to a fine not exceeding £100.

Article of police uniform means 'any article of uniform or any distinctive badge or mark usually issued to police forces or special constable or anything having the appearance of such article, badge, or mark'. 'Calculated' is not free from ambiguity but legal opinion is that it means 'likely'.

This section of the Act was introduced to control the wearing in public places of uniforms having a great similarity to police uniforms due to the increase in the use of uniformed personnel employed by security service companies. When going to and from their assignments they have to pass through the streets and as so many of their uniforms could easily be mistaken for that of a police officer something had to be done to reduce that possibility. One result of the new law is that the companies referred to have made changes in or additions to the uniform of their staff so there is now no doubt as to their identity.

With that in mind, if a uniform worn by security staff of a commercial or industrial concern is likely to be mistaken for that of a police officer it must not be worn outside the premises at which the wearers serve. If their duties require this, for example escorting money to or from a bank, or traffic control at factory entrances, the uniform should be of some distinguishing colour – dark green, maroon, or grey are practical as alternatives. If navy blue is still preferred it must bear some identifying badge and/or shoulder flash with the word 'security' on it. A cap badge incorporating the firm's name or its initials would also help in that direction. Stripes of rank should be inverted and for the same purpose metal bars should be substituted for stars on shoulder straps. The police wear black-and-white chequered hatbands and, although some ambulance personnel have started to wear green-and-white hatbands, security must keep to unbanded hats.

Apart from security there are other organizations whose uniforms approximate to that of the police – fire officers, bus inspectors, water and gas officials, but they are distinguished by their badges. It should be noted that prosecutions have been almost invariably based upon the actions of the wearer which were intended to convey the impression of being a police officer with the accompanying powers and authority. The difficulties which arise occasionally with the police seem to originate from a failure to reconcile themselves to the knowledge that pressures upon their time prevent them now from taking responsibility for everything in the public security field. This may cause resentment in individuals to accept that other non-police bodies have the same objectives and that there is a role for both in ensuring the common good. The essential facts about uniforms are that they should not be intended to deceive the public and do not. One of a security officer's best tools is his uniform. It should be, and can be, an excellent symbol both of identification and authority and assists him greatly in dealing with the public, other employees and anyone contravening the law as it affects his employer.

Uniforms must be of good quality and fit so as to encourage the wearers to take a pride in their appearance and be a good advertisement of the firm to callers. The use of second-hand uniforms purchased from outside sources is deplored as having a bad effect on the morale of the staff, and this will obviously reduce efficiency. Uniform should be renewed after specified periods of wear. Trousers for example have a shorter life than overcoats. The older issues of uniform can be worn for night duty with subsequent issues worn on day work when a good appearance is more important.

## *JOB TITLE*

No uniformity has been reached yet in naming the rank-and-file security staff, and indeed much of the terminology used is traditional to the individual firm or the site. 'Works police' is found in the older establishments; 'security guard' is favoured in Scotland and seems to be growing in popularity; 'security warden' is usually used in commercial premises and office blocks; 'patrolman' or 'guard' is frequently used by companies supplying security services; 'gateman' or 'gatekeeper' may be applied purely for static duties at entrances – these incidentally are probably on lower pay scales; finally, the most conventional title is 'security officer', which will be used hereafter. The name matters little; the performance is what counts.

## *STATUS*

If security officers are regarded as a force directing and supervising personnel in various ways, enforcing works rules, exercising discretion as occasionally they must, together with those private powers of arrest that probably only they in the organization know they have, then it is logical that they should be of staff status with all that that implies. This is increasingly the case; where there is opposition there are also likely to be found old-fashioned dogma and antipathy to the conception of security at a management level, which can obstruct the change. The main business of security officers is apt to be with payroll; so there are obvious advantages in having them of different status and in different unions.

Not least in importance is the added attraction that staff status has for the good-quality applicant who is in such short supply.

### *Promotion*

*Within the security department* – Whether there are ranks within the department depends on the number of people employed. Where each shift performs duty in rotation this can mean that usually at night and at weekends they do it without the direct supervision of the CSO or other person in charge of security. Any decision in circumstances calling for further action is usually taken by the person on duty who is senior in service.

To give each shift someone, irrespective of service, selected for that responsibility, the creation of NCO-type ranks should be considered. This would provide promotional opportunities for the staff, leading to better pay. The rank could be shown on the holder's uniform by a chromium-plated bar on the shoulder straps or by three inverted stripes on the sleeve.

*Outside the security department* — When selecting individuals who have shown the right potential for training for supervisory positions the security staff should not be overlooked. A reputation for fairness and consistency in the performance of duties should make a security officer acceptable to the general workforce in a supervisory capacity after any necessary training in the required skills. Persons passing Institute examinations will have shown their worth.

## *FIRE AND FIRST-AID QUALIFICATIONS*

An otherwise entirely acceptable applicant should not be disbarred because he lacks knowledge of fire training or first aid. Instruction in both is readily available, often in crash-course form taking two or three days of concentrated practical and theoretical study. A uniformed man must have these skills, for his credibility in the eyes of his fellow employees will diminish to nothing if he attends an incident involving either and perforce has to remain a bystander. Fortunately, virtually all employers demand a first-aid ability, and a considerable majority can see the virtue in having their ever-present security staff trained to combat fire.

### *First aid*

The British Red Cross Society (9 Grosvenor Place, London SW1), the St John Ambulance Association, Brigade (1 Grosvenor Crescent, London SW1), and the St Andrew's Ambulance Association (98–108 North Street, Charing Cross, Glasgow 3) are the main bodies providing training in which security personnel can freely participate; they hold their courses at centres throughout the country. Details can be obtained from their central offices if there is local difficulty. Certificates of training are issued after the successful conclusion of each course, and these are accepted by the Health and Safety at Work Inspectorate as evidence that the holder's training complies with the requirements of the Health and Safety (First Aid) Regulations 1981. These Regulations are dealt with in detail later (see p. 518).

The virtue of having all security staff trained is obvious; they provide a continuous presence that obviates the need for rotas and checking that designated persons in the workforce are on correct shifts, etc., and have up-to-date certificates. Frequently a gratuity is paid by employers for having the qualification; but really, as this is part of most security job descriptions, it should be weighed in assessing salary. Where an ambulance room is maintained, the security staff can be trained in the use of its equipment, so as to be competent to stand in for the trained nurses in an emergency.

False injuries have been known to be alleged as a basis for a diversionary incident to cover theft or breach of works rules. This possibility should not be overlooked when the casualty is not behaving in a manner consistent with pain.

Contract security firms would do well to regard their ability to supply first-aid trained personnel to clients as a distinct advantage when trying to sell their services, and those in search of employment in security work should appreciate that this is a skill which enhances their value to a prospective employer. Far too many applicants are found to have allowed their qualifications to have lapsed.

### *Fire*

There is no better source of training than the local fire brigade, who are invariably most helpful and periodically hold basic and more advanced training courses.

It is essential that security staff are competent in this since an uncontrolled fire can do infinitely more harm than any theft. In cases of difficulty the Industrial Fire Protection Association (36 Ebury Street, London SW1) may be able to assist, and membership of the Fire Protection Association (Aldemary House, Queen Victoria Street,

London EC4) will lead to a constant supply of specialized information on experiences with fires and up-to-date fire protection advice.

## *HOURS OF DUTY AND SHIFT ROTAS*

The old concept of security envisaged an almost static role whereby long hours of duty were thought workable without detriment to either staff or objective. This view unfortunately persists in some firms who can be easily identified by the appearance and calibre of the staff they employ and by the low efficiency they expect from them. Long unbroken spells of duty are psychologically and physically detrimental to those performing them on a regular basis. The optimum average weekly hours, in conformity with industry in general, is currently accepted as being 42, which lends itself to the optimum four-gang three-shift rota, and security staffing should be geared to this.

What is not recommended is a permanent night shift whose members may, by the regularity of the duty, be attracted to part-time day work for another employer which can affect their efficiency in performing their security duties. The arrangement is also undesirable where there is a permanent night shift of the general workforce. This can lead to familiarity which is not a good feature of security practice in any circumstances.

Reserve strength to cater for holidays and sickness will probably be kept to an absolute minimum so that, on occasions, considerable overtime becomes unavoidable. This is economically more desirable than to add extra wages and overheads for staff, to allow purely for emergencies. During such times, no shift should exceed twelve hours; by then the person's level of performance has sunk to a degree where his capability to deal with unforeseen situations must be suspect.

When the required coverage cannot be achieved without exceeding reasonable hours, the availability of people from the professional security firms on a temporary basis can be investigated. It cannot be expected that they will have the special skills required in some circumstances, but they will have had generalized training, sufficient to bridge an emergency. What must strongly be opposed is any suggestion of temporary transfer of members of the general workforce to security duties. They could thereby have access to information concerning their fellows which should remain confidential. They could gain an insight into the manner in which security patrols and duties are carried out which they could subsequently use to the detriment of the firm. Their only qualification would be familiarity with their surroundings. The fact that it was possible to use them on such a basis lowers the status of the selected and trained man in the eyes of other employees. It could also be anticipated, particularly where this was a staff-grade job, that there could be staff union objections.

*Shift duties*

These are necessary to provide the continuous coverage required. A proportion of applicants for security employment may not have had experience of the advantages or disadvantages of this form of working. Rather than have a person who finds that it is unacceptable to him when considerable expense has been met in his training and equipment, the full implications of shift working should be made quite clear to him at his initial interview so that he is in no doubt what to expect. This is also true of holidays where he will not be able to coincide with the rest of the community and he will have to accept the possibility of inconvenient overtime at short notice.

*Four-gang, three-shift duties*

This is the normal form of coverage. It can be applied in a number of different ways, as shown in Appendix 5. In every four-week period, 168 hours of duty are performed by each person giving an average of 42 hours per week. This requires a minimum of four people or multiples of four according to the number of staff required. The respective shifts performed over periods of four, eight, twelve or more weeks will balance out so that each person does an equal number of each according to the type of rota worked.

The normal roster caters for seven days on early shift (6 a.m. to 2 p.m.) followed by two days' leave; seven days on late shift (2 p.m. to 10 p.m.) followed by two days off; then seven days on nights (10 p.m. to 6 a.m.) followed by three days off; this completes a full cycle. A variant is to work 21 days continuously, performing three shifts in sequence followed by a full seven-day leave. Some men will prefer this since holidays should preferably be taken to coincide with the early turn week and this enables, in effect, several holiday fortnights to be taken each year by coinciding a rotational week of leave with seven days' annual leave taken in lieu of the early turn week.

*The continental system of 3 × 2 × 2 shift system*

An objection to the foregoing three-shift working lies in the complete absence of normal evening social life during the afternoon and night shift periods. By using this variant there are frequent changes of shift, so that some evenings are free each week. An example is shown in Appendix 5 and the common feature of this system is a rotation of 3 morning shifts, 2 afternoon shifts, 2 night shifts; 3 rest days, 2 morning shifts, 2 afternoon shifts; 3 night shifts, 2 rest days, 2 morning shifts; 3 afternoon shifts, and so on.

This system does not usually appeal to the older staff who prefer a settled routine which is less likely to cause disturbed sleep and digestive complaints. Longer breaks between shifts are included but with reduced leave days.

All these systems can be varied in respect of the starting day thereby modifying the break periods to coincide with weekends. Whenever it is intended to change a rota system, it is absolutely essential that the staff should have full opportunity to discuss it; if a change is made, all objections have then been ventilated and thoroughly examined so that there can be limited cause for subsequent complaint. With increased unionization, there is little doubt that those bodies will wish to become involved where their members are concerned.

*Special arrangements*

Conditions may appertain where it is necessary for more staff to be on duty at certain times of day than others. The shift rota previously mentioned cannot cater for this and special arrangements must be made, either to devise an alternative acceptable system, or to superimpose upon the shift rota other staff carrying out regular duties over the period of greatest demand.

This can particularly apply where a large number of gates are required to be manned during daytime and the necessity lapses when the main body of the workforce leaves, allowing some of the gates to be closed. The same can apply to commercial and retail organizations where there is a necessity for control of members of the public during the day reverting to control of limited staff after normal closing hours. These day staff will not, of course, have wages enlarged by shift premiums but the work will attract people who are averse to shift working and it should be possible to still demand a high standard. Some clerical ability may well be desired and there may be a different job description.

On occasions, however, the desirable extra cover may be needed during the particular period 6 p.m. to 2 a.m. A variation for this on the normal rotas is also shown in Appendix 5; this utilizes ten men as an alternative to employing twelve on a four-gang three-shift basis; cuts overtime in connection with payment of wages on Thursday and Friday of each week and provides a man to cover the 6 p.m. to 2 a.m. period each day. A further advantage of a system of this nature is that the necessity for this particular duty is progressively reduced during the holiday months so that the staff performing it can become holiday reliefs with minimum detriment to the coverage – they are available during the winter months when they are most wanted.

Whenever economies in manpower are desired, the possibility of variations of this nature are well worth considering.

### *Annual and public holidays*

Security staff should enjoy the same holiday privileges as their equivalent grades among other employees. It is quite obvious that they will not be required to be absent at the same time but days in lieu should be logged and taken to mutual convenience of the individual and the requirements of security manning. Where it is possible to allow extra manpower to be off at public holidays, this should be done.

### *Shift allowances*

Security, by its very nature, normally requires a presence 24 hours a day and seven days a week unless the premises in question can be adequately covered during 'shut-down' periods by electronic alarm systems or other means. If other employees have hours regarded as outside normal daytime working, they will inevitably be paid compensatory allowances; if this is so, similar payments should be applied to the shifts worked by security officers.

It is customary to pay a small premium for early turns (6 a.m. to 2 p.m.) and larger premiums for late turns (2 p.m. to 10 p.m.) and nights (10 p.m. to 6 a.m.). In addition to this, weekend working often attracts higher premiums; for example, a special rate may be paid from 10 p.m. Friday until 1 p.m. Saturday and an increased one from 1 p.m. Saturday until 6 a.m. Monday. Added to these, in some industries there is a 'disturbance allowance' or 'inconvenience allowance' related to the regularity of weekend shifts. In order to avoid variable pay cheques, the total salary payable over the complete period of a person's rota may be calculated and divided by the number of weeks in the rota, so that the person has a constant salary irrespective of the shifts that he is working on a particular week.

Where the security wages are linked into salary scales of other employees these premium payments can lead to excessive increases when the basic rates are improved. It may well be found desirable in such cases to substitute a compensatory grant rather than premiums for shift working to avoid wages becoming totally uneconomic.

### *Overtime payments*

In most industries this is at the rate of 1½ times the basic hourly rate. However, this is not universal, and the practice in respect of security staff must not differ from that of other employees; otherwise industrial discontent will be caused.

This overtime factor should not be influenced by the day of the week. In the event of a person who is being paid a balanced weekly wage, as mentioned above, volunteering to work on a rest day, there are grounds for arguing that his repayment should be not at overtime rate but at that of a conventional working day. This will no doubt have to be subject to local agreement.

### *The carrying of weapons by security guards*

High Court judgments have ruled against security guards carrying weapons such as truncheons or their equivalent unless they are in immediate danger of attack. A warning was given that security guards or their employers should not come to regard the carrying of a weapon on any occasion as a matter of routine or as part of the uniform. This was clearly stated in *R* v *Spanner, Poulter & Ward* (1973 1 WLR 488), where security guards at a dance hall carried truncheons as part of their uniform ostensibly for deterrent purposes – they were held to have no reasonable excuse in law.

The relevant Act is the Prevention of Crime Act 1953 which makes the carrying of an offensive weapon in a public place without lawful authority or reasonable excuse (the proof of which must lie on the carrier) an offence which entails a fine and/or imprisonment which can be up to two years on indictment and forfeiture of the weapon. 'Public place' includes any highway, premises, or places to which at the time the public have access whether on payment or otherwise. 'Offensive weapon' is anything made or adapted for use for causing injury or so intended by the person carrying it, a definition which was extended by the Public Order Act 1986, Schedule 2, para. 2 to include an article intended by the person having it for use by himself or some other person – which caters for the person who carries and passes a weapon to another to use. Only police have the power of arrest.

## *CODE OF CONDUCT*

A recommended code of conduct for security staff is given in Appendix 6, this has the blessing of the International Professional Security Association.

## *SECURITY ACCOMMODATION*

Ideally a security office should be purpose-built, but frequently it will be a conversion of an existing building or rooms adjacent to the entrance to the premises.

This is the first contact point for visitors to the firm and must convey an appearance of tidiness and efficiency. There is also a psychological aspect affecting the staff themselves; dingy and insufficient surroundings will be reflected in the attitude of the staff to their work and in the manner in which they carry it out.

If conversion or construction is to take place, careful planning must ensure that no essentials or desirable attributes are overlooked. Once a capital expenditure proposal has been approved, supplementary ones to cover additional items will reflect upon the competence of the senior practitioner whose advice should have been followed. The essentials may seem obvious but will bear codifying and are equally applicable to industrial and commercial premises – or for that matter, public buildings, hospitals, etc. where security is enforced.

1 The security office must have a full and uninterrupted view of the entrance.
2 It should be as near as possible to the entrance so that no incoming personnel and/or vehicles can avoid passing it. However, this proximity to the entrance should not be such that obstruction of any main road will be caused by vehicles waiting to enter whilst others are identifying themselves to the gate office.
3 The entrance and area immediately in front of the security office should be floodlit, with provision for some form of emergency lighting in the event of mains failure – a safety as well as a security measure.
4 Drop-arm barriers, electrically controlled from the office, should be considered as a means of the discretionary stopping of incoming and outgoing vehicles.
5 If the nature of business is such that a weighbridge is used, serious consideration should be given to positioning it outside the gate office as this will eliminate some of the frauds periodically encountered. It also permits the weighbridge to form part of the security duties, which also has desirable factors.
6 The office layout and windows should be such that the staff inside have an unrestricted view of anyone approaching from any direction. This is advisable because on occasions gate offices have been attacked before the staff knew what was happening, thereafter giving the intruders complete freedom in the premises.
7 The interior layout should be such that visitors are segregated

from the main office area by means of a continuous counter. This ensures the privacy of the office and its contents and reduces the tendency of other employees to congregate and waste time in the office.

8 Seating accommodation should be available for visitors to the premises who may be asked to wait for any purpose, for example collection by the person whom they are visiting.

9 Toilet accommodation should be available, either in or immediately adjacent to the office, as a facility for both staff and visitors.

10 The internal layout and siting of telephones should be such that the staff, whilst writing or telephoning, maintain a full view of the inner entrance and approaches. If circumstances permit the siting of the main clocking stations within this field of view, this would provide a useful deterrent for the practice of 'double clocking'.

11 Keyboards should be sited out of clear view of visitors to the offices.

12 Separate rest room accommodation should be available for staff and contain cooking facilities so that they do not need to leave the premises for meals, thus remaining available without presenting callers with the undesirable sight of food being eaten amongst the office books and records.

13 Provision should be made for lockers in which staff may keep personal clothing and other belongings.

14 If a separate room is available for lockers, provison should be made for drying clothing if the staff have to operate outside in inclement weather. The rooms as a whole should of course have adequate heating and lighting.

15 Depending on the size of premises and the degree to which the security staff are responsible for first aid, consideration should be given to providing a small room adequately equipped for the purpose. If this is unnecessary or not feasible, an adequate first-aid kit must be kept in the office.

16 At least two internal telephones and one external should be available. The latter should not be used by visitors or employees as this would interfere with the routine of the office and encourage time wasting. Outside the gate office is probably the best place to site an external pay telephone.

17 Except in the largest security offices it will be necessary to site radio telephone equipment and alarm system terminations in the main office. If this is so, they should be sited as unobtrusively as possible – especially the latter.

18 If money is to be held in the office at any time, a suitable safe should be installed, preferably not in the main office.

19 It is advisable to site an alarm siren on top of the lodge with push-button control at the desk, in case of attack upon the staff or if for any other reason it is necessary to immediately sound an alarm.
20 Controls for the external lighting of roadways, perimeter, and buildings should be contained in the office.

The factors listed above are primarily structural ones, but the office could with advantage be made an emergency communications centre for the entire premises. This would mean that it should contain:

1 The terminal panels for all fire, burglary, or specialized types of alarm, for example those fitted to boilers.
2 Tannoy equipment for the transmission of messages to all parts of the premises, supplemented by loudhailer equipment.
3 The radio control equipment and independent means of operating a 'bleeper' system.
4 Possibly a separate small GPO switchboard to enable the security staff to transfer incoming calls out of hours to internal departments.
5 Dependent upon senior management's attitude, closed-circuit television monitoring screens.

*Books and records*

A considerable amount of data will be required in the security office of a large firm, so that incoming queries can be answered promptly and there can be quick reaction to incidents. Some of the documents and books may contain matter that should be restricted from the knowledge of the general workforce, and these should be kept where not immediately in sight of casual callers. Appendix 8 lists books, forms, records, and equipment conventionally found in a large security office.

# 6

# *Duties of security staff*

The broad requirements of security work are too diverse to be adequately shown in the normal job description form, which must perforce be couched in concise terms. In addition the security officer needs a datum to which he can refer if his carrying out of a particular duty is subjected to query or, for that matter, if he himself is in doubt as to the extent of his responsibilities and discretion. Though the title savours of regimentation, there is nothing better for all purposes than the formulation and agreeing of 'standing orders' which can be added to as the job content changes – they are also invaluable at job re-evaluation times!

## *STANDING ORDERS*

Apart from initial generalities relating to clothing, standards expected, confidentiality, overtime and holiday arrangements, etc., the matters included will vary from employer to employer in accordance with the nature of the business and the trust reposed in the department's training and efficiency. A specimen of such orders, similar to one in use by a progressive employer, is shown in Appendix 7. It is unlikely that all sections can be universally applied, but many will be common, particularly in engineering and heavy industry. It will be noted that several items are of a nature that is helpful to other departments or have welfare implications. These represent opportunities to create co-operation and goodwill; the more the security staff feel that their function is a contributory factor in the firm's efficient running, the wider their involvement, the better the morale of the department, and the more valuable they are to their employer.

Some duties that a security department is asked to accept are peripheral to its main job, but unless they diminish the satisfactory performance of that job the 'hidebound' attitude of 'that's not security' should not be adopted. It has been said that weighbridge operation in emergency comes into that category but consider the monumental frauds that have occurred in connection with this – upwards of £750,000 in 1979/80 at one UK factory alone over a single weighbridge. Not security? Someone must be joking. Again, telex inspection and passage of messages: if security has to visit a telex room during shutdown periods to ensure absence of fires etc., a glance over the messages and intelligent interpretation of what is read could be invaluable to a firm with an urgent overseas order requiring immediate reply, and the sense of job importance given to the security officer doing it is worthwhile. A simple step from there is to changing a teleprinter roll when the room is unmanned during a holiday break – a matter of seconds. Not security? Perhaps not, but in one instance where a machine reached the end of a roll an unknown number of calls, estimated at 50, were all printed superimposed and were never satisfactorily sorted out. The cost to the firm might even have been that of the entire security department for a year, and the need for the roll change might only arise once in that time. If additional tasks such as these require additional skills, the time comes when a re-evaluation of the main job can be legitimately requested, and the job itself becomes the more indispensable to the employer – a factor not to be overlooked in these days of calculated economies in manpower.

Having made these comments on the diversity of duties, there are some that should be most strongly resisted, which usually arise in firms where security has been regarded as an imposed necessity or been traditionally inefficient. Cooking meals, preparing tea, sweeping offices, and the like, occupy time totally unjustifiably and create lengthy periods where it must be obvious to anyone criminally minded that there is no threat to their intentions. Employers pay and therefore dictate the job content, but they also get the calibre of employee they deserve. What they should not do in such circumstances is to have the effrontery to apply the term 'security' to the function.

### *Issue of standing orders*

It is advisable that the company secretary or a legal representative should glance through the orders before they are issued as they do authorize action on behalf of the firm which might result in claims against it. No alterations or additions should be made thereafter except by management responsible for their issue and then only after acceptance by the staff of the variation. A copy should be kept on file in the security office for ready reference by any member of the department, all of whom should have been given a copy during their induction. Whilst standing orders are generally termed 'confidential', management and union representatives must have access to them during some negotiations as they may have to be referred to at times when there is dispute about the carrying out of jobs. Hence they really should not contain information which is thought inadvisable for others to know.

*Confidentiality*

This is a most important matter for security staff as during the course of their work they will learn things about their employer's business which they must keep to themselves. No less important is the information that they may acquire about employees and anything of a personal nature that may be confided to them by employees. If that trust is abused, a great deal of trouble, both industrial and personal, may ensue, and a long time may elapse before relationships are stabilized again. A fundamental need of inadequate persons is to acquire a sense of importance, and they may seek this by showing, in gossiping, that they have knowledge that others do not; security staff cannot afford to indulge in this luxury, and the department cannot afford to keep members who cannot be so trusted. On the other hand, gossip should be listened to because there is often a grain of truth in what is being said that should be stored for future reference; it is also a means whereby a fellow employee with a conscience tries to pass information that he has without becoming involved himself.

In this connection a point which has caused, and no doubt will continue to cause, great annoyance concerns the process of making observations on some undesirable activity. Often when these have been found fruitless, it is discovered that some 'trusted' supervisor has been given an inkling of what was happening and he, not wanting to have any of his staff involved, has dropped a discreet word where it would do most good to that end. The business of the department must not be discussed with outsiders, and this should be specified in the standing orders as in Appendix 7, paragraph 6.

*Discretion*

A test of efficiency has been said to be the ability to deal with the unexpected and unpredictable. Mistakes will be made in well-intentioned attempts to do this, and this has to be accepted lest the philosophy of 'If I don't do anything, I won't do anything wrong' may be applied. An employer with any sense, or any commonsense for that matter, will acknowledge that discretion. At the risk of being repetitive, paragraph 20 of the specimen standing orders is worth quoting here:

> These instructions do not touch on all circumstances which may call for the attention of the security staff. Where a situation arises and no specific instructions have been issued which apply to it, the members of the security staff will be expected to use intelligence, imagination, and discretion to ensure that it is dealt with satisfactorily.

## *OFFICE DUTIES*

The various books used in the office are indicative of the duties expected of the security officers working therein. These are listed in Appendix 8, but some of the more important points are dealt with in detail here.

It must never be forgotten that the security office may be the first point of contact that visitors have with the company and that their opinion of it can be coloured by the impression that they form at that time. Always try to get customers dealt with promptly, politely, and efficiently. There is nothing worse than security officers talking together in an office whilst a caller stands patiently awaiting their convenience; an untidy office falls very nearly into the same category. Efficiency is also reflected in the answering of telephone calls. An elementary point? Both authors have frequently encountered the meaningless 'Hello', 'Yes', 'Eh?', and even deep breathing only! It is equally simple, and much more informative and correct, to say 'Security office' for internal calls, or 'Security office, Durham Metals, can I help you please?' to an outside caller.

Briefly, some other matters that bear repetition: do not allow other employees to loiter in the office; do not leave confidential papers where they can be read by callers; do not relax standards of dress in the office, for example tie off and jacket undone; and do not spread your meal out over the office desk unless there is absolutely no more private place to eat it. During quiet periods check that you are familiar with all of the emergency instructions that should be kept in the office. If the office controls an external door or gate and for any reason you have temporarily to go elsewhere for the odd minute, lock that entrance if this can be done – particularly after normal working hours.

*Registers*

*Occurrence (or log) book* – This is used to record all events related to security responsibilities which are not entered in a separate register, for example the search register. The occurrence book should be seen daily by the person in charge of security and signed by him. He will be responsible for any further action which might be necessary such as informing the manager of the department concerned of any incident which has received the attention of the security staff and the action which has been taken.

As an alternative to the occurrence book where a large security staff is employed, printed report forms can be completed by the security officer concerned in any occurrence and filed in a binder. If a shift of staff is under a supervisory member, he must inspect and approve the report of anything which happened during his period of responsibility, adding his signature.

*Search register* – In this must be recorded every instance where an employee or vehicle is searched when entering or leaving the premises, likewise hauliers, contractors and their employees who have accepted conformity with search conditions in their contracts. The date and time of search, the name of the person concerned, plus clock number where applicable, registered number and owner of vehicle plus driver's name, name of person carrying out search, should all be allocated columns in a permanent book register which should have additional spaces for 'Comments' or 'Results' and for signature by the person searched.

A written record that searches are carried out is much more useful to management than a verbal claim when dealing with a complaint associated with a search. Research through the records can establish whether complaints have previously been made and assist in placing the current one in perspective. For this latter reason, records are suggested to be kept for, say, three years.

*Lost and found property* – Meticulous instructions must be laid down for the handling and documentation of 'found' property as it can represent a source of temptation to security staff and of complaint by disgruntled finders who consider that they are entitled to it. 'Lost' property should be equally well documented to ensure easy cross-reference with property found.

Though reports of either lost or found property may be relatively rare they should not be treated casually, and an efficient system produces goodwill. It can be in the form of a separate book kept for the purpose or as a series of duplicated forms. Specimens of lost and found property forms are shown in Appendices 9 and 10. In most circumstances the book entries will suffice, but it is strongly advised that the person in charge of security should periodically (weekly) check the records against any property held in the security office and initial accordingly. All moneys found should be lodged in a safe with restricted access, or transferred to the employer's cashier for retention. Descriptions of property lost or found must be sufficiently detailed to identify. Where property is found the entries should be made in the presence of the finder, either in an officer's notebook or directly into the 'lost and found' property book, and the finder should be asked to initial that the details are correct. Where property is handed over to a bona fide loser he should be given the finder's name, and in turn the finder should be informed of who the loser was.

Legally, where property is found on premises the ownership, other than the priority of the loser himself, is vested in the person or firm tenanting the premises. In practice, except where high value is concerned, for all purposes it is better that after a suitable lapse of time it should be returned to the finder, who should, however, be asked to sign a form of indemnity to protect the firm against any subsequent claim from a loser. Whenever property is handed over to either a claimant or a finder there must be a signature and a witness to the signature.

*Car register* – The keeping of a register is not an imposition on the car owners in a workforce, it is a means of providing goodwill in that those with defective lights, flat tyres, leaks in radiators, etc. can be immediately traced – also those badly parked who are contributing to general traffic chaos. It also provides a means of checking the bona fides of vehicles parked upon company land which may be there unlawfully or for a hostile purpose.

Too often, car registers are made unwieldy. All that fundamentally is required is a card compiled by each car owner which shows the registration number, owner's name, department in which he works, and the telephone extension at which he can be contacted; the make of the vehicle can be added if desired. All that is then needed is a means of identification which, if necessary, can also show a parking area to which the vehicle is allocated. Windscreen stick-on discs, of a different colour for each park, surrounding a white smaller central disc with an identification number upon it can be used. These can be affixed adjacent to the licence disc or behind the inner mirror so as not to obscure the driver's view but yet be immediately visible to the security staff. A special docket with the company's name can be used if so desired. The requirement is to be easily seen and identifiable but not obtrusive. Any such scheme must be discussed with employees' representatives before implementation, and a speedy way of introduction is to fix a registration card and explanatory note beneath the windscreen wiper of each vehicle parked on the premises on a favourable day and get the stragglers later. The index can take the form either of the cards filed in numerical sequence from the registration numbers or of the details abstracted to make a book register when required.

*Borrowed tools register* – Some employers permit employees to borrow tools to do work at home. Their authority to take them off the premises is usually given on a form called a 'pass out' signed by an authorized person. This pass out is produced at the gatehouse when the employee leaves and the facts are recorded in this register. Sometimes the pass out is prepared in duplicate, one copy to be given up at the gate and the other to be retained by the employee.

When the employee is returning the tool he reports this and produces it at the gatehouse when he enters and it is noted in the register. Occasionally the register should be checked with the departments of the employees concerned to verify the tools have actually been returned; and also to see whether there are tools which have been outstanding for an unreasonably long period of time. If there are, and enquiries from the issuing department show they have not been returned, it is that department's responsibility to ensure they are. It is not unusual for an employee who is leaving to 'borrow' tools and 'forget' to return them – and tools have included calculators and desktop computers. A reminder by personal visit or official letter may effect recovery with the ultimate option of reference to the police.

*Key register* – Keys in the gatehouse should be kept in a cupboard which can be locked, the key of which is held personally by the security officer on duty. The list of identifications of the keys must not be visible to anyone entering the gatehouse. There have been instances where important keys hanging on open key-boards or in unlocked cupboards have been removed by unknown persons when the attention of the custodian has been distracted, used for an unlawful purpose, and later returned, again without the security officer knowing.

Only persons authorized to hold keys should be issued with them for a day, a shift, or for a specific purpose. Issues and returns are timed in the register and the entries initialled by the issuer. When the key is of special importance, the person receiving it signs for it.

Requests for keys at unusual times, without previous notice from an authorizing person, should be closely questioned and, if possible, the person wanting the key should be accompanied to the place concerned. An instance of when this was not done occurred at a factory where a woman who was known to be a canteen worker called on a Sunday, when the canteen was closed, and asked for the keys of the canteen. She was given them without question and allowed to proceed alone. Because she had not returned the keys after some time a call was made at the canteen where she was found to have gassed herself in an oven. Similarly, a worker with explosives on a pretended call-out, collected the keys to the store and proceeded to blow himself up over the adjoining area.

Undue delay in returning keys must be investigated in all cases.

*Telephone message pads* – When a telephone message is received at times when the official operator is not on duty it should be entered on message pads in duplicate. The name and initials of the caller, his telephone number, the person for whom the message is intended, and whether it is urgent or not should be included. The message will be completed with the time and date of receipt and the name of the person receiving the message. The original copy is available for delivery and the second copy is filed – this can be useful if the original copy containing important information is lost.

*Vehicle access passes* – This is an essential security duty in any circumstances where vehicles enter company premises as the availability of transport facilitates theft, fraud, or deliberate damage by either outsiders or employees. The existence of an efficient recording system in itself is a deterrent to illegal intention. The means of doing this varies from a simple note in a book of vehicle number plus time 'in' and 'out', to the issuing of an individual duplicated slip giving a permanent and comprehensive record of the visit.

A specimen used for this purpose is shown in Appendix 11, which is dual purpose in this instance in so far as it includes weighbridge particulars. The form is in triplicate. The third copy remains as a permanent record. The first is handed to the driver, who in the case of weighbridge usage will also be given a print-out ticket; on leaving the premises he must surrender the original, duly signed by an authorized person of the department to which he has delivered or from which he has collected. The details shown are self-explanatory. The signed original and second copy are sent to the department handling documentation and to that receiving or dispatching the material; so there is a complete and early cross-check. In certain instances these forms may be made of assistance to a firm's distribution department, in so far as it may be advantageous for a record to be made on them of the invoice numbers for all consignments of goods dispatched from the premises by road transport. It is conventional for copies of invoices and advice notes to be destroyed after a limited period, and it might well then be found that the only record for a delayed enquiry is that contained by this form of log which shows at least that the documentation left the dispatch point. The original slip goes with the driver's consignment notes, and the duplicate expedites the compilation of bonus sheets etc.

The onus for the accuracy of these records must rest with the security officer making them, who should visually check that the information that he has been given of vehicle number, carrier's name, etc. is correct. If circumstances arise where an internal weighbridge can be resited near the works entrance, the mere presence of independent scrutiny will dispose of a great number of opportunities for fraud, and the operation is a possible ancillary duty for security staff.

*Visitors' passes* – In the natural order of things there will be a great divergence in attitudes towards visitors, ranging from 'the more the merrier' for retail establishments to an almost complete ban for those with high-security defence commitments. The first priority is therefore to find the degree of restriction desirable, bearing in mind the necessity for review in the event of circumstances changing – the current IRA and other terrorist activity providing such an example. Within any complex the restriction requirement may be variable, and in these instances this can be recognized by using a form of authorization with defined limits. This may be done by issuing passes of different colour only permitting access to appropriate areas. Some are in a form that can be clipped to, or stuck on to, the clothing of the carrier, so that he is clearly and immediately identifiable as authorized or not authorized in the area where he is.

The main objectives of a visitor's pass system are:

1 To exclude undesirables and maintain the privacy of both employers and employees.
2 To control the access of legitimate but unwanted callers.
3 To maintain a record of all visitors in case of need.
4 To check that visitors have been dealt with in accordance with the purpose of their visit and have left.

For this to be done efficiently and economically, access points for visitors must be limited in number, and where there is no security present each should be in clear view of a surveillance point – for example, a receptionist, or a designated employee in an office with a clear glass window with an appropriate inscription such as 'Enquiries' or 'Visitors' reception'.

A form of visitor's pass found suitable for most general purposes is shown in Appendix 11. The procedure for using this would be for the issuer to ascertain: the name of the caller, and his company if applicable; the individual or department that he wishes to visit; and the purpose of the call. A check should be made by telephone to ensure that the caller is acceptable, which will also have the virtue of enabling the requisite courtesy to be extended and will allow the dispatch of an escort to collect the visitor, which in many instances will be desirable – especially in office blocks. Where a visitor is collected, the visitor's pass system need not always be applied, but a note should be kept. There are advantages in numbering passes so that a periodic check can be made to ensure that they are returned to the point of origin; this also provides a check upon departmental conformance with practice in so far as passes should be initialled by the person who has been visited.

Visitors' books are used in many instances, but their main value is in small establishments, and over a period of time there would inevitably be omissions.

*Telephone numbers of senior and key personnel* – These are necessary to assist the security staff to get in touch with such personnel in an emergency. They must be kept up to date by the persons concerned.

*Lifts* – The security staff should be conversant with the emergency action in the event of lift failure, have equipment and keys to release persons trapped, plus details of own staff or lift installers to be called out. Claustrophobic and nervous passengers may panic or become hysterical if there is delay so that everything which can be done or said to reassure them should be.

*Pass outs*

These are forms which can be used to authorize hourly-paid employees to leave the premises before their normal finishing time – for medical or dental treatment or on compassionate grounds, for example. The employee hands it to the security officer on gate duty and it is afterwards sent to the personnel or wages department so that any necessary adjustments can be made to his pay.

The form can also be used to authorize the removal from the premises by an employee of small amounts of material of no value to the company, for example short lengths of conduit pipe, or firewood, for his own use at no charge or on payment of a nominal amount. The pass out must show whether any charge has to be paid and whoever is authorized to receive that payment stamps it accordingly when it has been done.

In some cases the pass out for property is made out in duplicate, one copy to be handed to the security officer at the gate and the other retained by the employee to be produced should he be questioned by police or a member of the security staff. Material pass outs should be returned by the security staff to the persons authorizing them for them to check for any alterations and to deal with as required. Quantities should be written in words on such forms, figures not being used.

### *Staff sales*

The removal of company products through sales to the staff has to be strictly controlled. The principle to be observed is that employees must not be allowed to take their purchases into their places of work. A staff sales shop close to the exit from the factory has great advantages in that connection. All purchases, which to save time might have been previously ordered and wrapped, must be collected as the purchaser is about to leave the premises.

If the shop is not in such a favourable position one form of control is for the purchaser, having paid for the goods, to be given a ticket bearing the same number as one attached to the wrapping of whatever they have bought. The parcels are later removed by the shop staff to the gatehouse. On surrendering his ticket an employee is given his parcel immediately before leaving the premises.

Where there are no shop facilities or where products, scrap, or other materials may be bought from the firm at concessionary rates, a set documentary procedure should be established whereby an order is laid, then priced, the goods are made available, cashiers are paid according to the pricing and issue a receipt, and the goods are handed over against the receipt which is shown to security when they are taken out. Care should be taken, and instructions given, to ensure that only stock items of standard lengths etc. are sold and that 'scrap' has not been deliberately created for cheap purchase.

Staff sales in retail premises are mentioned in detail elsewhere (see p. 343). These are a special hazard in all types of shops, none more so than in supermarkets and if there is a loophole in the procedures it will be exploited. Recognizing this, a tight system should be laid down and enforced even if it causes some delay and inconvenience to employees.

### *Contractors' book or register*

Contractors are deemed a risk sufficient to justify a separate chapter about them in this textbook (Chapter 26). A fundamental step towards limiting that risk is to accurately record their comings and goings, and numbers of persons working on their behalf. The amount of data will depend on the conditions of contract under which they are operating – these could include a requirement that their staff individually sign in and out. If no such obligation has been enforced, and it is obviously one that should be recommended by the security head, the times of arrival and departure, together with the name of the firm, and number of persons should be entered into a separate book for easy reference.

### *Suitcases and large parcels*

Suitcases and large parcels carried in and out by employees provide an easy means of stealing especially where no 'search clause' is in force. Nowhere is this a greater risk than in retail premises where, in any new planning, it is important to provide cloakroom plus bag storage facilities immediately beside the staff entrance and to make it clear in rules, or conditions of employment, that it will be a disciplinary matter not to use those facilities.

In an industrial environment, the problem is less likely to arise but on the odd occasion it does, a temporary inconvenience of allowing the owner to leave a case or parcel in the gate office may both show a helpful attitude and prevent a risk.

## *PATROLLING DUTIES*

### *Objectives*

The objectives in patrolling premises are to:

1 prevent and detect fire;
2 prevent and detect damage from other causes, and waste;
3 ensure company rules are observed, for example no smoking in non-smoking areas;
4 prevent and detect offences against the company's interest;
5 prevent accidents.

### *List of duties for patrolling security officers*

The duties to be carried out when patrolling premises have been compiled and are shown below. These should be read in conjunction with the further duties outlined under 'Fire Patrols' in Chapter 30.

1 Examine roofs for holes through which water or sparks from

adjoining premises could fall to damage goods or machinery. Any necessary protective action should be taken immediately.

2 Note any other defects in roofs, gutters, walls, or windows through which rain water, smoke, etc., is entering or could enter the premises, and take any necessary action.

3 Check:

(a) locks on warehouses, stores, and offices and mark padlocks to detect any subsequent substitution;
(b) all perimeter fences for breaks, ensuring that goods or other materials are not stored directly against them;
(c) all vehicles of employees in car parks to detect lights or engines left switched on, and punctures – inform owners as necessary;
(d) seals on loaded lorries, railway vans, etc.;
(e) insides of empty railway vans waiting to leave the premises, to detect stolen property;
(f) that goods in basements are not stored directly on the floor whereby damage might be sustained from water accumulation;
(g) that any property is not exposed to the weather whereby it is or could be damaged;
(h) that all strong-rooms and safes are locked, removing to safe custody any keys found in locks.

4 Be on the alert to notice:

(a) any hazard which might cause or produce an accident and inform appropriate person;
(b) unauthorized persons on premises;
(c) to prevent and detect offences at clocking stations.

5 Turn off all taps or valves through which liquid or steam is escaping.

6 Visit men's dressing-rooms and cloakrooms to prevent theft and other offences.

7 Visit laboratories and other similar places to check on the operation of any equipment in use, particularly when concerned in an active process at times when this is not supervised. (The security department should always be notified in writing of such operations and given directions respecting action to be taken in any emergency. A suggested form of notice to be attached to the equipment in use in such circumstances is shown in Chapter 31.)

Alertness, interest and thoroughness must be displayed. A suspicious mind must be cultivated and anything that appears other than normal must be looked into.

A security officer cannot be expected to acquire more than a superficial knowledge of all the manufacturing and associated processes but it is essential that he should do that as early in his employment as possible. He must learn the identities of the persons in charge of each department and section so as to be able to communicate with them in case of need. To call for assistance in any emergency when on patrol the security officer should carry a police whistle and a truncheon to protect himself from attack (private premises).

*Frequency*

The frequencies of the patrols will depend on a number of factors including the length, complexity, and vulnerability to one or more of the risks which patrolling is designed to prevent.

Duration and main purpose will of course differ between time of normal working and those times when the premises are unattended other than by the security officer. The times of patrols must be irregular so that the arrival of the officer at any place cannot be anticipated. To prevent an interested person estimating with accuracy where he will be at any time he should retrace his steps occasionally and vary his route.

*Clocks and clocking stations*

There are divergent views on the requirement that a patrolling security officer should carry a 'watchman's clock' with which to register his attendance at clocking points located on the premises. One view is that if the calibre of the person is of the standard required he will patrol conscientiously and thoroughly and will not require the evidence of the clock to prove it. The other view is that the fact that he can prove his attendance at a given place at a specific time is valuable to him should subsequent events, for example an undiscovered fire, cast doubts on his claim to have been in the area at a relevant time.

If clocking points are used the number must not be excessive otherwise the officer will be required to spend his time making a round of them to the detriment of his attention to other matters. It is recommended that the first round should be a full one and subsequent ones at the discretion of the officer. Clocking points should be sited near places of high risk – the safe, stores, canteen, distant perimeter fences, and so on – which should receive proportionately more attention than the less vulnerable places. Examination of the tapes of the clocks will show how the patrolling has been carried out and whether a regular pattern of visits to certain areas emerges. The older mechanical 'clocks' are obsolescent and may now be replaced by electronic ones which are more convenient and have extra facilities.

### *Keys*

To save the patrolling officer having to carry a large bunch of keys to premises he is required to visit, the locks to them can be on what is known as a 'suite' which will be opened by one master key (see Chapter 38). Keys to the locks of dry goods and wine, spirits, and tobacco stores should not be held by the patrolling officer nor should they be on a suite of which he has a master key. The reason for this is that should shortages of stock occur suspicion could be attached to him if he had a key in his possession. However, provision has to be made for him to inspect the interior of stores – for example, through windows – and in the event of an emergency the keys should be within immediate reach. It is suggested they be sealed in a strong envelope and kept under strict security supervision at the security gatehouse. After use the envelope with the key should be produced for resealing.

## *COMMUNICATIONS*

### *Telephone*

Where particularly large areas have to be patrolled this will be done more efficiently if they are divided into named or numbered patrol sections, similar to the beats of the civil police. Where a security officer leaves the gatehouse in these circumstances he should tell any colleague he leaves behind which section or sections he proposes to patrol.

If there is no other security officer on the premises, and some process work is being carried on by the general workforce under supervision, he should note in a register where he has gone in case the manager or supervisor on duty requires his assistance. If he has a colleague at base, the patrolling officer must telephone him at intervals as a mutual safety precaution. The times should be recorded. If a person fails to communicate or return within a reasonable time, possibly due to accident or illness, the area of search under the circumstances described is much reduced.

### *Radio*

The use of personal two-way radios (transceivers) to provide constant and immediate communication between the gatehouse or other security centre and the patrolling officer has become recognized as a great assistance to good security (see Chapter 35). Some of the advantages do not require elaboration but one of them deserves emphasis. This is the ability, where the patrolling officer is out of touch with a telephone or is unable to reach it, to call for assistance in the case of accident or illness to himself or others or when discovering intruders carrying out crime or vandalism.

The value of this is illustrated by the experience of one patrolling security officer who found three men on the premises in suspicious circumstances. Before interrogating them, and without their knowledge, he switched on his personal transceiver and through the microphone in his breast pocket his colleague at the security centre heard his questions and their answers. These roused his suspicions so he immediately arranged for assistance to be sent. The three men were detained and subsequently charged with criminal offences.

One method of calling for a particular person is the radio call system in which a receiver carried in his pocket is operated by radio signal. This emits a bleeping sound and he then telephones the central switchboard to ask who requires him.

Another recent advance is the carrying of a device the size of a cigarette packet which is operated by pressing a button. This gives out a radio signal which operates equipment in the gatehouse to activate a recorded telephone call to police for assistance.

### *Closed-circuit television*

This sophisticated equipment can play a vital part in security and is especially useful in maintaining observation on particular areas from some distance. Its use will be developed in Chapter 35. It is now one of the most valuable aids in controlling theft from shops.

### *Dogs on security duty*

This will be dealt with fully in Chapter 37. Suffice it to say here that where a lone security officer is employed on a site where a high risk of theft exists and where he may be in danger of personal attack the employment of a dog to accompany him has great advantages.

### *Intruders*

This description is given to all persons unlawfully or irregularly on the premises. Problems associated with carrying a truncheon as means of defence against personal attack have been mentioned in the previous chapter, but the legality was only held in question when this was done in a public place. There seems no objection to doing so on an enclosed private site in an area where there is real risk – indeed, it would be commonsense to do so. Firearms are permissible under licence in some parts of the world for security purposes, i.e. Hong Kong, but are, hopefully, most unlikely to be either allowed or necessary in the UK in the foreseeable future.

On taking up employment on a security staff a person must accept that this may involve bodily injury in defence of his employer's interests. In circumstances where he feels he cannot cope single-handed with intruders it is better that he should obtain assistance rather than indulge in heroics. A security officer who can identify intruders or thieves who have escaped is more useful than an unconscious one unable to recognize his attackers.

The amount of violence a security officer may use against intruders is the minimum necessary to arrest or to prevent escape. Violence used in self-defence must have a reasonable relationship to that suffered or to that which is reasonably anticipated. Trespassers who have not committed any offence but who refuse to leave the premises when requested may be ejected using as much force as is necessary (see Chapter 12 on 'Trespass').

### *Vulnerable places*

When visiting warehouses, stores and offices where there is a high risk of loss by theft, care must be taken not to give notice of approach by heavy footsteps or the flashing of a torch. A minute or two of silence outside the premises listening for unusual sounds within are worthwhile and rubber soles should be worn.

The need not to announce one's presence is illustrated by the experience of one security officer who, without regard for the noise he was creating, entered a building where a safe containing a large sum of money was located. As he entered the office, which was in darkness, he was struck unconscious by a blow on the head from thieves who had forced open the safe and were preparing to leave. He was quite unable to assist the police investigation in any way.

### *Property found in suspicious circumstances*

It is common practice for a thief to hide things he has stolen from his employer on the premises or close to the perimeter fence to be removed later. Sometimes the property is thrown over the wall. Should a security officer come upon such property, which from its nature and the place where it is hidden has clearly been stolen, it must not be moved. Nothing should be done to show anyone who may be watching that the property has been noticed. Observation should be kept on it from a concealed position. No enquiries should be made of anyone outside the security staff.

When anyone is seen taking possession of the property he should be allowed to do so without interference. If this occurs within the premises he should be followed until the opportunity arises for him to be stopped in the presence, preferably, of a member of the managerial or supervisory staff, alternatively another member of the security staff. He should be told what was seen, asked to produce the property and explain his possession of it. If this clearly shows that he has stolen it, the policy of the employer must then be followed. If the property is seen outside the perimeter it should be kept under observation as described and police assistance sought.

### *Diversionary incidents*

Security staff must be on the alert to detect incidents which have been arranged to divert and occupy their attention whilst something unlawful or irregular is being carried out somewhere else. When any incident occurs which requires action from a security officer he should telephone his base and report the matter saying whether he has any suspicions. This is a circumstance where personal radio communication is particularly helpful.

## *MUTUAL AID SCHEME*

Security officers working alone on premises are obviously vulnerable to attack by intruders who are prepared to use violence and as a result incidents periodically occur in which they receive injuries sufficiently serious to have caused some deaths. Because of this, a scheme whereby they can readily obtain assistance when this is needed has been devised. It is called a 'mutual aid scheme' and is arranged between the security staff of a number of premises. The details must be kept confidential and the scheme works like this: at a pre-arranged time the security officer (SO) at factory A telephones the SO at factory B, then the SO of B telephones the SO of factory C who afterwards telephones the SO of factory D, he telephones the SO at factory A which completes the cycle. This is repeated at pre-arranged times.

If a security officer fails to answer a call it is repeated after ten minutes and if this is unanswered the SO making the call telephones the police. If a security officer expecting a call does not receive it, he telephones the factory from which he expected the call. If that call and another, ten minutes later, go unanswered he telephones the police. No time is wasted explaining the circumstances to police who should have been apprised of the plan. If a security officer has had an accident or has become ill and is unable to reach a telephone, by this arrangement assistance is brought to him within a reasonable time.

If a security officer is under restraint by intruders at the time of a call and they wish him to answer the telephone without disclosing what has happened he would use a previously arranged phrase which would sound quite innocent to the intruders but would inform the caller that assistance is required. The introduction of such a scheme between two or more factories where people are employed in similar conditions of personal isolation to those described is a morale boost to them.

Chief officers of police will not agree to police stations receiving routine reports directly from security officers nor allow telephone calls to be made to them for that purpose. This would unduly engage telephone lines and operators at the stations required for more urgent purposes.

One of the services which can be supplied by security companies is visits to or telephone communication with security custodians of premises.

# 7

# *Training security staff*

There is today an enormous capacity for the security industry to increase its effectiveness in crime and loss prevention and to gain the respect and confidence of the public who are seeking some form of extra protection.

Negative attitudes in all aspects of security must be replaced with a positive approach for the future and that in short means training.

Clearly, security employees should not be considered professional until they have received proper training, and management must accept this in view of the serious consequences that could arise from either misaction or inaction on the part of security staff. Supervisors and indeed management need instruction to deal effectively with the problems caused by operational personnel.

All too often firms apply high standards in their selection of security personnel, pay good salaries, and then do very little to improve the calibre of the person employed. The intake is rarely more than one or two individuals at a time, and this complicates in-company training, which usually depends on such time as can be devoted to it by the chief security officer (CSO). If he is already working to near capacity the consequence is that the recruit, if he learns anything about his trade, does so in the course of day-to-day contact with his colleagues. Any faults in performance or misconceptions of powers, or more usually belief in security's lack of powers, are therefore perpetuated so that the level of efficiency remains one of placid mediocrity. A company is entitled to expect due return for its investment, and it does not receive this from part-trained security officers. It is incumbent upon management to ensure that (a) the CSO is competent to the desired level and kept there by updating refresher courses, and (b) his staff are progressively instructed from their time of induction by both in-company training and attendance at organized courses with a broad spectrum of specialist lectures. The test of efficiency is the ability to deal with the spontaneous unfamiliar incident, not the routine day-to-day occurrence which sometimes seems all that some naive managements can visualize as confronting their security staff.

## *SPECIFIC TRAINING*

The present training arrangements in this country are minimal, being mainly provided by the larger companies for supervisory personnel. An effective future security education programme must encompass a large number of interrelated subjects. However, factors such as time, budget, manpower considerations and legal limitations have all to be acknowledged.

Any education programme related to security must be wide ranging to cater for all potential clients. Some of the potential students who need to be catered for are as follows.

*Security industry*

1. Guards – initial basic course, 16 hours; intermediate course, 16 hours.
2. Supervisory – leadership and administration.
3. Junior manager – communication, leadership and motivation.
4. Specialist – alarms, store detectives, loss and crime prevention, fire prevention, accident prevention, health and safety.

*Commercial*

1. Managers – security responsibilities.
2. Supervisor – security awareness.
3. Shop managers – loss prevention.
4. Directors – risk management analysis,

etc., etc.

*Chamber of Commerce*

To assist members.

*International*

Specialized courses for overseas delegates.

## *TRAINING PROGRAMME*

### *Aims*

Several positive results can be achieved through a good training programme.

1 Higher quality of job performance results once training completed.
2 Training should improve relationships between the trained individual and fellow staff.
3 Trained personnel get into the routine of work faster, because they are not in a 'trial and error' situation.
4 A pool of trained personnel for promotion possibilities.
5 Misfits are identified and can be removed from the organization before problems occur.
6 A better understanding of the total security concept of the industry and its work is developed and a better understanding of their individual roles and functions.
7 Increased flexibility and the capacity to adapt more readily to changes comes as a by-product to the overall programme.

## *OBJECTIVES*

To initiate a course of training or study in the security or loss prevention fields, certain objectives should be stated which would include:

1 To establish the operational role and function of security in society.
2 To identify the preventative role which can be established by the industry in greater measure.
3 To develop an understanding of the process and techniques of the security operation.
4 To develop an understanding of present-day problems in the security world, and to look at the symptoms, causes and possible solutions.

The first basic course given must therefore include loss prevention and security education and measures used by individuals or groups to retain and protect their own (and others) property and possessions.

## *IN-COMPANY INSTRUCTION*

Much of what is required is of course specifically of an in-company nature and cannot be given elsewhere. Knowledge of persons, places, processes, procedures, policies, etc., comes into this category. Part of this may be gained in a general induction course for new employees, but there are initial matters for which the CSO should accept responsibility:

1. Full clarification of the job description and conditions of employment.
2. Detailed explanation of the contents of standing orders for security staff.
3. Analysis of the works rules, with special emphasis on those items on which security will be expected to act or report.
4. Instruction in the firm's policies on security matters – for example, employee arrest, notification to police, searching, enforcement of rules.
5. The manner in which he will be expected to report and record incidents.
6. The limits on the discretion that he will be allowed to exercise when dealing with incidents of a security nature.
7. Any sensitive areas of employer and employee relationships where clumsy security action could have exaggerated repercussions.

## *FAMILIARIZATION WITH COMPANY AND SECURITY ROUTINE*

The next step is that of familiarization with the premises and activities of the firm. If circumstances allow for temporary attachment to departments with which there is close involvement, this should be done, if only for a morning or afternoon at each. Inside his own department it is advisable that the first process of familiarization should be in the office routine, which is the quickest way to learn about the business of the firm and people that he should know. It will also give him the chance to learn, by reading reports and records, how the department works and what is regularly expected of it and to read up on various emergency plans and procedures and other checklists which should be held in the office. The opportunity should arise during this spell for a read through of any security manuals or other instructional matter provided by the employer and to ascertain how he should act in purely internal disciplinary matters.

Use should be made of the training facilities of the organization for specialized instruction where this exists – first aid, fire, industrial relations, any of the risks peculiar to the business and premises. All too often security departments seem to keep aloof from this participation, and the importance of industrial relations can hardly be overstated nowadays.

## *EXTERNAL TRAINING*

The matters mentioned so far are primarily concerned with the company function as opposed to the more general security knowledge that has to be acquired. 'Basic' training courses designed to ensure the recruit knows certain essentials of a legal and practical nature are held in many parts of the country and at regular intervals. A minimum of two days (three if at all possible) is required to give basic training based on the following programme.

*First day* – Registration; History of security; Responsibilities and duties of security personnel; Attention to instructions.

Ethical standards for security personnel; IPSA Codes of Conduct; Personal conduct – Appearance – Demeanour.

Report writing; Pocketbooks; Dealing with incidents.

Site orientation requirements; Client expectation and requirements; Company policies.

Locks – Lighting – Alarms – Radio – CCTV – Fences – Doors.

*Second day* – Fire prevention; Fire detection; Fire equipment.

Health and safety; Accident prevention; Patrolling reports.

Administration of justice powers of arrest (Basic); Civil tort; Law.

Industrial relations; Environmental problems; Strike duties; Arrest; Searching; Responsibilities.

All including relevant films.
Each session – 1 hour.

There is no question of instruction to regular police standards; the time devoted to the Theft Act 1968 in speaking to security officers would be possibly as many hours as the police would spend days in their courses. It is not training of law enforcement officers, it is ensuring that those privately employed to protect have such information on legal aspects of incidents they are likely to encounter in their work as is necessary for them to do it efficiently. These essentials are dealt with in detail in subsequent chapters and also in the pocket-sized *Security Manual* (Oliver and Wilson, Gower Press), which was compiled specifically as a 'bible' of information and ready reference for security guards to carry during their duties. Examples of some subject headings would be:

*Making an arrest*

What is legally an arrest, and how should it be made?
What general powers has a civilian (security officer) to make an arrest?
What constitutes an 'arrestable offence', and what are those most likely to be encountered in security work?
What degree of force is permissible in making an arrest?
Is there power or justification to search a person who has been arrested?

*Action to follow an arrest*

Is it necessary to inform the police and commence proceedings against a man who has been arrested?
What will the police expect at the 'handover' to them of the prisoner from the person arresting?
What may happen to the prisoner after he has been placed in police charge?

*Notes, statements, and reports*

What are the rules which have to be complied with to make things said by an accused admissible in evidence, or to allow the officer to refer in the witness box to notes that he has made?
How should he record statements and versions from fellow employees, and how should he keep a notebook in connection with his duties?
How should he make out reports of arrests or other incidents?

*Court procedures*

What is the likely sequence of events during a prosecution?
What are the types of evidence that may be introduced?
How should he deport himself in the witness box?

*Constituents of serious offences*

> What legally is 'stealing' and the variants of it that are likely to be encountered in the course of security work?
> What legally are the main points about other criminal offences prevalent in commerce and industry – for example, criminal damage, indecency, inflicting injury, committing fraud, falsifying records, bribery?

*Miscellaneous legislation*

> What action can be taken in respect of trespassers and suspected persons on premises?
> What matters in safety and fire legislation or under the Health and Safety at Work Act should be known to recognize and prevent unintentional offences by the firm or hazards to employees or premises?

*Loss prevention*

> What are the essentials about 'security hardware' – locks, safes, fencing, lighting, alarm systems – for their value to be appreciated?

Such basic courses usually include a talk on the general duties of security officers which allows discussion of any procedures not clearly understood, and a series of practical problems is posed to alert students to the necessity of preconsidering incidents that they may encounter. There may also be a session devoted to practical instruction in the use of fire equipment.

'Intermediate' courses are designed for security officers regarded as having a supervisory potential or recently given that responsibility – also as 'refreshers' for well-established people. The content is similar to that of the basic courses but has emphasis more on the legal than on the practical aspects. Drug abuse, industrial relations, and administrative principles are usual additions, and discussion problems are appropriately more complex.

'Advanced' courses are intended for senior security personnel or managers with security responsibilities; it is a waste of time and money to send others to them as the content will be at too high a level. The emphasis is on the managerial and decision-making aspects, introducing problems associated with insurance, computers, company fraud, terrorist action, industrial espionage, disaster control, and matters of current legal importance.

Basic and intermediate courses are of three days' duration; advanced are of three or five days and usually residential, using university or adult education premises for the purpose.

*When should training occur?*

Training should occur whenever the need is apparent. Training can be categorized into three areas:

1. Pre-service or orientation training before carrying out direct security services.
2. In-service training – updating a person's knowledge at any given time in his career.
3. Specialized in-service training when a change of existing policy is necessary or where some advanced knowledge is required.

For two-day courses the middle of the week would be best to avoid weekends but exact times will be dictated by needs.

*Where should training occur?*

The presentation of training programmes must be made in the best facilities available. The ideal setting would be a well-designed classroom with the appropriate audio-visual support. Resources such as tapes, security books and hardware and related teaching aids should be easily accessible. It is the provision of a pre-designed room to be used solely for teaching students that is most important.

## *TRAINING OF EMPLOYEES OF COMMERCIAL SECURITY FIRMS*

Commercial firms face difficulties, both practical and financial, in training their officers to a desirable standard. They operate in a highly competitive market which means limitations upon the reserve of manpower that can be kept available to ensure commitments are met whilst training is being undergone. The larger firms have their own training establishments to which employees are sent once it appears likely they will remain in the job – high rate of staff turnover is a problem which may lead to wasted expenditure if the training is carried out immediately. All the reputable firms however endeavour to make an effort which is becoming increasingly necessary as customers query the value of the service they are being given or are promised (see p. 583). Providing the firm can call upon enough expertise from its local resources, training can be progressively given on a part-time basis of several hours per week which allows the performance of contracts without the increase in manpower that would be required by taking the individuals away completely.

Standards vary considerably and in the worst cases a new 'security officer' is simply a person who has been put into a uniform, taken to premises he is supposed to guard or supervise for security purposes, shown round for perhaps an hour and then left to it. It is a matter of record that some years ago, late one weekend evening, the son of a senior security officer was on the phone for nearly 45 minutes – the caller was a university friend ringing from the manager's office in a large Birmingham store – he was doing a part-time security job with a round-the-clock stint for a national firm. When quizzed by the father, he said that all he had been given was a route to follow round the building and a telephone number to ring if anything unusual happened; he had become lost, bored, and was spending his time ringing round all his friends! Of course, part-time employees are yet another training problem; few firms can do without them, though in fairness they are usually required for the weekend guarding of premises which requires least specialized knowledge. However, grounding should have been given on fire prevention, detection and use of appliances.

Training should follow the same general pattern as that given to in-company security, particularly if the contract is one to provide complete cover and the security officer can anticipate a regular assignment at one spot. If the firm concentrates on simple overnight guarding, or spot visiting, obviously there will be little concern with the customer's internal policies and procedures but familiarity with the premises will be required. Cash-carrying precautions need both explanation and demonstration before a newcomer is entrusted with the job; similarly, it would be most dangerous to turn a prospective store detective loose without making sure he was aware of the limitations on his powers as well as the ploys of store-thieves.

Should registration and licensing eventually be introduced, there is little doubt the legislation will make adequate training a requirement. Currently, it could be most embarrassing if a customer asked questions of the person supplied and found he had been grossly misinformed about the capabilities and instruction given to that individual.

Adequate minimum training necessary for assignments should be given in the first place, followed up by regular progressive training or refresher periods by competent instructors who can talk at a level understood by their listeners. It is highly advisable that all new employees be given an instruction booklet to which they can refer, and increase their knowledge by reading when the opportunity permits. The cynical comment of an ex-security guard supervisor is revealing: 'Everyone wants trained security guards. Untrained employees are the cause of wasted money and lost business and are a danger to themselves and others. But, the demand for manpower is so great, the wages and bids so low, that training standards have to be altered, with a little misrepresentation to salt it.'

## *SOURCES OF EXTERNAL TRAINING*

With the exception of an infrequent function organized under police auspices, practically all communal training for in-company security personnel is carried out by the International Professional Security Association (IPSA) on a regional basis. Senior members pool their expertise in providing lecturers, which gives an essentially business rather than public bias to the content. Both police and fire brigade give invaluable co-operation in dealing with specialist subjects, and similar assistance has been freely provided on request by the Factory Inspectorate, the British Insurance Association, the Royal Society for the Prevention of Accidents, the Trades Union Congress, the Institute of Personnel Managers, and a variety of other bodies which recognize the value to the public as well as to employers of security forces being as efficient as possible.

Commercially organized seminars are held periodically, mainly in the London area and for senior management, dealing in the main with specific topics: terrorist action, computer and company frauds, etc. These are advertised by countrywide circulation and in the security press, and the fees are in accordance with the promoters' assessment of the current importance of the subject matter.

Details of all the IPSA courses can be obtained from the International Secretary, Mr P. Rabbitts, 292A Torquay Road, Paignton, Devon TQ3 2ET. Telephone 0803 554849. They are held between September and May. The number is variable, approximating to 60 a year of all types. In addition a correspondence course is available from the same source, and an International Institute of Security formed with qualifying examinations twice a year for security officers who wish to study to improve their knowledge and promotional prospects. These examinations now enjoy City and Guilds approval and recognition.

## *PROFESSIONAL BODIES*

The IPSA's objectives are built round the concept of promoting mutual co-operation, professional standards, and efficient performance. It is the only organized non-trade body representing security personnel in the UK, and as a major part of its effort is associated with training this is an appropriate place to give some details of it since its activities have had considerable bearing on the development of in-company security conduct and duties.

The Association was formed in the Midlands in 1958. Regions have been set up for Scotland, North, NE, and NW England, Midland and North Midland, South Wales, the SW, the East (East Anglia), London and South East, Southern Home Counties, Retail (shops, stores and supermarkets), Overseas (worldwide), Northern Ireland, Eire, Hong Kong, and with the possibility of an affiliated European branch not excluded – in other words, a comprehensive cover. Representatives of each region meet four times a year as a governing body, with the permanent secretary at a registered office with supporting clerical and administrative staff; chairmanship is held for a two-year term only and rotates round the regions.

The Association is non-political in nature and has no trade union affiliations or aspirations. To support the central concept mentioned, its constitution provides for studies, surveys, conferences, forums, discussion, research, and indeed every step calculated to improve the status and ability of security practitioners and the exchange of information. There are four categories of membership which are as follows:

1 *Member Company* – Available to:

(a) Commercial companies, partnerships, individuals and other such organizations (which have been trading for a minimum period of 12 months) engaged in the supplying of security services, supplying/installing of security equipment, and the manufacture of security equipment.

(b) Commercial companies, corporations, nationalized industries, public utility undertakings, banks, Government departments and local authorities, who employ their own security staff (in-house).

2 *Member* – Available to persons who are wholly employed in industrial/commercial security.

3 *Associate Person* – Available to those not employed full-time, but who intend to achieve a career in industrial/commercial security (this includes serving police officers, members of HM Forces, students, etc.).

4 *Associate Company* – Available to those as indicated under 'member company', with exception of (b) who have traded less than 12 months.

This membership is available for one year only, and on having completed this period of membership 'Associate Company' members will be required to apply for the 'Member Company' category and meet the requirements as laid down by the governing body of the Association.

Overseas membership is available in all four categories listed above.

Membership of all categories entitles participation in all regional activities and the attendance at any training course organized by IPSA (by payment of fees).

The Institute of International Security was formed to provide an examination structure leading to the acquisition of professional qualifications. The ruling board of governors comprises equal numbers of nominated IPSA members and prominent businesspeople, lawyers, and others with keen interest in security matters. There are three grades of membership — 'Graduate', 'Member', and 'Fellow' – with again provision for 'Honorary Member' and 'Honorary Fellow' though these are sparingly given. Graduate and Member qualifications are gained by written examination.

Fellowship status is gained by submission of a thesis by an existing Member on a subject approved by the board which will be published in full in the official journal of the Association, if successful. Several firms now regard the passing of the examinations as a distinct advantage for promotion, if not a necessity, and many others make awards or merit salary increases for success. Such proof of knowledge and ability provides a counterbalance for the career security officer against the experience of older retiring police officers in contesting the more responsible and remunerative posts in the profession although police applications are now declining.

Overseas there are a number of other associations with like objectives and concentration on training. The larger and better known are:

American Society for Industrial Security (ASIS); 1655 N. Fort Myer Drive, Suite 1200, Arlington, VA 22209, USA

Canadian Society for Industrial Security, 336 Whimby Avenue, St Lambert, Province of Quebec, Canada

Security Association of South Africa, 203/4 De Korte House, De Korte Street, Braamfontein 2001, Republic of South Africa.

# 8

# *Security records and reports*

If security officers are to keep notebooks, they must do so in such a manner that their contents will automatically be accepted as accurate, should the need arise to refer to them. The police service has learnt by experience that the defence in court proceedings will take any advantage afforded by badly kept notes as a means to challenge and distort the validity of prosecution evidence. For this reason, a simple and logical set of police rules has been laid down which can well be copied by their industrial counterparts.

## *NOTEBOOKS*

Irrespective of criminal matters, there are convincing reasons why security officers should maintain their own personal records of day-to-day duties – in addition to a communal occurrence book. An officer on patrol does not inspire confidence or impress with efficiency if he has to fumble to find a scrap of paper on which to make an essential note or record an important message.

If used for no other purpose, a notebook would justify its existence purely as an easy means of recording items, no matter how trivial, best not left to fallible memory. It is never possible to tell when a query may arise in the future as to whether a security officer has acted correctly in respect of a matter reported to him – an appropriate brief note could provide an immediate answer either to management or to a court.

Where prosecutions are concerned, it is likely that the conversation between a security officer and an accused person will have a direct bearing upon the issue of a case. What is said by the accused may constitute an admission or an explanation, either consistent with innocence or implying guilt. It is beyond human credence that such conversations could be remembered verbatim for repetition in a court weeks after they have occurred. If such evidence is to be given the only manner in which a court would be fully prepared to accept it would be by the witness referring to notes which the court is prepared to accept as a true record.

Dealing first with the type of book that should be used; obviously it must be of a convenient size for carrying in a pocket, with either stiff or semi-stiff covers to protect the pages. Preferably the pages themselves should be numbered in sequence to preclude any suggestion of tearing out and redrafting entries. For legibility they should be white and lined, and a margin on the left is an advantage for the insertion of times.

Each book taken into use should be date-stamped at the time of issue and have the name of the user inserted. The practice followed in certain police forces of actually numbering the books and keeping a record of allocation is hardly a necessity in the case of security officers.

Entries by the user should be in chronological sequence and an entry made for each day of duty. This should take the form of lining off the previous day's notes and inserting the day and date, followed by the tour of duty for the current day, the actual time of commencing work, and any specific instructions that are given at the time of parading. Incidents of the day should follow in sequence with the corresponding time shown in the margin. Entries should be made as legibly as possible – it must be remembered that the book is the property of the employer who may wish to refer to its contents.

It must not be thought necessary to fill the notebook with trivialities in order to prove that the writer has been active. If there is a spate of incidents to record there is something wrong with the security arrangements – particularly if there is a sequence of thefts or arrests to be shown. The true test of industrial security is whether or not the precautions are adequate enough to strictly limit incidents and dishonesty, a notebook void of everything but dates and times of parading and dismissing may be the best testimonial.

There are certain matters which must be correctly and fully recorded in a notebook. Examples are:

1 In the case of arrests, the full name, age, occupation, and home address of the arrested person must be shown, together with the

time of his apprehension and the reason for which he was arrested. Any questions and answers should be clearly set out; it is as important to record the question as the answer and in all matters of consequence the precise words that are used should be quoted.

There is no reason why an abbreviated narrative on the circumstances of an arrest should not be entered, but this should be kept short since, whereas recollection of words may be vague after a period of time, actions of any kind should remain clear and should not be the subject of reference in giving evidence.

2 In the case of lost and found property, full details of time, place, and individual losing or finding must be inserted, as well as a full and identifying description of the property. Where a person has handed over property to the patrolling officer that person should sign the officer's note underneath the description. In the case of found property being claimed during the course of a patrol, the officer must, in addition to establishing ownership beyond all doubt, obtain the signature of the claimant together with details of department or home address.

3 Complaints of theft should include full name, address, and means of contacting the complainant; the time the property was last known in order and by whom; when the theft was discovered and by whom; location of the theft; full description of missing articles with their value; and a note of any information bearing on the identity of the culprit.

Equally, there are certain things which must not be done in connection with the notebook. Perhaps the most important of these is erasure. No entry should be erased or altered in any way. There must be no additions written between lines, and pages must not be torn from the notebook in order to make fresh entries. Whilst these matters may not be of importance in respect of normal security duties, any sign of this in a notebook produced before a court would result in a most severe criticism of the user and, undoubtedly, the striking out of the evidence to which the entries referred. The correct procedure to follow is to draw a line through the erroneous entry, initial, then record the correct matter.

The question of chronological sequence is of importance when records of conversations are in dispute. The validity of these entries is sometimes contested on the grounds that they have been made so long after the occurrence that their accuracy must be deemed suspect. All entries therefore must be made as soon as possible or convenient after the happening has taken place. It would be ridiculous to assume that an officer, in the course of making an arrest, stopped to make up his notebook whilst escorting a prisoner to custody!

Where two security officers are acting in unison in the arrest of an individual, there is no objection to their notes on the incident being compiled jointly, always providing, if there is a discrepancy in their recollections of what was said, both will record as they remember.

It is particularly important that the instructions concerning keeping notebooks are observed by those who are engaged in the prevention of theft in stores and other large premises where regular arrests and appearances before a court would be anticipated. Persons so employed will be well advised to meticulously follow police procedures, both in making their notes and in the manner and occasion in which they produce these notes before a court. It should be borne in mind that a defending lawyer cannot demand production and examination of a book unless a witness has referred to it in the course of the court proceedings.

## *REPORTS*

The books of record referred to in Chapter 6 provide only information in a static form for reference purposes. Reports must be submitted to the management on matters of importance or where the management requires information on which to make decisions. All incidents which necessitate reports should be committed to paper as soon as possible after the actual occurrence. This should be irrespective of whether full details were available at the time the report was made. A brief note of some happening still under investigation could well save embarrassment to senior management, who might otherwise be approached with questions concerning a matter of which they had no previous knowledge. There is no objection to submitting an incomplete report, if it is made clear that a more comprehensive one will follow when fuller information is available.

The object of any report is to convey full and accurate information without ambiguity. The amount of detail which is contained in it will be governed by the known familiarity of the recipient with the matter being reported upon. For example, the head of a department would not need details of the precise function of individuals employed in his department, which might be required by a director who is not conversant with personalities and responsibilities in that department. Similarly, an individual not conversant with a locality or a procedure would need the inclusion of extra details to make the matter comprehensible to him.

Persons for whom a report is intended should be clearly shown on the front sheet in a circulation list, preferably on the left-hand side at the top. It could be advantageous to direct it to a particular person or persons who will have to act on the contents, with copies to those whose status or responsibilities make it advisable that they should be informed. Where the contents are of a confidential nature, or could be construed to reflect upon a named individual, care should be taken to ensure that it is not read in transit by unauthorized persons. In these circumstances, the report itself must be headed 'confidential' and enclosed in an envelope addressed and similarly marked. The onus of deciding what degree of restriction is applied rests with the originator.

No one likes to wade through a quantity of written matter without first knowing what it is all about. Therefore, the first essential of a report is a very brief heading, indicative of the content.

If, for instance, a theft had taken place in Number 2 Warehouse of the Gorbals Works by breaking a window in the supervisor's office and forcing the safe during the evening of Friday 26 Feburary 1988, a suitable heading would be: Office breaking and theft, Number 2 Warehouse, Gorbals Works, 26 February 1988. The reader at once knows what it is all about.

The body of the report should begin without any courteous, old-fashioned preamble by setting out bluntly the time, date and full nature of the occurrence:

> At 21 00 hours on Friday 26 February 1988, Security Officer Jones found the supervisor's office in Number 2 Warehouse had been entered by breaking the back window and climbing through. The back of the safe had been blown off by explosives, £21 cash was later found to be missing. Glasgow CID were notified and attended; the Fingerprint staff have carried out their examination but ask that there should be no interference with the office and contents until Forensic Science staff have visited – expected at 10 00 hours on 27 February 1988.

Having conveyed the gist of the report, more details can follow and everything will fall into perspective without difficulty. The contents of the report should remain terse and to the point. No one particularly wants to know that Security Officer Jones was patrolling in such and such an avenue examining property, when he saw that a window in Number 2 Warehouse had been broken and he therefore looked inside. The material point is that Security Officer Jones found that the Number 2 Warehouse supervisor's officer had been entered by breaking a window.

The language should be plain English of the type which does not allow any misunderstanding; technical terms should not be used unless they are certain to be understood by those reading them. Slang terms must be avoided – they are a confession of a poor vocabulary and reflect adversely on the writer. The report should be as factual as possible and if opinions are quoted they must be soundly based and reasonable and the person expressing them must be sufficiently conversant with the subject to be competent to express such an opinion.

Where statements are attached they should not be recapitulated in the narrative but should be referred to and abbreviated down to the barest detail necessary to convey their content. Naturally, there are a wide variety of reports ranging from those relating to everyday minor occurrences being drawn to the attention of the departmental head to lengthy reports to a senior executive or director outlining or analysing a particular problem and making recommendations for its solution.

With trivia of everyday occurrence a stencilled form might lead to clarity and economy in time and writing. Notification of things such as doors left insecure, lights left burning, or electric fires left switched on are all appropriate for a form which would only need the addition of some five or six words and a signature to fully convey what information was required. The more lengthy type of document will be compiled almost invariably by chief security officers and their equivalent. Many will be of the type where one need simply state the facts of an incident and if necessary make recommendations based upon that particular set of facts for action to be taken.

It is obviously not feasible to give guidance on all types of report of a complex nature, but where problems are being analysed and recommendations made there are few better ways of doing this than by following the military type of format. No attempts should be made to set out an analysis or complicated report in full in the first instance. If this is done the mind is likely to become immersed in the intricacies of grammatical construction to the omission of points which it was intended to make.

The first step should be the selection of main headings in logical sequence under which detailed rough notes should be prepared, followed by a formally constructed report, after all the implications of what is being proposed have been considered. Unless thoughts are noted down to a pattern it is not possible to review all the factors involved in a complicated situation with complete impartiality. A system of standardized headings, if only in the preparation of notes, facilitates this process and guards the writer against omissions. The following sequence is a logical one for setting out preparatory notes:

1 *The aim* or objective of the report.
2 *The factors* which affect the attainment of the aim or which surround the objective of the report.

3 *The courses* opened which may be followed.
4 *The recommendations* or the proposed action which is suggested.

The whole value of a report depends upon a clear definition of its purpose which must be stated simply and concisely. Care must be taken to avoid any expression of ambiguity and to ensure that the definition of the aim is in accordance with instructions and responsibilities of the writer.

'Factors' are the facts, circumstances, or evidence from which deductions may be subsequently drawn. Those not bearing upon the main purpose of the report should be discarded as being superfluous to it. Each factor will be followed by a deduction and the word 'therefore' is invaluable; for example, in preparing a report recommending an increase in establishment:

> The works area is over 100 acres in extent and, during after-work hours, only one man is available to patrol it *therefore* it is not possible to make an effective patrol of this area more than once every four hours.

A tendency to note down a number of items of information and then draw a composite deduction must be avoided.

The factors to be considered will vary with every situation and it will be found there are usually a few which dominate the others. It is by noting these down and listing deductions from them that a clear and reasonable report will eventually be compiled, minor ones should not be neglected, since a critical examination by a third party may raise these and the answers could then be immediately forthcoming to the credit of the writer.

Only practical courses should be considered with the salient points for and against each. It is valueless to suggest lines of action merely to discard them again as being void of any merit. The course it is intended to recommend should be normally stated last. If recommendations are required they should be clear, unambiguous, and given in sufficient detail for a course of action to be taken upon them.

With any long and complicated report it is advantageous to number paragraphs for easy reference and indeed an index might be helpful. Tables and schedules should be attached and referred to in the body of the report. If statements are attached they should be placed in the order of reference in the text. Maps or plans must have the scale shown upon them and photographs should have a caption to indicate what is shown.

## *Vehicular accidents on private property*

When an accident occurs on a firm's premises involving a vehicle belonging to an employee it is not unlikely that an effort will be made to claim that the firm has a degree of liability. If a solicitor becomes involved, the customary letter nonchalantly suggesting that the firm will accept liability will undoubtedly arrive. A firm may be liable if it permits the existence of an obvious hazard which causes the accident. If there is a danger that such a condition will be created either temporarily or permanently, security should report upon the fact with a view to elimination.

Where no vehicle or property of the firm is directly involved, security may play an advisory role and furnish means of communication between the parties. This could be by effecting an exchange of names and addresses and details of insurance, giving such assistance as might minimize the affect of the incident. However, for subsequent reference a brief note should be made of the persons and vehicles, the circumstances, and the comments made by the drivers. Where a firm's transport or property is directly concerned fuller details should be taken, including names of witnesses and statements if they are prepared to make them. A sketch should be made of the scene showing the positions and paths of the vehicles. If a visiting driver is involved he should be asked his version; and if he wishes to make a statement, this should be taken.

If the premises are such that vehicle accidents are frequent, a standard form can be used to ensure that all the necessary details are obtained. An example is shown in Appendix 12. Similar forms can be used for recording fire incidents or accidents on business premises.

## *Summary*

To sum up, the main essentials of report writing for the average security officer are:

1. Make the report as soon as possible and amplify later if necessary.
2. Indicate clearly for whom the report is intended.
3. Provide brief explanatory heading with the date and time.
4. Be clear, brief, and to the point.
5. Do not pad out the report with superfluous material.
6. Use plain unambiguous English.
7. Limit technical phraseology to those who will understand it.
8. Do not use slang.

*PART TWO*

# *LAW AND PRACTICE*

# 9

# *Arrest and search*

The fundamental freedoms of this country are considered to be inherent in the rule of law, which clearly states that no person will be deprived of his liberty without his consent unless by the authority of a warrant of a court of law, or when a court of law has passed a legal sentence of imprisonment.

The main exception to this rule is that in some situations the law recognizes that it may be impracticable to wait for a court order before detaining someone, so in such cases a power of arrest without warrant is provided through Acts of Parliament, provided that the person arrested is taken before a court as soon as possible.

Here will be described what constitutes an arrest, what is required before it is carried out, and what has to be done at the time and immediately afterwards. The powers of search of a person and property will also be dealt with. Police action following an arrest by a private person will be given in Chapter 10.

The criminal offences most concerned in arrests are theft, fraud, criminal damage, including arson, and different types of assault against the person. There are other powers of arrest which have been provided by statutes and which are sometimes called preventative, in other words they secure the detention of a person reasonably suspected to be about to commit a crime, before he actually does it.

The Police and Criminal Evidence Act 1984 which came into effect on 1 January 1986, made sweeping changes in the law relating to arrest. It redefined the meaning of 'an arrestable offence', abolished most powers of arrest by statute exercised by a Constable, provided a new conditional power of arrest, outlined procedures which must be adhered to when a person is arrested and, finally, makes statutory provision for persons attending police stations voluntarily to be interviewed.

## *ARREST*

An arrest is the taking or restraint of a person from his liberty in order that he shall be forthcoming to answer an alleged crime or offence. It is not necessary to touch or lay hands on a person to arrest him; it is sufficient that it has been made clear to him by words or actions that he is under restraint and cannot go freely as he may wish.

### *Arrestable offence*

The term 'arrestable offence' was redefined in the Police and Criminal Evidence Act 1984 and now relates to offences where the sentence is fixed by law, for example, murder, and where a person 21 years or over (not previously convicted) may be sentenced to five years or more imprisonment.

A person under 21 years cannot be sentenced to imprisonment. It is the offence which is arrestable, therefore persons of any age can commit it and be liable to arrest and, subject to the seriousness of the offence, can be dealt with at a Magistrates Court or the Crown Court. Juveniles (under 17 years) will be dealt with at a Juvenile Court.

### *Serious arrestable offence*

This same new Act introduced a new classification of offence, the 'serious arrestable offence'. It is of the utmost importance in the investigation of crime. In simplistic terms, whether an offence is a serious arrestable offence will determine the police powers available and may affect the direction an enquiry must take.

Serious arrestable offences are shown in Section 116 of the 1984 Act. They relate to the very serious cases of treason, murder, rape; kidnapping, Explosive Substances Act – causing explosions; some sexual offences; possessing a firearm with intent to injure, or carrying them with criminal intent; causing death by reckless driving, taking hostages, hijacking aircraft.

This section also in fact creates circumstances when 'an arrestable offence' may become a 'serious arrestable offence', but does so only if the commission of the offence has led to, or it is intended to lead to, or is likely to lead to, any of the following consequences:

1 Serious harm to the security of the State or to public order.
2 Serious interference with the administration of justice, or with the investigation into a particular offence or offences.
3 The death of any person.
4 Serious injury to any person.
5 Substantial financial gain to any person.
6 Serious financial loss to any person.

It follows that a case of causing actual bodily harm to someone may not be serious in itself, but could become so, if the reason was to stop that person giving evidence in a court of law.

Similarly, whilst theft, robbery and burglary are not serious arrestable offences in themselves, they could be so regarded if any one of the above circumstances are found. In this respect 'serious financial loss' is not synonymous with 'serious financial gain'. Loss for the purposes of this section reflects how serious it is to the person who has suffered it, taking into account all their circumstances. 'Injury' is defined as including any disease and any impairment of a person's physical or mental condition.

It is therefore the circumstances of the offence which can reclassify it as 'serious', as well as those very serious matters which have always been regarded as serious crimes.

### *Manner of arrest*

An arrest will be made as quietly as possible and without unnecessary violence. An accused person arrested without a warrant must be told the reason upon which he is being arrested unless the attendant circumstances are such that he cannot be in any doubt, or where he resists arrest by force.

Section 28 of the Police and Criminal Evidence Act 1984 gives statutory effect to the decision in *Christie* v *Leachinsky* 1947 (1 All ER 567) on what constitutes a lawful arrest. This section now clearly provides that where a person is arrested, he must be informed he is under arrest or the arrest is not lawful, unless the person is told he is under arrest as soon as practicable after the arrest. Not only must the person be told that he is under arrest but he must also be told the grounds for that arrest. The simplest way is to say 'I arrest you for. . . .' For all practical legal purposes 'detained' and 'arrested' have the same meaning, but perhaps the former is better used for arrest by private persons because of the obligation to hand the accused person to the police as soon as possible. Both in making the arrest and thereafter, the accused person should be treated with consideration; should an arrest prove unjustified, complaint from the injured party is less likely if he has been reasonably treated.

Once under detention the suspect must be kept under constant supervision. If he asks to go to the lavatory he must be accompanied – a female, of course, would be escorted only by a female. Stolen property has often had to be recovered from lavatory pipe bends and drains where it has been disposed of by a suspect to avoid detection. Further, there have been instances of property being hidden in lavatory cisterns to be recovered later, so when searching, nowhere must be neglected. Suspects allowed to visit toilets have also been known to cut their throats or wrists, or use a tie or belt to attempt suicide. If a vehicle has been used to transport the suspect to an office, ensure the vehicle is searched where the suspect has sat, as often stolen property is pushed down the rear of seats etc.

### *Arrest by authority of a warrant*

An arrest can be with or without a warrant. If the statute concerned with the offence requires a warrant to be issued before a person can be arrested it is applied for either by the police or a private person from a Magistrates Court. Statements in writing or proofs of evidence taken from all the relevant witnesses, to prove a prima facie case against the person named, are supplied to the magistrate. A nucleus of witnesses sufficient for that purpose appear at the court in person and swear on oath that the contents of their statements are true. On finding that there is a case to answer the magistrate signs a warrant calling on the police to arrest the accused and bring him before the court where the warrant was issued to answer the charge described thereon. When police find the accused he is told that there is a warrant in existence for his arrest and told the charge which has been made against him. It is not necessary for the arresting police officer to have the warrant in his possession but if he has he shows it to the accused.

### *Bail*

The warrant will state whether or not the accused can be released on bail before appearing at the court concerned. Where bail is to be allowed – the sum of money stipulated will vary with the seriousness of the offence – the accused enters into a written recognisance or undertaking to owe the Crown that sum of money if he fails to attend the court at the required time. If he fails to do so and is subsequently arrested the court can order him to forfeit the whole amount. This is known as estreating bail.

In the more serious cases, friends of the accused, known as sureties, may also be required to agree to pay the Crown similarly, should the accused not attend the court and they would likewise lose all or part of their money if he failed to do so. In England and Wales it is not necessary for the accused and/or the sureties to deposit the equivalent amount of money when entering into recognisances.

Where the arrest has been made without a warrant, bail may be allowed at the police station with or without sureties depending on the circumstances, such as the seriousness of the offence and the character and address of the accused.

Quite a different attitude has been adopted by police and courts during the past few years towards bail. Overcrowding of jails and the work of pressure groups led to the Bail Act 1976 which many feel provides safeguards for criminals rather than public. Be that as it may, employees handed over to the police for serious offences, and in confident expectancy that they would be kept in custody for at least a nominal time, are likely to be released at once on bail with the prospect of indefinite delay before the final disposal of the charges against them. There is therefore little prospect of a breathing space before a disciplinary decision on dismissal is needed.

### *Private person's powers of arrest*

Until the passing of the Criminal Law Act 1967 the power of the private person to arrest without a warrant was derived from common law, and had reference only to felonies, which included theft in various forms. That Act created the original term 'arrestable offence' now redefined in the Police and Criminal Evidence Act 1984. This also defines the powers of arrest of a private person, although they remain identical with the old Section 2 of the Criminal Law Act 1967. Powers of arrest in this respect are divided into two categories: those which any person may exercise and those which can only be effected by a constable.

The Act provides that *any person* may arrest without a warrant:

(a) anyone who is in the act of committing an arrestable offence;
(b) anyone whom he has reasonable grounds for suspecting to be committing such an offence.

Where an arrestable offence has been committed, any person may arrest without a warrant:

(a) anyone who is guilty of the offence;
(b) anyone whom he has reasonable grounds for suspecting to be guilty of it.

Where a constable has reasonable grounds for suspecting that an arrestable offence has been committed, he may arrest without a warrant anyone whom he has reasonable grounds for suspecting to be guilty of the offence. A constable may arrest without warrant:

(a) anyone who is about to commit an arrestable offence;
(b) anyone whom he has reasonable grounds for suspecting to be about to commit an arrestable offence.

The powers of arrest entrusted to 'any person' obviously extend to a constable.

A security officer obviously falls also within the category of 'any person' and makes an arrest as a private citizen. Where a person is arrested by a constable for an offence, or is taken into custody by a constable after being arrested for an offence by a private person other than a constable under section 30, Police and Criminal Evidence Act 1984, he must take the accused person to a police station as soon as practicable. The constable can release an accused person before he reaches the station if he is satisfied there are no grounds for keeping that person under arrest, this means if a person can show his innocence, or the constable intends to proceed by way of summons, he must release that person immediately. The constable can also take the accused person to another place with a view to establishing his guilt or innocence, and immediate enquiries are essential. However, any delay in taking an accused person to a police station must be recorded by the officer. This is particularly important where a security officer or store detective hands over an arrested person to a constable. Once again the Act gives statutory effect to the decision of *Dallison v Caffery* (1974 Crim LR 548), which recognized that in some instances immediate enquiries are necessary, especially if a verbal alibi is offered at the time of the arrest, which might clear him.

*Miscellaneous powers of private person arrest*

The vast majority of occasions when a security officer will find it necessary to exercise his private person power to arrest are associated with stealing, committing criminal damage, and inflicting bodily harm on persons. The constituents of these offences are dealt with in detail in Chapters 11 and 12. There are powers also given by a number of statutes most of which have little bearing on security work; persistently importuning for an immoral purpose by a male in a public place is one of the unusual ones.

A citizen's powers of arrest under statute include powers to arrest: for arrestable offences (s. 24 of the 1984 Act); a person going equipped for burglary, cheat or theft (Theft Act 1968, s. 25(4) ); a person making off without payment (Theft Act 1968, s. 34).

Breach of the Peace – a private person may arrest on view of a breach of the peace, whilst it is happening, and if he has reasonable grounds to believe that it will continue or be renewed (*R* v *Nicholls* 1967 51 Cr App. R. 233 (C of A) ).

Vagrancy Act 1824 caters for offences such as indecent exposure and being found in or on enclosed premises for an unlawful purpose. It should be noted that the offence of 'suspected person loitering', better known to the media as 'the sus Act', has been deleted by the Criminal Attempts Act 1981.

Vagrancy Act 1824 offences can be committed on premises and in private car parks; the powers here could be better known to advantage and are dealt with in Chapter 10.

The Prevention of Offences Act 1851, section 11 lays down that any person may arrest any person found committing any indictable offence in the night, which is between 9 p.m. and 6 a.m. Indictable offences include all types of theft, criminal damage and offences against the person. Section 12 of the above Act is of considerable interest to security officers as it says that if the person arrested assaults the person arresting he commits an additional offence to the main one for which he may be sentenced to three years' imprisonment.

The important requirements which have to be satisfied before a private person decides to arrest anyone are that he knows that an arrestable offence *has* been or is being committed and that he has reasonable cause to believe that the person he intends to arrest is responsible for it.

The powers of a police officer extend to arresting a person where he suspects an arrestable offence has been or is about to be committed whether one was committed or not.

### *Use of force when making an arrest*

A security officer is covered by section 3(1) of the Criminal Law Act 1967, which says: 'Any person may use such force as is reasonable in the circumstances in the *prevention* of crime or in effecting or assisting in the lawful arrest of offenders or suspected offenders or of persons at large'. Chapter 12 deals with the use of force by security personnel to evict trespassers. There are practically no legal precedents to guide on 'such force as is reasonable', and the commonsense interpretation 'the minimum necessary to achieve the objective' seems advisable. This was reaffirmed by the Police and Criminal Evidence Act 1984.

*Codes of Practice*

Persons other than police officers charged with the duty of investigating offences and charging offenders shall comply with the Codes of Practice, Police and Criminal Evidence Act 1984, concerning the interrogation and apprehension of persons. These Codes are primarily directed towards the police, but it is important for security officers to remember that the decisions in *R* v *Nicholls* (1967 51 Cr App. R. 233 (C of A) ) and *Walmsley* v *Young* (1974 Crim LR 548) required them to comply with the Judges Rules. A security officer therefore, until those decisions are changed, will be subject to the new Codes of Practice which replaced the old Judges Rules. The Codes of Practice are fully explained in Chapter 13.

*Unlawful arrest/imprisonment*

The possibility of placing oneself in a position where a claim for damages may ensue if a mistake has been made is often raised as a deterrent to those who wish to take advantage of their lawful powers in the prevention of crime or the arrest of offenders. That contingency must be put in perspective: actions in the civil courts for damages for unlawful arrest are relatively rare, taking into consideration the total number of arrests which are made for every sort of offence and those which are not brought to conviction. Some claims, of course, are settled out of court.

If there is *reasonable and probable cause* to believe that the arrested person was responsible for an offence which had been committed he should be detained until he can be handed into the custody of the police. Should their investigations subsequently show that either the offence had not been committed or, if it had been, that the accused person had not committed it, a claim for damages will not be likely to succeed. Courts dealing with such claims are sympathetic towards those defendants who honestly believed they were acting in accordance with the law and *without malice*.

## *THE RIGHT OF SEARCH*

No private person, which includes security officers, has the right to search anyone. Consent to a request to do so must be given at the time. The agreement of an employee to a clause in the conditions of employment requiring him or her to submit to a personal search when thefts are suspected does not remove the necessity to obtain consent to the search of the person immediately before doing so.

To put one's hands on the person of another without his consent is an assault and could be followed by a charge of that offence and, although extremely theoretical, a claim in a civil court for damages for trespass of the person.

A refusal to be searched is not by itself justification for detaining a person until the police arrive, or attempting to do so. However, a refusal where there is credible evidence that the suspect has stolen property in his possession would be reasonable cause to suspect he was in the act of committing an arrestable offence and he could be lawfully detained until handed to police.

The extent and thoroughness of personal searching depends on the type of property likely to be concerned. The searching of employees, for example engaged in the refining of precious metal or in the manufacture of jewellery, would not have been efficiently carried out unless it required the stripping of the clothes to the skin. Indeed this is actually done in certain countries where native unskilled labour is used, but it is hardly an acceptable practice in a progressive environment where metal detector doorways or similar devices are used. On the other hand, where, say, foodstuffs are concerned it might be sufficient to inspect the contents of bags or packages carried and to 'frisk', that is, to run the hands over the outside of the clothes. Cycle clips have been known to support stolen property inside trouser legs.

It is generally the practice at production units, particularly where the raw materials or the finished products are attractive and can be secreted on the person, for the conditions of employment to include a clause referring to personal searching. The employer offers to engage an employee providing he agrees to the conditions and, from the employees' viewpoint, he knows what acceptance entails. On the employee accepting the offer a contract has been made between them. If the employee fails to comply with the conditions he breaks the contract and is liable to disciplinary action.

The words of the searching clause differ from firm to firm but in essentials are the same. Some examples are:

1 The company reserves the right to search all employees leaving *or entering* the premises and to inspect any parcel, package, handbag, or motor vehicle.
2 Should pilfering be suspected the company reserves the right to require employees to submit to a search. Any employee found removing company property from the premises without proper authority will be dismissed and may become liable to prosecution.

3 When employees join the company they sign an agreement that while on company premises any authorized official of the company may question them concerning company property and may examine any article in their possession or any vehicle used by them and that they will submit if required to a personal search.
4 The company may require an employee to submit to being searched and may take such precautionary measures as it considers necessary.

It may be advisable to add a further clause such as:

> At intervals random checks are made and it should be made clear that selection for search does not imply suspicion and it is hoped that employees will appreciate the need for such checks to be made in the interests of all.

The condition must be shown to apply to management, staff, and labour force alike and without discrimination. At one company the chauffeur of the managing director was using his car to remove company property from the premises. The absence of a searching clause does not preclude the owner of property or his agent from asking anyone to account for the possession of property believed stolen.

*Search procedure*

It is essential to carry out searches in as friendly and routine a manner as possible to avoid giving offence and causing complaints which might jeopardize continuance of the preventive operation. This happened in one instance where the search was resurrected after an inexplicable lapse and with full union agreement – at the outset! It was carried out like a military operation with the selected few ushered through into an inner room for individual search away from their colleagues; no words of appreciation for co-operation were spoken; the combined service with the firm of the first three people selected was 130 years. Tactless to say the least! Security/workforce relations became unnecessarily strained immediately.

A fundamental principle is that a woman must only be searched bodily by another woman. The local working environment and relationships with the female employees will dictate whether this should also be extended to examination of handbags and carrier bags – there have been objections about the former on the grounds of the personal nature of what might be carried therein.

Recommendations in respect of searching are:

1 Do not treat searching as an operation which is anything out of the ordinary.
2 Make the selections at random, but try to avoid bringing any individual in too regularly. This could cause him to think he is being 'picked on' and lead to complaints.
3 If there is a positive suspect whom it is wished to include, do so in a manner which will not bring suspicion of special selection that might point at a person upon whose information the search has been based.
4 Unless a really detailed physical search is entailed – as might occur with jewellery or rare metals – there is no reason why it should not be carried out openly in the security office, preferably with two or three employees there at the same time. Apart from giving an impression of routine this will show that everything is being done fairly and openly.
5 The employee should be asked to step into the security office 'for routine search'. If he professes not to understand or where, for that matter, as a new employee he may not fully understand, it should be pointed out to him that acquiescence to searching is included in the conditions of employment. This should be done quietly and politely without too formal an attitude.
6 In the event of a refusal it should be pointed out that this is a breach of a condition which he has accepted in being given the job and that it could result in disciplinary action likely to take the form of dismissal. He should be asked his reason for refusing. In other words give him every chance to agree whilst making sure he is in no doubt of the consequences of refusal.
7 If he persists, request details of name and department and inform him that the matter will be reported to his manager for disciplinary action. If there has been information of a reliable nature that the individual is in possession of stolen property, the action that should be taken will depend on the circumstances since the refusal adds credence to the information. For instance, if he has been seen surreptitiously putting an electric drill in a carrier bag and shortly after is leaving the works with such a bag which obviously contains something heavy and which he declines to let be examined, this could be an instance for detaining and informing the police.

8 Before searching in the office ask if the employee has any company property in his possession; that is, give him the opportunity to produce and explain anything that he has with him. If he does produce something, his explanation must be either theft, permission, or mistake, and the latter two possibilities should be checked straight away.

9 If he denies possession of property, search quickly and thoroughly to the extent necessitated by the materials likely to be stolen; articles of any bulk, for example, can be located by a quick tap of the pockets and check of anything carried.
10 If the result of the search is negative, thank the employee, ask him to sign the search register, and let him leave immediately. Date and time of the search should be entered in the register plus the signature of the officer(s) carrying it out.
11 If articles are produced or found upon the person and are identified as belonging to the firm, then should there be no reasonable and acceptable explanation for his having them the employer's policy and instructions for theft should be implemented (see Chapter 2).

Requests for another employee to be present during a search are rare and can be avoided by calling several employees in at one time for the purpose.

A very firm line must be taken by employers with instances of refusals to be searched, otherwise a mockery will be made of the requirement which, among other things, will encourage more employees to refuse and this would lead to an intolerable situation. Where such breaches of the conditions of employment occur, dismissal is justified and indeed recommended. There has been no occasion, known to the authors, where this has been carried out and objected to by representatives of organized labour. Industrial tribunals appear to be in full agreement that a strict approach is necessary and this is confirmed by their judgments in the few appeals for unfair dismissal that have been reported. In *Riley* v *Dingwall Blick* (25 November 1976) an employee refused to be searched and there was no search clause in the contract of employment. Dismissal was found to be unfair, on the grounds that in the absence of a contractual right there is no implied obligation on an employee to be searched by his employer and refusal may not itself justify dismissal. On the other hand, in *Edwards* v *F. W. Woolworth* 1977 *there was a contractual right* included in the conditions of employment and Edwards's refusal to be searched was held to be adequate grounds for fair dismissal. Both were Merseyside Tribunal hearings and neither judgment was appealed against.

When disciplinary consequences are so stringent, the security action must be scrupulously fair and patient so that the refusal is patently deliberate to anyone considering the facts and indefensible by the offender. Where a security officer told an employee, hurrying for a bus, to step in the office and see 'the boss', following this up by 'you'll be in trouble and get the sack if you don't', this hardly constituted anything remotely like a request in accordance with company rules and would have brought the department into ridicule if reported to official notice.

A complete and thorough examination of motor vehicles, though desirable occasionally, is dependent on the time available. When the search is carried out the driver and any passengers should be included. Likely places for the secretion of stolen property are the boot, in tool boxes, the spare tyre, behind, in and under seats, under the bonnet, and fixed to the chassis. This problem can be removed or reduced by excluding the cars of employees of all grades from within the premises. If they are admitted, security will be improved if they are required to be parked in a clearly marked area, not on factory roads and as far away from the production and storage areas as possible. The parking area should be well illuminated during the hours of darkness. An additional security measure which might be adopted is to lock the gates of the parking areas except for short periods at the beginning and the end of work periods.

### *Contractors' vehicles and staff*

The agreement in writing should be obtained from building and haulage contractors, whose vehicles and staff enter the premises, to their being subjected to the same searching conditions as company personnel and vehicles. The authorized removal from the premises of the materials and plant of contractors, which have a similarity to company property or of other contractors, can present difficulties. One way to overcome this is to provide the contractors with duplicated forms to be completed with the name of the firm, the registered number of the vehicle, the date and time of the removal, what is being removed, and the signature of an authorized member of the firm. A copy of all signatures would be supplied to the security department. The pass would be handed to the security officer at the gate and retained.

These forms could provide useful information should claims be received from contractors of the loss of material and plant. It is customary for contractors and other outside persons working on company premises to agree to conditions under which the company is indemnified against common law claims due to the negligent behaviour of their employees. It is suggested that the searching conditions should be included. Contractors on premises provide an important risk factor, the implications of which are dealt with at length in Chapter 26.

*Searches at exits*

When deciding to carry out searches at the exit from premises these will be more effective if small groups are picked out *en bloc* to be dealt with one by one rather than making individual selections as employees leave. The fact that searches are taking place will soon become known to the later leavers who, if they are carrying stolen property, will try to get rid of it. Therefore after searches have been carried out it is recommended that places where such property could have been hidden on the way to the exit should be examined.

Where small items are likely to be carried, places which might be overlooked are: a rolled or folded newspaper; a container for tea or sugar in a food box; a vacuum flask; a tin of tobacco; the shoulder pads and linings of a coat or jacket, or the waistbands of trousers; strapped or stuck between the shoulders, the back of calves or thighs; tops of stockings; under the inside sole or the lining of shoes; inside the peak of a cap or the lining; in piled-up hair.

Depending, of course, on the nature of the property at risk from theft, containers of all descriptions and other possible places where property can be secreted, such as umbrellas, briefcases, and the frame tubes, panniers and handlebars of cycles, should not be neglected when carrying out searches.

*Searching women*

Some industrial concerns have a high proportion of women employees, either full or part time. The material they are concerned with is often domestically attractive and searching procedures have to be introduced to deter them from stealing. If there is no woman on the security staff it is often the case that the searching of women employees is carried out by a nurse from the health centre or a female member of the personnel staff. This is not a popular duty.

The problem can be overcome by the engagement on a retainer or part-time basis of a woman used to search females, such as a retired nurse, midwife, policewoman or court matron. When she carries out searches they should be at her discretion within the terms of her engagement. The fact that such an experienced woman has been engaged will be a most effective deterrent to stealing.

*Random selection units*

Complaints may be made of discrimination in selection for search, or management may feel that total impartiality should be proven by an automated choice. A number of firms supply electrically-operated random-selection devices which are hand-held and can be set for say five out of every 100. Alternatively, and much cheaper, an ordinary mechanical counter of the type frequently used in warehousing can be used by the security officer and every, say, twentieth person chosen.

### *Establishment of a search clause*

In the present industrial-relations climate the introduction of a new search clause into works rules or contract of employment in even the mildest terms is a delicate operation, and the reasons for doing so must be so well founded as to be obvious to all concerned. Ideally, when a new place of business is set up this is a condition that should be included for acceptance by the new workforce from the outset. Later inclusion is much more difficult, and the necessity for it would perhaps depend on proof in the form of numerous arrests by police for pilfering or of the introduction of small tempting articles of high value into the business of the firm. The point must be made at all times that the objective of searching is not to catch but to deter employees who might otherwise lapse into dishonesty. Workforce representatives would have to be convinced of this and that the system would be fairly implemented. Ample warning of commencement must be given, and this should also apply where the procedure has not been enforced for a period.

### *Search upon arrest*

The Police and Criminal Evidence Act 1984 restricts statutory power to constables, but the common law power to search may not be restricted to constables, although of course in the vast majority of cases it is a constable who effects an arrest and carries out a search. The interesting result might be that when a constable arrests, whether under statutory or common law powers, the statutory powers of search will apply, but when a citizen arrests the common law powers of search apply.

It is clear also from the 1984 Act that the search depends on the fact that the person has been arrested and not on the existence of the power to arrest, confirming a previous judicial comment that: ‘The right to a personal search is clearly dependent not upon the *right* to arrest, but the *fact* of arrest, and that at the time of search the person is in custody?

*Search after arrest*

(See also Chapter 10.) It might be as well to mention here that when an accused person is taken into police custody it is the practice approved by the law that he is personally searched, where there are reasonable grounds for believing that the arrested person may:

1 have any weapon or anything else by which he may cause danger to himself or others;
2 attempt an escape from lawful custody;
3 have on him evidence of the offence concerned and of other offences.

These reasons are arguably equally cogent where a private person has made an arrest, though they have not been spelled out in the same way – there would be little difference in the end result between a hysterical prisoner cutting his throat in a police station or in a security office. An eminent solicitor made an apposite comment: 'If the original arrest is wrong, searching in a reasonable manner isn't going to make matters much worse, especially if there are good reasons for the searching.' This is a matter for commonsense and the circumstances, and a polite request such as 'Do you mind emptying out your pockets please?' will be accepted in most cases.

A divisional court of the Queen's Bench Division discussed the common law powers to search after arrest in *Brazil* v *Chief Constable of Surrey* 1983 3 All. ER. 537 and stated:

> That a personal search by police officers imposed a restraint on a person's freedom to which he should not be required to submit unless he knew in substance the reason for it; that a police officer who had decided to carry out a search should inform the person concerned of the reason for it unless the circumstances rendered the giving of reasons unnecessary or impracticable.

This was stated to apply to the common law powers of search on arrest. It may well mean that if a citizen has a power to search on arrest, he must, to comply with common law requirements, inform the person why he is being searched.

*Entry and search after arrest*

A constable may enter and search any premises occupied or controlled by a person who is under arrest for an arrestable offence, if he has reasonable grounds for suspecting that there is on the premises evidence, other than items subject to legal privilege, that relates to that offence, or to some other arrestable offence which is connected with or similar to that offence. The constable may seize and retain anything for which he may search.

The citizens power of entry has not been repealed by the 1984 Act, but it is a restricted power. The Royal Commission on Criminal Procedure which later produced the Police and Criminal Evidence Act 1984, summarized them as follows:

> A citizen may enter premises –
> (a) to save life and limb, or
> (b) to prevent serious damage to property.

In practical terms the 1984 Act will probably make little difference. It is of course going to be easier to know whether a police officer is acting lawfully or not when he enters premises.

*Search Warrants*

There are circumstances in which, on an application by a constable, a Judge may issue a warrant authorizing a constable to enter and search premises.

All entries on and searches of premises under a warrant issued under *any enactment* are unlawful unless they comply with the prescribed provisions of the 1984 Act and the Codes of Practice.

*Police powers of personal and vehicle search*

> *Police and Criminal Evidence Act 1984*
>
> A constable, if he has reasonable suspicion that he will find stolen or prohibited articles, may search any person or vehicle (and anything in or on it) and detain that person or vehicle for that purpose, in any place where the public have access, on payment or otherwise, or in any other place where people have ready access at that time.
>
> The constable may seize anything he finds which he has reasonable grounds for believing to be stolen or prohibited articles.
>
> Prohibited articles fall into two categories:
>
> 1 Offensive weapon – which is any article made or adapted for the use of causing injury to persons, or intended by the person having it with him for such use by him or by some other person.
> 2 An article made or adapted for use in the course of, or in connection with any burglary, theft, taking a conveyance, or deception, or intended for such use by him or another person.

A police officer exercising these powers has strict Codes of Practice with which he must conform, and records of such searching that he must make, and have available for inspection by the person searched for a period of 12 months thereafter. The obvious objectives are to prevent misuse of the authority given, pinpoint instances leading to complaints and, of course, provide statistics for reference. Security, on the other hand, carry out what are in effect contractually authorized searchings, not normally in places of public access, and routine rather than based upon suspicion though they may lead to arrests. Nevertheless, an adequate search register must be maintained and kept – in practice for a period of years rather than the 12 months stated (see p. 98).

# 10

# *Arrest by a private person – police procedure*

When police assistance is called for, following an arrest by a private person, the officer attending will ask to be told, *in the presence and hearing of the person detained*, what is the evidence that he has committed an offence. Police procedure may differ in unimportant details from one force to another but basically the actions and requirements are the same. The person responsible for stopping the suspect and who recovered, say, company property from him, would then tell the officer what he saw, what he did, what he said to the suspect, and what the suspect replied. The identification of any property allegedly stolen and its value would be required by the officer from a competent witness.

The officer then will ask the suspect whether, having heard what has been said, he wishes to say anything. He will caution him that he is not required to say anything unless he wishes to do so but what he does say may be given in evidence. The officer will record in writing anything he says. On the assumption that the evidence shows a prima facie case against the suspect the officer will then tell him he proposes to arrest him and for what offence and take him to the police station accompanied by the witnesses.

At the police station the custody officer on duty will ask to be told the evidence and this will be described by the arresting officer *in the presence and hearing of the accused* or, according to practice, the witnesses will repeat to the custody officer what they told the arresting officer. Before accepting the charge, the custody officer has to decide what offence appears to have been committed and if there is credible evidence that the accused was responsible for it. He does not have to be convinced of the guilt of the accused.

The offence is written out in plain language on what is known as a charge sheet with a reference to the Act and section which describes it. To ensure that the accused is required to attend court to answer the charge a prosecutor is required. This can be either the owner or his authorized representative who will sign the charge sheet, or the police officer arresting. It has become the practice that where there is clear evidence that the accused person has a case to answer the police officer arresting will sign the charge sheet. The accused is given a form on which the offence with which he is charged is written. This is to give to any legal representative he may wish to instruct.

After this, and when other administrative procedures have been completed and the address of the accused has been verified, he is usually released on bail to appear at the next hearing of the local Magistrates' Court though increasingly a longer period is being specified.

## *STATEMENTS*

The person arresting will be required to make a full statement to the police, as will all material witnesses. These statements will be taken down in writing by a police officer in the form that he needs. But it must be remembered that this is the deponent's statement; and if he disagrees with anything that is being recorded or feels that ambiguity is being created by the phraseology, he is the person who will have to substantiate it in court and should insist accordingly. Possibly the increased granting of immediate bail removes the element of urgency but a new tendency to defer taking of witness' statements by the police is manifesting itself. This may be a matter of convenience that reflects the pressures upon them but it does not help the accuracy or completeness of statements if there is a substantial time lapse. It is suggested that the whole process could be expedited by the person in charge of security causing statements to be prepared, in accordance with police format, as soon as possible from his own staff and other employees involved. These to be tendered to the police who can clarify the contents later if they deem this necessary. This is widely practiced already in retail circles and accepted elsewhere when the security staff have been accepted as competent. It is also a good idea to prepare statements when a serious offence has been committed and remains undetected; they can be kept on file and will prevent the uncomfortable situation arising where an arrest is made say 12 months after an offence and statements are then required.

As will be mentioned in Chapter 14, statements can be produced and accepted in evidence in court without the attendance of witnesses – this is a regular practice at committal proceedings to Crown Courts in accordance with the provisions of the Criminal Justice Act 1967, section 1 and subject to the conditions of section 2 of the Act. A variation of the original statement to make its contents acceptable to the courts may therefore be produced for the witnesses to sign – for committals it will include the declaration of liability to prosecution for facts known to be false therein. Do *not* sign it if anything is wrong.

The statements thus obtained pass through police channels to the Crown prosecution service for the presentation of the case. During this process they will be checked against each other, and any contradictions or discrepancies will be clarified.

At busy courts it can happen that a charge on the list of those to be heard that day is not reached before the court closes. The accused is then readmitted to bail to surrender at the court at a later date. This is called a remand. Congestion at courts can lead to very lengthy delays which, if the case is a complicated fraud, may even exceed a year. If this is likely, the witness is entitled to ask the police for a copy of his statement to refer to before giving evidence (this does not apply to Scotland).

Any property involved is retained by the police from the time of arrest until the case is completed, when it is normally returned to the owner. Where, despite a plea of not guilty, the accused is convicted it is customary for the police to retain any exhibits, that is, the property in the case, until the specified time for him to appeal against conviction has passed.

Occasionally there may be delay at the police station; this will be a result of new mandatory procedures and it is to be hoped a rare occurrence. The person detained may demand legal advice in which case the duty solicitor has to be called – the time factor of his arrival is unpredictable. If for some reason security officer witnesses have been requested to wait and the delays become excessive, they should ask to leave. Their statements can be taken later if the delay is in some way connected with those. They have no obligations towards a duty solicitor and should not be drawn into conversation with him, or anyone he may have with him, about the matter.

Such waiting about is frustrating and an appreciation of the cause may at least mitigate any annoyance against the police. The role and records to be kept by the custody officer are shown in Appendix 14. They are of interest value only to security personnel.

# 11

# *Law of theft and related offences*

The purpose of this chapter is not to train management and security staff to become lawyers or law enforcement officers. In the performance of security duties, which are fundamentally concerned with the protection of property against being stolen, it is vital to have adequate knowledge of what the law describes as stealing, what evidence is required to justify a prosecution, and what powers exist to arrest the alleged offender.

The law concerning theft is surprisingly similar throughout the world. Differences occur mainly in the manner of enforcement and the court procedures and punishments that follow. All over the English-speaking world even the private person's powers of arrest for the offence are almost identical, to the extent that it would seem that countries have studied each other's legislation before finalizing their own. This may be particularly true of the Theft Act 1968, which is almost unique in that hardly any new case law has accumulated around it during its years of existence and only one section has been seriously in dispute.

### *Scotland and Northern Ireland*

The Act applies only to England and Wales. Scotland has its own legislation which is very similar with the same presumptions of ignorance of law not justifying commission of crimes/offences, innocence until proved guilty; but the age at which a child can be guilty of an offence is eight years – not ten years as in England. Both theft and criminal damage are common law offences and embezzlement, as an offence, is not restricted to servants and employees. The evidence of a single witness is not deemed adequate to convict without factual or circumstantial corroboration or substantially by a second witness.

## *EXTRACTS FROM THE THEFT ACT 1968*

### Section 1. Basic definition of theft

1 A person is guilty of theft if he dishonestly appropriates property belonging to another with the intention of permanently depriving the other of it; and 'thief' and 'steal' shall be construed accordingly.
2 It is immaterial whether the appropriation is made with a view to gain, or is made for the thief's own benefit.
3 The five following sections of this Act shall have effect as regards the interpretation and operation of this section (and, except as otherwise provided by this Act, shall apply only for purposes of this section).

To establish the offence the prosecution is required only to prove the four elements of: dishonesty; appropriation; property belonging to another; and intention permanently to deprive.

*Recent possession* – This is a doctrine of importance to security officers. Where it is proved that property has been stolen and that very soon after the theft the defendant has been found in possession of it, he can be convicted of theft (*R.* v *Seymour* (1954) ).

### *Section 2. 'Dishonestly'*

1 A person's appropriation of property belonging to another is not to be regarded as dishonest –

(a) if he appropriates the property in the belief that he has in law the right to deprive the other of it, on behalf of himself or of a third person; or
(b) if he appropriates the property in the belief that he would have the other's consent if the other knew of the appropriation and the circumstances of it; or
(c) (except where the property came to him as trustee or personal

representative) if he appropriates the property in the belief that the person to whom the property belongs cannot be discovered by taking reasonable steps.

2 A person's appropriation of property belonging to another may be dishonest notwithstanding that he is willing to pay for the property.

Section 2 is what was formerly known as 'claim of right made in good faith' which if genuinely held, even if mistaken, allows a defendant to be acquitted, but a jury is entitled to look at all the surrounding circumstances. Hence a security officer, when dealing with a person who he believes has stolen, must give him the opportunity to explain what has happened. Having said that, however, the logical way to refute an allegation of theft is to claim rightful possession if this has any chance of being believed; the facts will either give credence or render the reply farcical, which in itself is good evidence.

It is possible to steal one's own property despite the claim of right; for example, car taken to garage for repair, owner used duplicate key to take it away without paying when work finished – theft (*R.* v *Turner* (1971) ); but work done on a chattel, taken to owner who refused to pay, workman took it away again as security for the amount owed – not theft (*Harris* v *Harrison* (1963) ).

Section 2(1)(c) is the defence in instances of theft by finding, the criterion being whether there was genuine belief that no steps within reason would locate the owner. Identification marks on property or the circumstances and locale of the finding may rebut the belief.

*Section 3. 'Appropriates'*

1 Any assumption by a person of the rights of an owner amounts to an appropriation, and this includes, where he has come by the property (innocently or not) without stealing it, any later assumption of a right to it by keeping or dealing with it as owner.
2 Where property or a right of interest in property is or purports to be transferred for value to a person acting in good faith, no later assumption by him of rights which he believed himself to be acquiring shall, by reason of any defect in the transferor's title, amount to theft of the property.

This provides for theft by a bailee (lawful possession of property); by an innocent finder who knows or learns of the true owner; by someone who acquires property by virtue of a mistake of another (for example, goods delivered to wrong address but accepted); and by parents of a child under the age of criminal responsibility who appropriate property that it has brought home.

*Section 4. 'Property'*

1 'Property' includes money and all other property, real or personal, including things in action and other intangible property. . . .

With certain exceptions that are unlikely to be of security interest, a person cannot steal land or things forming part of it that are severed from it by him or by his directions. Wild creatures cannot be stolen unless they have been tamed and kept in captivity, or killed and their carcass belongs to another. Similarly, a person cannot steal mushrooms, flowers, or fruit growing wild on land unless he does it for reward or for sale.

*Section 5. 'Belonging to another'*

1 Property shall be regarded as belonging to any person having possession or control of it, or having in it any proprietary right or interest (not being an equitable interest arising only from an agreement to transfer or grant an interest). . . .

3 Where a person receives property from or on account of another, and is under an obligation to the other to retain and deal with that property or its proceeds in a particular way, the property or proceeds shall be regarded (as against him) as belonging to the other.

4 Where a person gets property by *another's mistake*, and is under an obligation to make restoration (in whole or in part) of the property or its proceeds or of the value thereof, then to the extent of that obligation the property or proceeds shall be regarded (as against him) as belonging to the person entitled to restoration, and *an intention not to make restoration* shall be regarded accordingly as an intention to deprive that person of the property or proceeds. . . .

A partner can steal partnership property provided the ingredients of the offence are present (*R.* v *Bonner* (1970) ). As above in *R.* v *Turner*, a garage proprietor can have possession and control of a vehicle left with him for repair, and the owner can steal it from him. When £106 was overpaid on a bet to the knowledge of the recipient who kept it, he was convicted rightly and as the property in the £106 never passed to him section 5(4) did not apply (*R.* v *Gilks* (1972)).

*Section 6. 'With the intention of permanently depriving the other of it'*

1 A person appropriating property belonging to another without meaning the other permanently to lose the thing itself is nevertheless to be regarded as having the intention of permanently depriving the other of it if his intention is to treat the thing as his own to dispose of regardless of the other's rights; and a borrowing or lending of it may amount to so treating it if, but only if, the borrowing or lending is for a period and in circumstances making it equivalent to an outright taking or disposal.
2 . . . where a person, having possession or control (lawfully or not) of property belonging to another, parts with the property under a condition as to its return which he may not be able to perform, this (if done for purposes of his own and without the other's authority) amounts to treating the property as his own to dispose of regardless of the other's rights.

If the taker does not literally intend to deprive the owner of the property permanently, but means to abandon it where the chances of the owner getting it back are remote or to leave it in such a condition that it is useless, that is regarded as theft. If there is a taking, using, and then voluntary return, that is not theft but a civil offence of trespass. If the return was not voluntary, as where the taker knew that his action had been found out and so then effected return, that would not purge his original intent to permanently deprive – hence theft. The taking of money with merely a hope or expectation of repaying in the future might be mitigation (*R.* v *McCall* 1971) but is not a defence to a charge of theft – just as well, since that could be a very easy excuse to give for almost all monetary thefts.

*Section 7. Theft*

A person guilty of theft shall on conviction on indictment be liable to imprisonment for a term not exceeding ten years.

*Section 8. Robbery*

1 A person is guilty of robbery if he steals, and immediately before or at the time of doing so, and in order to do so, he uses force on any person or puts or seeks to put any person in fear of being then and there subjected to force.
2 A person guilty of robbery, or of an assault with intent to rob, shall on conviction on indictment be liable to imprisonment for life.

Robbery is stealing aggravated by the use or threat of force. There must be proof of a theft, and the force must be before or at the time of the theft and used for that purpose. If force is used to effect escape after the theft, the offence will not be robbery. Where the force used was primarily that needed to get possession of the property, as in snatching a handbag or basket from the victim, the old defence-exploited technicality between this and that of actually using force on the victim personally has been over-ruled and the issue is now one to be decided by the jury on the facts (*R* v *Clouden* 1987 Crim LR 56).

Threats must be to a person, not against property, and must be of subjecting the victim or some other person *there* and *then* to force; threats of future injury or of damage to property may be blackmail under section 21. In robbery the force need not be a physical impact; the firing of a gun to intimidate would suffice though this could also come under the heading of 'threat'. In section 8(2) – assault with intent to rob – 'assault' does not of necessity mean actual violence (see Chapter 12).

### *Section 9. Burglary*

1 A person is guilty of burglary if –

(a) he enters *any building or part of a building* as a trespasser and with intent to commit any such offence as is mentioned in subsection (2) below; or
(b) having entered any building or part of a building as a trespasser he steals or attempts to steal anything in the building. . . .

2 The offences referred to in subsection (1)(a) above are offences of stealing anything in the building or part of a building in question, of inflicting on any person therein any grievous bodily harm or raping any woman therein, and of doing unlawful damage to the building or anything therein.
3 References in subsections (1) and (2) above to a building shall apply also to an inhabited vehicle or vessel . . .
4 A person guilty of burglary shall on conviction on indictment be liable to imprisonment for a term not exceeding 14 years.

This section of course greatly extends the former meaning of burglary, making it inclusive of all buildings and not just dwelling houses. Houseboats and caravans are also covered, and it is immaterial whether they are for the time being unoccupied. Persons legitimately in a building commit the offence if they 'trespass' with the essential intent into parts where they have no right to be – as would a guest in a hotel if he stole from the room of another guest.

The entry as a trespasser must be done deliberately and knowingly and without the true consent of the occupier. Where fraud has been used – for example, the pretence of being a local authority health inspector – the consent is not real and the ingredient of the offence is present.

It will be noted that the intent must be to do one of four things: to steal (not to obtain by deception); to inflict grievous bodily harm; to rape; or to cause unlawful damage. The best evidence of intent is of course for the act to have been carried to completion, but it can be deduced from the attendant circumstances.

*Section 10. Aggravated burglary*

1 A person is guilty of aggravated burglary if he commits any burglary and at the time has with him any firearm or imitation firearm, any weapon of offence, or any explosive; and for this purpose –

(a) 'firearm' includes an airgun or air pistol, and 'imitation firearm' means anything which has the appearance of being a firearm, whether capable of being discharged or not; and
(b) 'weapon of offence' means any article made or adapted for use for causing injury to or incapacitating a person, or intended by the person having it with him for such use; and
(c) 'explosive' means any article manufactured for the purpose of producing a practical effect by explosion, or intended by the person having it with him for that purpose.

2 A person guilty of aggravated burglary shall on conviction on indictment be liable to imprisonment for life.

*Section 11. Removal of articles from places open to the public*

1 . . . where the public have access to a building in order to view the building or part of it [or any of its contents], any person who without lawful authority removes from the building or its grounds the whole or part of any article displayed or kept for display to the public in the building or that part of it or in its grounds shall be guilty of an offence. . . .
4 A person guilty of an offence under this section shall, on conviction

on indictment, be liable to imprisonment for a term not exceeding five years.

Intent to steal is not included in this section, which is likely to be of interest to security personnel at museums and 'stately homes' opened to the general public and also to commercial security staff manning temporary exhibitions of works of art. It is not an offence likely to be charged very often; a collection assembled purely for the purposes of sale would be excluded.

*Section 12. Taking a motor vehicle or other conveyance without authority*

1 Subject to subsections (5) and (6) below, a person shall be guilty of an offence if, without having the consent of the owner or other lawful authority, he takes any conveyance for his own or another's use or, knowing that any conveyance has been taken without such authority, drives it or allows himself to be carried in or on it. . . .
3 Offences under subsection (1) above and attempts to commit them shall be deemed for all purposes to be arrestable offences within the meaning of section 2 of the Criminal Law Act 1967. . . .
5 Subsection (1) above shall not apply in relation to pedal cycles; but, subject to subsection (6) below, a person who, without having the consent of the owner or other lawful authority, takes a pedal cycle . . . shall on summary conviction be liable to a fine not exceeding £50.
6 A person does not commit an offence under this section by anything done in the belief that he has lawful authority to do it or that he would have the owner's consent if the owner knew of his doing it and the circumstances of it. . . .

This in essence is a re-enactment of the Road Traffic Act offence of 'taking and driving away', and the same statutory defences are retained. The difficulty arises where a driver has lawful possession of his master's vehicle in the course of his work but without the owner's permission uses it for his own purposes. *McNight* v *Davies* 1973 provided a High Court ruling; a homeward-bound driver had an accident, panicked, drove to a public house from where, after having a drink, he drove three men home in his firm's lorry before continuing to his own home and thence to work next morning. It was held that the driver had 'taken' the lorry when he left the public house, assuming control for his own purposes in a manner inconsistent with his duty to the employer to finish his tour of duty and return to the depot. From this it would seem that route deviations for personal purposes, not uncommon with haulage drivers, could be treated as criminal matters rather than by the more conventional and perhaps more appropriate internal discipline.

It is of interest that, though the implication is that the offence will always relate to motor vehicles, 'conveyance' also covers boats and aircraft.

*Section 13. Abstracting of electricity*

A person who dishonestly uses without due authority, or causes to be wasted or diverted, any electricity shall on conviction on indictment be liable to imprisonment for a term not exceeding five years. . . .

*Section 15. Obtaining property by deception*

1 A person who by any deception dishonestly obtains property belonging to another, with the intention of permanently depriving the other of it, shall on conviction on indictment be liable to imprisonment for a term not exceeding ten years. . . .
2 A person is to be treated as obtaining property if he obtains ownership, possession or control of it, and 'obtain' includes obtaining for another or enabling another to obtain or retain.
3 Section 6 (intention to permanently deprive) shall apply to this section with the necessary adaptation of the reference to appropriating, as it applies for the purposes of section 1.
4 'Deception' means any deception (whether deliberate or reckless) by words or conduct as to fact or as to law, including a deception as to the present intentions of the person using the deception or any other person.

The deception must operate on the mind of the person who is being deceived (*R.* v *Laverty* 1970 3 All ER 134); in other words, the deception is what causes the loser to act as he does. Hence if a driver calls at a manufacturer's premises claiming to have been sent to collect a load for a customer and a check with the customer confirms that his story is false, then if he is loaded up and subsequently stopped by the police as he leaves, the offence is not committed, since the deception was not believed and was not the cause for parting with the load. There would, however, be an attempt to commit, and the intention would obviously be confirmed by the driver's going through with the sequence of action.

The making of the deception must precede the obtaining of property; a reckless deception is sufficient, for example, drawing money against uncleared cheques believed worthless.

*Cheques* – In *R.* v *Page* 1971 2 QB 330 the Court of Appeal made a noteworthy statement about the drawing of a cheque, as implying:

1 The drawer has an account with the bank on which it is drawn.
2 He has authority to draw on the account for the amount entered.
3 The cheque is a good and valid order for payment of the amount shown; that is, it will be honoured when presented.

Drawing does not imply that there is money in the bank to meet the cheque; there may be authority to overdraw or intention to pay in enough money to cover it before it can be presented. The comment of the court was:

> The accused's evidence was that he hoped before the cheques became due to produce funds to meet them. Of course, if he succeeded in doing so, all would have been well, and indeed if he had any intention of doing so and any real propsect of so doing, of course, there would be no offence. There would not be any dishonesty. But if he had not any prospects whatever of finding this money and these cheques were drawn quite recklessly without regard to whether there was any chance of their being met or not, and indeed when there was every prospect that they would not be met, then they were drawn dishonestly.

This was very valuable case and authoritative comment of importance, especially to security officers in retail business and the hotel world.

*Section 16. Obtaining pecuniary advantage by deception*

This is the only section in the Act that has caused problems of interpretation; these resulted in the repeal of section (2)(a) and its replacement in the Theft Act 1978 which does not affect the other parts of the section.

1 A person who by any deception dishonestly obtains for himself or another any pecuniary advantage shall on conviction on indictment be liable to imprisonment for a term not exceeding five years. . . .
2 The cases in which a pecuniary advantage within the meaning of this section is to be regarded as obtained for a person are cases where:

   (b) he is allowed to borrow by way of overdraft, or take out any policy of insurance or annuity contract. . . .
   (c) he is given the opportunity to earn remuneration or greater remuneration in an office or employment, or to win money by betting.

The section 16(2)(c) is particularly important in a security sense since it makes the deliberate giving of false information to obtain a job a criminal offence (see p. 64 for Employment Protection Act implications). This could include lying in answering certain questions, for example denying have been convicted of any criminal offence for a job which demanded and required integrity.

The Theft Act 1978, apart from simply eliminating the troublesome subsection, created new offences in itself:

*Section 1 'By any deception dishonestly obtain(ing) services from another'*
An example of this would be an employee having his personal car repaired and charged to his employer's account by pretending it was a company vehicle. The element of 'deception' (as defined in section 15) has to be present from the outset, be practiced before the service is obtained, and be the reason why the service was given.

*Section 2 deals with dishonestly by deception*
Securing the remission of the whole or part of a liability to make a payment, for example falsifying bill or invoice; inducing creditor to wait with the intention of defaulting; getting an unjustified exemption from, or abatement of, a liability to make a payment.

*Section 3 dishonestly making off without paying when knowing that on-the-spot payment was required or expected*
This could be an offence committed on the spur of the moment on finding a bill was bigger than expected.

*Power of arrest* – Sections 1 and 2 each carry a five year maximum sentence and are therefore arrestable offences as described in Chapter 9. Section 3 only carries two years, but also provides that: Any person may arrest without warrant anyone who is or whom he, with reasonable cause, suspects to be, committing or attempting to commit an offence under this section. There must be an intention permanently to avoid payment or to avoid payment altogether and not merely an intent to delay or defer payment (*R.* v *Allen* (1985) 2 All ER 641).

*Section 17. False accounting*

1 Where a person dishonestly, with a view to gain for himself or another or with intent to cause loss to another –

   (a) destroys, defaces, conceals, or falsifies any account or any record or document made or required for any accounting purpose; or

(b) in furnishing information for any purpose produces or makes use of any account, or any such record or document as aforesaid, which to his knowledge is or may be misleading, false or deceptive in a material particular;

he shall, on conviction on indictment, be liable to imprisonment for a term not exceeding seven years.

2 For purposes of this section a person who makes or concurs in making in an account or other document an entry which is or may be misleading, false or deceptive in a material particular, or who omits or concurs in omitting a material particular from an account or other document, is to be treated as falsifying the accounts or documents. . . .

It will be noted that these offences are not limited to falsification to the detriment of an employer but are general in their application; they are also only applicable to documents used for accounting purposes. Hence 'clocking offences' could be charged under this section since a clock card is a document used for accounting. Though there will almost certainly be an accompanying theft or attempted theft, this charge could be advantageous since it would almost be self-evident once proof had been given that the defendant was responsible for the false accounting entry – it does not follow that because a defendant is not guilty of theft he is also not guilty of false accounting.

*Section 21. Blackmail*

1 A person is guilty of blackmail if, with a view to gain for himself or another or with intent to cause loss to another, he makes any unwarranted demand with menaces; and for this purpose a demand with menaces is unwarranted unless the person making it does so in the belief –

(a) that he has reasonable grounds for making the demand; and
(b) that the use of the menaces is a proper means of reinforcing the demand.

2 The nature of the act or omission demanded is immaterial, and it is also immaterial whether the menaces relate to action to be taken by the person making the demand.
3 A person guilty of blackmail shall on conviction on indictment be liable to imprisonment for a term not exceeding 14 years.

Demands must be made with menaces, which have been interpreted as including threats of injury to persons or property and of accusing a victim of misconduct; they may be made by writing, speech, or conduct. *R.* v *Clear* (1968) said that conduct could amount to menaces if: 'It is of such a nature and extent that the mind of an ordinary person of normal stability and courage might be influenced or made apprehensive so as to accede unwillingly to the demand.'

In *R.* v *Garwood* 1987 1 WLR 319 at the Court of Appeal the word 'menaces' was defined. The judge commented that:

> The word 'menaces' in blackmail under section 21(1) Theft Act 1968 is an ordinary word that will not usually need to be explained to a jury.
>
> In the present case the jury asked for clarification of the term and its application to the individual victim and their personal sensitivities.
>
> Where conduct appears to an abnormally sensitive person as threatening, although it would not also do so to the average person, the offence is complete if the perpetrator realizes the effect his actions are having.

*Section 22. Handling stolen goods*

1 A person handles stolen goods if (otherwise than in the course of the stealing) knowing or believing them to be stolen goods he dishonestly receives the goods, or dishonestly undertakes or assists in their retention, removal, disposal, or realization by or for the benefit of another person, or if he arranges to do so.
2 A person guilty of handling stolen goods shall on conviction on indictment be liable to imprisonment for a term not exceeding 14 years.

This is a most useful variation of the former legislation bearing on receiving stolen property. It applies to goods obtained by blackmail and by deception as well as by stealing and adds 'believing . . . stolen' to the old definition – which makes the defence of not having knowledge that the property was stolen much more difficult to sustain. It must be proved that the defendant was aware of the theft, or that he believed the goods to be stolen, or that, suspecting them to be stolen, he deliberately shut his eyes to the circumstances (*Atwal* v *Massey* 1971 3 All ER 881). Actual physical possession of the goods is not necessary to complete the offence; it is sufficient that he controls them. Hence a metal dealer who sent his lorry to collect metal which he knew to be stolen and gave instructions that it should be taken directly to a buyer could come within the ambit of the section. But if his driver were caught in possession of the load, it would be necessary to prove that he was acting with the knowledge and authority of his master for the latter to be prosecuted.

Where a child under the age of criminal responsibility (10 years) steals and gives the property to his parents or some other person who knows that it is stolen, there can be no handling charge since the child cannot steal, but a charge of theft may be preferred against the recipient.

*Section 23. Advertising rewards for return of goods stolen or lost*

Where any public advertisement of a reward for the return of any goods which have been stolen or lost uses any words to the effect that no questions will be asked, or that the person producing the goods will be safe from apprehension or inquiry, or that any money paid for the purchase of the goods or advanced by way of loan on them will be repaid, the person advertising the reward and any person who prints or publishes the advertisement shall on summary conviction be liable to a fine not exceeding one hundred pounds. . . .

*Section 25. Going equipped for stealing etc.*

1 A person shall be guilty of an offence if, when not at his place of abode, he has with him any article for use in the course of or in connection with any burglary, theft or cheat.
2 A person guilty of an offence under this section shall on conviction on indictment be liable to imprisonment for a term not exceeding three years.
3 Where a person is charged with an offence under this section, proof that he had with him any article made or adapted for use in committing a burglary, theft or cheat, shall be evidence that he had it with him for such use.
4 *Any person may arrest without warrnat* anyone who is, or whom he, with reasonable cause, suspects to be, committing an offence under this section.
4 For the purpose of this section an offence under section 12(1) of this Act of taking and driving away a conveyance shall be treated as theft, and 'cheat' means an offence under section 15 of this Act.

'Cheat' is not defined but may have its common law meaning where 'any article' could include devices or tokens, or even documents, to support a false and fraudulent statement. Where an article has been specially made for the criminal purpose, proof of possession is acceptable evidence that it was carried for that purpose; but if it is an item of ordinary use (gloves, torch, screwdriver, etc.), the circumstances will have to indicate what was the intent.

*Section 26. Search for stolen goods*

1 If it is made to appear by information on oath before a justice of the peace that there is reasonable cause to believe that any person has in his custody or possession or on his premises any stolen goods, the justice may grant a warrant to search for and seize the same; but no warrant to search for stolen goods shall be addressed to a person other than a constable except under the authority of an enactment expressly so providing.

The Police and Criminal Evidence Act, 1984 clarifies police powers to enter and search premises when a person has been arrested and taken to a Police Station. Generally the written authority of a Police Inspector is first required but, if after the arrest, the effective investigation of the offence requires the action before the offender can be taken to the Station it can be done provided the Inspector is informed as soon as possible thereafter.

## *RESTITUTION AND COMPENSATION IN CRIMINAL PROCEEDINGS*

### *Compensation*

The attitude of courts to awarding compensation for victims of crime has changed during recent years under pressure of public opinion; in addition, companies have become more vocal in expressing their desire for repayment, perhaps because they realize that action of this kind is dual purpose in both mitigating loss and providing a deterrent. The fact that an immediate order can be made by a criminal court after conviction is calculated to make a loser explore any possibility of inducing a court to exercise its powers. It is infinitely preferable to the alternative – painfully extended civil proceedings which are unpredictable and frequently of financial benefit only to the legal personalities engaged.

The power given to the courts is now obtained by the Powers of Criminal Courts Act 1973, section 35, as amended by s.67 of the Criminal Justice Act 1982:

> A court by or before which a person is convicted of an offence, in addition to dealing with him in any other way, may, on application or otherwise, make an order (in this Act referred to as 'a compen-

sation order') requiring him to pay compensation for any personal injury, loss or damage resulting from that offence or any other offence which is taken into consideration by the court in determining sentence.

The amendment (s.67) simply states that a compensation order may be imposed as a sentence in itself and gives it payment priority over any fine. Courts will not make awards where the circumstances are contentious but will leave losers to obtain a civil remedy.

It will be noted that the reference is to 'an offence'; this vastly extends the field of matters for which compensation can be awarded, one of which could be a firm's losses occasioned by evacuation after a malicious bomb threat message. There are certain exclusions; offences under the Road Traffic Acts are the main ones, and no compensation order can be made to the dependants of a dead person. However, where a vehicle has been damaged as a result of an offence under the Theft Act, that damage can be the subject of a compensation order. This power is extended to any other offences under the Theft Act where the property has been recovered in a damaged condition.

A further point of importance is that the order may be made 'on application or otherwise'. Suffice it to say that a court's attention is much more likely to be directed to the issue if it has written notification of the complainant's desire for it. This proviso allows a prosecuting solicitor to mention the matter on behalf of the loser; he can be made aware of the wish that this should be done by an inclusion in any statement given about the offence by a firm's representative, on the lines: 'I am authorized by my employers, Wilson and Oliver Ltd, to ask that, should any person be prosecuted to conviction for this offence, the court passing sentence should be requested to consider exercising its powers to award compensation in respect of its loss.' Experience shows that individual police officers may not like this addition to a statement that they are taking; however, it is accepted procedure in many forces and the deponent can insist on it – it is *his* statement.

Prosecutors vary. Some do not want any additions to their normal presentation; others are plain forgetful. Hence if the loss is substantial, no harm would be done by also writing to the Clerk of the Court hearing the charge with a copy to the Chief Constable of the area concerned who would then pass it with the file to the Crown Prosecutor (see p. 227). There is an advantage in written application, even before the advent of the Crown Prosecution Service it could not be guaranteed that a complainant would be notified of the date and place of the court hearing in which he was interested so that he could attend or send an observer. In the event of an indicated 'guilty' plea it is even less likely now unless procedures are changed.

The Criminal Law Act 1977, section 60, established the power for a court to award compensation after the conviction of an offender; this has now been raised to a maximum of £2,000 in respect of each individual charge of which convicted.

There is no restriction, other than commonsense, upon the number of charges that can be laid against an offender, but the Act allows compensation to be ordered for those offences which are admitted and taken into account by a court when passing sentence (section 35(5) ). However, the total amount is linked to what could have been awarded on the substantive charges.

Suppose, for example, a man is convicted on five charges, the total value of which totals £2,000 but he admits 10 further charges which total £15,000. The gross compensation which can be ordered is £10,000 of which £2,000 must go to the substantive charges, leaving £8,000 for those taken into account. It can be appreciated that confusion could arise with ten separate claimants in respect of that £8,000; this would very likely be an instance where the court would not exercise its powers, and who could blame it?

*R.* v *Grundy* 1974 1 All ER 292 said in effect that complicated orders must not be made unless unavoidable. Could a loser therefore be blamed if he insisted that the police treated his loss as a substantive charge if provable?

An order is of course suspended where an appeal against conviction is pending, and will lapse if it is successful. Compensation now takes pleasing precedent in order of payment from monies received over costs adjudged to be paid to the prosecutor in summary proceedings. Civil action can supplement the compensation granted by a criminal court and any damages assessed will be without regard to the order, but the damages actually awarded will be reduced from the assessed value by the compensation already paid and no further enforcement of the compensation order will be made without leave of the court.

Complainants should prepare an itemized account of the loss or damage sustained and the accompanying cost of making good, not just cost of materials, and have documents at hand to prove their claim to the satisfaction of the court. When children are involved, all is not lost – a high percentage of all theft and damage is attributable to children and young persons (under 17) – a court must order a parent or guardian of such offenders to be responsible for fines, damages, costs and compensation (Criminal Justice Act 1982). There are limits on the maxima that can be ordered for fines and compensation combined for under-14 and under-17-year-olds. Another snag is thrown up by section 35(4) of the Act: the court shall have regard to the defendant's means when making an order, which is equivalent to saying that if the defendant can convince the court that he is a man of straw, he will be safe from a compensation order, irrespective of how rich he is.

*Restitution*

The Theft Act 1968, section 28 is that most likely to be invoked; it has been amended by the Criminal Justice Act 1972 to include offences taken into account. Where goods have been stolen and a person is convicted of *any* offence in connection with the theft, for example, the handling of the stolen property, a convicting court *may*:

1 Order *anyone* having possession or control of the goods to restore them to any person with a legitimate claim.
2 *On application* of a person entitled to recover stolen goods from the person convicted, the court may order anything those goods have been converted to by sale, exchange, etc., to be handed to the claimant in lieu.
3 Order payment out of money in possession at the time of his arrest of the person convicted to any person who would have been entitled to return of the stolen goods.

An order can be made under both (2) and (3) but the total shall not exceed the value of the original goods. A purchaser or lender of monies in good faith may be a beneficiary under the provisions of (3) at the convicting court's discretion.

As in compensation, orders will not be enforced pending the hearing of any appeal against conviction.

Whilst the ethics of so doing may be challenged by defending solicitors and the courts may well refuse to accept the application, where an offender is known to have been arrested in possession of a sizeable sum of money, to prevent dissipation of this money there is nothing to prevent the loser notifying both police and court clerk of his intention to apply under (2) above for an order and express the hope that the monies will be retained for that purpose, or to enable the court to exercise its powers under (3).

A further section of the Powers of Criminal Courts Act 1973, useful for losers, is section 43 – if a person convicted of an offence punishable on indictment with two years or more (all theft and damage matters included) has anything in his possession, or under his control at the time of arrest which has been, or was intended by him to be, used for the purpose of committing, or facilitating the commission of any offence, the convicting court may make an order depriving him of his rights, if any, in the property which will be taken into possession of the police. The provisions of the Police (Property) Act 1897 then apply with a limit of six months on claim applications and the necessity for a claimant to satisfy the court that he had neither consented, knew, or suspected the likely use to which the property was intended. In Scotland, if the conviction is on indictment, the court may order the property forfeit and dispose of it as it pleases.

The Police (Property) Act 1897 applies where the police have come into possession of property in connection with any criminal charge and the ownership is not resolved in court. The police or any claimant can apply to a court for an order to deliver the property to the person appearing to be the owner – or make such other order as they think fit; the police often are the instigators to get goods off their hands.

# 12

# *Other offences against security*

## *CRIMINAL DAMAGE*

The Criminal Damage Act 1971 has consolidated and made obsolete all previous legislation connected with damage caused deliberately or by completely reckless behaviour with the exception of certain special matters concerning railways and shipping. Its provisions do not apply to Scotland and Northern Ireland.

Where fire is the means employed to cause damage, the offence is called and charged as 'arson'. Where, for example, an employee, impelled by grievance or other motives, damages machinery or destroys records, that is termed 'sabotage' but the charge is not correspondingly named.

### *Criminal damage act 1971*

All offences under this Act make those responsible liable to imprisonment. None of the maximum sentences is less than ten years, indeed, arson and acts likely to endanger life carry a potential life sentence. It follows therefore that all are arrestable offences and therefore any person, which of course includes a security officer and employees generally, may carry out an arrest.

The basic offences created by this Act (not verbatim) are:

1 Without lawful excuse, destroying or damaging any property

belonging to another, intending to do so or being reckless as to whether it was so damaged or destroyed.

2 Without lawful excuse destroying or damaging any property, whether belonging to himself or another, intending to do so, or being reckless as to whether any property would be destroyed or damaged *and* by so doing intending to endanger the life of another or being reckless as to that risk.

To illustrate 2, there could be the deliberate blowing-up of premises with an intention to kill someone therein; alternatively, say for insurance purposes, a separately occupied part of larger premises could be set on fire with total disregard of the certain spread to other sections with inevitable risk to the life of those in them.

3 Without lawful excuse, making to another a threat, intending that the other person would fear it would be carried out, to:

(a) Destroy or damage any property belonging to that or a third person, or
(b) To destroy or damage his own property in a way which he knows is likely to endanger the life of that other or a third person.

It is possible to visualize in embittered industrial disputes that threats under 3(a) could be made.

4 Having anything in his custody or under his control intending without lawful excuse to use it or cause or permit another to use it:

(a) To destroy or damage any property belonging to some other person or
(b) To destroy or damage his own or the user's property in a way which he knows is likely to endanger the life of some other person.

Possession of petrol bombs might come under this section.

*'Without lawful excuse'*

The following 'lawful excuses' do not apply to any circumstances involving threatening or endangering the life of another. They are statutory excuses peculiar to this Act and in addition to normal defence 'excuses'. They are that:

1 At the time of committing the acts, he believed they had been consented to, or would have been consented to, by those responsible for the property if they had known of the actions and the circumstances.
2 His actions were designed to protect property, rights or interests belonging to himself or another and at the time he honestly believed:
   (a) That they were in immediate need of protection and
   (b) The actions used or intended were reasonable having regard to all the circumstances.

These excuses are obviously intended to cater for mistakes and incidents such as breaking down of doors and demolition to stop spread of fire.

*Compensation*

This is now catered for under the provisions of the Powers of Criminal Courts Act 1973, as amended by the Criminal Law Act 1977. By these a court may order compensation not exceeding £2,000 in respect of each charge upon which convicted. The court may require itemized proof of cost of work and materials to make good the loss or damage to property (see p. 170).

## *INFLICTING PERSONAL INJURY*

The vast majority of incidents will be spontaneous fights where neither party wishes to pursue the quarrel subsequently by involving the police. There is a question of company rules which will certainly lay down what disciplinary action employers would wish to take; but it is the employees themselves, as the potential complainants, who decide what transpires *vis-à-vis* the police – that is, if the damage done is trivial. If an employee is too badly injured to express an opinion, it is incumbent on security to obtain medical aid and inform the police.

All the ascending degrees of injury are covered by the long-standing Offences against the Person Act 1861, which includes an essential definition – that of 'assault'. The term is conventionally taken to include what are two separate offences: 'assault' and 'battery'. The former is an act which intentionally or recklessly causes another person to fear immediate and unlawful violence to himself; this is in effect the preliminary stage to 'battery' which is the actual application of the unlawful violence, no matter how slight. If an employee strikes at another and misses, it is an assault; if he hits him, a 'battery'. Court actions without some injury are extremely rare, and for all intents and purposes 'assault' can be accepted to have the inclusive meaning.

'Common assault' is the lowest charge under the Act (s. 42) and usually arises out of fisticuffs, the injured party taking out a summons himself for the offence though the police may assist; a cross-summons by the other party is by no means infrequent.

'Assault occasioning actual bodily harm' (s. 47) follows, with a maximum penalty of five years which thereby confers a private person's power of arrest if needed. The injury is more substantial, and security might become involved with an unprovoked attack that is knocking or kicking a victim into insensibility.

'Wounding' (s. 20) involves an actual breaking of the skin either with or without a weapon and the phrasing is 'unlawfully and maliciously wound or inflict any grievous bodily harm'. The essence of the offence is a direct assault that the perpetrator is bound to know would cause bodily harm. Injuries in the shape of broken legs sustained through jumping out of a window to escape an attacker, without blows being struck, have led to a conviction. Again the potential sentence conveys a power of arrest.

'Grievous bodily harm' (s. 18) involves a deliberate intention on the part of the attacker 'by any means whatsoever' to do really serious bodily harm or to resist or prevent a lawful arrest.

Security action will have to be dictated by circumstances, the use of a weapon, the extent of the injuries, and whether the victim is able to express his own wishes. If hospitalization is required, the police will attend in any case, and every assistance should be given to them. Unless an injury is really serious the actual detention of a known employee is unlikely to be necessary; but if an unfamiliar non-employee is the attacker, then unless there are good reasons to the contrary the powers of arrest should be exercised to prevent him from absconding or later challenging his identification as the offender.

Weapons used in violence should be kept with minimum handling for transfer to the police. Any scene of serious attack where there is likely to be evidence available should be isolated. The clothing of injured and accused, if in security possession, should be preserved; and if the attacker is arrested, he should not be allowed to change or clean it. Every effort should be made to obtain details of witnesses to the occurrence or anyone who can give any relevant information about it. In other words, help the police as much as possible. The brawl might turn out to be manslaughter or murder with a great deal of attendant publicity, and lack of co-operation would merit a headline.

### *Indecent assaults*

This is an awkward area for security officers. Again the action depends on the wishes of the offended party, and the firm will certainly not benefit from any unnecessary publicity for the incident. If the complaint is made, there must be investigation and action of some kind even if this is confined to internal disciplinary measures at the request of the complainant. The offence is the subject of the Sexual Offences Act 1956, which also provides that a girl under the age of 16 and a mental defective cannot give consent to an indecent assault upon herself. The maximum sentence of two years does not convey a private power of arrest. Proof is needed of an assault in circumstances of indecency. Indecent exposure coupled with an approach towards the complainant with an invitation to intercourse has been held indecent assault, as has kissing against a girl's wishes with a similar invitation – *R.* v *Kilbourne* (1972).

False accusations are not uncommon for a variety of malicious or semi-hysterical reasons. Note should therefore be taken of the manner in which the complaint is made and the complainant's demeanour, together with anything about her appearance and clothing which may confirm her story or otherwise and any inexplicable delay in reporting as it is not always possible at that stage to be certain how far the complaint will be taken. She should be asked for possible sources of corroboration and whether she has done anything that could induce the man to think his advances would be welcome. More often than not the complainant will realize that she will incur unwanted publicity if the police are notified and a charge is made, and she will want internal action by the firm. In a union environment she will probably have been advised by her shop steward on this point, and that person may be present at the time of complaint. In these circumstances a detailed report to the appropriate personnel or other disciplinary source is needed, and the accused must be given an opportunity to give his version. If the girl is under 16 a great deal of trouble may be avoided by ensuring that her mother has been informed and has the chance to advise her. If the police become involved, again assist as much as possible.

### *Sexual harassment*

Allegations of this type are attracting increasing attention and publicity. They are made in relation to employment in commercial, retail or industrial workplaces and involve men making unwanted advances to fellow female employees extending frequently to acts which technically are indecent assaults. Not only can this cause embarrassment but it can make the atmosphere in which the woman works intolerable, though the man may at least pretend that it is 'a bit of fun'. If the practice is regular it would be naive to think there were no sexual intentions. Trade unions accept that there is a problem; they have arranged counselling sessions, and have represented female members in court and tribunal actions in which it has been alleged conditions had become so bad they could no longer work for their employer and 'constructively dismissed' themselves. The suggestion is obviously that the employer has shown negligence in allowing such a situation to develop. In America, many companies have been successfully sued for exactly this – which does them no good at all either financially or in reputation. A newsworthy topic which would not help recruitment of good female staff.

Complaints are more likely to be made to female shop stewards than security officers and the disposal of the complaint is a matter for the personnel and/or welfare department. If the security officer is the recipient, he should 'offload' as quickly as possible by reporting verbally and in writing to the personnel department.

Managers and directors are not immune from indulging in this practice and unfair discrimination has been alleged, based on refusal of advances, or on a competitor's acceptance, when an expected promotion has not materialized. It has been known for chief security officers, who have received confidences of such affairs developing, to have dropped hints in the appropriate place of having heard rumours and this has caused an abrupt cessation.

## *CRIMINAL INJURIES COMPENSATION BOARD*

Though this scheme came into operation on 1 August 1964 it has not been utilized as it should have been. It is of great importance to security personnel, many of whom believe that, should they be injured in the course of resisting criminal attack on their employer's property, they will be compensated by one or other of the company's insurance policies.

This is not necessarily so. Indeed it will appertain only in a minority of instances unless a special risk cover has been negotiated, which could be a precedent that management might prefer not to create. 'Negligence' on the part of the employer is an essential ingredient for most claims, and where an outsider is concerned this factor obviously will not apply; several leading insurers have already made it clear that they would dispute liability and only consider *ex gratia* payments made without obligation. In one instance, a drunken employee returned to premises after finishing an early shift and assaulted the chief security officer. Neither the firm nor its insurers recognized any liability, but as a criminal offence had been committed a claim was made under the compensation scheme with a speedy and satisfactory result.

*Nature of scheme*

Payments of compensation may be made for:

> personal injury directly attributable to a crime of violence (which would include arson and poisoning) or to arrest or attempted arrest of an offender or suspected offender or to the prevention or attempted prevention of an offence or to giving of help to any constable who is engaged in arresting or attempting to arrest an offender or suspected offender or preventing or attempting to prevent an offence.

Applications are to be made to Criminal Injuries Compensation Board (10–12 Russell Square, London WC1B 5EN; tel. 01–636 2812/4201) and may be made by the injured person, or a spouse or dependant if deceased.

There are conditions which have to be met:

1 The injury is one for which compensation of not less than £400 is appropriate. (Courts have been directed to use their powers to compensate in cases where the injury is such that likely compensation from the Board would be less than this figure.)
2 Causation of the injury has been the subject of criminal proceedings or it was reported to the police without delay.
3 The Board has been given all relevant information, including medical reports.

The main exclusions from the scheme are injuries caused by traffic offences, except where the vehicle has been deliberately used as a weapon – that is, where a more serious charge than a normal offence under the Road Traffic Act would be preferred.

The amount of compensation will be assessed on the basis of comparable common law damages, and the amount will take into consideration any other sum awarded by a court.

The existence of this scheme is of benefit and reassurance to employees other than security who are at risk, that is, those handling wages or driving valuable loads.

The Board issues very detailed information about its function and how to claim – it usually does so very promptly and those contemplating claiming should contact the office as soon as possible after the incident.

## *BRIBERY AND CORRUPTION*

The action to be taken with offences of this nature probably depends on a board level decision for, regrettable though it may seem, in some trades gifts for preferential treatment by buyers are almost an accepted fact of life. Where there is proof a firm may prefer to dismiss rather than report to the police; but if it decides otherwise, as ethically it ought, an appointment should be made with, if possible, a Fraud Squad officer away from the premises to outline the facts. Acquiring the evidence is anything but easy since neither the giver nor the recipient of the bribe is likely to wish to talk about it. Where firms themselves are put under pressure to provide bribes as an essential step to acquiring contracts, they should consider the adverse publicity given to such practices internationally in recent years before deciding their action. If tempted, they may find that the provisions of the legislation make them subject to prosecution.

Security officers may be approached with bribes by scrap purchasers or contractors defrauding on time and materials contracts or, for that matter, by employees who want a blind eye turned to reprehensible activities. This should be brought to the notice of the security head without alerting the offerer, who will try elsewhere if turned down in this instance. In acquiring the evidence of attempted bribery, the means are also achieved of legitimately excluding the person from the premises or, better still, prosecuting him for the attempt to commit the offence for which the consideration is offered. Prosecution for bribery needs the fiat of either the Attorney-General or Solicitor-General, and any alternative charge will be favoured by the police. Note that the charge will be 'attempt' if the offender is caught actually defrauding or stealing when the sequence of events has been allowed to go to completion so as to provide sound evidence. Care should be taken that no suggestion can arise of *agent provocateur*, that is, inciting to commit the offence.

## *FORGERY*

Fundamentally, forgery is the altering of a genuine document, or making a false document so as to appear genuine, with intent to defraud; it goes beyond the mere forging of a signature. Using the forged document is referred to as 'uttering' which of course is an offence in itself. Legislation is contained in the Forgery and Counterfeiting Act 1981. This lists a series of offences which, paraphrased, are (main ones):

1 Making a false instrument with the intention that it should be used to induce someone to accept it as genuine and, by doing so, act to their own or someone else's prejudice (s. 1).
2 Copying, or using either the copy or the original of the false instrument with the like intention (ss. 2, 3, and 4).

Special provision is made to cater for forging money orders, share certificates, passports and the like (s. 5), the schedule for the section including things more likely to be of security interest – cheque and credit cards, cheques and travellers cheques. Having possession of the means of making the forged instruments, that is special paper and printing plates, with intent to make them, is also covered.

'Instrument' is defined by s. 8 – any document, formal or informal; GPO or Inland Revenue stamps; discs, tapes or soundtracks upon which information may be recorded or stored.

The penalty where intent is proven is a maximum on indictment of ten years which confers an 'any person' power of arrest.

A similar penalty is imposed for counterfeiting bank notes and coins with the intention of passing them off as genuine; knowingly having possession of such counterfeits or offering them or tendering them with that intent is also similarly penalized.

There will always be those amongst employees who will use printing or duplicating facilities to produce 'banknotes' out of curiosity or for some frivolous purpose without any real intent to use them as genuine – also moulds for coins. These in themselves are offences which carry a potential two-year penalty, the very mention of which will be more than adequate to stop further misuse of the employer's equipment!

Forgery is often a constituent of offences charged under the Theft Act ranging from submitting false production bonus sheets, false expenditure receipts, to the signing for another employees wages – the latter being a most annoying incident since the automatic reaction is to assume the loser is 'trying to pull a fast one', and, if it is not resolved, the doubt always lingers and the employee becomes disgruntled.

Documents controlling the loading of goods or the passage of them through any gate control are susceptible to being altered with intent to get stolen goods through. An example is where the practice was to summarize the individual deliveries of goods on a lorry on to a loading sheet to assist the loaders. Following shortages of stock an investigation showed that some of the summaries had been forged by adding the figure 1 in front and 0 behind others. This had led to the loads being increased over the correct totals and the drivers, who were parties to the fraud, sold the excess goods. It is recommended that if a document is used to authorize the loading or dispatch of goods it should be typewritten in duplicate and one copy retained by the authorizing department. Spot checks should be made on the loading bays to see that the original compares with the copy, if offences are suspected.

Obtaining stores of any description by falsely increasing the amount to be drawn or adding additional items with intent to steal them is obtaining by means of a forged instrument. Requisitions should be checked from time to time with the authorizers to see whether any alterations have been made. The fact that requisitions are subject to check is a deterrent to altering them.

## *TRESPASS*

Certain forms of trespass have long been criminally punishable by Acts of Parliament, but these are restricted in number. It is necessary to have a wider knowledge of this subject, the dictionary giving the following definition:

> To interfere with another's person or property; to enter unlawfully upon another's land; to encroach; to intrude; to sin.

The legal definition is derived from the above and states:

> The act of entering or being upon land or premises without any right to be thereon.

Notices which state 'Trespassers will be prosecuted' are commonplace, but most are empty threats because with those few exceptions, trespassing is not a criminal offence. The exceptions are trespassing on the railway (British Transport Commission Act 1949), trespassing on war department land (Manoeuvres Act 1958), trespassing in pursuit of game (Game Laws Amendment Act 1960), and on land enclosing explosive factories or magazines (Explosives Act 1875). A minority of security officers will be involved in these; apart from the offences created under the Criminal Law Act 1977 their main concern will be the civil wrong of trespass.

Most legal references interpret trespass as:

(a) the innocent action of someone who unwittingly is on private premises without a right to be there; or
(b) the wilful action of someone who enters land, or remains on private land or premises as an encroachment of the private rights of the owner (or occupier) of that land or premises.

In the present time of industrial espionage, bombs, cash office raids, thefts and other criminal and malicious activity, the presence of a stranger within the company boundaries must be regarded with suspicion and he should be questioned for his reasons for being there. Only when satisfied beyond reasonable doubt that there is no unlawful intent should he be regarded as a mere trespasser and asked to leave.

Deliberate and recurring trespass is best dealt with by seeking a court injunction to prohibit it.

Certain persons are entitled to enter premises without permission. Police with search warrants, and Health and Safety Inspectors are typical examples of officials to whom one common factor will apply; they will carry proof of their authority which they will produce on demand. Others entering premises as 'guests' – representatives and the like – or as implied invitees – customers of retail establishments – become trespassers if the permission to be on the premises is withdrawn. From the security point of view, such circumstances should be handled with the proverbial kidgloves, or an action for damages may ensue against the occupier and the security staff.

### *Security action*

After establishing, by watching, that a person is simply a trespasser and without any criminal intent, the first step is to make the fact known to him and ask him to leave. Refusals will be rare if the approach is polite and firm. Use of force is a last resort after all requests have been refused; and it is lawful then only to apply the minimum necessary to effect the ejection, which certainly does not include the use of a staff or weapon other than in self-defence.

If there is every indication that violent reaction is going to be encountered or there are a number of recalcitrant trespassers, the police should be called – as they should if there is any suspicion, by actions or unsatisfactory replies, that dishonest motives are attached to the trespass. A trespasser's name and address should be requested for reference purposes; but refusal to give them does not, in the absence of any other factor, justify either calling the police or any use of force and provides no basis for legal action.

One fearsome weapon which might be encountered on dispersed sites is the crossbow, which has achieved the distinction of having a special Act of Parliament named after it – The Crossbows Act 1987 which came into effect on 15 August 1987. It creates offences relating to the sale or letting on hire of crossbows or parts of crossbows to persons under 17 and for offences by that person of buying, hiring, or possessing. Section 3 is that which may concern security – a person under 17 who has with him a crossbow capable of discharging a missile, or the parts which can be assembled to make one, is guilty of an offence *unless* he is under the supervision of a person 21 or over. A constable having reasonable grounds for suspicion of committing or having committed an offence under that section can search the person or a vehicle for the crossbow or parts and can detain for that purpose; he has powers of seizure but not arrest. The obvious security action is to contact the police as soon as the youth with the bow is seen. Of course, to complicate matters, the Act only applies to those bows which have a draw weight in excess of 1.4 kilos! It must be remembered that this is a lethal weapon which can kill by accident and incidents should be treated with due care. Circumstances may arise which put the bow into the category of 'offensive weapon' (see p. 151).

Clearly it is the evidence of the intruder's intentions which is important, and unless he is caught in the act of doing something illegal this can only be obtained by questioning, or by evidence of property in his possession. This could include skeleton keys for doors or vehicles, wirecutters with which to cut through compound fences, etc.

Factors which should influence your judgement as to whether a person is a mere trespasser or something less innocent should be:

1 time of day or night;
2 the part of the premises where found;
3 information in your possession as to his previous movements, or conversation with other persons before he was challenged;
4 frankness in answering your questions;
5 willingness to co-operate in turning out his pockets, allowing you to look into any bag he may be carrying, or any vehicle in his use;
6 obsequious, over apologetic and anxious to leave – why?

*Public using company premises for access*

Where members of the public form the habit of crossing part of company land as a short cut to adjoining premises or to another street, action must be taken to stop them because to allow habitual use might create a 'right of way'.

Barriers or notices must be placed at both ends of the route indicating that people are trespassing if they cross the land without the consent of the company. An occasional stopping or refusal of access by security staff would diminish a claim of right by a habitual user and it is conventional to make a complete and supervised stoppage on one day each year.

Civil action in a county court may be taken against a trespasser, but mere trespass does not make him liable under the criminal law unless the land is protected by Act of Parliament, as is virtually all Crown property and that of nationalized undertakings.

*Legal matters*

The Criminal Law Act 1977 does, however, create a number of criminal offences, but the real target is the trespasser into residential premises – the 'squatter'. Excluding such premises, the point of main security interest is s. 6 which says that:

1 Any person who, without lawful authority, uses or threatens violence for the purpose of securing entry into any premises for himself or any other person is guilty of an offence, provided that:
   (a) there is someone present on these premises at the time who is opposed to the entry which the violence is intended to secure; and
   (b) the person using or threatening the violence knows that is the case.

The fact that a person has rights or interests in the premises that he threatens does not constitute 'lawful authority' for himself or anyone acting on his behalf. There are statutory defences provided where the premises are residential.

Section 8 provides that 'a person who is on any premises as a trespasser, having entered as such, is guilty of an offence if, without lawful authority or reasonable excuse, he has with him on the premises any weapon of offence'. 'Weapon of offence' means any article made or adapted for use for causing injury to or incapacitating a person, or intended by the person having it with him for such use.

Imprisonment not exceeding three months or a fine of up to £1,000, together with a power of arrest only for a police officer in uniform, are specified. It is difficult to see these circumstances occurring in a security situation where some other offence with a private person's power of arrest would not also arise – for example, under the Vagrancy Act, Theft Act, or for Breach of Peace.

## *PUBLIC ORDER ACT 1986*

*Police powers*

A new police power to evict trespassers in certain limited circumstances is now in force and gives the police the power to deal with mass invasion of land, such as hippy convoys.

If the senior police officer believes that two or more persons have entered land as trespassers for a common purpose of residing there for any period, and that reasonable steps have been taken by or on behalf of the owner or occupier, he can ask them to leave, *and* (a) if any damage is caused, or if they have used insulting, abusive or threatening words or behaviour towards the occupier, a member of his family, an employee or agent, or (b) if they have brought 12 or more vehicles on to the land, the senior police officer can direct those persons to leave the land. Vehicles include caravans.

Failure to leave, or returning to the land, is an offence and the person(s) can be arrested without warrant by the police. (This act was not designed to deal with 'gypsy squatters'.)

## *OCCUPIERS LIABILITY ACT 1984*

This Act deals with two important areas of responsibilities by the occupiers of land towards people who come on to their land. It has long been the law that occupiers owe a duty of care to persons on their land as their guests or with their consent. In 1972 the House of Lords held in *British Railways Board* v *Herrington* that occupiers also owe some duty to trespassers and others who enter their land without consent, as commonsense and common humanity dictate. The Occupiers Liability Act 1984 confirms that decision, and defines the duty to be shown. It also deals with the circumstances in which a person who occupies land for business purposes, can allow someone to come on to his land while at the same time *not* incurring any liability for damage or injury to that person, because of the dangerous state of the premises.

People can therefore now be allowed to go on to land for recreational or educational purposes without the owner of the land being liable for any injury they may suffer. This permission must be a management decision at the highest level.

## *VAGRANCY ACT 1824*

Though a very old Act, its provisions under s. 4 still have importance and carry a 'found committing' power of arrest for a private person (s. 6). There are two main offences which security staff may encounter:

*Indecent exposure* – The offender 'wilfully, openly, lewdly and obscenely exposes his person with intent to insult a female'. This offence is rarely if ever carried out by an employee to fellow employees, for obvious reasons of identification; it is usually by an outsider to the female office staff, either in sight of their windows or as they leave. These are more often than not persistent offenders. The police can be informed for the necessary action; but there is no reason why security men, in circumstances which allow no doubt, should not arrest and hand over.

*'Being found on enclosed premises for an unlawful purpose'* – In most instances there will be more formidable alternative charges to be preferred. 'Premises' includes a dwelling house, warehouse, etc., or any enclosed yard, garden, or area, and the offender must be found *in* or *on*. The unlawful purpose must be the commission of an offence which if completed would lead to a criminal prosecution. A man trying doors in an enclosed yard at the rear of a retail store who could not give any reasonable explanation for his presence and actions would come under this offence.

## *CRIMINAL ATTEMPTS ACT 1981*

This is an appropriate point at which to mention this Act which repealed the invaluable 'suspected person loitering' offence contained in the Vagrancy Act and also took the opportunity to rectify some judicial decisions, absurd to the layman, which had been reached in connection with offences it was claimed impossible to commit. For example, an intending thief took a wallet from a drawer meaning to take money out of it – it had no money in so he was acquitted of any attempt to steal.

The focal part of the Act is the section 1, 'Attempting to commit an offence'. Section 1 (4) defines 'an offence' so far as this section is concerned – 'any offence which, if it were completed, would be triable in England and Wales as an indictable offence' other than:

(a) conspiracy (at common law or under any enactment);
(b) aiding, abetting, counselling, procuring or suborning the commission of an offence;
(c) offences under s. 4(1) and s. 5(1) of the Criminal Law Act 1967 – assisting offenders or accepting or agreeing to accept a consideration for not disclosing information about an arrestable offence (see p. 33).

*Comment* The offences with which security work is mainly concerned – fraud, theft, criminal damage, inflicting serious injury, etc. – are all indictable and therefore fall within the scope of the Act.

*Section 1*

(1) If, with intent to commit an offence to which this section applies, a person does an act which is more than merely preparatory to the commission of the offence, he is guilty of attempting to commit the offence.
(2) A person may be guilty of attempting to commit an offence . . . even though the facts are such that the commission of the offence is impossible.
(3) In any case where (a) apart from this sub-section a person's intention would not be regarded as having amounting to an intent to commit an offence; but (b) if the facts of the case had been as he believed them to be, his intention would have been so regarded, then, for the purposes of s. 1 (1) he shall be regarded as having had an intent to commit that offence.

*Comment* Section 1 (2) and (3) abolish the very awkward common law rule that one cannot intend or attempt the impossible (3) caters for intent formed on the basis of totally incorrect information. For example, a gang are told a vehicle loaded with whisky will be left overnight on a lorry park whereas in actual fact delivery is made during the afternoon and it is parked empty. They are caught attempting to break into the vehicle with intent to steal the whisky – an offence never capable of being carried out.

Where the dividing line comes between an act of preparation and the commencement of an 'attempt' will depend on the particular circumstances and evidence. Perhaps, for practical purposes, the safe criteria would be – if the act in itself is carried through, would it result in the commission of the full offence? The purchase of a case opener with the object of using it to force a warehouse door would be a preparatory act, carrying it through the streets to his target would still be preparatory (though an offence under s. 25 of the Theft Act 1968 – 'going equipped for stealing'), but when he applies it to the warehouse door, the attempt begins.

*Section 2. Procedural and other provisions*

This section lists ways in which 'attempts' shall be regarded as if the full offence had been committed. Of main interest are that the same powers to arrest or search, institute proceedings within statutory time limitations, and seize or detain property are conferred. Similarity of penalites for attempt and full offence are confirmed by s. 4.

*Section 3. Specific offences of attempts*

Some enactments have specific offences of 'attempting to commit'; this section simply extends to those the provisions of s. 1 (1), (2), and (3) so that impossible offences are covered and the distinction between preparatory act and attempt is preserved.

*Section 5. Conspiracy*

Again, this takes care of the impossible offence and also provides a new definition of conspiracy:

> If a person agrees with any other person or persons that a course of conduct should be pursued which, if the agreement is carried out in accordance with their intentions, either:
>
> (a) will necessarily amount to or involve the commission of any offence or offences by one or more of the parties to the agreement, or
> (b) would do so but for the existence of facts which render the commission of the offence or any of the offences impossible he is guilty of conspiracy to commit the offence or offences in question.

*Comment* Those planning and organizing the intended theft of whisky mentioned earlier would be caught by this section.

*Section 9. Interference with vehicles*

This creates a new offence and appears to be recognition of the fact that abolishing the 'suspected person' offence left a dangerous vacuum in powers to restrain thefts from and of motor vehicles. Although the concept will be a matter for court interpretation, it would appear that acts which fall short of 'attempts' will be actionable. Section 9 (1) defines the offence – interfering with a motor vehicle or trailer or anything carried therein or thereon with intent that he or another will commit one or other of the following:

> *Section 9(2)*
> (a) theft of the motor vehicle or trailer or part of it;
> (b) theft of anything carried in or on the motor vehicle or trailer; and
> (c) an offence under s. 12(1) Theft Act 1968 (taking and driving away without consent).

*Comment* Power of arrest without warrant is only conferred on a constable, as opposed to the 'any person' power of the Vagrancy Act;

this is regrettable since possibly the main security involvement with that Act arose from intended thefts from vehicles on works car parks. However, the limitation is of restricted security importance since the actions may be covered by other statutes conveying an 'any person' power: forcing a car bonnet or boot lid with a screwdriver can only be criminal damage or attempted theft; trying door locks with a bunch of keys could be an offence against s. 25 of the Theft Act; undoing wheel nuts can only be attempted stealing of the wheel or malicious damage; undoing the roping of a load on a trailer is surely reasonable grounds for suspecting an attempt to steal from the load etc.

## *OFFICIAL SECRETS ACT*

This Act usually comes to notice when the media produce 'revelations' based upon confidentially-rated papers which have come into their possession via a Government employee who had a duty not to disclose them. 'Freedom to know' arguments inevitably ensue as do the political overtones and criticism of whichever party happens to be in power. Repeal of this part of the Act, section 2, has been threatened almost since it was first on the statute-book and it is fortunate that normal security work has little to do with it.

Until now the Official Secrets Act 1911 (the subsequent Acts of 1920 and 1939 are both construed as part of it) has been largely of academic interest since Service and MOD police have been responsible for the premises to which it has applied. However, privatization of naval dockyards, tank and munitions factories, and the like, provides a scenario which is still not entirely clear and makes it advisable for the powers given by the Act and the places to which it applies to be at least summarized. Those who find themselves working in such places must clarify with their superiors the action they will be expected to take if foreseeable incidents occur, a necessary precaution for a highly newsworthy and thereby highly sensitive subject. One author, full of pre-war zeal, 'picked up' a suspect photographer near a tank factory and spent a long time thereafter licking severe verbal wounds, superior-inflicted – and still does not know what happened to the photographer!

Section 1 is the most important security-wise and relates to actions carried out by 'any person for any purpose prejudicial to the safety or interests of the State'. Case law has established that this is not limited to spying as such. Section 1(a) – 'approaches, inspects, passes over, or is in the neighbourhood of, or enters any prohibited place' (as defined by the subsequent s. 3). Section 1(b) 'makes any sketch [which has been held to include photograph], plan model or note which is calculated to be, or might be or is intended to be directly or indirectly useful to an enemy' (enemy has been held to include potential enemy). Section 1(c) 'obtains, collects, records or publishes or communicates to any other person any secret official code word or password, or any sketch, plan, model, article or note, or other document or information which is calculated' etc.

It is not necessary to show that the suspect was guilty of any particular act prejudicial to the state if his conduct or known character makes it appear his purpose was so prejudicial. Offences committed against this section carry a 14-year maximum sentence which of course entails an 'any person' power of arrest (see p. 139).

'Prohibited place' is given a very wide definition under s. 3 – any work of defence, arsenal, naval or air force station, etc. occupied 'by or on behalf of Her Majesty' and similarly any place used for building, repairing, making or storing any munitions of war or plans, etc. relating thereto. Factory is also included in this section (a) but presumably on privatization will cease to be; however (b) reinstates it as a place, not belonging, but under contract to make repair, etc. In addition the Secretary of State can make orders declaring a place to be 'prohibited' – notices to the effect are always displayed.

The contentious and complicated s. 2 deals with wrongful communication etc. of information by anyone holding or has held office under the Crown, or holds or has held contracts on behalf of the Crown or is/has been employed by the contractor. It is legalistically worded so as to be all-embracing and includes an interesting offence seemingly rarely pursued – that of failing to take reasonable care of, or so conducting himself as to endanger the safety of documents etc. These offences are rated as misdemeanours with a two-year sentence; and consent for any prosecution has to be obtained from the Attorney-General for all offences under the Act.

There are a few basic commonsense points for security staff employed in 'prohibited places' which merit special emphasis:

1 Take the job seriously, do not relax vigilance because incidents are rare, when they do occur they are likely to be of significant importance.
2 Do not let curiosity lead you to tamper with papers or files which do not concern you – your intentions could be misinterpreted.
3 Similarly, do not use master keys or other means to enter rooms or other areas which you are not supposed to – unless there is good reason for doing so – then make a record of the fact.
4 Do not gossip about your job, the work that is being done, or specialists who may be employed there.

5 Err on the side of being over-cautious in reporting and logging what you deem is suspicious activity.
6 Remember that if a genuine incident does occur, you will be dealing with a professional who will do everything he can to get rid of or destroy anything incriminating he has with him, for example expose a camera film, and he must be watched like the proverbial hawk, later checking any car he may have been in and the route along which he has been escorted.

Those carrying out gate duties should note offences include those of unauthorized use of naval, military, air force or police uniform – or one that resembles an official uniform – to get access to a prohibited place, and for the same purpose forging or tampering with passports etc. or impersonating officials.

A final item of interest: a police officer has been held a person holding office under the Crown and he commits an offence by communicating information entrusted in confidence to him by his superior officer (*Lewis* v *Cattle* 1938 3 All ER 368) and the person receiving the information may also be guilty unless he can show 'it was contrary to his desire'. This case has been interpreted in its widest sense and those who might be tempted to exploit personal friendship, or an 'old boy' relationship, to obtain information a police officer should not disclose, should note that prosecutions are increasing in number both here and overseas.

The other aspects of the Act are really outside the security remit though 'soliciting and inciting to commit' is specifically mentioned as carrying the same penalty as the major offences.

## *DRUGS*

It is only possible to touch briefly on this very important subject which in the industrial and commercial world can have unpredictable consequences. The cost to society is very high, with drug-related deaths, the cost of imprisonment for offences, medical expenditure, administration of treatment programmes and crime and damage related matters. A recent survey in the US put the total cost at some 47 billion dollars per year. In the UK it is a growing problem.

In the workplace, people under the influence of drugs are a very serious problem for management, over which they initially have little control. In 1985, 348 kilos of heroin, 79 kilos of cocaine and 20,424 kilos of cannabis were seized by Customs and Excise officers.

The effects of taking drugs can include a diminished sense of responsibility for the protection of employers' property against loss and for the observance of safety rules.

Drugs create an increasing reliance on them and the addict often becomes unable, from normal resources, to pay for them. He is then compelled to obtain money from some other source, for example stealing cash and company property which is convertible into cash with which to purchase what is required. But for the craving, such actions would be unthinkable to the addict.

While under the effect of a drug the addict is unstable and his behaviour could vitally harm the interest of his employer or himself.

It is therefore essential that indications of drug abuse must be identified as early as possible for action to be taken. These include:

1 A dramatic increase in pilferage.
2 Empty pill or medicine bottles and used syringes in rubbish bins or wastebaskets.
3 The smell of burning rope in washrooms or hallways.
4 Sudden behaviour changes in otherwise stable employees.
5 Erratic changes in emotion or activity by formerly stable employees.
6 Degeneration of performance level.

Addiction is not confined to any particular class of employees; at a CBI conference in March 1987 it was said to be an increasing problem amongst highly-paid staff in the City with heroin as the main cause for concern. Astronomical figures were quoted for possible losses and it was suggested that some young executive types spent as much as £300 per week on drugs. Causation was thought to be 'stress' but an impartial observer could be excused for feeling more likely explanations would be the usual trendy aping of others exploring new sensations, coupled with an excess of money and boredom with conventional sources of pleasure. Irrespective of reason, in a computer-orientated environment CBI alarm is understandable.

Management must also be aware that employees may be 'pushing' drugs for personal gain, or to help pay for their own habit. A House of Lords decision in 1987 (*R* v *Maginnis*) is of interest; it involved a person in possession of drugs, which had been left with him by a friend. The Judges stated the ordinary meaning of 'supply' was not merely to transfer physical possession of an article, but implied provision of an article for use towards a particular purpose of the recipient. Deposit of drugs for safe keeping is not therefore a supply, within the meaning of the Drugs Act 1971.

It is important that management and supervisors are aware of the physical symptoms stemming from taking drugs. A checklist is included in Appendix 13.

An indication of a drug-using group of employees might arise from their conversation and inclusion of words associated with drugs. 'Fix' or 'joint' refers to a single dose; 'stoned' or 'on a high', to be well under the influence of drugs; 'skins', the papers used for rolling marijuana cigarettes; 'roaches', the cardboard cylinders to hold the cigarettes; and 'gear', syringes or any other form of equipment for injecting or any other way of using the drug. 'Weed', 'crack', 'hash', 'snout', 'snow' are amongst the multiplicity of names given to drugs used.

Cannabis is sometimes regarded more leniently than other drugs but its addiction leads on to the harder ones and gives a sense of euphoria, dangerous amongst machinery and equally dangerous to decision-making. It is the most common in use, and as a 'starter' to the others can save a lot of trouble if it is recognized at an early stage. Fortunately, the smell is persistent and distinctive; at work it will be smoked frequently in toilets where it may be noticed by patrolling security officers. Most will not be familiar with the smell but if during their training a demonstration can be given (hopefully as part of a full lecture on drugs) by the local police drug squad, it will certainly never be forgotten.

There is very great interest in prevention; such a lecture was arranged for management on a large site during a strike-enforced shutdown. Not only did everyone from the managing director downwards attend but senior shop stewards from the picket line asked permission to attend! This is an interest which can be exploited for the good of all.

Some of the symptoms for managers and supervisors to be instructed to recognize are as follows:

1 The drug user may wear sunglasses at inappropriate times to cover dilated or constricted pupils, and wear long sleeves at all times, even on hot summer days, to cover marks caused by intravenous injection of drugs.
2 The abuser of depressants such as the barbiturates usually shows most of the signs of alcohol intoxication – slurred speech, staggering, etc. – with one exception: there is no odour of alcohol on the breath. This type of abuser may fall asleep on the job or appear confused.
3 Abuse of amphetamines (stimulants) such as diet pills or 'pep' pills brings about excessive activity, excitability or irritability, and heavy perspiration. The pupils may be dilated, even in bright light. These drugs may also cause unusual bad breath unlike that

resulting from garlic or alcohol, and chapped or cracked lips caused by drying of the mucous membranes. The amphetamines also cause complexion problems. Finally, excessive nervousness may lead to itching, chain smoking or talkativeness.

4 The most obvious symptoms of abuse of heroin are injection marks and constricted pupils. Sometimes the addict may make frequent or prolonged trips to his locker or to the restroom for privacy to inject the drug. He may become excessively secretive about his belongings. After injection the user may become drowsy and lethargic.

5 Marijuana cigarettes rolled in brownish paper and crimped at the ends are usually smoked while the users are gathered in groups. The smoke of marijuana is recognizable by its odour, like that of burning rope. Smoking marijuana may produce talkativeness, bursts of hilarity, and distortions of time and space. There is some loss of co-ordination, and the eyes may become bloodshot. As the drug wears off, drowsiness and lassitude follow. (Cannabis and hashish are derivatives from the same source the hemp plant and have like affects.)

6 Although it is unlikely that an employee will use hallucinogens at work, it is possible. Under the influence of hallucinogens like LSD, the individual appears to be in a trance or dream-like state. The most common symptom is dilation of the pupils. Mental effects are unpredictable; the user may experience exhilaration, an urge to self-destruction, or sheer panic. 'Flashback' drug episodes may occur weeks later, perhaps while on the job, without further ingestion of the drug.

7 Glue sniffing is mainly confined to juveniles and teenagers and is therefore less likely in workplaces. Other solvents and aerosols may be used. Users will stink of what they have used; they may be violent and irrational or disorientated, 'not with it', drunken appearance and may collapse. Lavatories (and out of the way corners) are likely places of usage. Hopefully this is becoming less of a menace.

Supervisors should be certain that there are no other explanations for the symptoms before concluding that the employee is a drug user. Simple fatigue, for example, can cause a number of symptoms similar to those listed.

### *Drug Trafficking Offences Act 1986*

This makes provision for the recovery of the proceeds of drug trafficking, and for new offences of assisting another person to retain the proceeds of drug trafficking or disclosing information which would be likely to prejudice a drug trafficking investigation.

The Act also amends the Police and Criminal Evidence Act 1984 to classify certain drug trafficking offences as 'serious arrestable offences'.

No one involved with an in-company drug investigation should treat the matter lightly, it may have wider implications, and the police should be involved.

It should be carefully noted that section 21 of the Misuse of Drugs Act 1971 states:

> If an officer of a Company which has committed offences against the Act has connived in the offence, or has allowed the offence to occur through neglect, he, as well as the Company, shall be guilty of the offence.

Security officers may incur a direct responsibility with regard to 'controlled substances' which the Misuse of Drugs Act divides into various categories:

(a) Hard drugs – heroin, cocaine, morphine, LSD and mescaline.
(b) Soft drugs – cannabis, amphetamines and codeine.
(c) Others – those considered addictive but not as dangerous as those listed above.

Section 10 states that where a person is lawfully in possession of a controlled drug (that is, seized from an employee) he must take adequate precautions for the drugs safe custody, must supply information to the relevant authorities, and may be responsible for transporting the drug if so directed. Effectively this means the police must be informed immediately, that any substance should be locked away in a safe, and a record kept of who has handled the substance for evidential purposes. It is a good idea to place it in a large envelope and all persons handling it should sign the envelope with the time and date and purpose for doing so.

Any other items, for example, syringes, connected with drugs should be treated in the same way (s. 9).

### *Company policy*

At the beginning of 1985 there were some 6,000 registered addicts in the UK and the Drug Advisory Committee estimated an additional 300,000 regular users of cannabis. The figures will have escalated since then and inevitably any sizeable company will have employees who are 'drug-abusers'. Drugs, like theft, is a matter on which a policy must be formulated and included in company or works rules, company policy manuals or whatever other means are used for codifying discipline and penalties. Drink addiction, now called 'alcohol abuse', should be lumped under the same conditions; it is now generally accepted as being a more widespread threat than drugs. As always, a policy must be applicable to all – a drunken or hallucinating shopfloor worker might injure himself or workmates but just imagine the damage a similarly afflicted purchasing officer or sales director could do to a firm's prospects.

If the drug problem does manifest itself within an organization it should not be swept under the table. A deterrent campaign should be mounted using posters or any other means suitable to the environment to publicize the danger and the firm's attitude. Since safety is a paramount factor, unions are apt to be more than co-operative and perhaps more rigid than management – it has been known for a union to insist on a member's dismissal in the face of intended management leniency. The problem does not bring good publicity and some companies, including multinationals, have their personnel departments covering up by producing the anonymity of 'fitness to work requirements'.

Opinions vary as to what action should be taken when a user is found amongst employees; there is a school of thought that says that a responsible firm will take a benevolent interest in 'drying him out'. It is a matter of policy which must be consistent. There are two cases, however, which must not be debatable: if the user is employed with machinery, he must be kept away from it; if a supplier is pinpointed, it is a public duty to notify the police and give them every assistance to obtain a conviction – the worst possible publicity would arise and deservedly so otherwise.

## *CONFIDENTIAL MATERIAL/INFORMATION*

Not infrequently problems arise when senior executives change employment and in doing so take with them confidential papers, data, or information which they then proceed to use to the advantage of their new employer, or personally in setting up their own business in competition with the previous employer. Indeed, one departing sales executive is said to have used cartons from, and the services of, the firm's postal department to remove his 'goodies'.

Civil action to restrict the use of the material seems the norm in such cases and the law is complex. The Court of Appeal has done its best to provide clarification by giving a very detailed legal analysis when adjudicating on two consecutive hearings: *Roger Bullivant Ltd and another* v *Ellis and others* 1987 IRLR 491 and *Johnson & Bloy*

*Holdings Ltd and J. & B. Ltd.* v *Wolstenholme Rink plc & Fallon* 1987 IRLR 499. The two Law Reports should be compulsory reading for any senior security officer asked to give an opinion in such a contingency; their contents are more important than the judgments. In the first case, an index of trade contacts had been taken away with many other documents, all later reclaimed. The executive had a standard directors' contract limiting his responsibilities to one year after termination and that was the restriction laid upon him. In the other, documents were taken containing details of a special drying agent in print work and the ban was made permanent.

A cynical security officer might well preface his opinion by asking whether the departee 'had a claim of right made in good faith' to what he had taken (Theft Act 1968 S. 2. page 158).

## *REFERENCE SOURCES*

Whilst the offences mentioned in this chapter are those most likely to be encountered under normal security conditions, a group security officer or security manager would be well advised to requisition specialist reference books to have at hand to elucidate any of the more obscure offences brought to his notice. Two standards to cover the whole range from the trivial to the most serious are *Stone's Justices Manual* and *Archbold Criminal Pleading Evidence and Practice*.

A basic briefing about drug abuse (DM3) is available from Department DM, DHSS Leaflets Unit, PO Box 21, Stanmore, Middlesex HA7 1AY. The leaflets, prepared by the Institute for the Study of Drug Dependence, are a more detailed guide to the effects, dangers, prevalence and legal status of all major drugs from tobacco and alcohol to heroin and other opiates.

# 13

# *Questioning and Codes of Practice*

It is difficult to visualize any set of circumstances in which a security officer can act without indulging in some form of questioning. For the most minor breach of works regulations, where he may have to request the name and works number of the employee and point out the breach, his mode of approach may be subject of complaint by the person to whom he is speaking unless he behaves in a reasonable manner. It is, unfortunately, a characteristic of persons in trouble that they are apt to find fault with the individual who is responsible for their predicament. This is sometimes seen in its ultimate phase at the criminal courts in the form of making unfounded allegations against the conduct of prosecution witnesses, with a view to diverting the attention of the jury or magistrates away from the essence of the matter – the guilt of the accused. Apart from the precautionary nature of so doing, it is commonsense to adopt the reasonable approach which one would expect if the positions were reversed. No excessive use of authority should be made or even implied. However, many offenders regard attack as being the best form of defence, so neither should there be any question of backing down in the face of an aggressive approach from someone who is in the wrong. Politeness, coupled with firmness, soon makes such a person see that he is making a fool of himself to no purpose.

Unlike the duties of police officers, the ordinary security officer should keep questioning to a minimum, consistent with establishing that no mistakes have been made and that the individual being spoken to has had every opportunity to clarify the circumstances. The only security officers who need to develop a genuine technique of questioning are limited to those chief security officers or security managers whose companies direct their own internal enquiries into major defalcations. However, both are now subject to the criminal law and codes of practice.

The Police and Criminal Evidence Act 1984 came into force on 1 January 1986. Prior to this date, persons engaged in questioning suspects were given guidance under the Judges Rules, which were fundamental requirements to ensure the admissibility of evidence, especially statements, written or oral, made by suspected persons. These were rules which left the courts to finally decide the admissibility of any evidence in the course of the proceedings, even if the rules were not followed properly. The introduction of the Police and Criminal Evidence Act 1984 made these rules into a lawful requirement, and everyone must comply with this Act of Parliament and the Codes of Practice which are part of it. Section 67(9) of the Act says:

> Persons, other than police officers, who are charged with the duty of investigating offences *or* charging offenders shall, in the discharge of their duty, have regard to the relevant provisions of such a Code.

Different interpretations can be given to who are such persons under this section of the Act, and following questions to the Home Office a letter in reply concerning security personnel clearly gave further guidance. The letter stated:

> This Act does not alter in any way the position of such people under law. They have no special powers of arrest, search, entry or seizure. If they make an arrest, it is a citizen's arrest which involves no power of detention for questioning and they must bring the person arrested before a Justice of the Peace, or a police officer as soon as reasonably possible. Security personnel and store detectives are not charged with a duty of investigating offences within the meaning of section 67(9) of the Act. However, the Codes of Practice represent good practice in the investigation of offences and it is clearly desirable that people who may take it upon themselves, whether by virtue of a contract of employment or otherwise, to investigate a crime, or a category of crime, should have regard to the standards set in the Codes so far as in commonsense they are applicable to the work they do and the limited powers they have. For example, they should follow those standards in matters such as cautioning and the treatment of young or handicapped suspects, but they should be most cautious that they do not, in so doing,

> overstep the clear limitations on their powers both under the statute and at common law.

Whilst the majority of this Act is clearly for the police alone, this advice appears to reinforce the requirement for all security officers to abide within the Codes.

This matter has still to be tested in a Court of Appeal, when final clarification will be forthcoming. Certainly, under s. 78 of the Act, the position of the court is clear. The section says:

> In any proceedings, the court may refuse to allow evidence on which the prosecution proposes to rely to be given, if it appears to the court that, having regard to all the circumstances, including the circumstances in which the evidence was obtained, the admission of the evidence would have such an adverse effect on the fairness of the proceedings that the court ought not to admit it.

This contradicts a House of Lords decision in 1980 (*R* v *Sang* 1980 AC 402) which stated that in general terms all evidence not being evidence of admissions, or evidence amounting to self-incriminating material arising after the commission of the offence, is admissible in subsequent proceedings, however obtained.

It is evident that a security officer, who is employed to safeguard his employer's interests and apprehend those who attempt to steal company property, has a full duty to comply with the Codes if he arrests anyone. Likewise, store detectives (previously stated in *R* v *Nicholls* 1967 51 Cr App. R. 233, as a person charged with the duty to investigate offences) must also follow the Codes, whilst other persons, for example a store manager whose duty is to the store, would not be considered in this category. Failure to follow this Code may make certain evidence inadmissible in court. Security officers are likely to be regarded more leniently by a court than would be their police officer equivalents if in contravention of these Codes. Circumstances would be the deciding factor. Sometimes what has been said will be so vital to the prosecution that the defence may well try to have it erased completely from the proceedings. There is little doubt that this would be successful where an employee who was questioned by a senior security officer later claimed that he was not afforded his rights under the Code of Practice in full.

Suffice it to say then, that if the matter under enquiry is discovered to be of some gravity and the replies to questions to be put are vitally important to the truth of the matter in issue, it is necessary to administer a 'caution' to a person against whom there are grounds to suspect has committed an offence. This person must be cautioned before any questions about it (or further questions, if his answers to your questions have given the grounds for the suspicion) are put to him for the purpose of obtaining further evidence which can be put before a court or tribunal.

It is essential, therefore, for anyone in authority who has to interrogate persons in connection with a criminal offence, to know the material points raised in the Codes, relevant parts of which are shown later.

It is true to say where the facts speak for themselves, any conversation with an offender will be relatively unimportant if there is overwhelming evidence against the person, so the less said in such circumstances the better. It is re-emphasized that the Codes of Practice are lawful requirements and must be obeyed.

## *PRINCIPLES AND PRACTICE OF OBTAINING STATEMENTS AND QUESTIONING SUSPECTS*

Real ability in ascertaining the truth from a confused witness or an obdurate suspect cannot be acquired by memorizing a checklist of points to be observed. Much will depend upon the personality of the questioner and his correct assessment of the person to whom he is speaking. Technique is only acquired after years of practice in talking objectively to a diversity of people to an extent which security officers are rarely likely to experience. This can become an intuitive skill rarely appreciated by any layman or even perhaps the questioner himself. Principles to be observed are as follows:

1 The object is to learn the truth, not to induce a pattern of deceit or simply to obtain answers satisfactory to the questioner.
2 Under no circumstances whatsoever should any implied threats or promises be used to obtain answers.
3 Do not indicate (except in answer to direct question) what action will be taken in event of a person answering questions, making a statement, or refusing either. If such a direct question is asked, it is permissible to inform the person what action is proposed *provided* that action is proper and warranted.
4 If a person has a possible explanation for his actions, he must be given every opportunity to state it. If, having been given the opportunity, he fails to take advantage of it, a court may be subsequently entitled to comment on this and he would encounter little sympathy for complaints of inconvenience arising from his failure to do so.
5 Questions must be asked in a language and phraseology which is clearly understood by the person being spoken to.

6 Ambiguous questions must not be asked. If the answers are ambiguous, they should be clarified by further questions so that there is no misunderstanding on either side.
7 Questioning must be done methodically to cover all the information that is required. A person who is not telling the truth can put up a much better performance if he has intervals in which to collect his thoughts – there is usually a reason for inexplicable hesitancy.
8 Under no circumstances must the person who is asking the questions lose his temper, irrespective of provocation. Aggressiveness, abuse, ridicule, complaint, unwarranted argument, and 'talking down to' are all means employed by an 'experienced' suspect to discomfort a questioner, and a reluctant witness may employ some of these in a lesser degree.

Within the limitations imposed by these principles, there is scope for the questioner to display his virtuosity in playing on the personality of the individual to whom he is speaking. Formal approach, sympathy, incredulity, and even a degree of flattery may be needed to persuade the witness to divulge what he knows. There are several guidelines which can be given, but these do not replace practice or make a person into the 'good mixer' which is necessary to establish confidence.

The conditions under which the interview is taking place are important. It is a waste of time being in a place where people are constantly passing to and fro, diverting attention and distracting by noise and conversation. Equally, if the continuity of the conversation is continually broken by telephone calls, queries to either party, or any other form of interruption, this is fatal to any complex discussion, particularly if dealing with a suspect. In the further interest of the privacy of the person being spoken to, which is desirable, a quiet, separate room is preferable, with refreshments available if needed, but not alcohol, during a long session. Rooms should be adequately lit, heated and ventilated.

In dealing with employees, there may be established company rules which allow them the presence of a representative, should this be practicable in the circumstances. If dishonesty is the subject, the probability is that an employee will prefer to keep it between himself and the interviewer; if the implication is disciplinary, it is more likely that the meeting will be prearranged and the attendance of a shop steward or other 'friend' asked for. It is worth remembering and pointing out, that if prosecution follows, the 'friend' may find himself in what may be the invidious position of being required by the police to give evidence.

When taking written statements of any kind, the best way of producing a thoroughly bad one and perhaps causing a witness to close up completely is to pick up a notebook or sheet of paper, then sit down and start writing immediately. The first step is to get a clear picture of what has happened before committing anything to paper; this will make the subsequent screed more readable, sequential, coherent and comprehensive. It will also give an opportunity to draw out the witness on any points on which he is reticent. Having achieved this end and, perhaps, having jotted down a few rough words for sequence, an attitude of 'let us' or 'should we' get this down in writing will make the witness take a keener interest in what is being recorded and be more personally involved in ensuring absolute accuracy, since he himself has had time in the preceding discussion to marshal his thoughts.

Do not hurry the process of making a statement as you may omit matters that should be included. Although everyone at some time or another has done this, it reflects adversely on the capability of the notetaker and should be avoided if at all possible. Keeping to a chronological sequence helps to prevent omissions.

### *Technique*

The amount of information that can be obtained from a witness or a suspect is dependent upon the rapport established by the questioner. He has to recognize the individual's probable emotional reaction to the situation, mentally put himself in the witness's position, and then modify his own personality and approach accordingly to establish the desired relationship.

Correct recognition of the witness/suspect type is necessary. The talkative are easy to handle but need coaxing back gently into the areas from which information is required. The boastful need nudging in the same direction but at the same time their vanity can be played upon so that in the process of self-aggrandizement they may throw out information they really do not wish to give. Nervous witnesses are difficult and their nervousness may be mistaken for a desire to conceal facts, or for stubborness – this type needs to be dealt with sympathetically. The taciturn have to be coaxed and persuaded to talk, perhaps even by questions unconnected with the main line of enquiry. The aggressive have to be cooled down by a total lack of reaction to their attitude and by patience. The self-pitying need a good listener, but one who takes every opportunity to channel the conversation on to the subject he is interested in. The casually evasive may react to a direct and formal approach to bring them down to earth. Outright and blatant liars can sometimes be embarrassed into truth by unconcealed and somewhat hilarious disbelief. Of course these are generalizations and there may be occasions where a person does not clearly conform to any group, possibly due to embarrassment of the situation in which he is found. With such circumstances a dispassionate approach, as if this was an everyday matter, is advisable. The cardinal characteristics to acquire may be listed as:

1. Be tactful – do not cause unnecessary friction or resentment.
2. Be a good listener – try to create an atmosphere of interest to encourage the witness to talk confidently.
3. Be patient – above all avoid any display of boredom.
4. Do not interrupt unnecessarily, except to keep the conversation to the point.
5. Do not show annoyance or make either verbal or physical threats.
6. Do not use ridicule or abuse.
7. Do not make promises or offer inducements to obtain answers.
8. Do not be diverted from your objective by any antics of the person being spoken to.
9. Persist until you are satisfied that you have found out as much of the truth as possible.
10. Do as you would be done by, if the positions were reversed.

Remember that if any of your assumptions have been wrong, any annoyance felt by the person to whom you have been speaking will be mitigated by a pleasant and courteous approach. If you are found totally wrong, an apology should be given.

## *CODES OF PRACTICE*

### *The Police and Criminal Evidence Act 1984 (s. 66)*

This Act of Parliament and the Codes of Practice which are part of it were drafted after extensive public consultation and reflect the views of the Royal Commission of Criminal Procedures. The Act provides clear and workable guidelines for the police, balanced by strengthened safeguards for the public. The Act covers all aspects of police officers powers to stop and search, search premises, seizure of property, the identification of suspects, and the detention, treatment and questioning of persons in police custody.

However, this Act and Codes of Practice also covers many items of relevance to security personnel overall, who although said by the Home Office not to be persons charged with the responsibility of investigating offences, have been held by Judges to be such persons, at least in their observance of the old Judges Rules. The Police and Criminal Evidence Act 1984 and the Codes of Practice completely replace the Judges Rules which are however repeated almost verbatim. It should be noted that the Judges Rules were *guidelines* to ensure the admissibility of evidence, whilst the new Codes are *legal requirements*.

The foundation stones of the present Codes do not affect the following longstanding principles:

1. Citizens have a duty to help a police officer to discover and apprehend offenders.
2. Police officers, other than by arrest, cannot compel any person against his will to go to, or to remain in, any police station.
3. Every person at any stage of an investigation should be able to communicate and consult privately with a solicitor. This is so even though he is in custody, provided that in such cases no interference will be made to the proper administration of justice.
4. When a police officer, who is making enquiries about an offence, has enough evidence to prefer a charge against any person for the offence, he should without further delay cause that person to be charged, or informed he may be prosecuted for that offence.

*Interview technique*

A police officer enquiring into an offence may question anyone from whom he thinks useful information could be obtained, irrespective of whether the person to be questioned is in custody or not, so long as he has not been charged or told he would be prosecuted in connection with the offence being enquired into.

*When to caution*

Section 10 of the Codes of Practice clearly lays down when a person must be cautioned:

> A person whom there are grounds to *suspect* of an offence must be cautioned before asked any questions about it (or further questions if it is his answers to previous questions that provide the grounds for suspicion) are put to him for the purpose of obtaining evidence which may be given to a court in a prosecution. He need not therefore be cautioned if questions are put to him for other purposes, such as to establish his identity, his ownership or, or responsibility for, any vehicle or the need to search him in the exercise of powers to stop and search by police.

This is a significant change in the law; before it was necessary to *have evidence* giving reasonable grounds for suspecting a person has committed an offence, now, all that is required is to have *grounds to suspect* a person has committed an offence.

The caution given should be in the following terms:

> You do not have to say anything unless you wish to do so, but what you say may be given in evidence.

Minor deviations do not constitute a breach of this requirement provided that the sense of the caution is preserved. A person must be cautioned upon arrest for an offence unless:

(a) it is impracticable to do so by reason of his condition or behaviour at the time, or
(b) he has already been cautioned immediately prior to the arrest when grounds to suspect the person became apparent.

When, after being cautioned, a person is being questioned or elects to make a statement, the time of commencement and termination of such questioning and statement shall be recorded together with the place and persons present.

*Charging with an offence or informing a person they may be prosecuted*

When an officer considers that there is sufficient evidence to prosecute a detained person, he should without delay bring him before the custody officer at a police station, who shall then be responsible for considering whether or not he should be charged. Custody officers are usually police sergeants (see Appendix 14). Where a person is to be charged, or informed he may be prosecuted for an offence, he shall be cautioned again. This time in the following words:

> You are charged with an offence shown below. You do not have to say anything unless you wish to do so, but what you say may be given in evidence.

At the time a person is charged he shall be given a written notice showing particulars of the offence with which he is charged and including the name of the officer in the case, the police station and the reference number of the case. Charges should be stated in simple terms, but they have to show the exact offence in law with which he is charged.

Questions relating to an offence may not be put to a person after he has been charged with that offence, or informed that he may be prosecuted for it, unless they are necessary for the purpose of preventing or minimizing harm or loss to some other person or the public, or for clearing up any ambiguity in a previous answer or statement, or where it is in the interests of justice that the person should have them put to him and have the opportunity to comment on information concerning the offence which has come to light since he was charged, or informed he would be prosecuted. Before any such questions are put it is necessary to caution the person again,

### *Written statements after caution*

Until now, the taking of statements under caution from offenders has been almost entirely restricted to police officers. The exceptions have been in instances where firms have decided to exercise complete control over an investigation and have a discretion on whether to prosecute if the findings justify that course. This they would not have if the police did the job – it may be that they wish to avoid adverse publicity, or come to some financial agreement to cut their losses. If they have a chief security officer of accepted competence, they may allocate the job to him; otherwise a private investigator or commercial security firm, not all of whom would have the necessary capability, would have to be engaged.

For the chief security officer, the taking of a statement would be a rare occurrence but one in which his reputation would either be enhanced or badly damaged if, for instance, he took statements which proved inadmissible. With increasing pressures on police time and the duration of an involved fraud enquiry, it seems likely that the UK will slowly follow the trend in North America of firms utilizing their own resources. A by-product for them would be the expediting of the enquiry since those they engage to do it would devote all their time to it – the police are unlikely to be able to do so.

Statements made after caution which amount to admissions, even if not used in criminal proceedings, might be invaluable to a company in any subsequent unfair dismissal proceedings against them. Indeed the existence of such a statement might preclude the inconvenience of such proceedings being instigated. This means the taking of the statement must allow for its validity to be questioned, and this involves conformity as far as possible with the Codes of Practice under the Police and Criminal Evidence Act 1984. These Codes of Practice are outlined in detail in the Appendices: Appendix 15 deals with interviews and shows interview record forms; Appendix 16 deals with the taking of statements under caution, and includes specimen statements.)

Careful note should be made of s. 76 of the Police and Criminal Evidence Act 1984 which states:

> In *any* proceedings a confession made by an accused person may be given in evidence against him if it is relevant to the proceedings and is not excluded by the court in pursuance of this section.
>
> A confession made by an accused person must have been obtained properly. It will be excluded if
>
> (a) it is obtained by oppression of the person who made it, or
> (b) it was obtained in consequence of anything said or done which was likely, in the circumstances existing at the time, to render unreliable any confession which might be made in consequence thereof.

A Court will not allow a confession to be given in evidence unless the prosecution proves beyond reasonable doubt that it was not obtained improperly; that is, not by oppression, making false promises or by deception. This veto stands although the confession may be true. In this section 'oppression' includes torture, inhuman or degrading treatment and the use or threat of violence whether or not amounting to torture (*R* v *Fulling* 1987).

'Oppression' ought now to be given its ordinary dictionary meaning. Its third definition in the *Shorter Oxford English Dictionary* is 'exercise of authority or power in a burdensome, harsh, or wrongful manner, unjust or cruel treatment of subjects, inferiors, etc.; the imposition of unreasonable or unjust burdens'. It was likely that such oppression would involve some impropriety by the interrogator. In the present case, although the police officer had acted unkindly in revealing the information to the appellant, his words were not oppressive.

*Confession obtained by deceit (R* v *Mason*, 1987 TLR (23 May 1987) )

The appellant had been arrested in connection with criminal damage by fire to a motor car. A confession was obtained after police officers falsely pretended that the appellant's fingerprint had been found on a material object. The same false information was given to his solicitor who then advised his client to give his version of what happened. The Court of Appeal held that the trial judge had a discretion as to the admissibility of any evidence including confessions. The Police and Criminal Evidence Act 1984, s. 78, restated this common law power. The confession had been obtained by a deceit practised by the police on the appellant and his legal adviser and was not admissible in court.

### *Court of Appeal*

Care should be taken to avoid any suggestion that a person's answers can only be used in evidence *against* him, as this may prevent an innocent person making a statement which might help to clear him of the charge.

### *Statements of persons jointly charged or told they may be prosecuted*

If at any time after a person has been charged with, or has been informed that he may be prosecuted for an offence, a police officer wishes to bring to the notice of that person any written statement made by another person, who in respect of the same offence has also been charged or informed that he may be prosecuted, he shall hand to that person a true copy of such written statement, but nothing shall be said or done to invite any reply or comment. If that person says that he would like to make a statement in reply, or starts to say something he shall at once be cautioned or further cautioned.

This could apply if a second person involved makes a statement, after the first person is charged, and it is necessary to serve that statement on the first person.

### *Investigators other than police officers*

As shown at the beginning of this chapter, this section lays an onus upon security officers to comply as far as possible with the Codes and they should familiarize themselves with their contents.

It is essential that any written statement after caution is taken properly to comply with the Codes of Practice. There are plenty of occasions, especially for store detectives, where such a statement is vital, either for the firm to take out a private prosecution or to justify the action taken if necessary.

It is helpful if the 'cautions' are written on a card and kept in the back of a notebook, used at work.

### *Summary*

Most of the sections of the Police and Criminal Evidence Act 1984 are almost wholly applicable to police officers. The main points a security officer should know relate to the following:

1 methods of investigating offences;
2 obtaining evidence correctly;
3 interviewing suspects;

4 requirements to 'caution';
5 completing interview records;
6 taking statements under caution;
7 taking witness statements;
8 reporting offenders to the police;
9 dealing with cases internally;
10 preparing correct reports;
11 preserving exhibits;
12 giving evidence required.

## *USE OF INTERPRETERS*

A person must not be interviewed in the absence of a person capable of acting as interpreter if:

(a) he has difficulty in understanding English;
(b) the interviewing officer cannot himself speak the person's own language; and
(c) the person wishes an interpreter to be present.

The investigator gives guidance to the interpreter of what is required of him, but he does nothing to influence him. The interpreter is an independent medium.

Questions are asked in the language the person fully understands and both questions and answers are recorded in that native writing. The notes and/or statement are read by, or read to, the maker and he is asked to sign them. The interpreter certifies the accuracy of the record.

The interpreter translates the records into English, attends the court/tribunal and gives evidence and produces the records and explains them. He carries out verbal examination for the court/tribunal, or a court interpreter can also be used.

It is imperative that any person used to interpret in official matters is of good character and is fully conversant with the duties.

In the case of a person making a statement in a language other than English:

(a) the interpreter shall take down the statement in the language in which it is made;
(b) the person making the statement shall be invited to sign it; and
(c) an official English translation shall be made in due course.

Both statements (a) and (c) must be produced to the court by the interpreter.

## *HANDICAPPED PERSONS AND JUVENILES*

It is important to bear in mind that, although juveniles or persons who are mentally ill or mentally handicapped are often capable of providing reliable evidence, they may, without knowing or wishing to do so, be particularly prone in certain circumstances to provide information which is unreliable, misleading or self-incriminating. Special care should therefore always be exercised in questioning such a person, and the appropriate adult involved, if there is any doubt about a person's age, mental state or capacity. Because of the risk of unreliable evidence it is also important to obtain corroboration of any facts admitted whenever possible.

A mentally ill or mentally handicapped person or juvenile must not be interviewed or asked to provide or sign a written statement in the absence of an 'appropriate adult'. In the case of a person who is a juvenile, mentally ill or mentally handicapped, 'the appropriate adult' means:

(a) a relative, guardian or some other person responsible for his care or custody;
(b) someone who has experience of dealing with a juvenile, mentally ill or mentally handicapped persons but is not a police officer or employed by the police; or
(c) failing either of the above, some other responsible adult who is not a police officer or employed by the police.

Where the parents or guardians of a person at risk are themselves suspected of involvement in the offence concerned, or are the victims of it, it may be desirable for the appropriate adult to be some other person. A person connected with the investigator, a company official, or any person who is possibly biased towards the investigation can never be an appropriate adult.

### *The deaf*

If a person is deaf, or there is doubt about his hearing ability he must not be interviewed in the absence of an interpreter unless he agrees in writing to be interviewed without one.

## *CONCLUSION*

The Codes are more liberal in ensuring that an accused person is advised to be careful in what he says and against making an admission. Interrogating officers must be fair to the persons being questioned and scrupulously avoid any method which could in any way be regarded as underhand or oppressive.

### *Application of the Codes to industrial tribunals and civil matters*

The application of the rules which are now contained in the Codes is clarified by the Employment Tribunal case of *Morleys of Brixton* v *Minott* which has been repeatedly confirmed (1982 IRLR 220).

The Codes are not applicable and, in consequence, if the objective of questioning is purely confined to disciplinary or civil issues, non-observance will not automatically make admissions unacceptable. Nevertheless, the principles and practice advocated earlier should be adhered to and, if there is a possibility that a prosecution decision may follow questioning, it would be advisable to caution in accordance with the Codes. The situation could arise where a perfectly sound dismissal could be made based part on evidence unacceptable to a criminal court, even if the subsequent prosecution failed because of this.

One must also bear in mind s. 67 (11) which reads:

> In all criminal and civil proceedings any such Code shall be admissible in evidence; and if any provision of such a Code appears to the court or tribunal conducting the proceedings to be relevant to any question arising in the proceedings, it shall be taken into account in determining that question.

After nearly two years in operation there is yet no indication this section is relevant to evidence in industrial relations proceedings. The Act clearly shows in its title that it refers to criminal evidence.

# 14

# *Elementary rules of evidence and court procedure*

Individuals will always have conflicting interests and it is the function of the law to settle any dispute in a just way. For the effective dispensation of justice, the law must also be efficiently administered. Therefore, it is not only important to have a structure of courts, but also that their administration should not be corrupt. In the British constitution this has been attained by completely separating the functions of the judiciary from the other powers within the State.

Each court is bound by and operates according to rules of procedure. These regulate the preparation of legal disputes for a court hearing, the actual conduct of the court trial and the enforcement of the court's decision. The rules of procedure differ greatly according to whether the dispute is a civil or a criminal matter. It is essential that each court of law should be subject to procedural rules which enable each party to have a fair chance of stating his case and answering the allegations that are made against him.

## *ENGLISH COURT SYSTEM*

The House of Lords is the highest court in the country and all cases that come before it are on appeal from a lower court. It has virtually no original jurisdiction; in other words, it will not be the first court to which a dispute is taken.

*The Crown Court – Full-time Judges*

Section 1 of the Courts Act 1971 states: 'The Supreme Court shall consist of the Court of Appeal and the High Court, together with the Crown Court established by this Act'.

The Crown Court has jurisdiction throughout England and Wales in criminal matters. All trials on indictment, in other words trials with a jury, must be brought before it. It deals with all serious offences.

*The Magistrates Court*

A Magistrates Court consists of Justices of the Peace who have been appointed by a document known as the Commission of the Peace. Every county and many boroughs have a separate Commission of the Peace for appointing the JPs to serve in their area. Justices of the Peace need no legal qualifications as they are appointed for their qualities of integrity and understanding and are persons who are broadly representative of the community which they serve. They deal with lesser offences.

A Juvenile Court is a special sitting of the Magistrates Court. It hears the majority of summary and indictable offences committed by children (persons under 14 years) and young persons (persons between 14 and 17 years old). The court will usually consist of three magistrates (at least one must be a woman) drawn from a panel of Justices who are specially qualified to deal with juvenile cases. The procedure in court is more informal than the ordinary Magistrates Court and the court is not open to the general public. The court has power to place the juvenile on probation, impose a fine, commit to an approved school or detention centre or place him in care of a 'fit person'. They have no power to send him to prison but may commit him to the Crown Court with a recommendation for training. It can deal with any offences except murder.

In certain areas of the law there is a need to treat minors, that is persons who have not yet reached adulthood, in a way that is different from persons who are considered to be of full age and capacity. The age at which an individual becomes an adult is, for legal purposes, 17 years. Any person who has not yet reached this age is a juvenile or young person.

A child under 10 years of age is not criminally liable for any offence and therefore will not be prosecuted. A child between 10 and 14 years old is presumed by the law to be incapable of forming an intention to commit a crime, but if the contrary is proved, he may be found guilty of the crime. Over 14 years of age a young person is considered old enough to be liable for his criminal behaviour.

Security officers who find persons on their premises, who are under 17 years of age, must understand the difficulties of dealing with such persons. The offence committed by the youth(s) may be clear, but there are juvenile panels who decide whether or not to take the young person to court, and often the police will give the person a 'caution', to guide him out of crime.

If the person under 17 years old is with an adult (over 17 years) he can be taken with the adult to the Magistrates Court, but again in practice the young person is remitted to the Juvenile Court for sentence.

*Summary*:
Under 10 years – not responsible in law
10–14 years – a juvenile
14–16 years – a young person
Over 17 years – an adult

There is little which can be done with the under 10 years youth. Hopefully his father will deal with him properly.

### *The Administration of Justice*

It is desirable that anyone concerned in the investigation of an offence against the criminal law, or in preparation of a prosecution, should know what would be acceptable to a court as evidence and what would be rejected. Moreover, a knowledge of what happens in a court, may prevent a security officer making a fool of himself to the detriment not only of himself, but also his profession – and possibly permitting an undeserved acquittal.

Industrial tribunals have greater discretion in the evidence they will accept, considerably more hearsay latitude is allowed, but in preparing a disciplinary case it would be well to look at the evidence to see whether it would stand up to the criteria of their criminal counterparts.

## *TYPES OF EVIDENCE*

### *Law of evidence*

This determines (a) what facts may be proved in order to ascertain the innocence or guilt of the accused person, and (b) how and by whom those facts may be proved.

The word 'evidence' means the facts, testimony, and documents which may be legally adduced in order to ascertain the fact under inquiry.

### *Direct evidence*

Evidence is direct when it immediately establishes the very fact sought to be proved and is normally what a witness saw or heard connecting the accused directly with the offence.

### *Circumstantial evidence*

Evidence is circumstantial when it establishes other facts so relevant to or connected with the fact to be proved that they support an inference or presumption of its existence. For example, if *A* was seen in a cloakroom by *B* to put his hand into the pocket of a coat which was not his and take out some money, later proved to have been stolen, what *B* could say would be *direct* evidence. If, however, *B* saw *A* standing beside the coat from which money was later reported stolen and *A* was unable to account for his possession of a similar amount to that stolen and denied his presence in the cloakroom at the relevant time, what *B* could say would be *circumstantial* evidence which in association with the other evidence of finding the money in *A*'s possession, and what he had to say, would go towards proving his guilt.

### *Hearsay evidence*

Where a witness cannot give direct evidence of a fact. For example, *B*, who saw *A* steal the money, tells *C* what he saw; *C*'s evidence of what he was told would be hearsay and would not be allowed to be given in court. The evidence can only be given by *B*.

Another example is where a document has to be produced to prove a fact in a prosecution. It is not acceptable for it to be done by someone who did not prepare it. For example, the fraudulent addition of figures on a consignment note (see Chapter 28) would require evidence from the actual clerk who wrote the original figures and who would say that since he did so they have been altered by the addition of other figures thus inflating them.

### *Opinion*

As a general rule a witness must give evidence of facts within his knowledge and recollection and not of his opinions. It is for a court to draw opinions from what he has to say. The exception is where specially-qualified experts are called to give their opinion on, for example, medicine, art or foreign law. Fingerprint officers giving evidence must satisfy the court as to their experience and expertise before they are allowed to be heard.

### *Corroboration*

By English Common Law the evidence of one competent witness is enough to support a verdict in either criminal or civil proceedings (not in Scotland). However, by statute, corroboration is demanded in a variety of contexts, and is also considered desirable by the common law in broad categories of situation where a court receives evidence from what are traditionally considered the 'suspect' classes of witness – namely, accomplices, sexual complainants, children and persons of admitted bad character. Corroboration has been authoritatively defined as:

> Independent [evidence] which affects the accused by connecting or tending to connect him with the crime. In other words, it must be evidence which implicates him, that is, which confirms in some material particular not only the evidence that the crime has been committed but also that the prisoner committed it (*R* v *Baskerville* [1916] 2 KB 658, 667 *per* Lord Reading, CJ).

or, more simply by a later authority:

> Nothing other than evidence which confirms or supports other evidence. That is in short evidence which renders other evidence more probable (*DPP* v *Kilbourne* 1973 1 All ER 440).

### *Character*

In general terms, evidence that an accused has been previously convicted of a criminal offence is not admissible in evidence until after he has been found guilty. It is then helpful to the court to know his background in deciding on the most appropriate punishment. However, if a person of bad character gives evidence of his good character the prosecution may, in rebuttal, give evidence to the contrary *before* the court arrives at its decision. Also, at the discretion of the trial judge, if the character of prosecution witnesses is attacked by the defence, then the defendant, should he give evidence, can be questioned about his character.

## *GIVING OF EVIDENCE*

A witness may not read his evidence from a document. He may, however, refresh his memory by reference to any memorandum if made at or about the time of the fact to which it refers and either made by himself or seen by him while the facts were fresh in his mind and then recognised to be correct. If this reference is done in the course of giving evidence, it will usually be from a notebook which the defending solicitor will have the right to inspect and to cross-examine on, in so far as it relates to the evidence for which the reference was made. Hence the importance of keeping a personal notebook in a manner which will render it acceptable to a court (see Chapter 8).

## *WITNESS SUMMONSES*

Where a witness required by either the prosecution or defence is not willing to attend court to give evidence, an application is made to the court for a summons. This is served personally on the witness and calls on him to attend the court issuing the summons at a specified time, pointing out that intentional failure to do so might lead to an issue of a warrant of arrest under which he will be detained until the next hearing of the court.

If a person required to give evidence attends court but refuses to give evidence this amounts to 'contempt of court', which can be followed by his committal to prison until he is prepared to give evidence. These circumstances rarely happen but the power to deal with recalcitrant persons is there if required.

The most common instances where witness summonses are used are where bank managers are required to give evidence of the accounts of their customers.

## *EXHIBITS*

These are tangible objects produced in court which go towards proving the guilt, and sometimes the innocence, of the accused person. They are evidence in themselves but they have to be produced in court by witnesses who speak of how and when they were found and explain their connection with the accused.

An exhibit can be a weapon alleged to have caused an injury which is the subject of a charge, for example a blood-stained knife in the case of murder or causing grievous bodily harm. Other exhibits could be hair of the injured person found on a weapon in the possession of the accused; safe ballast found on gloves or other clothing worn by a safe-breaker at the time of his arrest; the written statement alleged to have been made and signed by the accused; or, the most straightforward of them all, the property concerned in the charge of theft subsequently recovered in whole or in part.

### *Electronically-produced exhibits*

Sound and video recordings are now acceptable in evidence (Police and Criminal Evidence Act 1984, s. 68). Tape recordings are treated in the same way as documents; the original should be produced if possible and proved by the person making it, copies will only be accepted in exceptional circumstances (*Grant* v *Southwestern and County Properties Ltd* 1974 2 All ER 465). Video recordings likewise to be proved by evidence from a competent witness (*R* v *Fowden and White* 1982 Crim LR 588) though later cases have shown more latitude (see p. 579).

These methods are going to be invaluable to prosecutors, and the publicity accorded closed-circuit television in dealing with soccer hooliganism has resulted in vastly increased use. They have proved their worth many times in the conviction of bank and building society raiders.

Documents in the form of computer printouts are also acceptable subject to conditions establishing validity (Police and Criminal Evidence Act 1984, s. 69) and correct use of the computer. Photographs too may be exhibited provided they are verified on oath by some person, not necessarily the photographer, who can testify as to their accuracy.

## *STATEMENTS OF WITNESSES*

In Chapter 8, reference is made to obtaining statements or proofs of evidence from witnesses who may be called upon to give evidence in the prosecution of the offender. The statements may contain hearsay which, as explained, would not be admissible in a court. This is sometimes necessary to explain the actions of a witness in relation to other persons to make a complete story of what happened.

### *Statements in lieu of oral evidence*

Since the passing of the Criminal Justice Act 1967, a revolutionary change has taken place in the administration of the courts. Whereas until then only a witness in person could be heard, evidence can now be adduced through the reading of a written statement from a witness, subject to certain conditions. This requires the editing of written statements to remove hearsay or other passages which would not be admissible if the witness were giving evidence orally.

Section 2 of the Act says that in committal proceedings (that is the hearing in a Magistrates Court of an offence which, if a prima facie case is made out, will be committed for trial at the Crown Court) a written statement of any person shall, if the following conditions are satisfied, be admissible to the like extent as oral evidence:

1 The statement purports to be signed by the person who made it.
2 The statement contains a declaration by that person to the effect that it is true to the best of his knowledge and belief and that he made the statement knowing that, if it were tendered in evidence, he would be liable to prosecution if he wilfully stated in it anything which he knew to be false or did not believe to be true.
3 Before the statement is tendered in evidence, a copy of the statement is given, by or on behalf of the party proposing to tender it, to each of the other parties to the proceedings.
4 None of the other parties, before the statement is tendered in evidence at the committal proceedings, objects to the statement being so tendered under this section.

The following provisions have to be complied with:

1 If the statement is made by a person under the age of 21 it shall give his age.
2 If it is made by a person who cannot read, it shall be read to him before he signs it and shall be accompanied by a declaration by the person who so read the statement to the effect that it was so read.
3 If it refers to any other document as an exhibit, the copy given to the other party shall be accompanied by a copy of that document or by such information as may be necessary to enable the party to whom it is given to inspect that document or a copy thereof.

Although a written statement may be admissible, on a court's own motion or on the application of any party to the proceedings, the person making it may be required to attend the court and give evidence.

Section 9 says that, in any criminal proceedings in a Magistrates Court where it is competent to proceed to a conviction or dismissal, a written statement of any person shall, if conditions which are basically the same as required by s. 2 are satisfied, be admissible in evidence to a like extent as oral evidence.

There are certain administrative requirements which are not likely to affect others than the police or court officials.

If a witness wilfully included in his statement anything he knew to be false, or did not believe to be true, he would be liable to prosecution under s. 89 of the Act and on conviction would be liable to a term of imprisonment not exceeding two years, or a fine, or both.

### *Statutory declarations*

These are of importance in cases where charges are based upon transit losses from firms. Under the Theft Act 1968, s. 27(4) and subject to certain conditions, in any proceedings for the theft of anything in the course of transmission (by post or otherwise) or for handling stolen goods from the theft, a statutory declaration made by any person that he dispatched, received, or failed to receive any goods or postal packet, or that any goods or postal packet when dispatched or received by him were in a particular state or condition, shall be admissible as evidence of the facts stated in the declaration.

This is a valuable facility for the saving of witnesses' time at court on matters unlikely to be seriously contested.

### *Refreshing memory from statements*

Home Office Circular No. 82/1969 to Chief Officers of police says:

> . . . the Secretary of State is able to commend for adoption a revised practice which has the approval of the Lord Chief Justice and the judges of the Queen's Bench Division. This is that, notwithstanding criminal proceedings may be pending or contemplated, the Chief Officer should normally provide a person on request with a copy of his statement to police. . .

This procedure was confirmed in *R* v *Richardson* (Court of Appeal) 1970 2 WLR April 1971 where witnesses were allowed to see their written statements to police made 18 months before they were called to give evidence in accordance with them. Richardson appealed against conviction on the grounds, *inter alia*, that as the prosecution witnesses had been allowed to refresh their memories through this means their evidence was inadmissible.

The Court of Appeal held, dismissing the appeal, that there was no general rule that prosecution witnesses might not refresh their memories before trial from statements which they had made non-contemporaneously but near the time of the offence.

This is an important privilege and where a witness of fact is uncertain through passage of time what he said when his statement was taken he should ask to be allowed to refresh his memory before giving evidence.

## *PROCEDURE IN COURT*

### *Examination-in-chief*

After taking the customary oath or making affirmation the witness gives his evidence, in which he will be assisted by the lawyer representing the side for which he is appearing. This is called examination-in-chief.

### *Cross-examination*

The lawyer appearing for the other side then has the right to ask the witness questions in an attempt to negative or diminish his evidence by casting doubts on his memory, hearing, or accuracy, etc. This is called cross-examination. After any cross-examination the lawyer representing the witness's side then has the opportunity to re-examine in order to explain or remove ambiguity in any answers given by him in cross-examination.

### *Leading questions*

Leading questions are not as a rule permitted. They are questions so framed as to suggest to the witness what answers are required. For example: 'Did you see *A* take some money from the coats in the cloakroom?' will be disallowed. The proper question would be: 'What did you see *A* do?' The exceptions are where lawyers, to expedite the giving of non-controversial evidence, agree that the one tendering the witness shall, with the permission of the judge or magistrate, make use of leading questions.

### *Perjury*

This offence is committed by a witness who gives, under oath or affirmation, evidence which he knows to be false or does not know to be true and which is material to the proceedings.

*Practical advice on deportment and the giving of evidence*

It has been said that the degree of credence that a jury or judge places upon a witness's evidence depends on his:

1 knowledge of the facts that he gives;
2 impartiality;
3 apparent truthfulness;
4 appearance of integrity;
5 acceptance of the binding nature of the oath or affirmation that he has taken.

These are points that any witness would do well to bear in mind, but the value of his evidence can be jeopardized by the manner in which he gives it and the way he reacts to questions designed to confuse and occasionally annoy. The role of a defending solicitor or barrister is to secure the acquittal of his client or minimization of any sentence that might be imposed. To do this, in the case of a 'not guilty' plea, he must start from the premise that his client is innocent and his explanations, no matter how ridiculous, are true – which occasionally has caused an advocate presenting them to squirm physically as well as mentally!

An old dictum still remains valid – 'Stand up, Speak up, Shut up'. Security officers are regarded as disciplined and responsible individuals; they should therefore appear in a court tidily and quietly dressed, standing up straight while giving evidence and not leaning over the edge of the witness box. 'Straight' does not mean rigidly to attention, but upright and at ease.

Practically all evidence will be given in the form of answers to questions; do not over-expand when replying, confine yourself to the subject matter about which you are asked; do not withhold information which may be in favour of the accused; if in doubt about a question, ask for it to be repeated or clarified. During cross-examination, remember that in very many cases, yours will be the evidence it is essential for the defence to disprove or at least cast doubt upon. If you can be irritated into arguing, something may be said which allows the defence to pick on a triviality and magnify it into a massive red herring to divert attention away from the facts that really matter. If you are led into making a mistaken reply and challenged, at once admit the error; be scrupulously polite in giving your evidence, no matter how unfair you think the line of questioning.

There is an element of the theatrical in the approach of some advocates – designed to intimidate and worry. A stern accusation, followed by a 'well we shall see' when refuted, is more likely to portend that that line of attack is to be abandoned as non-productive, rather than pressed; a period of silence spent apparently looking into infinity is indicative of wondering what to say next, rather than being about to administer a *coup de grâce*! If your evidence is true, straightforwardly given and without embellishments, it will not be shaken and it is for the defence to worry.

Since October 1986 when the new Prosecution of Offences Act 1985 was commenced, all prosecutions in England and Wales have been conducted by the Crown Prosecutors. These are lawyers who decide all cases. The police no longer have any right to decide whether or not to prosecute a case, the decision is made by the independent prosecution service. Before deciding to prosecute the lawyer takes into account: the sufficiency of the evidence available; and the likely penalty, the staleness of the case, the complainants attitude, and the personal circumstances of the accused persons infirmity or mental stress. The Crown Prosecutor can decide to discontinue with any proceedings, he can select the charges to be proceeded with, and he can advise the court regarding the mode of trial, either at Magistrates Court or at Crown Court. He is bound by a Code of Practice formulated under this Act.

## *DECIDED CASES AND APPEALS*

Few contentious issues are brought before a court of law which have not been the subject of previous legal decisions. Broadly speaking those precedents, if made by a sufficiently authoritative court, are binding in similar circumstances thereafter and are therefore frequently quoted by lawyers during court hearings. They are also well documented in textbooks such as *Archbold* to which a security chief would do well to refer if faced with a tricky legal point.

If the decision of a trial court is disputed, the defence may appeal to a superior court against the sentence or conviction. The appeal may be lodged even if the defendant has pleaded guilty, if it can be shown that he did not appreciate the nature of the charge or pleaded guilty by mistake or that the facts were such that he could not legally have been convicted. Otherwise the application can be based on questions of law or mixed law and fact. The prosecution has limited powers of appeal based on points of law.

Whilst the Court of Appeal considers itself bound by its own decisions in civil matters, in criminal appeals it may, on rare occasions of misunderstanding or misapplication of the law leading to the earlier decisions, reverse them. The House of Lords may be the final adjudicator in matters of general public interest.

Case law or decided cases therefore become an interpretation of the law of the land. In legal reports and textbooks they are shown in the form *R* v *Rogers and Tarran* 1971 Crim LR 413. In this *R* stands for Regina (or Rex); and the names of the co-defendants, the year of hearing, and the journal in which the case is reported – in this instance *Criminal Law Review* – are shown.

## *LEGAL PHRASEOLOGY*

Certain legal terms and phrases of Latin origin are in regular use in law books and reports, verbally in the courts, and occasionally in letters from solicitors claiming damages in industrial accidents. It is as well to have knowledge of at least the most important of these, omitting those which are now in everyday usage, for example bona fide, ex gratia, etc.

| | |
|---|---|
| *actus reus* | guilty act |
| *ad hoc* | for this purpose |
| *de facto* | in fact |
| *inter alia* | amongst other things |
| *ipso facto* | by the mere fact |
| *mens rea* | guilty mind |
| *modus operandi* | method of operation |
| *non sequitur* | it does not follow |
| *obiter dictum* | comments of a court which do not form the basis of its decision (not binding as a precedent in themselves) |
| *per se* | in itself; taken on its own |
| *prima facie* | on the face of it |
| *ratio decidendi* | the reason for a judicial decision |
| *res gestae* | the facts surrounding a transaction |
| *sine die* | indefinitely |
| *sub judice* | under judicial consideration |
| *ultra vires* | beyond the powers |

*PART THREE*

# *SECURITY IN OFFICES AND SHOPS*

# 15

# *Wage protection*

Hard cash needs no criminal disposal avenue, being portable, immediately usable and, more often than not, impossible to identify. Wherever it is held or handled in bulk it must be regarded as a priority risk to be protected against the possibility of planned and violent attack.

As in many other departments there is a tendency for cashiers and wages-handling staff in general to become complacent about their role. They constitute perhaps the most logical target for both internal dishonesty and attack by outsiders, and this is not brought home to them until a colleague is found to have been systematically stealing or defrauding or until laxity in arrangements has permitted a sudden violent raid which has resulted in severe loss, possible injury, and certainly a long-lasting shock to those involved.

There are several facets to this risk which should be separately examined for weaknesses and loopholes. In terms of sheer quantity, wage protection must be given the priority.

The collection and handling of wages are matters in which insurers have a direct interest; premiums, and indeed the availability of insurance, will depend upon their approval of the methods used. If it is practice or intention to use company staff to fetch the wages from a bank, this must be made known to the insurers so that they can have the opportunity to impose any conditions that they think appropriate. If the sum involved exceeds what they deem to be an acceptable risk they may possibly require special precautions by way of escort and regular route and vehicle change. For sums grossly in excess, cover will probably be refused. Obviously, from their point of view the use of a commercial cash-carrying company has major advantages, not the least of which is that the danger period for wage thefts in the past has been 'in transit' during which the carrying company's insurers would be responsible. However, modern trends are to attack as or immediately after wages have been delivered to premises, though the frequency has fallen off as firms have become alerted to the risk. Cash in bulk is rarely recovered and more often than not there is a considerable time lapse before the criminals are caught; professional gangs, operating well afield from their bases have been responsible for many raids.

Attackers in these instances are almost invariably armed with sledge-hammers, shotguns, pick-axe handles, or other dangerous weapons which by and large they will not hestitate to use since prison sentences for armed robbery are long in any case and there is no deterrent death penalty. Wage protection measures must take this into account as employers cannot disregard their legal requirements where there are foreseeable risks connected with the health and safety of their employees engaged in wage handling. Indeed, an improvement notice under s. 2(1) of the Health and Safety at Work Act 1974 was served on a Midlands Building Society to provide anti-bandit screens at one of its branches and was upheld at an industrial tribunal, the grounds being that 'the staff were not protected so far as is reasonably practicable'. However, an appeal to High Court, Queen's Bench Division (*West Bromwich Building Society* v *Townsend* 1983 IRLR 147) quashed the tribunal finding on the grounds that there was no contravention of a statutory duty – 'that a precaution is physically possible does not mean it is reasonably practicable'.

If despite all the arguments to the contrary the company decides to use its own staff, or where the amount involved is too small to justify a carrier, consideration should be given to using specially designed and alarmed bags (which shoot-out telescopic arms, emit coloured fumes, siren noises, etc.) or waistcoats adapted to conceal what is being carried. Times, routes, vehicles, etc., should be varied and only able-bodied youthful persons used, preferably in pairs. They should be instructed to avoid side streets, keep away from kerb edge, etc. if part of the journey must be made on foot. Local police crime prevention officers will be more than willing to advise in detail on the various precautions that can be taken.

## *COMMERCIAL CARRIERS*

There are several stages to wage protection, the first of which is deciding how the money shall be brought to the premises. Wherever possible the services of a cash-carrying company are advisable because it is equipped and geared to the function with specially-constructed vehicles which have full radio communication and also, as mentioned earlier, a major risk for company employees and own insurers is thereby eliminated. Competitive quotes should be obtained from the commercial companies operating in the area which are of nationwide repute. Their quotations will vary only a little but may depend upon the local availability of their services and depots. In addition some may not be able to conform to the required delivery times or have the ancillary service of recollection and overnight storage which may be required at holiday periods or in other special circumstances. They often provide the additional facility of wage packeting to reduce the necessity of calling in other departmental staff to supplement wages office staff for making-up purposes. The carriers have special counting equipment and staff trained for the purpose, with the result that charges per packet are usually quite reasonable. This service, of course, also limits the length of time for which the money is readily to hand during counting and make-up.

In arrangements with the cash-carrying company, ensure that insurance responsibility at all stages is quite clearly established and that your own does not commence until the money has been actually signed for – that is, not simply on delivery to the firm's premises. The agreement should in fact show the actual delivery point. It is perhaps trite to say that no signature should be given before checking amounts, but this is known to happen. Each commercial firm will have its own system of identifying employees for cheque receival or cheque submission to the bank, and it can be anticipated that this aspect will be adequately documented. If possible the carrier should collect the cheque from the firm's premises rather than have an employee take it to the bank where there may be a chance of attack involvement. The internal and external delivery route should be discussed with the cash carrier's representative to note any special precautions that are necessary. They are fully aware of the dangers, and several have in the past refused to accept contracts where the recipients were not prepared to take measures to reduce risk to the carrier's staff.

## *WAGE OFFICES*

Certain principles must be borne in mind in designing a wage payout office or modifying an existing one. It is most important to plan on the supposition that an attack *will* be made on the premises and not to regard this eventuality as a remote possibility. The following principles should be considered:

1. Forcing entry must be made a noisy and difficult operation.
2. There should be no clear view from the outside of the occupants, where they are, and what they are doing.
3. There should be no opportunity for an attacker to threaten occupants from outside to cause them to open doors or windows.
4. Windows, particularly external windows, should be fully protected against entry by force, for example by sledge-hammers.
5. A silent electronic alarm system should be installed in the wages office, connected to a constantly manned internal point or to the police if they so agree. Staff should be fully conversant with how it operates and what it does.
6. A sequence of action should be laid down to follow alarm operation.
7. Apart from wages office doors having substantial locks, the doors themselves must be adequate to withstand sledge-hammer attack, for example steel faced.
8. Payout windows should be as small as acceptable and protected by overlapping bullet-resistant glass so that the clerks cannot be threatened. Packets can be fed out via a 'scoop' cut in the wood of the base so that the glass gives complete cover.
9. Inspection devices for callers may be fitted in the outer door, or a system may be established of prior notification at a window before access is allowed.
10. Do not overlook employee amenities in any new building, for example ventilation and heating.

## *ALARMS*

There is a division of opinion as to whether a silent or an audible attack alarm should be used. It has been said that the audible will scare away attackers without harm to the employees. However, this prospect is no consolation to those who might be killed or injured – the coincidence of a slight movement by one of them with the sounding of an alarm might well induce immediate reprisals from an attacker, and this is just not worth risking. Apart from that, any planned countermeasures are likely to be aborted by the alarm sounding, and set plans to prevent escape may be nullified by the taking of hostages or very hurried departure.

It will be noted that a sequence of action should be laid down to follow alarm operation; in formulating this a first consideration must be the safety of employees of all kinds. Apart from prompt notification to the police it may be possible to lock doors or gates on an escape route and to deposit vehicles so as to block off escape roads. There is an inevitable delay in police attendance which will depend upon the location of the works and the involvement of their other resources, so that any hampering measures will be welcome and may result in attempts to escape upon foot or even abandonment of the proceeds. Such plans must be discussed with the police so that they are fully aware of what is likely to happen independently when they are notified, and undoubtedly they themselves will have guidance to give on the subject.

## *RECEPTION OF WAGES*

Certain steps should be taken before the arrival of wages. Those recommended are:

1 Before the arrival of wages on site the external delivery point, passages, toilets, etc., leading to the wages office should be checked to ensure that there are no intruders about and no suspicious circumstances.
2 During wage delivery, if undue interest in the proceedings by the occupants of any vehicles or bystanders is noticed, a note should be taken of pertinent details about them, especially vehicle numbers. There should be no hesitation in passing this information to the police even if it is subsequently found worthless.
3 At the time of arrival, if there is the slightest cause for suspicion, the carriers must be warned to stay in their vehicle until doubt is alleviated.
4 If a works site is concerned, once the carrying vehicle has entered the perimeter then, if at all feasible, the gates should be locked behind it until the payroll is known to be safely in its destination.

Of course, these are points which should have been discussed with the carrier's representative in the initial discussions.

## *PAYOUT OF WAGES*

The wages office staff have a prime responsibility for getting the wages to the respective employees correctly and as expeditiously as possible, but they should also be indoctrinated with the necessity for security measures with emphasis not to disregard them on grounds of expediency or inconvenience. This should not be difficult; security can be made part of a routine which they will appreciate is designed for their own safety.

Matters of internal practice that do not always receive the attention that they merit are:

1 Wage packeting should be such that an employee can check the contents without opening, and it should be clearly laid down that a cash shortage complaint will not be entertained if the seal has been broken. (Note that with some packets it is possible to 'roll' notes on to a thin pencil or similar cylindrical object and abstract them without outwards signs; the solution is to staple through the notes and packet.)
2 A wage packet must not be issued other than against a signature or other accepted form of receipt – for example, pay slip, authorization note from sick employee, or clock card. Bulk collection for a department or shift must be checked and signed for against a list of those concerned, and individual signatures or the equivalent must be returned in proof of delivery.
3 Packets pending issue must be out of reach of everyone but the wages clerk. It must not be possible for an employee to reach over and steal packets if the clerk's attention is momentarily distracted.
4 Entry into the wages office by any level of employee should be discouraged to the point of their being banned from wage reception time up to payout time. Doors must be kept locked and/or bolted throughout.
5 Any complaints of shortages or wrong collection should be investigated at once if at all possible, not deferred until a long payout is completed.
6 Wage payout should be made as expeditiously as possible. If it is necessary to retain money overnight or pending distribution to say a night shift, a safe or other container approved by insurers should be used. If the amounts are above the acceptability rating of the safe (insurers have a table of types against their permitted cash-holding capacity), consideration should be given to recollection and redelivery by carriers. (Note that purchase of new safes must be discussed with insurers.)
7 Custom-and-practice bulk collections which cannot be revised

should be checked to ensure that there are no security loopholes and that persons of appropriate status and repute are used.

8 If there is a handover between clerks during payout, the person accepting the boxes or envelopes must check and sign accordingly.

9 Safe key or combination holding must be kept to an absolute minimum, and this must not be extended on the grounds of temporary convenience since arrangements like that are apt to become permanent. It is good practice to lodge duplicates at the company's bank.

## *UNPAID WAGES*

For a variety of reasons, packages containing unpaid wages may well accumulate, and these constitute a hazard in so far as they may put the cash holding of a safe beyond what is acceptable. The longer they are retained the greater the temptation to the staff handling them. It has not been unknown for IOU's to be put inside wage packets awaiting collection (which hardly helps to improve evidence of intent to steal!), and false signatures have appeared in unpaid wages books. Whilst money is resting in packets for an indeterminate period it is money dormant and unproductive. This is a factor which should be discussed with a view to rebanking the money in the firm's account and making subsequent payment from petty cash against record and signature in an unpaid wages book. Retention for one month should be a mandatory maximum.

## *SECURITY CASH AUDIT*

Periodically in every group or organization it is advisable to carry out an audit of what exactly is being done in connection with money, held or obtained for a purpose. This is also of advantage to accountants, cashiers, and internal auditors and can result in minor malpractices being cancelled out before losses are sustained. A specimen *aide-mémoire* and report form are shown in Appendix 17.

It is essential that such an audit results in a report showing either security-acceptable practices or explanation why shortcomings in particular features have to be deemed acceptable risks. Copies must be served on departmental heads with one to the director responsible for finance on which is shown the circulation list. The object of this is to inhibit any subsequent changes of procedure without discussion and notification to the director. An instance of what might happen occurred when a chief cashier removed a check on the payment of supplier accounts which enabled a junior clerk to quickly pay at least £89,000 into an account for herself and her husband. The change was ostensibly for economic savings! The chief cashier's advancement came to an abrupt halt.

It will be noted that one of the points made in Appendix 17 relates to the loss of wage packets or alleged wrongful collection. It must be made quite clear in works rules or by notices at the pay stations that the responsibility for a wage packet is the collector's. Where a wrongful signature is alleged it must not be forgotten that collections for a person alleged to have suffered in this way often exceed the amount missing – a fact which has not been lost upon persons whose gambling or other propensities have accounted for the wage packet before it has actually been received!

# 16

# *Cash handling and cheques*

## *CASHIER'S OFFICE*

In most organizations the offices used for payout purposes will not be those used by the cashier's department in its day-to-day routine. In the cashier's office different requirements of accommodation, filing, clerical support, etc. apply, and there will be the need to maintain a ready, though much reduced, constant supply of cash. There is less necessity to be readily accessible to the workforce, and a degree of safety can therefore be incorporated by its location well inside an office block. Despite the reduced amount of cash held, the fitting of attack buttons also in the cashier's office would help staff indoctrination to the existence of risk, do their morale good, and give them confidence. There should be at least two such alarm points, well apart and placed near the employees who are least likely to panic.

The best location for the office will be a compromise between the needs of administration, access by users, general office availability, and security considerations. Ideally it should not be on the ground floor nor directly located near stairs, lifts, or access doors from the outside. The objective is to complicate matters for an intending intruder without constructing a Fort Knox complex. The compromise between convenience and security must be accepted as a legitimate risk to be mitigated.

Provision should be made for installing an adequate safe within the office area – there is no sense in having the main cash storage elsewhere, involving carrying money backwards and forwards when the office opens or closes. This postulates another requirement: the floor strength must be adequate to take the weight of the safe, not just in its intended position but also along the route needed to get it there. At the risk of repetition it must be emphasized again that the insurers must be consulted about the choice of safe in the light of the maximum sums of money intended to be placed in it. If burglar alarms are to be fitted to the office or to the safe itself, the protective value of these will be taken into account by assessors in fixing the maximum amounts permitted to be held.

The actual siting of the safe will to a large extent depend upon the general layout and floor strengths, but it is sensible that it should be in full view and not in a separate small room – unless there is an inbuilt strongroom or fire-resistant room. Whilst it should always be kept locked, inevitably during working hours there will be a tendency occasionally to leave it temporarily open; if it is where everyone can see it, the chance of an unauthorized member of the staff taking advantage of this lapse is much reduced. Whether there is virtue in leaving a light burning over the safe will depend upon a number of factors, the main one perhaps being whether the premises are being patrolled; if so, the light and a clear inspection panel in the door or window will facilitate checking by the patrolling man and be an embarrassment for anyone making a sustained effort to break into it. Types of safe, anchorage, and other matters relating to them are dealt with in Chapter 38.

A facility is needed for employees' queries to be dealt with and payments to be made without the individuals coming into the office proper – for example, by a hatch in the door, enquiry window in a wall, or segregating counter within the office.

The office is a likely target for after-hours intruders, and apart from its siting the precaution should be taken to locate it in a room(s) with solid walls and substantial doors with security locks. If considerations dictate that electronic alarms should be fitted to risk areas on premises, these offices should be accorded a high priority. There is merit in applying the shatter-resisting adhesive plastic sheet now available to the windows of the offices, partly to hinder entry and partly to obscure an outsider's view of the routine business of the office.

## *PETTY CASH*

This is held to meet a variety of contingency needs including expenses, advances, minor items of expenditure, stamps, employee cheque cashing, repayment of unclaimed wages which have been banked, etc. The danger is that a cashier will, for convenience sake, hold considerably more money than is strictly necessary. Payment is normally a matter for the cashier himself or a deputy in his absence; the actual recording is probably done by a member of his staff based upon various receipts submitted.

As always, where there is ready cash there is temptation for petty frauds or theft – if due care is not exercised in the claiming and recording systems and physical storage. Where there is an internal audit department their checks will throw up discrepancies, but not necessarily frauds arising from lax managerial control. The least practical number of senior management executives should be nominated to authorize the payment of expense claims. In a small or medium size factory, this would be the factory manager; in the case of a large departmental store, the store manager; in a large organization, ideally each director or executive in charge of a division or department. The responsibility for ensuring that claims are correct and reasonable lies strictly upon the person authorizing them, but he may be too far removed from the purpose for which the cost was incurred. It is advisable therefore that claims by an individual should be initialled by the person to whom he is responsible before being submitted for authorization. Particularly because of VAT returns, receipts and vouchers should be insisted upon; where this is not possible an itemized list should be required, not simply a bulk sum.

Amongst the frauds that have come to notice are mileage claims in respect of journeys not carried out, or grossly inflated; false petrol or parking receipts; forged meal chits; claims for entertaining for which a customer or supplier has paid. In fact the list is endless and limited only by the blank receipts which an individual can obtain – or the benevolence with which management is prepared to regard such petty dishonesty. Internal audit's main preoccupation must be with the reconciliation of figures, but a security chief should, on glancing through the monthly pile of expense vouchers and receipts, by his personal knowledge be able to pick up any pattern of deceit – for example, claims for 'out of hours' visits to premises which have not actually taken place (there is virtue in the recording of such visits to premises by senior personnel) and petrol receipts from garages outside the area of travel.

Auditors will lay down the precise manner in which the petty cash register should be kept. Basically each receipt should be numbered sequentially and separately recorded on a monthly sheet, both of which should be retained for an agreed number of years. The records should be checked against the cash balance at intervals of not more than one month, and the person responsible for the checking should not be either the person paying out or the individual keeping the records. In addition to this regular check and to spot checks by auditors the security manager, if he is of accepted seniority, may include this amongst his responsibilities.

Apart from expenses, all payments made from petty cash must be backed up by receipts of one kind or another, plus an authorizing departmental manager's initials where necessary.

Businesses of any size nowadays tend to use franking machines rather than purchase stamps; but in smaller units this is not financially advantageous, and money has to be made available from the petty cash. It is a managerial decision as to whether a postage book will be kept to validate stamp usage. This does take time, and a loss from potential petty dishonesty may be regarded as too trivial to justify the cost. However, there are advantages in keeping such a book since it does record the dispatch of letters which may be the subject of future complaints and queries. If kept, it should contain a record of cash expended on stamps, balanced against letters dispatched, and should be checked at monthly intervals.

### *Obtaining and holding petty cash*

There is no sense in making separate withdrawals from the bank for petty cash. Weekly amounts over a period should remain reasonably predictable, and the money can be obtained simultaneously with wages. Where possible an effort should be made to pay out the larger claims from petty cash before weekends so that a minimum amount is left in the office safe – that is, representatives' expenses and employee cheque cashing. It is desirable that the petty cash should be kept separate from other monies, and the person responsible for it should have a separate locked cashbox to be held within the main safe.

In smaller installations – for example, stores, shops, warehouses, and sales offices – divorced from the larger units the local manager must be responsible, and it will be his responsibility to ensure that any monies on the premises are actually at a minimum. In some circumstances it might well be that a sum of £100 would be more than adequate. This would not justify expenditure on a safe, but a cashbox would remain essential; this should be secreted in a locked cabinet and concealed as best possible. Alternatively, the standard form of small floor safe set in reinforced concrete is a reasonable buy and would prove more than adequate for the risk.

The overriding essential in respect of petty cash is that those who are responsible for dispensing it, keeping records, and requisitioning for it must know that their stewardship will be subjected to regular checking, augmented by spot checks. There is virtue in a regular scrutiny of monthly receipts since this can throw up a pattern of claims which, although apparently innocent and acceptable in themselves, become distinctly suspect when viewed as a whole.

Appendix 18 lists points of potential weakness in internal cash handling.

## *RECEIPT OF CASH AND REMITTANCES*

The nature of the business will dictate the degree to which precautions are required; and ready-cash situations, as in shops and supermarkets, are dealt with as a separate and perhaps more complicated problem.

Payments to industrial and commercial firms are likely to be made by mail, subsequent to invoice. It is refreshing that nearly all firms have specific arrangements, with reliable personnel, to ensure that mail is safely and quickly passed to the designated recipients.

The cashier's department is unlikely to be involved in the opening of mail, and indeed should not be so. It is advisable that at least two persons should be concerned, one preferably of supervisory grade. Letters addressed to the accounts department or another recipient designated on the invoice should be delivered there direct and unopened. Others found to contain cheques or other forms of payment should be transferred to the accounts department by hand, from those opening the mail. Again, it would be surprising if an adequate system were not in force since this constitutes the life blood of any company.

The conventional ways of paying invoices strictly limit the possibilities of frauds being perpetrated by the employees handling them. Nevertheless there have been instances of crass carelessness, not only at a firm but at banks also, whereby the former has accepted a cheque addressed to someone else and the bank has paid it into the firm's account without examining it. The routine in an accounts department should be that the cheques or other orders of payment are compared with the accompanying documents, usually the invoices, immediately after the envelope is opened. The date, name of payee, written amount and figures, and the fact that the cheque has been signed should all be verified and the documents marked in some manner to show that the correct amount has indeed been received. The cheques should be stamped with the company's name and removed from the accompanying documents.

The cheque amounts and those on the documents should be separately totalled, and the respective totals should agree. The cheques with the addition list should be sent to the cashier, and the documents with their respective list should be sent to the sales ledger section to be dealt with.

*Receipts*

Official receipts, if used, provide a limited control over remittances, by comparison between counterfoils, or duplicates and cash book entries. Receipt books should be serially numbered, and any spoiled ones must be retained to assist reconciliation. Of course, care must be taken of receipt books, and they must not be readily available for misuse.

*'Bouncing' cheques*

Security involvement in these will depend upon the nature of the business. In some instances, where goods have been obtained by cheques subsequently dishonoured, the security officer may be called upon to act as an intermediary with the police if the circumstances merit prosecution. There is inevitably a time delay during which the firm's accounts department or credit control re-presents the cheque and enters into correspondence with the customer. The really financially painful instance where a firm may suffer is when it becomes the victim of a long-term fraud (see Chapter 21). By and large, though, in commercial and industrial practice most dishonoured cheques arise through errors in drawing them.

*Presigned cheques*

These are printed by a customer's bank and held there in bulk to be drawn on as required in the course of the customer's business. As a security protection against their misuse they are normally overprinted with 'Account Payee Only'. Subject to that restriction they are valid orders for payment bearing the facsimiles of the signature of the payer(s) with no amount or payee's name entered upon them. A subsidiary account is usually created for their use, and this is 'topped up' as needed. The same service is offered by the post office Giro system with an endorsement up to the maximum amount for which they can be encashed – a similar restriction is also conventional upon the banks' cheques.

It is not always appreciated that insurers may not regard such presigned cheques as being covered in a cash loss policy. For example, a firm pays its retired employees by presigned Giro cheques validated up to a maximum of £100; for convenience it wishes to keep a reserve stock of 500 cheques to cater for misprint eventualities via the computer. The only steps required to make these cheques cashable are to type in the name and address of the payee, the post office at which payment is desired, and an amount not over the prescribed maximum. The potential loss value of those cheques is therefore £50,000, and it would be most unlikely that they would be kept in anything like the conditions of security that the equivalent sum of money would be held. The risk in respect of presigned cheques limited to the account of the nominated payees might be considered less, but it is not difficult to envisage circumstances whereby false accounts are opened and used before stolen cheques are made out and presented. It follows therefore that every precaution must be taken to ensure safekeeping and accountability if this means of making payments is used.

Fortunately, it is practice now to pay outstanding accounts on fixed days of the month rather than on a day-to-day basis. This means that the required numbers of cheques can be drawn from bulk at a specific time, for example on perhaps two days in a month, which will facilitate record-keeping and checking. There must be a strict reconciliation of cheques used, cheques spoiled, and cheques returned on each occasion, particularly where computer printing is used. Locked safes or containers must be used for their safekeeping, and keyholding should be restricted preferably to one single responsible person, ideally the cashier. Yet again the difficulty is ensuring that the risk is recognized, and here the co-operation of the insurers could apply the necessary pressure.

### *Cheque books*

The theft of one of these could prove a source of embarrassment as well as loss, and they must be held under even stricter standards of security than are accorded to cash. No cheques should be signed unless there are supporting and authorizing documents upon which to base them. Spoilt cheques must be cancelled and retained for internal audit scrutiny and reconciliation to prove that they have not been misused. Each cheque book should be used to completion before another is started. The danger attached to presigning by one of several authorized persons has been exemplified recently by the executive who made such a cheque out to himself, provided the second necessary signature, and then disappeared with sufficient cash to last him for the rest of his life – thereby confirming that senior position does not equate with greater integrity, only greater opportunity!

### *Cheque-signing machines*

These machines require two keys which must be separately held: one secures the signature block; the other permits operation of the machine. A meter records the total number of cheques signed, and at the conclusion of each use the figure shown should be noted by both keyholders and checked against the list of signed cheques. Again, spoiled cheques must be endorsed with cancellation and retained. Prior to each operation the previously recorded meter figure should be checked. Duplicate keys must not be kept on the premises but should be lodged in bank custody. The signature block should also be taken from the machine and locked away.

### *Cheque cashing for employees*

This employee amenity can grow beyond what was originally intended and cause excessive amounts of cash to be drawn from the firm's bank at short notice to make up the petty cash so used up. This is not a matter purely for the goodwill of the cashier's department; it should be the subject of an established policy which specifies who can have the facility, what limitations are placed upon persons and amounts, and when it can be done. The latter is important; otherwise there will be unnecessary interference with the work of the department, especially by senior staff who wish it to be at their convenience. The practice of a department sending a junior with a number of cheques should be discouraged, as it creates temptation and could cause embarrassment if a cheque were to be refused or queried.

## *INTERNALLY-GENERATED CASH*

There are a number of internal sources from which a cashier will receive money. Some of these will be subjected to a double book entry for transfer into petty cash. Others, on the advice of auditors, may be kept as separate accounts – the canteen, in particular, if receipts from it are substantial. The cashier should keep a vigilant watch on these monies as they can accrue to the point where his cash holding is greater than is covered by insurance.

### *Canteen receipts*

Where contract caterers are concerned it is likely that they will be given a facility by the cashier's department to hold their day-to-day takings in the safe, banking once or twice weekly. This may raise insurance complications; and there should be an understanding of what liability, if any, is being accepted. The best solution is for the caterers to have a separate cashbox to which they alone hold the keys and which is lodged in the company safe without liability, all records being kept by the caterers.

### *Vending machines*

Increasingly, these are installed in business or commercial premises, either purchased by the occupier or simply installed and maintained there by a commercial supplier who accepts full responsibility and takes any profits that may accrue.

In the latter circumstances the security interest is in seeing that the machines are sited where they are least likely to be broken into and that the contractor's employees who enter the premises to service them are identifiable and aware of the requirements imposed upon them to report on arrival and observe the host firm's rules – including search if so desired.

If the machines are owned by the employer, they should be operated under the aegis of the canteen manager and a regular system of emptying, checking, and replenishing established. Inevitably there would otherwise be opportunities for defrauding, especially in beverage machines. Care should be taken in selecting the persons who are responsible for these, who should be long serving and reliable; this is a good post for disabled persons. Preferably they should operate in pairs, though this is not always economically possible. Keyholding for the machines should be restricted to the person who is responsible for opening them; and any duplicates should be carefully locked away by the canteen manager, not held by security. Records should be kept of the ingredients and articles used in the machines against meter readings etc., so that a reasonable assessment can be made of what cash returns should be expected for comparison with what actually accrues. The proceeds from these machines should be kept by the canteen manager as a separate record within his main accounts.

### *Staff sales*

The amount of money involved in these will depend upon the nature of the business; in retail establishments, for example, it will be quite sizeable as employees exercise their right to a percentage reduction in normal selling prices (see Chapter 6). The prime requirement is that a receipt is given for all payments received, either as a cash till register slip or as an 'authority to purchase order' on which is shown the amount of the payment required and which is stamped 'paid' before the goods are taken from the premises. Spot checks are desirable upon these purchases to prevent goods which have been trivially damaged or shop soiled from being sold at ludicrously low prices as favours to the relatives, friends, or acquaintances of the person giving authority. Such checks have further value in preventing the taking out of 'scrap items' which are actually normal material upon which a great deal of conversion work has been done to prepare it for a special use for the purchaser. The staff purchase concessions are a valuable ingredient in stopping temptation to steal, but they can be abused to provide an avenue for theft and fraud.

### *Sales of scrap materials*

In some industries, particularly the metal-using ones, this represents an opportunity for fraud which can reach higher levels of management.

Realization of the maximum value of these sales can affect profitability, but because the term 'scrap' is applied there is a tendency to regard this as nuisance material to be got off the premises as soon as possible and for whatever financial return as may be immediately forthcoming. The sums involved may be very substantial and paid in cash, so that a specific procedure should be laid down from the first approach to purchase up to the payment into the cashier's department.

Disposal authority should be centralized, not given to the individual who will come in direct contact with the purchaser. A fraudulent approach commonly found in the metal industry may be worthy of mention; it is applicable to other industries also. The proposed purchaser claims an urgent need for the material (for example, terrazzo flooring or 'earthing' material); he offers above the conventional price; the offer is at once accepted. To show his appreciation he makes a present to the person authorizing the purchase and also tips those who are involved in the loading. What follows depends upon his assessment of the people with whom he has dealt but perhaps mainly upon the attitude of the person authorizing the sale, who may himself broach the subject of making excess material available for a consideration. In any case, if the persons involved in the loading are substantially tipped, it is a short step to overloading and not charging. Once a bribe has been given and accepted, there is no limit to the consequences.

## *BANKING OF CASH*

Canteen proceeds in the case of a large company, or other monies accruing, may be sufficiently substantial to merit regular banking though not perhaps of such magnitude as to justify using contract cash carriers. If that were done there would be few complications; either the carriers could deliver a sealed bag to the bank containing a statement of contents, or other acceptable arrangements could be reached with them.

If the firm's own resources are used, the degree of security will depend upon the amount involved. The arrangements should be cleared with the insurers, who may in a law-abiding area impose no precautions at all for sums up to say £20,000 but may be much more demanding where there is a history of robberies. In any case, however, for the protection of employees it is advised that no fixed date/time pattern be followed and that the cash carrier be an able-bodied man – females or young persons should not be asked to do this. Unless the distance involved is negligible the transfer should be in a vehicle, preferably an employee's or a company car, and apart from the driver there should be another able-bodied companion for the man carrying the money. Under current conditions of extreme violence the practice of chaining a bag to the carrier's wrist is certainly not advocated, as it could cause serious injury for which any extra security for the cash would be small consolation. Carrier and escort should be dropped outside the bank, and before leaving the car they should look round to satisfy themselves that there are no suspicious types hanging about the bank frontage who might intercept them. It is better to drive around the block to check, and possibly feel silly at the sense of unease, than to risk an attack.

The specialized carrying bags mentioned when dealing with collections from banks can of course be used for delivery purposes.

## *CARRIAGE OF MONEY IN BULK*

If the sums of money involved are such as to compel a firm to modify one of its own vehicles to provide additional security, it is unlikely that the insurers will look upon this favourably. It has been known for them to impose conditions far beyond what they require of commercial carriers. It may be, however, that the firm carries its own insurance or that the folio overall is so large that the insurers can be persuaded to relax their antipathy to the principle. Another possibility is the intention to use a converted vehicle not just for bank delivery, but also for transferring wages after make-up to a series of payout points where it will act as a mobile wages office – as is provided by some commercial companies. If this is so, the following precautions should be regarded as a minimum requirement:

1 Provision should be made for fixing the cashbox or cash bag firmly on to the frame of the vehicle – for example, by eyebolts and a chain to pass through the handle or by welding in a steel frame or a special steel box.
2 The vehicle doors must be lockable from the inside, and provision

must be made to immobilize the vehicle immediately and totally in the event of attack.

3 An audible alarm should be fitted to the vehicle that is capable of operation from an internal switch.
4 If there is a wireless system in use for either security, internal transit, or another purpose, a two-way communications set should be fitted in the cash-carrying vehicle.
5 In the event of its use as a mobile wages office, a van should be converted so that payout can be made from it without opening the doors, for example through a protected window.

The use of a company vehicle presupposes that it will be seen on sufficient occasions to attract attention to what it is. In these circumstances, on public roads the crew should take precautions as would the crew of a commerical cash-carrying vehicle:

1 Vary routes and times (this should be more possible with the company's own vehicle than with a commercial one).
2 Do not leave the vehicle if involved in what appears to be a minor accident.
3 Recognize the fact that police uniforms have been used by wage robbers. Do not therefore open the doors of the vehicle if stopped by police officers, but offer to drive to the nearest police station. If there appears to be no reason for a police officer to stop the vehicle, be ready to accelerate away immediately.
4 Watch for any vehicle apparently following; and if this is suspected, make a minor diversion or slow down to confirm. If suspicions cannot be alleviated, drive direct to a place of safety. In any case report the registration number of the vehicle and other details to the police – by radio if this is fitted.
5 Exercise extra care at the bank in scrutinizing the cross-pavement environment.

## *NIGHT SAFES*

These are useful for depositing shop takings to prevent excessive amounts being left overnight on unguarded premises. The danger is that closing times are easily noted, and intending thieves can watch the movements of carrying employees to the night safes or even loiter near the safes without preselecting their victims. As the use of these safes increases, attacks in their vicinity may well follow suit.

An advised solution is to prepare and deposit the bulk of the proceeds in the night safe at variable times before business actually ceases and whilst the streets are still busy, implementing as far as possible the precautions advocated for taking money to the bank – especially not permitting the use of a female employee for the purpose. Regrettably, some areas of industrial cities have become notorious for attacks upon small shopkeepers carrying their takings either home or to the night safe. It is strongly suggested that the relatively small expenditure on a floor safe should be incurred, so that deposits can be made during the busier parts of the day and the residual money can be reasonably protected.

# 17

# *Security in offices*

Security is difficult to enforce in an office block with numerous tenants unless the owners of the building are prepared to provide a commissionnaire service, with some form of after-hours cover by means of caretaker or watchman. During the day, without control at the entrance there is nothing to prevent anyone so minded from entering the building and wandering about at will. If the whole building is occupied by a single firm, the situation is easy: it either imposes the necessary supervision or accepts the risks that ensue.

## *ACCESS DOORS*

All access doors should open into areas subjected to scrutiny by one means or another. The fewer the doors the better, and one main entrance would be a security ideal. Any ground floor doors not used for access or for delivery purposes should be treated as fire doors, or kept locked if they are not part of a fire-escape route. Doors or roller shutters used for delivery purposes should only be opened as the need arises, being closed and relocked immediately the supplier has left, or when a meal break intervenes during unloading, or when the delivery area is vacated for any other reason. Crash bars are not recommended for fire doors, unless they are of the type fitted with an audible alarm to sound when they are opened without use of the control key. Redlam panic bolts can also be used and will at least indicate misuse of the door, but they must be checked regularly.

Occasions have occurred where small pieces of glass have been nipped out of an end of the glass tube so that it can be slipped off, releasing the bolt, then replaced, and that point turned towards the door so that the damage cannot be seen; to the casual glance the door has then never been opened. The special alarm-locking devices are more expensive than the other means but are well worth the extra cost.

If there is an alarm system to the building as a whole, it is possible to have all fire doors on a permanently live circuit with an internal-only audible warning to indicate interference. Where glass doors are used at entrances, lights should be left on after hours behind them so that patrolling police, either on foot or in cars, have a better view of the interior; and the illumination will in any case deter anyone from forcing those doors. The customary advice to use good security locks and substantial doors is as applicable to external office-block doors as to those of other buildings.

## *RECEPTIONISTS*

The nature and usage of the building will determine the form of surveillance that should be given at the main entrance. Multi-occupier premises are best served by uniformed commissionnaires or security personnel with a reception desk. They should have internal telephone communication facilities to all tenants and an outside line, either direct or through a switchboard, for emergency use.

A 'Reception' or 'Visitor Reception' sign with an arrow should clearly indicate the whereabouts of the reception desk. This should be sited in full view of the access doors and must be between them and the lifts or stairs, so that everyone coming into the building has to pass by it. Display boards listing the occupiers with their respective floors or room numbers are best placed beyond the reception desk and outside the lifts, so that a fraudulent caller is denied the opportunity of having a name or firm to quote glibly in answer to an unanticipated query from reception.

A set drill should be established as to the manner of dealing with callers so that their bona fides are established, as far as is reasonable, without causing offence. 'Can I help you? Whom do you wish to see, sir?' will induce a reply upon which assistance can be based or one which will indicate sufficient unfamiliarity with the premises to justify caution. If a system of notification to the persons or offices to be visited can be organized, this will allow unwanted representatives or other types of callers to be put off and genuine ones to be directed or collected by escorts.

*Duties*

If the offices are in single tenancy, reception duties may incorporate those of telephone operator or perhaps some clerical work, but the same principles must apply of capability to observe and challenge unknown incomers. In many instances it might be possible to site an internal tannoy microphone at reception to locate personnel as soon as possible. The use of a visitor's pass system is more feasible than in multi-occupier offices, and much better control can be exercised.

However the reception must be manned, whoever is there must have instructions on what to do in the emergencies which might arise – for example, fire, flooding, injuries or illnesses requiring urgent medical attention, suspect callers, bomb threats, incidents to which the police should attend (thefts, break-ins, wage snatches, etc.) The calibre of staff employed may be such that 'instructions' comprise only a printed list of internal and external telephone numbers for persons and authorities who should be contacted in the event of one of the anticipated contingencies. The manager to whose department reception is attached, should be responsible for seeing that this information is provided. All too often this is not done until a security survey pinpoints its advisability or until the utter confusion arising during an incident demonstrates the neglect. If a telephone operator's role is incorporated, the instructions should contain a specific drill for dealing with nuisance, obscene, and threatening calls.

Co-operation between tenants is the essential ingredient in multi-occupier offices. If this does not exist or is not enforced by necessity or under pressure from the fire authorities, each tenant will have to take such steps as it deems necessary to ensure the privacy and safety of its operation.

## *OFFICE SUITES*

No firm, no matter how small or inefficient, is indifferent to unwanted and unauthorized persons wandering into its offices at will during or after office hours. The best way economically to prevent this will depend on the layout of the areas occupied and the risk potential attached to the business.

Where a firm has a complete floor with a single main access from lifts and staircase (it should endeavour, as with the building itself, to keep the ways in and out to the absolute minimum), the reception or the office of the person designated to receive callers should be sited immediately at the entrance to the suite.

A number of firms, with international trading connections that are liable to attract attention from terrorists or the more aggressive pressure groups, have found it convenient to fit closed-circuit television to survey callers. These are provided with a push-button buzzer to attract attention and a 'speak through' facility whereby to introduce themselves. An electric remote-control lock release can then be used to allow entry into the office suite. The maintenance costs and breakdown possibility of a system like this are negligible, and the installation charges can be offset against the alternative of constructing enquiry windows etc. An alternative is to have buzzers and sliding windows in outer walls beside the entrance, with electronic lock release for approved visitors. Both systems have advantages in that no entry can be effected casually to infringe privacy of the offices – which, whatever the measures used, is the desirable objective.

Where offices front on to a communal corridor, one office at the point that is likely to be first approached by callers should be clearly marked as 'Reception' or 'Enquiries'.

## *RISKS*

The fire hazard has to be met by conforming with fire legislation requirements, which should not clash with anti-theft security measures in offices to any material degree.

Risks are predominantly those of sneak-in or break-in theft, theft after entry by deception, malicious damage, information loss, misappropriation of funds coupled with falsification of accounts by employees, and petty internal stealing.

### *Sneak-in and deception-entry theft*

An efficient reception procedure should inhibit these, but the best of systems falls down around lunchtime when there seems a general belief that thieves accept a compulsory break. Offices are vacated and left unlocked; personal property is left about; pocket calculators are left on desks, safe keys in drawers, and confidential documents freely available for anyone to see. If it is the practice for everyone to go out at lunchtime, dropping a yale lock should become an automatic routine even in an office block of single occupancy. In open-plan offices, calculators, personal property of value, and confidential papers should be slipped into drawers that should then be locked – a matter for personal routine.

At all times those with supervisory status should feel it incumbent on them to clarify the presence of persons unfamiliar to them in areas where they are responsible; and this may apply to window cleaners, maintenance men, and GPO engineers – all poses that have been used to get where strangers would not normally be allowed and which might permit access past an unwary receptionist.

Cloakrooms are a common attack point for the sneak-in thief. These should be sited within the office complex, not near to entrances, lift shafts, stairs, etc.; in particular, they must not be accessible to applicants waiting for employment interviews.

There was a widespread belief that employers could be claimed upon to make good any loss that an employee suffered from theft or damage to his property whilst at work; a belief that was encouraged by the obscure terminology used in some sections of the Unfair Contract Terms Act 1977. As an initial result of this Act, a flock of claims for losses from cars on company car parks were made; so far as is known all were rejected and to date no court decisions have been publicized that any such liability is conferred on employers. This is logical because a whole spate of spurious claims would follow such a decision, and the topic is one for which the Act was never apparently intended to cater.

### *'Disclaimers'*

These should be exhibited wherever employees leave their personal property and should be included in works rules or similar office policy documents. Suitable wording is:

> The management cannot in any circumstances accept responsibility for loss or damage to any property or vehicle of employees whilst on these premises. Such property can only be permitted to remain thereon at the owner's risk.

This is a ruling equally applicable to newly-received pay packets, and no sympathetic precedents should be created by grants where these are stolen or alleged stolen by virtue of the owner's carelessness.

### *Petty internal theft*

It is often difficult to differentiate between this and sneak-in stealing, particularly if the latter is done by someone from adjacent offices.

Cash, either company or personal, should not be left accessible; instruction handbooks should contain a specific warning that money accruing from football pool contributions, staff outings, charity collections, etc. should not be kept in desks or drawers. Facilities should be given for holding this in a safe or other more secure means than the employee has at his desk.

Calculators and other small items disappear from time to time in most establishments, and an enquiry should always be made as a deterrent to repetition. There is no reason why office equipment, if not over-expensive or fragile, should not be lent overnight or over weekends to staff in the same way as tools in the works. Calculators are the prime target and are easily disposed of; this applies to the larger and higher priced type as well as the pocket ones. If possible they should be locked away in drawers each night and some form of identifying mark made on the smaller ones, which often have no serial number whereby to prove ownership. An inventory should be kept of all office equipment of any consequence together with serial numbers, make, etc.; for insurance purposes additional details, such as date of purchase and price, may also be advantageous or required.

Occasionally, items of jewellery and watches are left at the side of washbasins in staff cloakrooms. It is advisable to display small warning notices besides these where they are used by a number of people.

There are other sources of loss which are not regarded by those responsible as 'theft' but can escalate unseen until they become excessive. The main ones are: misuse of telephones – a crossed line revealed that a receptionist was making regular long-distance calls to a former colleague in New Zealand, compounding her guilt by accepting reversed charges on occasions; duplicator – an executive printed the whole of his church magazine each month; postage system – the secretary of a countrywide association sent out all his material through company mail, fed in in batches and using his employer's envelopes. These are only odd examples which are probably duplicated in all firms who do not impose constraints. Gadgets are readily available to record/tape outgoing calls and telephone misuse will stop immediately this becomes known; a duplicator log of usage can be insisted on and the machine locked-off outside normal hours; post abuse can be limited by circulation of a positive directive and instructions to the post dispatch section to query any suspect batches of letters.

Massive timewasting can also occur, which may include all the foregoing matters, in the running of mail order clubs and football pools. These can, of course, operate during lunch breaks but managers should see they do not intrude as of right into working hours. The office bookmaker does not seem to be the problem he once was.

### *Detection*

Where persistent petty theft occurs, traps can be laid for the culprits by use of marked articles of the type being taken. Camrex, a Tyneside firm that manufactures security paints, also produces a kit of powders, greases, varnishes, etc., all of which will fluoresce to different colours under the ultraviolet lamp incorporated in the kit. An invisible permanent identifying mark can be put on articles with the varnish, and the other substances will adhere to a thief's fingers or clothing and remain traceable for up to 24 hours, despite washing, though this progressively reduces the amounts that are detectable. If articles are used for this purpose, then apart from any powder or grease put on them they should be given a positive individual mark, either with the varnish or by some other means.

### *Burglary*

If alarm systems are being considered, again the problem of sole or multi-occupancy arises. A single owner can have a conventional '999' or similar police-linked installation which would not be feasible for say the fourth floor of a building by itself. If a communal watching or security control is in being, there is no reason why a series of alarms should not be connected to an internal manned post; such an arrangement might be offered by a developer to prospective tenants. Alternatively, a firm could fit its own audible alarm purely to protect the part that it occupies. The guidance of the police crime prevention officers for the particular circumstances should be obtained.

Once inside a building an attacker can exercise such force as he wishes to break down doors, and the possible damage has to be weighed against what is to be protected in deciding whether really strong mortice locks should be installed or simply locks which will ensure privacy under normal conditions. A solid oak door to a managing director's office merits the former, but flimsy office doors are best fitted with a light 'night latch' or pin tumbler rim lock' whose weaknesses are detailed in Chapter 38.

During holiday periods when the offices will be vacated for several days at least, it is advisable to collect and lock away the more valuable items of office equipment as a precautionary measure, or at least to remove them from obvious view if there are no means of doing the job more thoroughly.

### *Locks, keys, and safes*

A perpetual danger in large office blocks, especially those built for leasing, is that door locks are on a common sequence with the keys readily obtainable. In those circumstances a master-keyed security lock system is well worth its cost, and key replacement can be made the responsibility of a single employee whose signature will be needed before the manufacturer will cut a replacement. By doing this, strict control can be imposed on the number of persons who can obtain access to particular offices. Desk locks are even more prone to be of a common type; in a large office it may be that 50 per cent of desks can be opened with the same key, which is unfair to the employees. Moreover, these locks have their number inscribed on the face, and duplicates can be easily obtained from any office supplier; if this duplication can be avoided in ordering furniture, it should be done, but this is a manufacturers' fault, difficult to overcome.

Very few filing cabinets have any security value. Some, if turned over, can be opened from beneath; a smart tap with a fine chisel can open others; and hardly any are resistant to brute force. The rather expensive fireproof cabinets are attack resistant to a degree, and these are recommended for documents of importance if no safes are available. Such cabinets usually have 'unique' locks, and the serial numbers can be obliterated; it may also be possible to acquire them fitted with combination locks, but these are likely to be a 'special order' which will be reflected in the cost.

So far as safes themselves are concerned, the comments made in Chapter 38 about their suitability and acceptability to insurers are applicable in all circumstances.

### *Locking up of premises*

Office staff leave at variable times and often unpredictably, which creates a situation where it is impossible to designate a single member with responsibility for final locking up unless a caretaker, commissionaire, watchman, or security officer is employed to give out-of-hours protection. A keyholder must be nominated for callout in emergency, and if the final door lock is of the type where the person leaving simply slams the door shut to lock it, the number of keys to be held can be kept to a minimum. The police should be given a keyholder address for the whole building (with a reserve), and he in turn should have a list applicable to all departments or tenants.

### *Cleaners*

These will have to have access to all offices either before or immediately after working hours, since their activity is hardly acceptable whilst the business of an office is in progress. They must be strictly instructed to lock themselves in and to admit no one unless they are quite satisfied as to identity and authority. What keys are held and how they should hold them must be agreed; quite frequently they arrive as the offices close and therefore can leave when their work is done, slamming the outer door locked as would any staff member. Experience shows that cleaners are suspected of far more petty thefts than they are actually responsible for; but if permanent cleaning staff are employed, references should be carefully checked before acceptance, and it may be found convenient to entrust keys to the most reliable to open up for the cleaning to be done before business hours.

### *Incoming mail*

Periodically there is an outbreak of thefts from letterboxes in office entrances; invariably either these boxes have not been of a type that can be locked or the locking practice has fallen into disuse. The annoyance, uncertainty, and interruption of business are far more important than any likely immediate financial loss, and precautions that can be taken to ensure that mail is not interfered with should be closely considered. If the reception desk is manned sufficiently early, no difficulties should arise.

### *Out-of-hours patrols*

Office occupants have the responsibility to ensure that no fire hazard remains when they leave and evening cleaners will provide an additional check. If a caretaker is employed, he should tour the premises before retiring for the night, and any alarms fitted should have an audible warning in his flat. Should watchmen or security officers be employed, especially where there is multi-occupancy, a clocking system is strongly advised. As the only persons in a locked building, the discretion which may be desirable to a factory area patrol is not necessary, and the inducement to sleep on the job is stronger. Clock points will therefore indicate whether patrols have been made and show tenants that the service for which they are paying has been carried out.

### *Information loss*

This is dealt with in detail in Chapter 19. Obviously, offices are the focal point at which information that is of value to outsiders accumulates. Often staff do not appreciate the importance and value of the documents they handle, and consequently inadequate care is taken of them. A 'clean desk' policy will improve the tidiness and appearance of the rooms and stop restricted papers from being left about.

### *Misappropriation of funds and falsification of accounts*

Incoming cash and cheques should be banked as soon as possible in accordance with a procedure which minimizes opportunities to intercept either (see Chapter 16). Petty cash must be kept to a minimum with regular scrutiny of records and checking of cash holding. The wide range of accounts that may be held precludes advising specific checks, but it must be made obvious to all who have opportunity to gain by falsifications that such checking will be carried out. Ideally, management should try to work out how the firm can be defrauded by internal account manipulation and then take preventive steps based on their findings.

'White collar' fraud and general dishonesty are said to be on the increase throughout the world. If such are suspected in an office, there may be some indication of responsibility in a change of demeanour on the part of one of the staff. A person of previous honesty who turns to crime often shows by lack of sleep, irritability, and a reduced efficiency that he has something on his mind – and of course he may also show signs of living beyond his known income.

## *CHECKLIST*

Appendix 19 lists points of security importance which are especially applicable to offices. It can be used for an appraisal of whether existing procedures and precautions are adequate or should be improved. Needless to say it is not fully comprehensive as matters raised in other chapters have bearing on its contents.

# 18

# *Explosives and other stores*

The importance attached to the safe storage and usage of explosives is shown by the scope of the Explosives Acts of 1875 and 1923 and the mass of Statutory Rules and Orders and Statutory Instruments based upon them. The safety aspects are obvious: there are few more instantaneous and devastating ways of creating havoc than by the ill-timed setting-off of an explosive charge. Reduction of incidents of this type was no doubt the prime initial objective of legislation but security has become increasingly important.

## *SECURITY OF EXPLOSIVES*

Old-fashioned safes offer only a token resistance to thieves with adequate cutting gear or the ability to get at the ill-protected backs. As improvements came in design and construction, thieves soon found that gelignite and detonators took all the hard work out of their operations and speeded them up. Thus their first need was to acquire these substances in adequate amounts. It is a sad reflection upon users that there never seems to have been any difficulty for thieves in this respect. It does not always necessitate breaking into explosives stores – checks on usage by employees are rarely so exact as to provide any knowledge of misappropriation. It is appreciated that where there is heavy consumption, as in coal mining and quarrying, opportunities will always exist for those so minded to steal the materials they are using.

It is obviously not within the scope of this chapter to deal with legal requirements in detail; for those who need minute knowledge, the *Guide to the Explosives Act 1875* (fourth edition), issued by HM Stationery Office will provide this. A number of summaries are also available from the same source (under Code 34/99). Summary Number 4 deals with 'Stores'; Number 6 with 'Registered Premises'; Number 7 with 'Sales'. In addition, the Nobel Division of ICI has a publication, *Explosives: Their Sale, Stores and Conveyance by Road*, which sets out the essential requirements quite clearly and indicates Statutory Instruments on which they are based.

The sale and purchase of explosives have less importance to security requirements than the storage; suffice it to say that all users of explosives, including gunpowder, all types of detonators, capped fuses, and safety fuses must have a police certificate or licence to cover their purchases. The only exceptions are government departments and licensees of magazines – the latter are controlled by the Secretary of State who has power to refuse a licence. It should be noted that where the storage of explosives is concerned, a local authority has no power to withhold a store licence, provided the conditions laid down for its construction, proposed site and the quantities held, are complied with; the discretion vested in the police to refuse a certificate is, therefore, the means of control to prevent undesirables having control over explosives. A standard form of certificate is used which certifies the holder 'is a fit person to keep during the continuance of this certificate at his store licensed for mixed explosives (or at his registered premises; or for private use) . . . the following explosives. . .'. A factor which any chief constable will take into account in considering whether a person is a 'fit person' is the manner in which he keeps the explosives in accordance with regulations or otherwise.

## *STORAGE OF EXPLOSIVES*

### *General information*

When mixed explosive is stored, gunpowder and explosive are reckoned weight for weight though in the lower amounts, if gunpowder alone is kept, provision is allowed for a higher quantity under the same conditions. For all types of storage, other than private use, the explosive equivalent weight of detonators should be calculated at 2.25 pounds (1 kg) per 1000. Cordtex is in general use and when storing it, its weight as an explosive should be reckoned at 16 pounds per 1000 ft (23.81 kg per 1000 m).

The police certificate runs annually and there are distinct advantages in making the renewal of this concurrent with that of a store licence or registration of premises (see below).

At the time of writing there are special restrictions on all aspects of purchasing, storing or using explosives in Northern Ireland. The Explosives Act 1875 is normally applicable there with minor differences in respect of sale and conveyance. It is likely to be advisable for some time to consult local police offices there for up-to-the-minute guidance.

*Private*

Small quantities up to a maximum of 10 pounds (4.5 kg) of explosives and 100 detonators can be stored for private use without legal restrictions on the method of storage. It is, however, recommended that the detonators be kept separately and both should be contained in strong boxes which are transportable (for the event of fire), appropriately labelled, and fitted with substantial locks. This labelling, which is necessary for safety reasons, could attract the notice of a thief so when the boxes are not in use, they are best kept out of sight and in locked premises.

While in 'private' storage 10 pounds of explosives may be replaced by 30 pounds (13.6 kg) of gunpowder.

*Immediate use*

If a user requires explosives for immediate purposes in larger quantities than can be acquired under 'private use', a Chief Officer of police may issue an 'immediate use' certificate to authorize this, provided they are not intended for resale. These documents only cover one transaction and must be signed by the supplier. It is recognized that the explosives will not always be used immediately but that there will be a necessity to store on many occasions. HM Inspectors of Explosives, now part of the Health and Safety at Work Inspectorate, advise that if explosives so obtained are not entirely used on the day of receipt, they should be stowed with minimum hazard and maximum security and the police, or local explosives officer, informed and given an indication of the safety security steps taken.

This type of certificate has been found extremely handy when, because of administrative delays, the renewal of licences has been delayed and a supplier has had no authority to release the needed explosives. The police will normally be prepared to issue the necessary certificates to enable work to continue.

### *In registered premises*

There are two types of registered premises:

*Mode B* caters for up to 15 pounds (6.8 kg) of mixed explosives. These may be kept in any type of building, provided that they are in a substantial container, which is used exclusively for the explosives, under lock and key to exclude unauthorized persons. Detonators must again be kept separately; a fireproof safe would be acceptable for gunpowder only (up to 50 pounds, 22.7 kg, if not mixed).

*Mode A* increases the amount up to a maximum of 60 pounds (27.2 kg) of explosives (up to 200 pounds, 90.7 kg, if not mixed). Dwelling houses are not acceptable and the building used must be substantially constructed of brick, stone, iron or concrete (fireproof safe again for gunpowder); provision is made for keeping in excavations. In all cases the building or excavation must be at a safe distance from highways or places where members of the public pass or work – never less than 15 yards (13.7 m) is considered to comply with this requirement.

Obligations are laid, as follows, in respect of the buildings for mode *A* and the receptacles for mode *B*:

1 The shelves and fittings must be lined to prevent exposure of iron or steel.
2 Precautions must be taken to observe scrupulous cleanliness to eliminate grit.
3 Water must be excluded where this may have a dangerous effect on the explosive in store.
4 All tools, locks, keys, etc., must be of a non-spark-producing metal.
5 All articles of a highly-inflammable nature must be kept well away.

### *Multiple registered premises*

In certain industrial concerns, where there is substantial usage of explosives, as in collieries, it is often impossible to keep to the safety distances required for licensed magazines or stores. Special provision is made so that, apart from the main central store, a number of individual points, each regarded as registered premises under mode *A* are established and are subject to the same conditions of registration and inspection.

### *Explosives stores*

These are divided into five divisions, referred to alphabetically as *A*, *B*, *C*, *D* and *E*, by capacities ranging from 60 to 4000 pounds (27 to 1814 kg), in excess of which magazine licences are required. Special limitations are placed upon the siting of these stores, according to their division and the proximity of 'protected works', which are themselves divided into two main classes. To illustrate the application of these regulations, in the case of a division *E* store, 2000 to 4000 pounds (907 to 1814 kg), this could not be sited within 352 feet (107 m) of a pier or jetty (Class 1), or within 704 feet (214 m) of a hospital or theatre (Class 2).

Under no circumstances may detonators be kept with other explosives in the main store; they must be in a separate annex or licensed building. Where a few hundred detonators only are concerned, special arrangements may be allowed, providing there is no simultaneous access to both detonators and explosives inside the store. All stores containing more than 1000 pounds (454 kg) of explosives must be fitted with an efficient lightning conductor.

### *General safety rules*

There are some general rules for stores which must be complied with. The more important are:

1 The permitted amount of explosive must not be exceeded.
2 The store must only be used for keeping explosives and tools and receptacles in that connection.
3 All tools, etc., shall be of some soft metal or wood to obviate sparks.
4 The interior must be lined to prevent the presence of exposed iron or steel.
5 The interior must be kept clean and free from any kind of grit.
6 The explosives must be taken out and the store thoroughly washed out before any repairs are made to it.
7 Provision must be made, by use of suitable shoes and pocketless clothes and searching or other methods, to prevent means of causing fire – matches, steel, or grit – being taken into the store.
8 There must be no smoking in the store.
9 A person under the age of 18 shall not be employed in or enter a store, except in the presence and under the supervision of some person of the age of 21 years or above; a person under the age of 16 shall not be employed in the store at all.
10 A copy of the general rules shall be affixed inside the store,

together with an extract from the licence showing the amount of explosive that may be lawfully kept.

Not mentioned in regulations, but nevertheless a practice that should be followed, is that of having a prearranged drill to be carried out in the event of an explosion. This should include designation of individuals with responsibility for specific actions and names and telephone numbers of those services and persons who must be informed. If possible, a list of those persons who might be expected to be working in the danger area at any time should be available.

Where any accident occurs by explosion or fire in connection with an explosives factory, magazine or store, notification must immediately be sent to HM Inspectors of Explosives; similar action should be taken in respect of registered premises when the incident involves any form of personal injury. In serious cases, notification in the first instance should be by the quickest available means and the debris should not be touched except to recover bodies or treat injuries.

It should be noted that explosives factories are included in the premises specified under the Fire Certificates (Special Premises) Regulations 1976. Guidance on these can be obtained from area offices of the Health and Safety Executive.

## *WORKSHOPS*

Any workshop used for making up charges must be sited at least 25 yards (23 m) from the store or 'protected works'. It must not contain more than 100 pounds (45 kg) of explosives, and the local authority must be notified of it. To all intents and purposes its upkeep and usage should be similar to that of a store.

## *STORAGE OF EXPLOSIVES: SECURITY ARRANGEMENTS*

A Home Office Circular No. 113/1972 contains an appendix outlining in detail the requirements of HM Inspectorate for the secure storage of explosives.

They express the firm opinion that well-constructed steel stores are preferable to those of brick, stone, or concrete with walls less than 14 inches thick. A basic premise is that attacks on explosives stores must be frustrated, either by the construction of the building itself, or by the existence of supplementary alarm arrangements and/or surveillance, or by a combination of the two. The store should resist attack with unpowered tools for a longer time than that necessary for security forces to react; a steel store is expected to resist entry for at least four hours. If a building does not meet this requirement, an alarm system should be fitted.

The circular does not list the manufacturers of safes conforming to recommendations, but the very useful ICI book *Explosives: Their Sale, Storage and Conveyance by Road* does. They are:

Hornsby and Goodwyn Ltd, Scunthorpe, Lincolnshire
James N. Connell Ltd, Coatbridge, Strathclyde
M. J. Turnbull, East Boldon, Durham
Tredomen Engineering Works, Ystrad Mynach, Glamorgan

If existing safes need to be modified, it is essential to obtain and study a copy of the circular which is very specific on the constructional requirements. All safes have to be ragbolted into a base of concrete at least 8 inches thick with the bolts to a depth of at least 6 inches. As opposed to the previous brass locks, two steel multilever mortice deadlocks are now decreed, and HM Inspectors will advise on acceptable types. ICI again comes to the rescue by naming the MD17 by Bramah, M/101 by Banhams, and E2553 by Erebus. It should be borne in mind that these reference numbers may be changed, but the approved suppliers persist.

Padlocks are not recommended because they can be 'blown' with safety by a small amount of explosive; nor should they be used to supplement mortice locks because the latter might then be left unlocked or the fittings for the padlocks used as an aid for levering open the door.

Advice is given that the stores should be regularly cleaned and painted in an inconspicuous colour.

Very detailed recommendations are made to ensure that the alternative forms of non-steel stores are adequately resistant – a copy of the circular is again essential. When alarms are fitted they must be such as to function when attempts are made to attack any part of the circuitry or controls; microwaves are not to be used where there are electric detonators.

## *GENERAL STORE SECURITY ADVICE*

If a daily balance of what should be in the store can be maintained by logging all incoming explosives and detonators, those passed out for use, and those returned unused, a quick check on the contents should reveal deficiencies and provide accurate information to the police in the event of a theft. A strict control must be kept on keys, which should be signed in and out daily.

Where explosive stores or registered premises are within a works or factory perimeter, the actual physical protection offers no more difficulty than any other important point. Alarm systems can be easily fitted and protection afforded against unreliable employees by making an enclosure with a padlocked gate and illuminating the whole area.

Quarry stores are particularly vulnerable, they are invariably well away from habitation and rarely visited after working hours. The heavy metal prefabricated type is preferable, sited in open view and illuminated if possible. If a compound of the Lochrin Palisading type can be erected, so much the better. Bushes and other obstructions in the near vicinity of the store should be cleared.

If a light is to be left over the store the police should be formally notified so that its absence will at once arouse suspicion. If steps are taken to make it absolutely obligatory for thieves to use considerable force to get into a store they will not try it. They have everything to lose if anything goes wrong.

### *Security powers in respect of explosives stores and magazines*

Anyone who enters without permission, or otherwise trespasses, on any explosive factory, magazine, or store, or land immediately adjoining and occupied therewith, can be removed by the occupier or his agents or servants, as well as by the police, and may be fined up to £5.

Anyone, other than the employee, doing any act which tends to cause a fire or explosion can be fined up to £50. (A wide power of arrest is given to a variety of persons for this offence under the Explosives Act 1875, s. 78.) Notices warning all persons of their liabilities shall be posted by occupiers, but their absence will not exempt from the penalties.

## *CONVEYANCE OF EXPLOSIVES ON ROADS*

To date, there do not appear to have been any instances of stealing explosives in transit in this country and the danger, which would attend any violent attempt to do so, limits the probability down to one of collusion with the driver. The following general regulations, imposed by a series of Statutory Instruments, are similar for both diesel and petrol-driven vehicles:

1 Two men must always be in attendance.

2 The cab must be separated from the body by a fire-resisting screen to within 12 inches (30 cm) of the ground, the whole of the exhaust pipe being in front of the screen, which must have a clear space of at least 6 inches (15 cm) between it and the body.
3 The petrol tank must be in front of the screen.
4 Adequate means of fire extinction must be carried.
5 There must be a quick-action cut-off to the petrol supply, near the carburettor, but not so close that it could be involved in a fire therein.
6 The vehicle interior must be lined with non-inflammable wood or asbestos, with a sheet metal exterior.
7 Electric lamps only should be carried.
8 Normally, the only opening in the body should be a door at the back.
9 The cab area should be rendered non-inflammable.
10 The engine must not be run either loading or unloading.
11 Speed limits must be rigidly observed.
12 Where possible, stops should not be made in built-up areas.
13 Smoking in the vehicle must be absolutely barred.

## *MISCELLANEOUS STORES*

Every variety of store holds a temptation for petty pilfering by its staff and by those employees who can gain access to it. Whether it is an engineering, general, consumable goods, stationery, canteen, or any other kind of store does not seem to matter. The reasons may be that useful items are there in a quantity and under conditions where a petty loss will not be noticed and that they belong to the employer, not to another employee. The very number of different items kept and sheer volume of stock may conceal shortages for some time. Even at a stock check it may be felt that discrepancies are due to bad documentation rather than theft, so that the losses become substantial before any positive action is taken.

### *Structural considerations*

The first essential is to exclude the possibility of outside intrusion, for those working in the stores will be deterred by knowing that any shortcomings will be attributable to them. If a storekeeper has responsibility for safekeeping stock, he must be given the authority to enforce it and to insist that nobody, whether of supervisory status or not, can take anything without documentation or, for that matter, be free to enter the stores at will.

Thefts from stores are not infrequent where night-shift working is carried out, particularly in engineering works where useful tools are held in quantity and a general stores will contain many items which are readibly usable for domestic purposes. Steps are needed to prevent such thefts from happening, or at least to make it known that they have occurred so that necessary remedial steps can be taken to prevent recurrence or detect those responsible.

If possible stores should be in rooms, buildings, or areas which can be locked and away from the rest of the activity on the premises. Rooms and buildings offer little difficulty as windows can be protected by bars or mesh and doors fitted with security locks. It is unlikely that employees will go to the extent of forcible break-in which causes considerable damage, unless the contents are of sufficient value to justify the risk.

If separate rooms or buildings are not available and areas on a workshop floor have to be allocated, the only economic solution is to erect a substantial cage which either has its sides extended up to roof level or has a separate roof fitted to the cage. Night-shift incursion normally takes the form of climbing over the top of a cage, dropping down inside, and leaving by the reverse route.

The matter of locks on stores is especially important. If a maintenance fitter can get hold of a key, even for a short period, he may be able to make a duplicate. This has happened on many occasions when suspicion has fallen upon the storekeeper's staff since there are no visible signs of entering. A number of locks are available on the market which offer resistance to quick duplication, either at a retail key-cutting shop or by an internal craftsman. The main essential is to have the strictest possible key control, with a single key to be collected daily and surrendered when the stores close and the briefest possible list of persons authorized to take it.

To prevent access into the stores by those not actually working therein, a counter should be built into the outer door or a full-sized counter constructed immediately behind it with a flap that has to be raised before entry is possible into the stores proper. Only storekeepers should 'look out' items being requested unless these are of such bulk that assistance is required in handling.

## *Documentation*

For economic reasons it is desirable to keep stock levels at the lowest possible, and this will entail accurate documentation for receivals and issues. Some issues of trivial value may be allowed without record if these represent acceptable risk and unlikelihood of fraud.

Stock records may be based on computer or on something like a

Kardex system supplemented by bin cards; they will progressively become computer-based. Stock checks on lines that are likely to show shortages should be made at irregular intervals, not just left to an annual stock check by which time the maximum damage would have been done financially. A set form of requisitioning and of authorizing requisitions should be laid down, so that a storeman will know from the document presented that the issue is bona fide and authorized. Where a requisition is handed to a subordinate for collection from stores it is advisable that a line is drawn underneath the entry to prevent additions from being made to what has been written. The form should identify both the person authorizing and the department to which the issue should be charged. A procedure may be necessary for urgent production needs whereby a temporary requisition can be made out at the stores, to be supplemented later by a formal one which can be processed in the normal way.

### *Out-of-hours issues*

Instructions should be given which keep these to absolute minimum; for example, tools or materials required for night shift or weekend working should be withdrawn during normal working hours, so that the stores are only entered in an emergency. Again a set procedure should be laid down for this, so that the key can only be drawn by a designated individual of supervisory status who is aware of the documentation needed, completes it at withdrawal, and leaves it for the storekeeper.

Where there is a security presence on the premises it should become their responsibility to accompany any person making a withdrawal from the stores. This should apply to every type of stores. The instigation of this practice in one particular firm resulted in an immediate reduction of 50 per cent in the apparent usage of soap and cleaning materials from a consumable stores, and there was a noticeable improvement in several other categories. This may be considered by some security personnel as a totally extraneous duty, but it has a distinct bearing on profitability and is a material deterrent to dishonesty by employees of all levels – supervisors are not exempt from temptation when the contents of a store are open to them; nor for that matter are department managers.

### *Requisitions*

Procedures should be checked to ensure that no one can requisition for personal usage, though this is difficult to ensure and may entail subsequent knowledgeable scrutiny of requisitions.

It is most important to check that all incoming purchases pass through a general stores for acceptance and recording. If it is possible for an outside purchase to be made and go directly to a department, with the requisition and an 'approved' signed invoice later being passed to the stores for recording, this may permit fraud by the person authorizing the requisition. A case in point would be a painter's department apparently collecting from a supplier for urgent use, not taking the purchases into the stores but presumably immediately beginning to use them; in these circumstances all the purchases might not be brought back into the premises and part of them could be diverted for personal use or personal business. Numerous incidents of this type have occurred, and the possibility is one that every firm should bear in mind.

### *Engineering stores*

Temptation to steal from these may be reduced by a scheme for lending tools. Such a scheme should incorporate an accurate recording system which is appraised at regular intervals to stop tools from being out on a virtually permanent loan akin to theft.

Where contractors are working it is highly inadvisable that they be allowed temporary issues of tools unless this is essential for the job to be done. Once they have left a site the company's chances of retrieving the tools fall off abruptly.

### *Oxy-acetylene equipment*

Cutting gear and bottles of gas are important to thieves who intend attacks on safes or strong rooms. The burners must be locked away, and circulars to this effect have been sent from the police to industries where this equipment is used. There is a risk attached to the cylinders which are themselves of value and are an increasing target for itinerants. They should not be left outside buildings but preferably put away inside and out of sight.

Gas cylinders more often than not are leased from the manufacturers who operate a replacement service for empties with their own transport. The frauds and thefts associated with this have caused a leading supplier to draw the attention of customers to the fact they are financially responsible for the safe keeping of the supplier's bottles; the notification being followed by a series of spotchecks by the supplier which produced many shortfalls in numbers. The cylinders are expensive and extremely easy to dispose of by a dishonest delivery driver, but losses are equally easy to prevent by enforcing simple receival procedures and sound documentation to show location, numbers, and usage of cylinders internally.

Delivering drivers are the main offenders – they have maximum opportunity and are in regular contact with potential sources of disposal. Even more than tanker drivers, they become accepted visitors and are increasingly left to their own devices. Empty cylinders may be stacked in an area ready for collection and replacement – a driver who is unsupervised has ample opportunity to under-deliver. Again, a storeman may be corrupted by a driver who is prepared to split proceeds of an under-delivery coupled with an under-collection which would show a constant number of cylinders but an apparent over-consumption of gas. How many firms' accounting systems would pick up an increase if such a fraud began?

### *Ladder stores*

A minor matter perhaps, but if contractors are employed on sites and good quality ladders are left about, there is a distinct tendency for them to disappear, having been carried out unnoticed on the roofs of the contractors' vehicles. There is no difficulty in constructing racking on to which the ladders can be adequately secured horizontally or vertically. If it is felt necessary, ladders can be marked with identifying paint or by other means.

### *Petrol stores*

These may be subject to fraud by internal drivers who are allowed to fill their own vehicles (see Chapter 28), but the main interest is what may be done by delivering drivers. Unfortunately, frauds have been prevalent in several parts of the country. On one occasion 19 drivers from a single depot were convicted of stealing from their loads which indicated repeated losses to customers.

Fuel receipt and issue should be made the job or part of the job of a single individual if possible, rather than rotate haphazardly amongst any staff who happen to be available. If the latter is done, there will be no real interest in precautions and no means of attributing responsibility for any shortage.

There is always the possibility of collusion between the storeman accepting the fuel and a delivering driver, and for this reason there should be independent scrutiny of the records at intervals and test dipping of the tanks. A delivering driver can, however, perpetrate frauds without collusion by the storeman provided the latter is not conversant with the manner in which he can be deceived. For example:

1 On a metered delivery the meter may not have been reset back to zero before delivery commenced, which allows the driver the opportunity to keep what was already recorded. (It has also been suggested that it is possible to falsify a meter by pumping from one tank to another on the vehicle.)
2 If the delivery hose is not emptied, it can retain a surprising amount of petrol.
3 Dipstick frauds are possible by 'bouncing' – the dipstick is dropped into the tank, and the impact bounces fluid up the tube indicating more present than is actually there. In a similar manner, a driver may dip two tanks of different size with the same dipstick, whereby an incorrect total, either up or down, will be recorded on one of them.
4 A shortened dipstick used after delivery may show excess fuel in a tank; one fitted with a ring or flange will show 'dry' by not reaching the fluid (driver taps tank to simulate stick reaching bottom).

These are the main frauds that a storeman can expect, and there is a routine that he can apply to stop them. He can check the gauges or sight glasses on the tanks receiving the fuel, both before and after delivery, or use his dipstick on them where this is applicable. In some firms it may be possible to weigh a tanker in and out, and from this it is easy to calculate what has been delivered. Delivery drivers become familiar figures, and all too often latitude is allowed to them in how they carry out their business. No signature should be given until the storeman is certain he has got what is stated on the documents, and he must not allow a driver to dictate how the delivery is made and checked.

Fuel stores are controlled by stringent safety regulations too detailed for inclusion here. Fire authorities exercise strict supervision and will give any guidance and advice needed.

## *Summary*

Satisfactory security standards for stores should entail:

1 Restricted access during working hours.
2 Structural strength to exclude intruders when unmanned.
3 Adequate documentation to account for all receivals and issues.
4 Adequate stock records to identify sources of loss.
5 Periodic stock checks to deter theft and reduce potential gross loss.

# 19

# *Commercial and industrial espionage*

This is an area of risk where the consequences are difficult to evaluate. A firm of ship repairers could not get quotations accepted until they excluded everyone from the estimators' office other than the staff working therein; they then began to acquire orders. At the other end of the scale a £35 million Middle East contract was lost when the tender was revealed to a competitor.

During the mid 1980s there have been repeated accusations of governmental sponsorship of industrial espionage made against several countries, diplomats being expelled and businessmen imprisoned. It is noteworthy that one extremely reputable international firm asserts in its employee's handbook that, if precautions for the security of information and research are not observed, in 20 years' time the firm will not exist.

There are grey areas where market research may fade into industrial espionage, and those in higher authority may be faced with the ethical problem of whether information of extreme value has been acquired by acceptable or unacceptable practices. If a firm faced with a shortage of orders and substantial redundancies were offered information which could gain for it a life-saving contract, how predictable would its reaction be?

How widespread is industrial espionage ('business espionage' might be a better name)? It cannot be quantified; its extent will be exaggerated by those selling their services ostensibly to counteract it and minimized by those who consider that they have no secrets to protect. The really successful efforts may cause suspicion but lack adequate proof of commission. Whilst there have been criminal convictions and actions for damages, the law in many countries is so vague that the farcical charge of stealing a piece of paper, referred to later, may be the only one which can be laid. However, there are indicators sufficient to cause misgivings. For example:

1 As far back as 1965 the French government saw fit to issue a pamphlet *Espionage: A Reality* to warn of East European industrial espionage. Later, the UK Institute of Directors, a sober body if ever there was one, caused *Industrial Counter Espionage* to be produced.
2 Electronic 'bugging' equipment is widely manufactured and advertised mainly in the Far East and America; a complete shop window at Hong Kong Airport contained such devices openly on sale, Directional microphones used by the media and others are readily available and seem increasingly evident in intrusions on privacy. Prosecutions have followed 'phone taps on GPO lines, one of which has been discovered during a major UK takeover battle in 1986/7.
3 A whole variety of electronic detection devices are now offered in the security press to counter 'electronic eavesdropping'; a number of firms offer their services to sweep rooms where 'bugs' are feared – at considerable expense to the client. Without due cause such facilities would find no market.
4 'Market information research' now appears amongst the services offered by some organizations which have hitherto been known as sources of private detective work and security consultancy. One leading UK investigator in the private detective field claims to have positively identified 20 such firms acting internationally with an interest in trade spying (more discreetly called 'aggressive market research'; the term 'offensive electronic surveillance' has been used to imply the use of bugs).
5 In 1976 a US National Commission survey report showed that, out of 115 private investigation agencies in seven cities, 42 said they would do electronic eavesdropping themselves or would recommend others that would; that proportion is still said to apply approximately.
6 The regularity of expulsions of foreign nationals and prosecutions for industrial espionage throughout industrialized countries is too frequent to attribute even a majority of them to actions of political expediency.

7 After Watergate, can anyone believe the facilities and manpower are not available if the incentive is great enough?

## *RISK ASSESSMENT*

Every firm which submits tenders for contracts will want all information about them to be restricted; otherwise the potential risk will vary according to the nature of the industry and the comparative state of the market. Unless management has had good cause for apprehension a somewhat naive view might be taken, but the person responsible for security must mentally place himself in the role of a competitor, look at his own company, see what he would want to know about it, and recommend precautions accordingly.

Ideally, at some stage the possibility of attack by industrial espionage should be considered at board level and a decision taken as to whether countermeasures are necessary. If it is thought that they are, a detailed security survey should be carried out to establish which activities a competitor would like to know about and how valuable that knowledge would be to him. This could then lead to: designation of areas to which access is to be limited; means of identifying personnel authorized to enter such areas; control over documents in every shape and form; and a form of security audit to be carried out to ensure that instructions have been implemented.

If an outside consultant is to be employed for the purpose, his bona fides must be established beyond any doubt and a binding agreement made for punitive compensation in the event of breach of confidence by him – he will, by nature of his survey, become privy to all the matters the firm wishes to protect.

### *Target areas*

These are established from the security survey and could include the departments working on:

1 Plans for production and sale of new products.
2 Plans for new advertising campaign.
3 Customer lists coupled with rebate and discount particulars.
4 Marketing projections.
5 Evaluations of possible acquisitions or mergers.
6 Sources and costings of raw materials and components.
7 Policy decisions on future activities, redeployments, closures, etc.
8 Contractual or trading agreements, and customer contracts.

9 Specifications and tenders.
10 Research and development projects.

Though of less interest to competitors but perhaps more immediately embarrassing are those matters which can cause industrial unrest, such as:

1 Rationalization and redundancy plans.
2 Personal files and confidential reports.
3 Wage-negotiation preparatory material.
4 Confidential instructions on industrial relations negotiations.
5 Salary structures and job-weighting factors.

Apart from any considerations of business espionage, human curiosity can be one motive for seeking out information just so as to be better acquainted with facts than colleagues; or the objective might be to convert knowledge to personal advantage or material benefit.

Excuses of moral objections to secrecy are currently the norm for capitalizing on material divulged to the media but the reason for disappearance of files relating to a national newspaper's negotiating policy with a print union is much more obvious. Protection is required equally against the inquisitive and the rogue.

## *CONTROL OF DOCUMENTS*

### *Document classification*

A system of classification should be agreed for documents whose contents need to be restricted in circulation; the simpler the better, and temptation to overclassify should be resisted. Responsibility must rest with the originator. 'Confidential' normally implies contents that it is wished should not be general knowledge but which the recipient can divulge at his discretion to those who need to act upon them or who should know of them. 'Secret' should mean not to be known to anyone other than those shown on a circulation list which should be marked upon the document or attached to it. To avoid drawing attention to the special nature of the communication, classification should not be shown on the envelope containing it, which should, however, be sealed. 'Personal', however, would be an acceptable notation which would convey little to a handler. It is generally accepted that personal secretaries will open all mail unless their masters decree otherwise; if an originator does not wish this to be done, a second envelope can be put inside the unclassified external one and endorsed 'Secret – to be opened by Mr . . . only'.

These precautions are not adequate for those projects vital to the profitability of the firm. An authorized circulation list should be drawn up for these and expanded as the project develops. A code word should be given to the project, and it should be referred to in this way in correspondence and telephone conversations. The specific number of copies of documents or plans needed should be printed and numbered sequentially with each recipient allocated an identifying number. It would be advisable, if the project demanded this degree of secrecy, for internal movement to be by hand with signature on receipt.

### *Safekeeping of documents*

All classification procedures can be invalidated by the careless executive who leaves papers strewn on his desk at night or the secretary who, having been given them to file, leaves them in an 'out' tray until convenient or locks them carefully away then leaves the cabinet key in an unlocked drawer, the top of her typewriter, or the pullout accessory tray in her desk. Cabinets adequate for the risk should be used and a check made that their locks are not on a common sequence. The range of such locks is more limited than could be wished, and for this reason the number of the key, conventionally marked on the lock surface, should be deleted. The fire-resistant cabinets available are sufficiently substantial to offer some security protection but are expensive; old safes should be considered for the purpose rather than disposed of for scrap.

A sound investment is a high security-rated, master-keyed suite of locks for offices containing restricted information, with strict control of keys and key replacement. A periodic task of the security head should be a tour of inspection of offices with reporting of those who offend by not observing a 'clean desk' technique.

### *Movement of documents*

This is almost invariably a purely internal risk, but fortunately it is the practice of most employers to allocate mail sorting and internal distribution to older and long-standing employees of established integrity and loyalty to the firm. Mailroom conditions can be easily created where any departure from routine sorting will attract notice by a colleague; experience suggests that possibilities of outside interference arise at the reception point where letters for a department are simply left in an office tray for sorting and distribution without anyone having been specifically designated to receive and deliver. In some instances, where several departments occupy the same building, a rack for letters exists inside the entrance doors for subsequent collection by departmental staff; this has caused unpleasant situations unconnected with industrial espionage where papers of personal or industrial relations importance have gone accidentally or deliberately astray. Where a department has classified correspondence the manager should insist on delivery to a particular person nominated by him; the security head should include this as an item to check.

### *Destruction of documents*

This is an often neglected precaution. The importance of a document does not of necessity lapse when the project to which it refers terminates, and nowhere is this more important than in the industrial relations field. Highly-confidential papers bagged for burning or shredding are given over to unskilled newly-appointed workers, and it is assumed that this is the end of the matter with no supervision needed. This is asking for trouble when small relatively cheap shredders are readily available for use in a department to ensure that nothing is decipherable when it leaves; the tendency with a machine to hand is that papers will not be allowed to accumulate as they would for collection. Special shredders can be obtained for computer printout, tapes, and discs and should be sited immediately adjoining the computer room for the convenience which will improve certainty of use.

All papers that are 'classified' as previously mentioned, and especially those originating in 'target areas', should be shredded; the secretaries of directors and key personnel who might handle the very sensitive material should have specific instructions to personally attend to this.

There are security companies who offer collecting and shredding facilities but their clients have to take on trust the integrity of those companies probably poorly-paid employees and may think it advisable to detail a responsible person to travel with the waste and oversee its destruction. Eventually, perhaps a firm will offer the service of a purpose-built vehicle carrying a large shredder which can do the job on the spot.

Burning is a further and most unsatisfactory alternative which is offered by some local authorities in incinerators. Paper in bulk does not burn easily and again the calibre of those engaged in the operation could be suspect. Pulping is yet a further possibility, not much used and subject to the same transit and handling risks.

Buying a big efficient shredder is desirable for convenience, economy, and peace of mind – and the proceeds may be used within the company for packaging purposes. The destruction of documents is one which the person in charge of security should monitor himself at intervals to ensure procedures are being followed and complacency has not set in.

### *Copying of documents*

This is hard to control, especially where a single original is to be circulated round a number of people amongst whom there will be those who will want to keep a personal copy for convenience or reference. Such documents should be stamped 'not to be copied'. No material of 'secret' classification should be copied without the agreement of the originator, and the task should then be given to someone who can be trusted.

### *Research and development*

Valuable information is not confined to the written word. Research is an increasingly costly business in a technological age; and a competitor, by simply knowing that a process is not feasible or a material is unsuitable for a purpose, can save wasteful expenditure of expertise and cash. On the other hand, knowledge of successful work may enable prior claiming of patent rights or wreck any market advantage that the researcher would have gained.

The same may be said of development in the designing and construction of new machinery where both the paperwork and the components will be informative to a skilled observer.

These are instances where there must be the strictest control on personnel entering the working areas – not just strangers but also members of the workforce who are not directly concerned. The size of the area may justify a constantly-manned reception desk to exclude the unwelcome, or electronic control of access may be more economic. This may be relatively simple or complex, ranging from coded lock, magnetic card-operated lock, or combination of card and code, to card and code linked to a mini-computer which monitors and provides a permanent record of who goes where. Cost increases rapidly with sophistication of device, and the cards may be adapted to show also a photograph of the holder; lost cards can invariably be blanked out by internal adjustment. A further variation is to issue each authorized person with a pocket transmitter which will cause an electronically-controlled bolt to be withdrawn when the carrier approaches the door. Photo-identity cards are often clipped to lapels for easy recognition;

colour variations in these can be used to denote the areas to which the wearer is entitled to have access.

## *PREVENTION*

Basic principles of countering business espionage are:

1. Survey the possibilities thoroughly.
2. Do not place anyone whose loyalty is in any way suspect, or whose personal circumstances might cause acute temptation, in a position of being able to capitalize on secret information.
3. Make sure that those handling secret information are aware of the importance of what they are doing and of the consequences of negligence. Apply the principle of 'need to know' to the passage of information, accepting the necessity of restricting information from trusted colleagues to prevent possibilities of innocent and unintentional disclosure.
4. Ensure that the originators of documents, designs, or plans are aware that it is their responsibility to establish the appropriate degree of protection.
5. Establish procedures, make sure that they are understood and implemented, and then monitor them at intervals to ensure that they continue to be applied.
6. Thoroughly investigate the slightest suspicion of a leakage to pinpoint the source and emphasize to everyone the importance of adhering to procedures.

## *FORMS OF ATTACK*

As with other forms of dishonesty, countermeasures should be based on an assessment of how the potential adversary will operate. In this instance the objective of acquiring information can be achieved by:

1. Exploiting the carelessness, boastfulness, or negligence of employees.
2. Corrupting employees, or forming an emotional attachment and inducing confidences, which can be extended into deliberate blackmail.
3. Inserting an agent into the workforce.
4. Deliberately recruiting an employee from a position where he holds the requisite information.

5 Holding a detailed interview of a knowledgeable employee for an advertised, attractive, financially wonderful, but non-existent job.
6 Electronically 'bugging' telephones and offices, or using other specialized forms of surveillance.
7 Entering premises to locate information either:
 (a) as a visitor, or
 (b) as an intruder by force.

Method 7(b) has rarely been encountered, but 7(a) in many establishments is the easiest way of all – just walk in and look if the target firm's procedures are so lax as to make this possible.

Since simplicity of operation and effectiveness go together in acquiring information, special interest should be shown in devising ways to stop the 'walk in and look' technique.

*Blackmail*

The risk of blackmail by emotional involvement cannot be more clearly demonstrated than by the subversion of the marine guards at the US Moscow Embassy in the early months of 1987. An executive whose womanizing or gambling leaves him open to such pressures, especially if he travels widely, should have his attention drawn to the risks – of course he will be indignant! The lessons of Moscow will have been widely noted but were almost certainly already widely in practice.

*Subterfuge*

Entry has been gained by persons claiming to be carrying out safety surveys or appreciations of equipment or posing as prospective customers with potentially large orders, telephone engineers, gas and electricity supply employees, laundrymen, window cleaners, etc. Obviously the permutations are as numerous as differing circumstances demand. Commonsense checking on the bona fides of most of these will suffice, but that this should be done is a matter for inclusion in instructions. Many will carry forms of identification which should be examined carefully. If regular visits are to be made by servicing experts or others, personal passes can be issued to them to minimize delays.

*Visitors*

An agreed procedure should be laid down for visitors which will satisfy security requirements without alienating customers or other genuine callers. A visitor's pass, as shown in Appendix 11, can be completed and a coloured lapel badge or similar means of identification issued which will show at a glance to which areas or departments the wearer is restricted. An escort from reception desk is always advisable regardless of any personal reasons or regularity of visits; the least suspected rogue always has the greatest opportunities. Special care should be exercised in dealings with contractors and suppliers who also deal with competitors.

### *Cleaners*

Cleaners have the most time of any to pry unsupervised, and there is much to be said for employing own staff in risk areas, recruiting them locally and having them vouched for by existing employees. Several commercial security firms now offer a 'security cleaning service'; too much should not be read into that title, however, though it can be assumed that extra care in selecting may be taken, some degree of supervision will be given, and a deliberate 'plant' cannot guarantee to be placed at the target firm. However, it might be advisable with contract cleaners to have the work done at times when the firm's staff are present. Set instructions should be given for the speedy disposal of the contents of waste-paper baskets which is one of the more prolific sources from which information can be gleaned.

### *Temporary staff*

On special projects it can be assumed that new employees will have been thoroughly scrutinized, especially if they have worked in the recent past for a competitor. However, lack of care creeps in when secretarial or typing assistance is brought in from agencies. Where possible these persons should not be utilized in a risk area, even if this means temporary, inconvenient, internal transfers of existing staff. If this is impossible, the agency should be acquainted with the fact that the post is a confidential one and asked to supply details of the person's service with them and the firms to which he/she has previously been allocated, so that there is at least an informed option to accept or refuse. If accepted, there is no reason why the desired confidentiality should not be impressed upon the incumbent; this will be appreciated very quickly anyway.

## *PRECAUTIONS*

With all these factors in mind the following precautionary steps are suggested, though they will not be necessary or applicable in all cases:

1. Define key posts and areas. Apply stringent selection and vetting techniques to the former, and do not use agency or other temporary staff. Provide means of entry restriction to the latter.
2. Establish classification, handling, copying, safekeeping, and destruction procedures for documents, plans, and designs. Provide adequate safe-storage facilities for them, as previously mentioned.
3. Destroy shorthand or other notes connected with the production of secret material, carbon paper used for typing the papers, the 'once only' plastic typewriter ribbons usually used with electric machines, and any spoiled or excess copies produced. All should be shredded if possible, and under no circumstances should they be only screwed up and dropped into a waste-paper basket. Similarly, wipe out all tape-recorded dictation.
4. Restrict the responsibility for copying to a person (secretary) already associated with the project, and ensure that adequate instructions are given.
5. Do not leave wall charts, graphs, or other visual aids hung up where they can be freely seen. This should also apply to screens of visual-display units if it is possible to produce a display of restricted data thereon.
6. Periodically, have offices visited out-of-hours to check that papers are not being carelessly left out and that no electronic listening devices have been affixed. The boardroom and conference rooms should be included in this survey, and they should be checked immediately after each usage in case papers, indented pads, notes, agendas, used blotting paper, etc., have been left. When vital board meetings are to be held, consider changing their venue at the last moment with minimum prior notice to those concerned.
7. Indoctrinate all staff with knowledge of the risk, the importance of the precautions, and the necessity of not relaxing vigilance with the passage of time.
8. In the event of a person who has access to secret information tendering his resignation for any reason or being dismissed, seriously consider termination of his services immediately with payment in lieu of notice, plus a check of his files and inventory before he leaves the building.

After all that advice, what recommendations can cater for the unworldly scientist who takes every precaution in his own environment, then writes a learned paper for some scientific journal or addresses a highly sophisticated symposium and, in doing so, carefully and in detail explains his firm's line of research? Of course he never believes that his learned colleagues could ever abuse his professional trust! Such individuals value their independence, and the only prevention is a strict board-level instruction to the contrary. However, the drunken boastful representative or employee and the naive susceptible secretary are probably the easiest sources for an information-seeker to tackle.

*Electronic surveillance devices*

Considerable emphasis has been placed on the physical and practical measures which can be provided to make industrial and commercial premises secure. Today a new threat has evolved, that of 'stealing' information through the use of electronic devices which are placed in sensitive areas, boardrooms, etc. and the spoken information recorded some distance away for future reference. Better known as 'bugging', the use of these has probably had greater publicity than they deserve, but that publicity might lead to temptation to use them.

There seems little doubt that they are an expected part of diplomatic life, the charge and counter charge of embassy 'bugging' is a regular media feature, even Denmark protested about the discovery of devices in its Warsaw embassy in February 1987.

Reported instances are increasing in the UK and it is noteworthy that in two recent cases professional agencies were responsible, though it should be said that both were phone-tap recording devices wired in outside the premises. The devices can take almost any form, even cigarette lighters, and may be very effective. The standard telephone 'bug' is a transistorized oscillator which can be fitted in next to no time, is powered by the telephone line current, and uses the line itself as an aerial. A variation allows the microphone to be active even when the handset is not in use, and the nastiest one of all is the 'infinity bug' which can provide a listening point at any spot in the world from which it is possible to dial the number of the tampered telephone.

Other forms have been encountered built into power plugs, or affixed quite independently with their own power supplies from miniaturized batteries. These are little radio transmitters in their own right and may be installed in any kind of artifact which makes a presentation gift for office use, sent from an unlikely source, distinctly suspect. Desk-top cigarette lighters, pen sets on stands, electronic clocks, calenders, and calculators were openly on display and shown for this purpose in a prestigious Hong Kong shop. What then may be available under cover? – very expensive fountain pens certainly are – beware such gifts made to executives.

The advent of microchip technology makes size even less important and it has been said that eventually a sheet of blotting paper may contain all that is necessary to listen in on conversations over the desk where it lies. Pinhead-type microphones have already been encountered and progressively more sophisticated gadgetry will be devised and utilized especially if the resources of governmental agencies are applied.

Detection equipment can be purchased; but the really efficient variety is expensive, and a capital expenditure case to justify purchase would not be easy in most instances. A limited number of the larger private-detective agencies will provide a checking service for offices. As in making a survey, if the outside agency is to be used, make sure it is reputable, insist that a keen and observant member of the firm is present during the 'debugging' exercise and that any gadget that is found is handed over to the firm. The latter requirement was a source of acute embarrassment to a practitioner locating a miniaturized tape recorder in a telephone linkage – he had 'found' it several times previously for customers who had thereon recommended him for his efficiency! It must be remembered that the searcher can also plant a bug if given the opportunity and the fact that the service has been required is confirmation in itself that something of potential cash value is *en train*.

Offices can be checked visually, but this must be done thoroughly and patiently with a knowledge of what to look for and where. Screwdrivers and a torch will prove useful tools, and it is not a job to hurry. The saving grace is that 'bugs' have to be planted; and if the operator is detected, he is in trouble, the victim is warned, and the whole exercise is aborted and may become an acute embarrassment to those sponsoring it – there are easier ways.

Directional listening devices have appeared sufficiently often in televised presentations to make the confidentiality of outdoor conversations suspect. The practicability of 'bouncing' laser beams off windows to record and interpret vibrations caused by conversations in the rooms behind has been the subject of several articles in quite reputable technical publications – the results are suggested to be of poor quality and obliterated by the drawing of curtains.

If in the course of a search a suspect phone or device is found it is advisable that the room be sealed off temporarily or steps taken to prevent any interference with it. A minimum number of people should know of it and a decision on action should be made at top managerial level – it is of course possible to retain the bug and feed false information through it. For example, a security manager found in the boardroom, which he was checking before a meeting the next day, a phone that responded to a detector being used for the first time. It would not dismantle and he contacted the office manager at home to arrange GPO engineer attendance first thing the next morning. The manager came in very early with an engineer who was almost full-time on the site; the phone was replaced, taken away, ostensibly checked and found innocuous, but then 'misplaced' amongst similar ones. The office manager said he had misunderstood what was wanted but that in any case he was responsible for having the boardroom equipment ready for the meeting. Had the office manager or engineer, or both, been engaged in 'bugging'? The doubt will always remain.

## *INTRUSION BY COMPUTER*

On more than one occasion students at educational establishments have claimed to have accessed the computers of firms with which they had no business connection and obtained data which was confidential – and produced proof that they had done so. These are deliberate and skilled acts but such access can be made accidentally by 'piggy-backing' where a line of communication into a central processor is accidentally left open and the next user finds himself into records which he would normally have no right to inspect. The problems associated with computers are discussed at greater length in the next chapter but there is a non-technical risk to users of the services of a computer bureau – that of an employee of that bureau offering confidential data of the user to a competitor. This has happened and its possibility is one of the items to be raised when entering into a contract with the bureau; the user cannot divest himself of responsibility simply by putting the job out to contract. The confidentiality of the material must be made clear and the precautionary measures agreed. A firm introducing a new product will take the greatest care to tie-down 'security-wise' advertisers, lawyers, component suppliers, and the like to ensure privacy but then overlook the avenues of leakage existent in computer records and output and, even more surprising, in telexed messages. An example of the latter occurred some years ago when an engineering firm received a most perplexing communication, obviously highly confidential and nothing to do with its business, equally obviously misdirected and could have been a potential source of acute embarrassment to the department of origin if its content had been released. Fortunately, it was 'fielded' by the head of security who was consulted and recognized its importance.

## *THE LAW*

The law in the UK provides little deterrent to those who indulge in information 'stealing', hence the rather pathetic charges of theft of a piece of paper albeit the value of what is written on the paper may be substantial. Where there has been theft of something tangible – for example, documents relating to a new heart drug, special moulds from a foundry, secret dog food ingredient, computer tapes – no difficulty has arisen. In addition to these, charges have been laid of conspiracy to corrupt and to effect a public mischief, but these are hardly likely to be increased by the complications of the new restrictions on conspiracy charges imposed by the Criminal Law Act 1977. The extremes that have been sought are exemplified by charges of using an unlicensed radio transmitter in telephone-tapping offences. It still appears to be no criminal offence for an employee to photograph, copy, or memorize information and pass it to a competitor for gain, unless it is possible to frame charges under the Prevention of Corruption Act 1906, s. 1. This Act relates to agents, but 'agents' are defined therein as including 'any person employed by or acting for another' – and unfortunately the fiat of the Attorney-General or Solicitor-General is needed for a prosectuion. It would be necessary to prove a corrupt act, a dishonest intent, and a consideration in the form of apparent gift or benefit. There have been very few prosecutions under this Act, and these have mainly involved public bodies for purposes other than business espionage.

If an outsider is found without permission in premises and it is suspected that he is there to seek restricted information, he should be detained, using force if necessary, in accordance with the powers of arrest previously mentioned, and the police should be called. The person must be given no chance to dispose of anything in his possession pending their arrival. They will no doubt wish to search him for evidence of stealing; and if he has a camera, the importance of what may be on the film should be drawn to the notice of the police. Charges might be possible in those circumstances under the Vagrancy Act 1824 – being on enclosed premises for an unlawful purpose – or under s. 9 of the Theft Act 1968 – burglary. For both there is a private person's power of arrest.

### *Civil actions*

Expert legal advice should be taken by an aggrieved firm before seriously considering whether to take civil proceedings of any kind. Action may only be possible against a previous employee of little substance, not the recipients of the information, and that is a waste of time and money. The hearing may result in the very dissemination of information that it is desired to avoid. Contractual agreements may have to be disclosed, system weaknesses spotlighted, and security arrangements compromised, and there is no certainty that the action will be successful. Cases have failed in the past through simple proof that the firm apparently did not regard its operations as being in any way secret since it took no steps to ensure privacy nor did it emphasize to employees that the operations were so regarded. In those cases it would have been cost effective to implement adequate precautions in the first place (see also page 200).

### *Comment*

Although only seven indicators were shown as confirmation that the risk of industrial espionage is genuine, the reaction of many of the multi-national firms is even more conclusive. Logical targets themselves, a number make no secret of the fact that they know they have been attacked, some have even sued, practically all have instructions dealing with ensuring the confidentiality of their operations. To quote the words of the head of one of the UK's most profitable and efficient groups, and one of the largest:

> Security of information is important to protect the present and safeguard the future of any company and its employees. You cannot show on the balance sheet a special item for security profit and loss but it is there, hidden under some other term, you neglect it at your peril.

## *THE CRISIS OF CONSCIENCE*

The ethical problem of involvement in business espionage was mentioned at the outset; in times of normality and acceptable return on capital, the senior executives of most enterprises would probably set themselves on the side of the angels and decline to have anything to do with it. But what happens if the opportunity of a substantial financial gain arises, or circumstances become so straitened that the firm itself is in jeopardy and personal futures of those making decisions are seriously threatened? Scruples may be found flexible at such times and mitigating reasons easily found for actions that might otherwise be deemed reprehensible.

There are of course many legitimate ways in which the activities of a competitor can, at least to some degree, be kept under scrutiny; a firm which failed to do this could find itself falling behind in research, efficiency, market share and consequently profitability. In an ascending schedule of information-producing methods, ranging from the innocuous to the downright criminal, the dividing line between the acceptable and unacceptable will vary, not only with the individual's conception of what is sound business sense, but also with external pressures which are acting upon him. A questionnaire to test reactions has been presented to many senior management seminars over a period of years by a highly experienced colleague and friend, Peter A. Heims, FIPI, FBIM, CPP. The answers, given in complete anonymity, showed a flexibility of conscience in times of adversity calculated to depress any high-minded purist. It is said that one paper had a '21' added to the list and marked 'X' – 'Murder'!

With Peter's permission a variant is shown hereunder; it does not purport to be comprehensive but all the methods which might be considered 'suspect' have actually been encountered. Where would you, or your executives, draw a line: (a) under normal trading conditions; (b) when the continuance of your firm and your own standard of living were in hazard? A supplementary question could be added for both contingencies: 'How would you respond to an offer of invaluable information about your competitor for suitable repayment?'

A final point to be borne in mind when a leak of information is suspected: senior status and previous good repute are no guarantee of innocence when pressing external factors are working on the individual.

*The 'information-gathering' questionnaire*

1 Study of competitors' annual reports, balance sheets and brochures.
2 Study of articles, trade publications, etc., discussing competitors' products and projects.
3 Market surveys and consultants' reports (authenticated trade sources).
4 Study of competitors' exhibits at trade fairs and the answers of their employees manning the exhibit stands to questioners.
5 Expert analysis of competitors' products to obtain specific information.
6 Assessment of inadvertent disclosures by competitors' employees noted and reported by own salesmen, purchasing agents, etc.
7 Establishing a policy that representatives, salesmen, etc., will quiz customers as to information tendered to them by competitors re pricing discounts, delivery schedules, production data, etc., and report in.

8 Similar instructions to discreetly quiz competitors' employees inadvertently met in the routine course of business, or at technical meetings and seminars.
9 Arranging attendance of specialists, posing as prospective customers, at trade fairs to put questions to competitors' employees on exhibit stands to induce desired information.
10 Taking advantage of genuine job interviews (for own firm) to question applicants who are past or present employees of competitor.
11 Advertising a specialist vacancy in the local press of the competitors' area, inflating the salary, perks and conditions to attract an employee with the desired knowledge from them, i.e. buying in the information.
12 Where specialist executive jobs are advertised by a competitor, arranging that a fake applicant with apparently attractive credentials should submit himself with a view to obtaining information during interview.
13 Advertising attractive but non-existent job, with the object of getting a maximum number of applicants from competitors' employees and drawing out as much information as possible during interview (using agency without disclosing identity of own firm).
14 Arranging attendance of agent, or trusted employee, in public houses or clubs near competitors' premises to listen to careless conversation – or to induce it.
15 Endeavouring to plant a trusted employee on competitors' payroll.
16 Direct observation of competitors' premises using binoculars, and/or cameras, or causing a trusted employee to seek access as prospective customer, supplier, contractor, Health and Safety at Work inspector, or local official.
17 Hiring a professional investigator to obtain a specific piece of information on a 'no questions asked' basis.
18 Finding that a competitors' employee is prepared to divulge information knowingly for payment – taking advantage of the fact, . . . bribery.
19 Knowingly participating in eavesdropping on competitor by electronic or other means of supplying information adequate to facilitate competitors' premises being entered for obtaining required information, i.e. copied drawings, documents, etc.
20 Conniving at the use of knowledge of compromising circumstances to put pressure on a competitors' employee to obtain desired information . . . blackmail.

# 20

# *Computer security*

Computers have proliferated during the past few years, becoming progressively more sophisticated, more costly and more essential to the user's day-to-day activity and profitability. Their intrinsic value is substantial but interference with them may have an adverse effect on the firm that is cataclysmic beyond any other form of loss or disaster.

For this reason it is proposed to deal with their protection in detail from the planning of a new installation to consideration of staffing and policy. Much of this is a direct responsibility of the computer manager, who may well not have had any experience in security aspects and will welcome discussion.

A great deal of publicity has been given in the media to the frauds which may be carried out by use of computers, but it is impossible to quantify any true total of losses. The malpractice may be skilfully concealed and evade detection or be terminated before suspicion is aroused, or an embarrassed board of directors, apprehensive of the consequences of disclosure, may keep knowledge of the crime to themselves. A CBI report estimates at least £30 million pounds will be disclosed as lost by such fraud in 1987 and that this will only be 30 per cent of the actual figure; a two-year study by an internationally-known firm of auditors put the annual estimate at £40 million. The 1980s have seen a virtual explosion in computer usage in every conceivable branch of industry and commerce and for every purpose where they can enhance efficiency and make cost savings – statistics, planning, production, finance, inventories, sales, etc., an endless list.

The amount of data now contained in programs and files is such that loss or interference could have unpredictable adverse consequences.

Whilst cases have occurred in the UK, the bulk of reported frauds are still of US origin, and considerable research has been done by the Stanford University of California to catalogue their nature and the status of the individuals responsible for them. As a numerical world-wide statistic the frequency is insignificant, but that does not reflect potential magnitude; for example, the Equity Funding insurance frauds were estimated at $2,000 million, and instances are freely quoted where valuable research and other data have been wrongfully acquired. There is strong suspicion that reported cases are only those where the perpetrator has made an error or has simply been unfortunate in that some accident has brought his actions to light.

The computer industry is producing increasing numbers of systems analysts, programmers, and operators, and it is too much to expect that the criminal percentage of the community will not be represented amongst them. Moreover these individuals in general are youthful, intelligent, energetic, and ambitious; and though it is reasonable that some will experiment with their skills to prove that a computer can be manipulated, success may be the first step to temptation to capitalize upon the knowledge. Nevertheless, at the moment it is significant that programmers and operators are a minority amongst the other categories of documented fraudsters. This is an emerging problem whose full extent is probably not yet recognized, and economically sensible precautions must be taken.

## *POTENTIAL MALPRACTICES BY COMPUTER STAFF*

1 *Systems programmers*
- deliberately writing in errors or logical oversights into the computer program providing weak points which can be exploited later.
- Disclosure of security measures to outsiders.
- Disabling or neutralizing the program security features.

2 *Computer operators*
- Deliberate substitution of altered programs.
- Disclosure of organizational and procedural security safeguards.
- Copying of files for sale to competitors or other buyers.

3 *Maintenance personnel*
- Use of test programs to examine or copy files or alter system programs.
- Disabling security hardware.

4 *Users*
   - Fraudulent impersonation of other legitimate users.
   - Falsifying own files to deceive third parties.
   - Penetration of operating system and alteration of object programs.

5 *Others*
   - Data processors who provide or handle input data.
   - Supervisory personnel with specialized knowledge.

## *STANDBY FACILITIES*

A first consideration when installing a computer is a rather surprising and pessimistic one, but one which the manager has to bear in mind from the outset – it is that of ensuring alternative suitable facilities in the event of a breakdown from fire, supply failure, or any other cause. In many concerns, such a breakdown can cause utter confusion in otherwise well-organized regimes because of the degree to which the computer has been integrated into the firm's activity.

Standby arrangements should be explored from the outset on a mutual-aid basis with owners of like machines within reasonable distances. This could have a bearing on which of two equally acceptable machines is selected for purchase in that an identical standby for one of them may be immediately to hand. There is little doubt that full co-operation will be extended, though it might be advisable to avoid potential competitors in making arrangements!

The manual or detailed instructions kept in the computer room should contain as a minimum of information:

1 The exact specification of each standby computer contact with details of time availability.
2 The identity and means of contacting those personnel at the other installations who can authorize use of the machines.
3 A list of contacts in order of preference to minimize delays in the event of the first choice being unsuitable.
4 The precise location of each contact with instructions on how to reach it.
5 A checklist of the minimum software, stationery and other equipment needed together with any particular personnel to be called out.

All senior personnel should be accessible by telephone at their homes and detailed instructions should be laid down for action in the event of a failure.

## *HAZARDS TO BE GUARDED AGAINST*

*Environmental disaster* – Fire, explosion and flooding can destroy not only the building fabric and the machine itself but also computer programs, information, data and records stored on magnetic media. Card and paper tape input, printed output and supplies of special paper are also at risk in this context.

*Machine failure* – New generations of computers should have increasingly greater degrees of reliability with their improved technology, but they are still pieces of equipment subject to mechanical failure and dependent on external power supplies which can fail.

*Deliberate or accidental loss of information held on computer files* – This is an area of human involvement and covers all those aspects outside environmental damage. The loss can arise from errors and acts of negligence on the part of the operating staff, faults in programs, the consequences of unauthorized persons or visitors gaining access to the computer room – or deliberate acts of sabotage.

*Misappropriation of company funds and assets* – The areas which effectively provide the best opportunity of using the computer resources for this purpose are those of programming and operations – they will be dealt with later.

## *FIRE*

The construction and usage of modern computers would, one might think, make fire a minimal risk; this has been effectively disproved by several disastrous fires which have originated in the computer rooms themselves and resulted in almost total destruction. Where a central computer processes information from ancillary 'slaves' in essential component factories, even a minor outbreak might have consequences affecting the whole production schedule. Preventive precautions must therefore be designed to detect and control fire at the earliest possible stage. The means of doing so must be built into the installation during construction, and the environment in which the computer is situated must not be neglected as a source of potential danger.

### *Housing of computer*

Although computers are usually inserted into existing buildings, the amount of work entailed may render this a false economy when all the requirements of the insurers have been complied with. It is better in every way to house it in separate purpose-designed premises. If this is not feasible, the whole computer should be enclosed in a fire-resistant compartment, protected both from horizontal and vertical spread – and from potential water damage if other parts are affected.

The isolation of the central area containing the vital computer equipment is most important. Here the devices for detection and protection must have maximum effectiveness. Segregation from the surrounding areas of data processing and ancillary offices must be by fire-resisting partitioning or solid walls. If other offices are sited above, the roof of the computer room must be fully resistant to water as well as fire. It must not be perforated by openings whether for lighting, ventilation, or any other reason. Rooms below are equally important and their occupancy should be limited to purposes which entail the least possible fire risk – there are advantages in locating the room at ground level.

These steps of segregation will meet a further essential purpose in that a reception area can be constructed to stop casual entry into the central rooms.

### *Contents of the central area*

It is elementary that all furniture and fittings in the central area should be of metal or at least of materials which are not readily combustible. This must be remembered during the ordering of such items when senior personnel, being consulted on their requirements, may be more concerned with comfort or possibly even prestige.

Tapes, cards, plans or files should be restricted to those needed for immediate use; fire-resisting cabinets should be available for their temporary storage. Any bins or containers for waste paper should likewise be of metal and regularly emptied. Smoking should be prohibited within the computer room and the ban extended to the stationery stores where it is even more important. Consideration should be given to incorporating a rest room where smoking would be allowed – this might limit abuse of the rule where it mattered.

### *Floor and roof voids: ventilation and wiring*

Temperature and humidity have to be controlled between relatively narrow limits. Extensive fluctuations adversely affect the computer and its tapes, and could cause operating failures. There must, therefore, be a full and reliable air-conditioning system with extensive ducting which, with the vast amount of wiring, has to be housed in roof and floor voids. A false floor is preferable to cable trenches in the structural floor as, among other advantages, it allows a construction whereby computer equipment can be easily rearranged or added to. The floor itself should be prefabricated panels of fire-proofed timber or metal, individual panels of which can be lifted by suction pads to give access to the services beneath; all such removable panels should be clearly marked. It is conventional for the false floor to be at the same level as the rest of the building, the void beneath it will then become a potential water trap in the event of flooding or fire in other parts of the building, causing danger from the presence of live cables. It is therefore advisable to fit a sill across the threshold of any door leading on to a false floor.

It is important to eliminate combustible materials as far as possible from these voids; materials for suspension units and brackets, insulation, sound and waterproofing panels and ducting must be considered with this in mind – ducting in fact should be of metal. A fire in these areas may not be immediately perceived and could be difficult for access with portable fire extinguishers. Detection devices in these voids are essential and they should be coupled to an automatic fire-extinguishing system.

The extensive ducting of the air-conditioning system provides an avenue of entry for heat and smoke, particularly if the computer room shares a ventilating system with the rest of a building. Sampling of the airflow is necessary to ascertain combustion and smoke content by means of probes and detectors, fire dampers held open by fusible links or electrically operated should be built into the ducts to stop smoke or hot gases reaching areas where they could cause damage. A manual means of operation should also be incorporated.

In addition to normal detectors, heat-sensitive cables can be laid among the wiring runs to detect abnormal temperatures before they reach the stage of creating smoke or flame.

### *Doors, hatches and windows*

In the computer room, partitions, doors, hatches and glazing should generally have a fire resistance of at least half an hour. The offices surrounding the computer room can be treated as a 'buffer' zone for fire defence purposes between it and the rest of the building by giving them a high degree of protection, though not equal to that of the central installation. Sprinklers could be used in the main areas of the building if there was no danger of water damage to the computer installation but smoke/heat detectors should be installed in the adjacent offices and hand appliances made available.

Doors in the rooms around the installation should be self-closing to function as 'smoke stops' and fitted with sills where required. Doors to the computer and plant rooms are especially important; substantial five-lever mortise locks should be fitted on these and strict control kept of keys. If these doors open internally into the rest of the building then they must be further shielded by self-closing four-hour fireproof doors operated by fusible links. This is a normal insurance and fire protection officer requirement, which can only be relaxed, to two-hour doors, when the rest of the building is office accommodation of negligible risk.

Where 'borrowed' lights are needed they should be of wired glass and fixed shut, any sliding or drop hatches should be self-closing and fire-resistant – wired glass if vision is required. If a viewing window is desired from the reception area into the computer room it must be expected that a solid fire-resistant, drop panel will be required as an emergency shield. This viewing window is a good idea since it permits visitors to be shown something of the computer at work while keeping them at sufficient distance to preclude damage or seeing restricted material. It also provides an opportunity for patrolling security officers to inspect the interior without entering – for the same purpose, doors in the installation offices should have clear, wired-glass panels. Having regard to the importance of the data held, this is an ideal situation for fitting special locks under a master key system to all doors.

## *FIRE DETECTION AND EXTINGUISHING EQUIPMENT*

Limitation of direct and consequential loss rests on the prompt detection of abnormal conditions such as may be caused by insulation failure and overheating. Detectors must be linked with a method of audible and visual warning and be capable of setting off an automatic extinguishing installation after a predetermined interval. The warning system should be linked to a repeater unit at a point which is permanently manned – either a security gate office, or a fire station in a similar manner to which a burglar alarm is linked to a police station.

The area surrounding the computer itself must have complete space protection. The usual practice is to halve the normal coverage which is allocated to a detector to double the certainty of immediate recognition of a fire source. Smoke-sampling detectors are recommended for general purposes and those of the ionization type are particularly suitable in underfloor spaces and for airflow sampling in probe units. The computer room, underfloor and ceiling voids should be covered by separate groups of detectors so as to define the fire area immediately. The detection circuit must be so wired as to be capable of automatically shutting down the ventilating system and activating a damper unit in the duct into the computer area to stop the inward spread of smoke and heated air; it should also automatically cut off all power supplies to the computer itself. An unnecessary automatic shutdown could cause serious consequential loss – hence the need of alternative manual operation when the computer is in use. Switches or buttons for this purpose should be located near the operator's console and the exit doors. A clear visual indicator is needed to show whether the system is on 'automatic' or 'manual'.

The extinguishing agent is provided by separate banks of gas cylinders for the plant and computer rooms. These are usually housed in the plant room with interconnecting delivery piping. Carbon dioxide is commonly used but vapourizing liquids of the BCF variety are tending to supplant it since they are less likely to cause the thermal shock to the computer which results from the cooling effect of a massive $CO_2$ release in a confined space. For testing purposes, a cut-off switch or mechanism for the bank of cylinders must be incorporated; this must have a prominent warning light or buzzer to ensure it is not left in the 'off' position making the cylinders inoperative after the test.

A predetermined delay on 'automatic' before the release has a positive value in that it allows an opportunity, albeit brief, for an immediate investigation to stop the sequence if the danger is one easily contained or the functioning is accidental – this could happen with a circuit fault on changing from manual operation.

For the purposes of dealing with the minor outbreaks, adequate hand-operated extinguishers of the $CO_2$ or similar type should be readily to hand in the computer room. Whilst the general opinion in this country is that water, either from hand-held equipment or through sprinklers, causes damage which outweighs its usefulness, in the USA there is an increasing tendency to fit sprinkler systems on the grounds of safety for personnel and because equipment can be subsequently dried out with minimal damage. A further alternative to a gas 'flooding' system is having large cylinders fixed to hose reels with nozzles whereby the gas can be directed precisely on to the source of fire. This is only possible where there is round-the-clock watching of the premises with efficient alarm devices to activate personnel for operating this equipment at a sufficiently early stage.

The recommended minimum for the computer room is:

2 × 5 kg (10 lb) $CO_2$ extinguishers
2 × 2.5 kg (5 lb) $CO_2$ extinguishers
2 × asbestos cloths 1.4 m (4½ feet) square.

*Humidity warning*

The damage that can result from substantial humidity or heat changes has already been mentioned; the suppliers of the air-conditioning system must build in adequate temperature controls but it is advisable to have a separate form of warning for dangerous humidity fluctuations. This can be equipment showing conditions in the form of a graph in the computer room itself, with linkage to an indicator unit in either the gate office or some other permanently manned point to show when a state of dangerous humidity is being neared. This unit should incorporate a means of testing that the equipment and link are in order.

*Indicator panels, manual buttons and switches*

The main panel should be sited in an open position adjacent to the central area and in clear view to everyone. By means of coloured lights appropriately labelled – or something similar which is obvious in meaning – it should show at a glance:

1 Whether the power supply is operative.
2 The location of a fire – with separate lights for computer room, roof void, floor void or plant room (these could be extended to other parts if desired).
3 The presence of a fault in any of the sections of the installation.
4 By separate lights, whether the gas cylinders are set to operate manually, automatically or have discharged.

A push button or switch should be fitted into the panel to stop the alarms, with manual operating buttons for computer and plant rooms to be used in emergency.

Other manual buttons should be sited inside the computer room itself, preferably beside the exit doors; a distinctive audible warning buzzer should also be there. Both $CO_2$ and BCF are said to be non-toxic, but in fire-extinguishing concentrations the staff must leave at once. There are obvious advantages in having the buzzer of a different variety to that which functions when a fire warning is given for parts of the building outside the computer area.

*Manual buttons*

Apart from entailing expensive refilling of gas cylinders, inadvertent discharge will cause time loss, potential damage and the displeasure of insurers if replacement cylinders are not readily to hand. The buttons should be such that the chance of accidental firing is at a minimum; glass-fronted alarm boxes are suitable but the buttons must not be spring loaded to fire on the glass being broken. Each of these buttons should be tested at specified intervals and the test recorded.

### *Manual/automatic switch*

Should a fire break out while staff are engaged in the computer room, they should notice it before an automatic release could operate and it would be invidious if some inadvertent action by any of them caused the system to fire while they were there. This can be obviated by coupling the alternator control which activates the computer to the release system, so that switching on the computer effects the change from automatic to manual operation and the reverse happens when the computer is closed down as staff leave.

### *Repeater panels*

The remote panels in the security office need not be as comprehensive as the main indicator. It would suffice if the 'manual' and 'automatic' conditions were shown, together with 'fault' and 'fire'. Similarly, the humidity repeater need only consist of red and green lights. A test switch to check the circuit should be incorporated in each repeater, and regularly be used.

## *ACTION IN THE EVENT OF FIRE*

All staff must be given instruction and have drill in fire prevention and the use of the hand appliances that are available for fire fighting. They must know exactly how the fixed installation functions, who to contact in the event of faults developing, and their own sequence to follow when leaving the premises, or when a fire occurs.

A notice should be displayed in the computer room on the following lines:

*In the event of fire*

1. Turn off master switches for machines, ventilation and ancillary plant.
2. Inform telephone operator who will call fire brigade.
3. Attack the fire with apparatus to hand (gas cylinder on electrical apparatus, asbestos blankets on waste bins, etc.).

A checklist of action to be taken when leaving the premises should normally be displayed beside the departure door. An example is:

*Check*
1 All doors and hatches between rooms are closed.
2 All waste paper has been removed.
3 Any soldering irons or heating appliances have been unplugged from sockets.
4 All master switches have been turned off.
5 Security office has been notified that you are leaving.

From the outset of building a computer installation, the local fire brigade should liaise closely and their advice sought on any matters of difficulty; it is not enough to decide on a series of structural measures and then ask their approval, they should be consulted at the draft plan stage when their suggestions may save time, inconvenience and money later. This liaison will pay off in their intimate knowledge of the premises if they have to attend an actual outbreak in or adjacent to the installation.

## *SAFEGUARDING THE INSTALLATION*

If the installation is sited in a part of premises away from immediate surveillance or in buildings unoccupied and unsupervised outside normal working hours, serious consideration should be given to incorporating an effective burglar alarm installation. The contents have little value to thieves, unless they are expert to the degree of being interested in the data content of the records, but the damage a frustrated intruder can perpetrate among that kind of equipment is frightening.

In other instances, patrols should visit and inspect the main indicator panel regularly, the first occasion soon after the staff have vacated. Lights should be left on in the computer room and approach passages to deter any unauthorized interloper.

### *Insurance cover*

Apart from structural and equipment loss from any cause, that of consequential loss might reach monumental proportions and could be induced by a simple prolonged power failure. The risk is almost impossible to assess, the extent of coverage is a matter for experts and will be related to the degree to which the firm's activities have become reliant on the computer. One thing that is certain is that the insurers will insist on all the aforementioned precautions as a minimum requirement.

### *Duplication of records*

The information stored on magnetic media, such as programs, master records or data for future reference, is more valuable by far than the media. Its loss or the cost of recreation could be a serious matter. It is conventional for all master file records on magnetic tape to be kept on the 'grandfather, father and son' principle. Master files on disc are usually duplicated on tape. The hierarchical system works on the basis that the 'grandfather' tape contains master records updated two processings before the current updated master file – the 'son'. The 'father' tape is that immediately before it. Both these preceding tapes, and the duplicated master record copies, should be kept in a purpose-built fireproof cabinet-type safe, away from any fire risks likely to affect the central area.

The compiled tape and disc programs will have source program-card packs backing them; these packs should also be separately and securely housed in a different part of the building, to be used if necessary to recreate the programs.

## *FRAUDULENT MANIPULATION OF COMPUTERS*

The most common problem for companies is not deliberate fraudulent or malicious action but simple human error. However, there is no doubt that the number of incidents in which a computer is used as part of a fraudulent operation is increasing – but then the number of computers in use is increasing in proportion and many of the offences would probably have been easier if the older manual systems of recording, documenting, etc. had remained in use. A disturbing statistic is that the bulk of the UK abuses of the computer have been by persons of managerial status and this is especially true where the department has been of limited size and the person responsible had unrestricted access and freedom from supervision. It is fortunate that professional institutions are setting high standards for ability and behaviour of data processing personnel but the suspicion is widely held that a tip of the iceberg situation exists with many frauds concealed in the complexities of computer operations with the attendant auditing difficulties.

Tampering with input data has proved the most popular ploy, adding false vouchers for payment authorization along with legitimate ones, expanding payments to an accomplice – as has happened several times in wages frauds. Data is altered to have payments made into a personal account – the possibilities are the same as existed before computerization – adding to, omitting from, or altering records for dishonest purposes. A UK example concerned some £50,000 over a period of years during which a clerk in a catering firm with a grocery supplier accomplice, fed the computer with false account numbers and invoices for food which was not delivered.

Altering an actual program has happened less frequently and can be worked in conjunction with input data falsifications. A manager set up accounts in the system for fictitious suppliers and then provided data to show equally fictitious goods received; a long-serving accounts clerk, in a brewing concern, found he could delete accounts from the computer so that only alternate deliveries to a tenant were charged – they shared the difference and he made £9,000 in a very short time. Both cases received barely a mention in the media. Other variations on the theme have included mis-posting transactions in the form of debits to large accounts where they would not be queried, increasing charges marginally on a large number of accounts and transferring the produce thereof to a personal account, by-passing credit limitation controls on personal accounts. Further examples are quoted below where a variety of means have been employed which have been the subject of successful prosecutions:

- A payroll clerk deducted a few pence off each pay cheque and added them to his own. (Note that the company's accounts balanced.)
- A bank programmer, whose program calculated savings account interest, instead of dropping off fractions of pennies, had them added to his own account.
- A computer consultant found a blank form used for adding a new employee to the company payroll. He added his own name to the payroll list and picked up his cheque as it came off the automatic cheque-writing machine. (Note the ease with which a visiting outsider gained access.)
- Two systems analysts set up their own company while working for another company that sold metal ores. Their own company bought ore from their employer and sold the same ore back to their employer at a profit. The entire transactions were accomplished by a computer which they controlled.
- A computer expert working for a bank typed his own bank account number on hundreds of blank bank deposit slips. He took actual deposits from various firms, replaced their deposit slips with his own, writing in the firm's account number and handing the deposit slips and money to the tellers. Because the computer thief used a

magnetic typewriter and because he knew that the computer was programmed to read the magnetic tape before any other notations, all the money went into his own account. He collected £20,000 in one day.

- A pensions superintendent defrauded his firm of £400,000 over a period of 15 years with the help of a colleague and the firm's computer. He diverted payments into at least 24 false bank and building society accounts only coming to grief when a clerk noticed his address on one of the accounts.
- A former employee kept his key to the building and the password giving him access to the computer which he instructed to pay cheques totalling £318,000 to a company he had established; he was not sufficiently experienced to cover his tracks.
- A clerk in a London branch of an American bank diverted payments into her own account and added bonuses to her salary. She made £279,000 before being caught.

An analysis of the 100 reported cases in 1983 showed the majority to involve sums in excess of £10,000 and two of £500,000. Of these, 63 per cent were simply manipulating computer input and source documents to inflate amounts, give illegal discounts, charge for services not provided, and for fund transfer without authority. Seven per cent exploited computer reports to get information to plan frauds; only 5 per cent related to the making of unauthorized program amendments, inserting private fraudulent information or writing off debts. It is noteworthy that remote terminals were used in 15 per cent of the cases.

There have also been instances of deliberate sabotage in which random errors have been introduced – more of these might exist and be treated as human shortcomings than are currently known. Output has been destroyed to conceal input tampering or at least to delay discovery, name lists have been stolen and substitions have even been made. Master files have been deliberately damaged, copied and sold, erased, and, as mentioned elsewhere, taken for ransom. Perhaps one of the most costly and least seriously treated items is that of stealing computer time; there was an instance some years ago, which was totally unreported, where to a biased outsider it appeared that the most important use of a major research establishment computer by its highly-qualified staff was in connection with football pools and racing! How would those individuals, now in very senior positions, treat subordinates who did likewise?

The police have recognized that they may be required to investigate fraud in a field where they lack expertise, and to cater for this they have created a month-long course on computer crime at their National

Police College – which in itself is tacit admission that offences are expected to become more frequent.

*Preventive measures*

Unless the security head is a computer specialist himself, it would be foolish to suggest he should advise on any technical measures to prevent misuse of the computer. He should however make his interest in the possibilities known and draw the attention of the manager in charge to cases mentioned in the media. This indoctrination of a computer manager has been known to be counter-productive in that one began to send papers on highly-technical aspects of the problem in the reverse direction to a security recipient to whom they were virtually double dutch!

Operators and programmers have the best opportunity to defraud and the precautions that can be taken will to some extent be dependent on the number of computer staff involved. If working is continuous with operators on the rotating shift basis, there is less chance of collusion because the job mix will change weekly and give little opportunity of an individual running one job on a permanent basis; operators thereby acting as automatic checks on each other. In writing programs, the work can be organized so that no one programmer is employed in writing a complete suite for a system.

A commonsense checklist of points in accordance with normal usage is suggested.

1. A library should be established for the safekeeping of programs and magnetic tape files. The librarian should keep an accurate record of usage and should not be a member of the computer operating staff; programmers should not have unsupervised access to the library.
2. There should be prior authorization of all computer usage by a senior operations officer. A tight control should be exercised on computer time; the job sheet for a program should be endorsed with the estimated running time, a run of unexpectedly long duration should be queried and an analysis of computer use should be made periodically.
3. A formal standards manual should be laid down and supervisors must see its standards are maintained.
4. No operator should be allowed to work the computer alone; a second operator should sign the log book which should contain times of starting and reasons for any delays.
5. Programmers should not be allowed to operate the computer.
6. When amendments are made to a program in use, it should then

be tested by an independent person. Program documentation should include a list of all changes.

7 Operators should not be involved in the preparation of any operational programs, nor should they be allowed to alter input data.

8 The preparatory work and verification of data should not be carried out by the same person; the work should be batched and controlled so that unauthorized batches cannot be inserted.

9 A log of all errors should be kept with a note of the remedial action and a copy of the printout.

10 With a main application to banks, master files should be printed out periodically and checked for accuracy. Any changes in information should be handled, as far as possible, by personnel other than those handling day-to-day transactions.

## *THEFT: 'INDUSTRIAL ESPIONAGE'*

The concentrated knowledge contained on computer tapes or discs represents a very real risk factor. It is reported that a single incident once resulted in a loss of some $800 million worth of research data. A tape containing the entire name and address list of a company's customers – or confidential plans for the future development of a company, its financial position, or other vital information – could be removed under a coat or in a briefcase. Apart from value to a competitor, at takeover times restricted information could be invaluable to those interested in stock manipulation.

There is a further hazard which has been pinpointed: that of actually stealing *all* copies of particular programs and data and using these for blackmail. Such an episode has had publicity, and others may therefore be tempted to emulate with better preparation, at least in the collection of the ransom! It is therefore logical to take steps to reduce, as far as possible, the number of those authorized to handle *both* original and back-up material. It may be objected that such restrictions will reflect upon the integrity of operators and slow down their work. Nevertheless, this is something which will have to be discussed, and it is particularly important where computer time is sold to customers who will require assurance of the security of their data. Another matter which should be borne in mind is that the copying of a program can be easily carried out by an operator, and the precaution should be taken of discussing with the computer manager how this can best be prevented.

The chance of knowledgeable outsiders entering a computer room to steal programs is remote, but commonsense dictates the strictest limitation on those so authorized. This is not purely on the grounds of theft but also because of the risks of distraction of the operators from their tasks, accidental damage, and the consequential general reduction of the sense of security that is necessary in the room. Indoctrination with this sense of security is essential as the high calibre of the individuals concerned tends to militate against disciplinary or compulsory measures. Restriction of access to the scene of operations may convey a feeling of élitism conducive to the objective, and can be achieved by having a sophisticated form of entry control. There are several varieties of these. One effective device is the small transmitter in the breast pocket with rechargeable batteries, which emits a signal as the carrier approaches the door and releases the lock; it is expensive, and it usually incorporates an additional cost in the form of an automatic opening mechanism. Alternatively, there are locks dependent upon magnetized cards which can be keyed to individuals and provide for the blocking-out of lost cards; a further sophistication is to couple this with a digital system whereby each operator has not only a card but also a personal number to tap out. This effectively precludes any danger due to lost cards, but checks are necessary to ensure that operators do not print their numbers on the back of their cards for convenience! Systems such as these can be linked to a computerized recording system whereby the times of door openings are tabulated together with the identity of those responsible. This does not preclude cardholders themselves from removing material, and a further refinement is security-linked closed-circuit television to overlook the reception area and the entrance to the computer room; this in itself would be a deterrent to anyone carrying material out.

The closed-circuit television can be dual purpose if the computer shift leader is given a monitor whereby he can identify callers at the outer reception door and admit them via a remote control lock without leaving his working position after the reception staff have left.

A further precaution can be implemented where a burglar alarm system is fitted to the computer suite – its keys to be held only by security so that the cardholding staff cannot make unannounced out-of-hours visits without this coming to notice. Security staff would have to set and turn off the alarm, but this could be incorporated in a patrolling routine.

There are sources of risk outside the suite itself, and lack of foresight and rank carelessness there can nullify all the internal precautions. Data, tapes, and discs may be transported backwards and forwards by personnel of low calibre and with total lack of interest in security; confidential printouts, no longer of use, may be left freely accessible instead of being fed into a shredder at first opportunity; current ones may be left casually about; where visual display units are installed, adequate precautions may not have been taken to limit the user's access to the computer and high status executives may well leave matters displayed on their unit which should have been cancelled off.

One factor is predominantly evident: that care should be taken in the recruitment of staff beyond that normally exercised. It should not be forgotten that even junior staff in systems and programming will have access to restricted information in the compiling of data. In the selecting and during the training of staff, consideration must be given to the qualities of loyalty and honesty as well as ability; no appointments should be offered before references and antecedents are satisfactorily checked.

If an employee does intend to pass on information that he handles daily, it is almost impossible to prevent this entirely, but steps can be taken to make matters more difficult for him:

1 A reception area should be created and manned at the entrance of the computer area.
2 Visitors, except with high-level authorization, should not be allowed in the computer room.
3 Unknown engineers, cleaners or others who might legitimately claim access should be asked for their credentials in reception.
4 Mutilated copies of printout should be destroyed and not disposed of with the ordinary waste paper.
5 Duplicating paper used in printouts must be destroyed by shredding or burning.
6 Staff who are discharged for any reason, or give notice under a cloud, should be paid off immediately in lieu of working notice. Consideration should be given to applying this rule to all leavers.

If the last suggestion would appear to be detrimental to a firm's rights or an employee's interests, remember that a disgruntled programmer has been known to wipe off a complete program and at least one individual leaving for another position has taken all details of his employer's current research with him.

Where a firm takes advantage of timesharing on a large commercial multi-access computer and has matter to feed which is vital to its functions it should query what controls are incorporated to ensure that other users have no access. Elaborate password systems are now commonplace and can be supplemented by other measures.

The future of fraud associated with the computer is not pleasant, at least from the standpoint of the prospective loser. Systems now exist to send data direct from a telephone into a computer and an employee with the requisite knowledge of passwords and the like will be able to use this means to get unauthorized access into records from a distance. Increased use of terminals means reduced paper output and more dependence upon controls in the computer – difficulties of efficient auditing are not likely to decrease. Perhaps most important has been the attention drawn to computer crime by documentary or fictional presentations on television and radio which have stressed the ease of performance and the potential rewards. A case in point was provided by a TV series in summer 1983 indicating in some detail not only how to manipulate a banking chain's central computer fraudulently, but also how to circumvent aspects of its physical security . . . and there was no triumph of virtue in the ending, only gloating over ill-gotten and prospective gains! It is of interest that the series also depicted the tactic of fake interview of an employee to obtain general information of procedures, and later that of employee bribery for specific facts required (see p. 293).

In 1987 the prediction made four years ago in the previous edition of this book has become fact: the word 'hacker' has come into the vocabulary to signify a person with computer expertise who makes use of it to access, without authority, computer information available via the telephone network and to interfere with it if so minded. This is good material for the media but the actual impact of hackers is not known. Banking institutions in particular are now tending to use cryptology (coding) for transmission of messages over public telephone systems. The equipment and procedures for this are matters for expert advice; several reliable UK firms market 'data encryption devices' and will readily provide information about them.

The legal difficulties of dealing with the intrusion of hackers into computer privacy have been clearly shown by the dismissal of charges against two men who accessed Prince Philip's personal computer files. In *R* v *Gold* and *R* v *Schifreen* reported in TLR 21.7.87 the prosecution tried to show that the impulses transmitted without authority in the form of the ten figure customer identification number and those for the password constituted in themselves a false instrument. Further, that the user segment upon which they acted, for the brief instant it required to store, check, and authorize access to the data, became a false instrument or device. Both submissions were rejected by the Court of Appeal who referred to an in-court procrustean attempt to force the facts into the language of an Act (Forgery and Counterfeiting Act 1981) to which it was not intended to apply. Their lordships said the appellant's conduct amounted in essence to dishonestly gaining access to the relevant Prestel databank by a trick. That was not a criminal offence. If it was thought desirable to make it so, that was a matter for the legislature rather than the courts – back to the drawing board, in other words.

## *DATA PROTECTION ACT 1984*

After a decade of discussion a Data Protection Act came into being in 1984 which requires data users, that is those who hold personal information records on computers and those who run computer bureaux, to register with the Data Protection Registrar and thereafter comply with certain standards within two years of registration. After that time the subject of the record can ask for a copy of what is recorded about him and can claim compensation if he can prove damage due to inaccurate data or damage resulting from the data user not employing adequate security precautions. The operator of a computer bureau commits an offence by disclosing personal data without the permission of his customer. A problem arises both for security and personnel in connection with suspicion or adverse comment to be recorded against an employee which they would obviously not wish him to see – the easy way is simply to not put personal information into the computer but to keep it on an ordinary record card system or personal file. Regrettably the Act does nothing that has bearing on the prevention of fraud or theft.

The Act does specify that personal data should not be retained for longer than is necessary for its original purpose in recording. In getting rid of it, the responsibility to the individual still continues so reliable means of destruction must be used. There are firms which offer a service for doing this but perhaps the most satisfactory way is internal utilization of efficient shredders or disintegrators which can deal with all kinds of recording material and are increasingly used in financial institutions and the armed forces.

## *OFFICE EQUIPMENT*

### *Desk-top computers*

Microcomputers and wordprocessors have become commonplace pieces of office equipment and as such might be expcted to rank with typewriters as desirable targets for thieves. Difficulty of disposal may be proving a deterrent and there is no indication of any epidemic of stealing them developing. Nevertheless, they are valuable and easily vandalized. They therefore need a reasonably safe environment. The discettes ('floppy discs') conventionally used may be adversely affected by outside influences, i.e. magnetism, and the manufacturers instructions should be carefully followed; the information on them may be sensitive and if so they should be housed accordingly and password controls used to prevent access to their contents. Cleaners and 'office idiots' may play with the keyboard and the obvious way to stop this, and prevent other misuse of the machine or unauthorized access to the information it holds is to key-lock it and for the operator to keep the key in his/her possession.

*Data safes*

The normal fire-resistant safes are not of necessity suitable for the storage of discettes or tapes whose tolerance of heat is significantly less than the paper such safes are designed to protect. Special types are available and it is feasible to use an insulated inner container within the normal safe. Under some conditions, condensation may occur in older safes and this can be damaging to the software. The discette covers again may give precautionary instructions and the preliminaries to the purchase of new safes/containers should include written assurance from the manufacturer that their product meets requirements.

# 21

# Commercial fraud

There is a strong body of opinion in the USA that major criminals are moving out of the sphere of violent crime into that of fraud. There would appear to be some evidence that the same may be happening in this country, and indeed throughout the world, with fraudsters not only mimicking what is being done elsewhere but travelling abroad to areas where they are not known to try their skills. The artistry of these internationals should not be underestimated – a purported Saudi pilot very nearly 'conned' his way into an agency for a firm with the tightest security vetting and only foundered on their 'when in doubt don't' principle which is much to be advocated. He was later found to have succeeded elsewhere.

UK statistics showed a steady increase throughout the early 1970s until stabilizing at about 120,000 offences of fraud and forgery per annum between 1977/79. From the beginning of 1980 a new system of recording multiple, continuous, and repetitive offences resulted in an apparent drop which, if it proved nothing else, at least demonstrated the unreliability of statistics. Frauds more than any other offences are likely to be repetitive and carried out on more than one sufferer at a time, hence their frequency was bound to appear less after such a change. However, the upward trend has continued and during 1986 police statistics recorded 133,431 reported cases of fraud or forgery; 88,828 of these (67 per cent) were detected; 1854 cases of false accounting were included, 1764 of which were detected.

It has been suggested by a criminology research establishment that only one in twelve of those committed are reported to the police; this may be true and pinpoints the gravity of the situation. There is obviously a variety of reasons for non-reporting, one of which for business-people is that of not wishing to be exposed to derision for being foolishly credulous. This is regrettable since criminals obviously rely upon this reluctance in planning their activities. In the higher echelons of management there may be reluctance to voice suspicion of colleagues or take action which might damage the corporate image – which is injured much more if the offender(s) have gone beyond UK jurisdiction when the full extent of defalcations become unconcealable. The 1986 statistics show only 25 frauds by company directors, all detected – it stretches credulity to think that these were all that were committed.

Under modern monetary trading and taxation procedures there are areas where there is a decidedly blurred dividing line between what is regarded as legitimate and what is punishable by law. Nowhere is this more evident than in some methods of carrying on business.

Statistics show an increasing number of small companies springing up and then failing within a matter of months, occasionally with substantial losses to suppliers or customers. Often these failures are sufficiently suspicious to be investigated by the Department of Trade and Industry but without adequate evidence to prove criminal acts. The extent of the risk is shown by the fact that in 1976 the Metropolitan and City Fraud Squad alone recorded £90 million of fraudulent activity during its investigations; in 1972 this was a relatively meagre £40 million. By 1984 it had risen to a staggering £867 million of losses or potential losses with a survey showing that 40 per cent of large UK companies had suffered at least one fraud costing more than £50,000 during the preceding decade. In 1986/87 there have been instances, still being enquired into, which each represent a substantial proportion of the 1976 figure. Even Lloyds and the Stock Exchange have been shown to be vulnerable when the 'reward' has been sufficiently attractive.

One of the many factors in favour of the large-scale fraudsperson is the delay in getting such cases before the courts. This results in part from their very complexity in documentation and from the difficulties in finding out where missing money has gone to. One of the hallmarks of a firm that is trading fraudulently is frantic activity in the movements of its money to the confusion of all interested parties who stand to lose.

The law keeps out of business fraud inquiries as long as possible, because their commencement can complicate the continuation of a firm's existence if the fact that an investigation is taking place leaks out. It follows that firms may trade on a borderline of dishonesty which is not always obvious to an unsuspecting and gullible manager intent on increasing business. There should be an element of self-interest and self-preservation in any manager's attitude towards fraud. His career can be jeopardized if he is a victim to the detriment of his company or his shareholders – he should realize this and take precautions accordingly. The law has provided a number of punitive and other safeguards which it is unnecessary to deal with in detail here. Those of main interest to internal security departments are listed in the Theft Act 1968 (see Chapter 11). The offences in question include the following:

1 Obtaining property by deception, s. 15.
2 Obtaining a pecuniary advantage by deception, s. 16.
3 False accountancy, s. 17.
4 False statements by directors, s. 19.
5 Suppression of documents for gain, s. 20.

## *MAJOR FRAUD*

The gravity and extent of major fraud is emphasized by government action to bring legal and accountancy skills into inquiries and to amend the law so as to reduce the legal loopholes and delays which wealthy fraudsters have exploited.

In 1985 a Fraud Investigation Group was established under the aegis of the Director of Public Prosecutions. This was followed by the Rosgill Commission on Fraud Trials which was critical of the number of bodies involved in investigation and prosecution – 43 police forces, the Department of Trade and Industry, Inland Revenue and Customs and Excise – and it suspected lack of full co-operation between them. It also criticized the customary short-term posting of police officers to fraud squads, normally three years.

The Commission's recommendations led to the setting up of a separate department, the Serious Fraud Office, staffed by lawyers, accountants and ancillary staff to work in conjunction with the other investigatory agencies and having wider powers than the police in that a suspect is to be denied the right to remain silent and has to produce documents – refusal to do so, or failure to answer questions, are now offences. It was established by the election-truncated Criminal Justice Act 1987 and came into being on 1 July 1987. It will be supervised by a Director of Serious Fraud who will be responsible to the Attorney-General. He is empowered to investigate any suspected offence involving 'complex or serious fraud' in conjunction with any other body, and to take over proceedings already commenced. Section 2 of the Act gives the power to compel attendance to answer questions, produce documents and answer questions about them – also to obtain search and seizure warrants to acquire those documents, by force if necessary.

Offences of falsification, concealment, or destruction of relevant documents are created for persons who do these knowing that an investigation by police or the Serious Fraud Office is being, or is likely to be, carried out; the onus is on an accused to prove such actions were innocent.

Proceedings may be transferred to a Crown Court without the usual committal before a Magistrates Court – with multitudinous provisions to ensure fairness. Enhanced powers are given to the judge to 'manage' the trial and safeguards are included in respect of the evidence given under compulsion.

The foregoing may be thought of technical interest only to security investigators but it should not be forgotten that 'from little acorns oak trees grow'. A matter originating within security competence may be the tip of an iceberg which ultimately calls for a major effort by the official bodies.

A useful clarification of the common law 'conspiracy to defraud' is made by the Act which legitimizes the laying of this charge in circumstances where substantive offences have been committed and there has been a conspiracy to commit them; the objective would appear to be to provide a 'backstop' charge in case the others are avoided on legal technicalities – an excellent idea. The wording in s. 12(1) is:

> if – (a) a person agrees with any other person or persons that a course of conduct should be pursued; and (b) that course of conduct will necessarily amount to or involve the commission of any offence or offences by one or more of the parties to the agreement if the agreement is carried out in accordance with their intentions, the fact that it will do so shall not preclude a charge of conspiracy to defraud being brought against any of them in respect of the agreement.

There has been a longstanding legal allergy to preferring conspiracy charges where the offences themselves could be charged; this would appear to be a most useful section to counter that attitude.

Other than in giving peripheral assistance it is unlikely that even the most senior security personnel will be actively involved in enquiries that are categorized as 'major fraud' in the Criminal Justice Act 1987. They will probably be:

1 *Market frauds* – In these, speculators or unprincipled directors

endeavour to manipulate share prices for personal gain by means of unethical practices.

2 *Insider dealing* – Using confidential knowledge of impending transactions in breach of professional standards for personal gain, possibly acquiring such knowledge by corrupt payments.

3 *Management frauds* – These are concerned with senior managers who use their authority for personal gain contrary to the interests of their shareholders by methods which range from borderline misapplication to actual theft.

Other offences become 'major' by virtue of the huge sums involved. The widespread multiple mortgage frauds perpetrated by estate agents/solicitors are a prime example of the almost incredible amount of money that can be obtained by exploiting a successful gambit.

Corruption by public officials may be a category which by magnitude or complexity is referred to the Serious Fraud Office whose powers of compulsion would be invaluable in producing evidence previously extremely difficult to accumulate. Security investigations have on occasions been part of the initial moves in such matters.

*Lesser frauds*

These may involve quite large financial loss without the complexities that accompany major frauds, and a security investigator may be called upon to make enquiries for his employer or client in some detail before reference is made to the police.

*Fraudulent trading*

This is a type of offence in which the victim is most likely to be a manufacturer or wholesaler. In *R* v *William C. Leitch Brothers Ltd* (1932), the judge said:

> If a company continues to carry on business and to incur debts at a time when there is, to the knowledge of the directors, no reasonable prospect of the creditors ever receiving payment of those debts, it is in general a proper inference that the company is carrying on business with intent to defraud.

There is of course a fine but proper distinction between a company which is trading fraudulently and one which is trying desperately to survive.

The credit manager should be able to spot the symptoms of the latter. Invoices will take longer and longer to pay, necessitating letters of complaint and repeated threats of legal action. This becomes a matter of 'teeming and lading', whereby commitments are increasingly deferred pending monies coming in to pay them, and at any given time the company is quite insolvent.

### *Long firm fraud*

This is dishonest in every aspect and one which every credit manager and supplier should be on guard against. The object is a simple one – to obtain as large a quantity of goods as possible without paying for them and then to disappear before retribution can follow.

For this to be successful, the criminal must have capital at his disposal, the patience to carry out a lot of preparation, and sufficient business acumen, particularly in the line of goods he wishes to acquire, to deceive his prospective victims into believing they are engaged in bona fide transactions. It follows that this is a field for intelligent and organized crime.

Having established the objective, which will invariably involve goods that are easily and quickly disposable through market stalls, cut-price shops, or an outwardly genuine chain, the next step is to create an apparently legitimate business. This can be done by purchasing any suitable established one which comes on the market cheaply for any reason, such as death or pending lease expiry. The trade connections are then immediately available and can be impressed by hints of capital being poured in for expansion with the prospect of considerably increased orders.

Alternatively, there has to be a start from scratch by leasing premises, going through the motions of legitimately setting up a firm, probably with an imposing name, and above all creating trade references. This can be done by opening simultaneously a number of one-room 'company offices' or even using accommodation addresses in different towns purporting to be suppliers with which the new firm has had satisfactory trading relationships. Enquiries are then placed with selected suppliers for goods and the credit terms allowed – naturally trade references are then asked for and those given are the prearranged ones, all of whom naturally answer in glowing terms and use impressive letterheads.

First orders may be small and promptly paid but as soon as confidence has been created, maximum ones will be placed in quick succession and the goods got away as quickly as possible until the inevitable happens and direct enquiries are made about non-payment. The organizers then decamp at speed leaving empty warehouses and little indication of their identity. Obviously this is a fraud which will come to light and a description will be available of the main negotiator. He will very rarely be the brains behind the scheme, but a 'front man', whose photograph may not be in police files. He may not even know who the organizers really are and in any case will put up a show of innocence if caught saying he is only acting on instructions and giving fictitious details of his 'employer'.

The temptation that is placed before representatives is that of the 'front man' allowing himself to be led into placing orders for slow-selling lines which the representative is elated to get rid of. Signs to look out for at premises to arouse suspicion are the absence of any apparent filing system, invoices and papers piled in trays, a secretary without knowledge of the business, whose main function is to send out letters inviting reps to call, who has to refer all enquiries to 'Mr Smith' who appears to be the only person with authority, and his invariable absence when needed in connection with complaints. At the warehouse itself, delivering drivers of suspicious mind can assist – and long-serving ones have a 'nose' for such things if they are consulted. Goods not checked off, delivery notes signed without scrutiny, an empty warehouse in dirty condition and no signs of order, carelessness in handling – in fact a general 'don't care' attitude and lack of familiarity with the job – all these will register with the intelligent driver.

There are two official sources of information which can be used to verify the background of any firm with which business is to be transacted. Where it is shown as a limited company then useful information as to its shareholders and standing can be found at the Registrar of Companies and Limited Partnerships, Companies House, 55 City Road, London EC1. If 'Limited' is not shown in a name that includes 'and Company', then lesser but useful information as to the registered owner may be obtained from the Registrar of Business Names at the same address.

Whether a firm can avoid being defrauded in this fashion depends mainly on its credit control measures and the enquiries it is prepared to make about new contacts. The services of the various trade protection associations or societies can be used but perhaps, even better, a call by an experienced representative would establish the true nature of things.

### *Subscription or investment frauds*

These involve offering shares in various speculative enterprises. Subsequently the money is put to the benefit of the organizers rather than the subscribers. This type of fraud uses the inducement of a promised quick and abnormally high profit on monies invested. Typical frauds of this kind are the inducement to invest in pig breeding schemes, where nothing like the stated number of pigs is ever purchased; sales of vending machines for cigarettes and other articles with a promise of excellent siting – rarely if ever fulfilled, and at an excessive price for the machines; and various plans involving the sending of money for placement in connection with bets. One common factor runs through all of these – it is difficult to imagine why any organizer, with such a profitable proposition at his own disposal would wish to openly invite so many others to participate. The Prevention of Fraud (Investments) Acts 1958 deals with offences of this kind.

## *MISCELLANEOUS MINOR FRAUDS*

### *Directory frauds*

These will be aimed at persons responsible for a smaller firm's publicity. They are less likely to be attempted at larger establishments. They are a regular occurrence and impartially distributed over the country. The fraudsman introduces himself as a publisher of trade directories, telephone book covers or similar means of advertising in which those wishing to make an insertion pay pro rata for the space used. He invariably has an assortment of excellent material to show, which may have been specially prepared for him, or may be that of genuine firms but without indication of origin. Payment or part payment is requested in advance but the directory is never published or is eventually produced in cheapened form after very prolonged delays and threats of reference to the police. Many of those defrauded do not complain, and if they do there is always a series of excuses to rebut varying from personal illness and delays by printers to alleged indecision on the part of clients as to requirements.

The only certain way of avoiding losing money in this manner is to deal directly with bona fide established firms and not casual callers.

A variation on the theme is that of requesting verification of entries in telex directories; these are compiled for sale to those needing them, not as a service to those to whom the entries relate. A caller visits the firm, producing a pretyped sheet showing the telex details of the firm and its branches and often also the directory said to be scheduled to be replaced. Confirmation is asked if the details still appertain, and a signature is requested to show that this is so or to approve any variations. If this is given, a sizeable invoice subsequently arrives based upon the pretyped sheet and the signature, the sheet having been worded as an order form. A simple way to get rid of this caller is to ask for a sight of the correspondence or order forms relating to the earlier entries. The printers attempting this fraud have been European, not UK, based.

### *Hire purchase frauds*

The Hire Purchase Act 1964 which provides for a four-day period during which a purchaser may revoke a contract after signature, has drawn the teeth of the fraudulent salesman who by persuasion, misrepresentation or intimidation has induced the buying of unwanted goods. However, legitimate commercial firms who accept payment by hire purchase as part of their normal method of trading are always at risk to the fraudulently-minded customer. Modern selling pressures induce a haste to dispose of goods which militates against checking the bona fides of the customer and his ability to pay. False names, temporary addresses, false references and papers are all employed to obtain delivery when the intention is to sell for cash as soon as possible thereafter and certainly not to complete the cycle of payments.

The motor trade and the finance companies who specialize in providing money for car purchase have particular problems and many of the latter employ their own investigators to trace and reclaim vehicles on which payments have lapsed. Systematic fraud can reach high proportions where payments are continued for a period after cars have been resold but this can be limited if the services offered by firms who keep records of hire purchase defaulters are consulted. Such a service is H.P. Information Ltd, Greencoat House, Francis Street, London, SW1. There are also many organizations which will make status enquiries as well as keeping records of individuals' credit status and transactions.

### *Pyramid selling*

Both this and its fellow fraud, 'inertia selling', which were once very prevalent, have now been made illegal, but both might reappear on a small scale. Perhaps it would therefore be advisable to know what they are in case they are brought to notice by an aggrieved employee; they affect individuals, not companies.

Pyramid selling essentially is that positions of varying authority inside a door-to-door sales organization are awarded to applicants on a pro-rata scale dependent upon their cash subscriptions or investment in the organization; experience or competence are disregarded. Inertia selling is simply the sending out of unwanted goods with subsequent demands for payment.

### *'Carbon paper' frauds*

Akin to inertia selling, these lapsed for a number of years, but instances now occur occasionally where an operator, usually in the

London area, 'tries his hand'. A telephone call is made to the prospective purchaser about lunchtime. The vendor asks for the purchasing officer or, more often, the office manager and in doing so apologises for having forgotten the person's name; he will almost invariably get the full name from the telephone operator, whom he will engage in conversation. If he is put through to the individual, he will endeavour to obtain a sample order for say carbon paper and, if successful, will send a much larger supply with an invoice referring to the verbal order. If he does obtain a name, he will follow the same procedure without speaking to anyone in authority and will subsequently claim to have been deliberately misled by someone to whom he has spoken at the firm – many invoices have been paid in the past on the assumption that this is true. The carbon paper, not surprisingly, is apt to be expensively priced.

*Holiday travel agency frauds*

These again concern individuals, or perhaps organized parties of employees. Small new firms with glowing literature and low prices solicit advance bookings accompanied by deposits. The firm collapses and the facilities offered are then found not to be available. The answer is to deal with established firms of known integrity, or those new ones which are subsidiaries set up by large concerns normally outside the holiday market.

*Distribution frauds*

Most frauds in this sphere involve employees but, if a firm's procedures are lax, it may be possible for an order to be rung in and accepted apparently from an existing customer asking to collect immediately and quoting an acceptable order number from a known sequence – this of course involves either a member, or former member of a customer's staff or one's own. Collection without documents follows and the fraud does not come to light until the customer refuses the invoice.

A simple way of stopping this is to insist on proper documentation being presented even in connection with a most urgent order. Also in connection with distribution, any telephone instructions to change a delivery point, unless there is no doubt about the identity of the caller, should be questioned by ringing back. As mentioned before, if any comments are made by drivers concerning the customers, they should be taken seriously and looked into.

## INTERNAL FRAUD

A malaise of trading in some types of industry is that of a firm's buyer virtually demanding a personal consideration from a supplier for inducing his employer to place an order. This may be reflected in the price which has to be paid. The 'consideration' may take the form of cash, on a percentage basis, or 'presents' in the form of goods. In spheres other than purchasing, a person in authority may be prepared to grant similar favours in the form of licensing permissions, or by employing certain firms to do work for his masters, with a percentage for himself, or by specifying the materials of a particular firm to be used in connection with some project. The same malpractice can happen in a reverse sense where goods are in short supply or subject to quotas, in this instance the customer pays a 'consideration' for the privilege of preferential delivery.

Suspicions and rumours of this type of 'fiddle' in a company may be prevalent, but it is extremely difficult to get evidence to substantiate. The only possibility comes when a supplier of services or goods who has been approached for an honorarium, is too principled to comply and is not only prepared to complain, but also to co-operate in trapping the offender. This pernicious type of fraud ought to be stamped out in the interests of both morality and efficiency but under modern trading conditions it is difficult to see how this can be achieved.

There is legislation under the Prevention of Corruption Act 1906 (see p. 290) to deal with any instance it is desired to prosecute, but unless a firm has suffered considerable loss by its employee's actions, there is a tendency to deal with the matter internally. If a decision is then taken to refer to the police, those involved should not be questioned, otherwise the police investigations may be impeded by prearranged stories.

Christmas traditional gifts in recognition and appreciation of services rendered or trading relationships can raise suspicions. A dictum has been laid down by an Association of Chief Purchasing Officers which could be widely applied with advantage – it is an absolute ban on any gifts being accepted by members of their staffs other than at the Christmas season – then there is no objection, but the presents and their origin must be declared to the chief purchasing officer.

### *Embezzlement*

The legal aspects of this, now covered under the general definition of theft, are dealt with elsewhere (see Chapter 11) and surprisingly enough, more often than not it involves trusted employees of long standing. This is probably due to the fact that anyone else's activities would be subject to closer scrutiny; falsification of records and the embezzlement of monies usually comes to light when the trusted servant is ill, or, against his wishes, has to take a holiday. It is then that irregularities in records or alterations and forgeries are noticed by the substitute.

### *Forgery*

Most internal frauds will involve forgery in some form or falsification of records. An alteration or erasure may, of course, be a simple correction, but it might also be a deliberate act, designed by the person making it to cover up a theft. Increases are easy to make to amounts shown on cheques and in records by adding a zero to the last figure and a 'ty' to alter, say, six to sixty. Wherever anything of this type is suspected, a thorough check is indicated and usually what has happened soon becomes evident.

Where there are no alterations but, contrary to usual practice, pencilled figures have been over-written in ink, these should be regarded as suspect. Forgery of signatures is most frequently done by taking a genuine one and either practising copying it or writing over it, then following the indentation in the paper beneath to form the forged signature. False signatures may seem evident to the eye but proof in court is almost always required by expert evidence to contradict denial of responsibility. Overwritten signatures may be spotted by a minute failure to follow the indentation or by the stilted appearance of writing which seems to have been done laboriously.

## *PETTY INTERNAL FRAUDS*

Ample scope exists in many industries for petty frauds by employees which are frequently dealt with by internal disciplinary measures rather than by reference to criminal courts. 'Double-clocking' is in this category; an employee punches a friend's time card as he himself commences work, the friend arrives say an hour later and is therefore paid for work he knowingly has not done. (Let the 'friend' clock-off before challenging so that he cannot claim he intended to draw 'the mistake' to the notice of his supervisor.) Bonus 'fiddles' likewise; these can be extremely annoying to security is so far as batch production, packaging, goods movement, etc. may be recorded in excess of what can be actually traced, thereby giving a semblance of theft.

Internal sales to employees – there are endless permutations such as items not charged, perfect ones sold as 'scrap', special jobs done and the end product treated as sale of a standard item, scrap may be deliberately created with the intent of cheap sale. Again, a firm which allows its employees to use its facilities to purchase from suppliers at reduced rates should take care that these are cash transactions: it is not unknown for such to be charged to a firm's account and not noticed until the recipient is virtually beyond trace.

If security becomes involved with these, or anything similar, in bringing them to the notice of senior management the stressing of criminal connotations will help to ensure positive action, which may in serious cases be reference to the police.

## *THE INTERNAL AUDIT*

The ideal deterrent in a large company is to have an internal audit department with wide powers of inspection without reference to departmental heads. It is important that they should not need to give courtesy notification of impending visits and should be responsible to a central high authority, preferably the finance director, so as to be unaffected by those in positions of authority who may be involved or who may be embarrassed by their findings.

It is essential that where there is an internal audit organization there should be the closest co-operation between that and the person in charge of security. He may have received rumours of malpractice by a particular individual or in a particular department which it would be inadvisable for him to investigate directly – for that matter he may not have the facilities or knowledge of the recordkeeping to investigate personally. He can however direct the attention of internal audit to this and they can apply their specialist abilities in what would ostensibly be a normal check. Conversely, internal audit may throw up discrepancies for which there is no apparent reason other than sheer carelessness. These should be made known to the security department where they may have relevance to other information that had been previously received.

There is a by-product from such a liaison; auditors are primarily concerned with records of all kinds, books, documents, vouchers, invoices, cheques and records in general, and some frauds will become evident purely from discrepancies in figures and perhaps alterations which cannot be justified. After involvement in inquiries of a security nature they would probably then start looking more closely to see whether receipts and other documents that they may be examining are in themselves genuine, by comparison of signatures, watermarks, number sequence and the like. This liaison is a form of cross-pollination which must be of value to both departments.

A classic example of what can be done by such a liaison has been demonstrated by an 'asset protection unit' set up by a nationalized industry. This created an official combination of security and audit personnel at top level, working together, with a broad remit, and reporting to the central board direct so that local opposition could not countermand investigations. This has proved one of the most effective bodies ever set up for security purposes and is to be commended to all large groups, industrial or commercial. Prosecutions arising from its activities have been numerous, mainly for corrupt practices, and involving many employees of senior status – which poses the question, 'Why had these matters not been actioned before?' with the logical *non sequitur* 'Would enquiries have been carried through to fruition if ultimate responsibility had been to other than the central board?'

### *Comments*

Internal frauds and thefts when successful become repetitive and make a persisting drain on a firm's profitability which may carry on for years – for 15 years in an example quoted in the following chapter which deals with the investigation of these matters. There is a technique for pursuing such enquiries so as to expedite them and provide optimum chance of success; it is recommended that the points raised be studied by anyone entrusted with responsibility for the task.

# 22

# *The investigation of internal fraud and theft*

The regrettable increase in major fraud in public life was referred to in the preceding chapter 'Commercial Fraud'; the extent of the measures now deemed necessary to deal with it (see p. 317) is a clear indication of its gravity. Senior managers and professional people are the offenders; if greed and ambition can warp integrity at that level, is it reasonable to think they are not even more widespread amongst lower echelons? There is reluctance to accept that persons in responsible managerial and executive positions will defraud and steal from their employers but the statistics make it clear that this is becoming commonplace. The 'staff' status of office workers and professionally-skilled employees can no longer be regarded as a near guarantee of honesty. The offences that these and their superiors commit are known as 'white-collar crime'.

Thefts and frauds carried out by such persons are less likely to come to immediate notice than those in which something simply disappears. They are more easily camouflaged, they can be longlasting and profit-draining to the extent they affect the very viability of a firm, their existence may be hard to establish, and their investigation calls for patience, expertise, and a methodical approach coupled with a disbelief in the good intentions of all concerned. Increasingly, investigations are being carried out privately by senior in-company security or specialized contract security firms and referred to the police only when suspicions have been confirmed on clear evidence. If the contract security option is to be exercised it is essential that a personal recommendation be obtained. Whoever does the enquiry, it is fundamental that they know how to tackle the job properly.

The police, not unnaturally, are often reluctant to undertake investigations unless there are clear and substantial criminal offences with a probability of successful prosecution. Enquiries can be time-consuming, tedious, frustrating by reluctant co-operation from the complainants and, eventually, produce evidence of bad internal stock-keeping, or tolerated malpractices which have got out of hand and an offender who, at least originally, thought he was exploiting an accepted 'perk'. Other disincentives can be the opinion that it serves the company right for displaying poor business acumen, that there is a civil remedy for recovering losses, and finally the difficulty of presenting a complicated case in a manner comprehendable to a jury. Occasionally, not just to a jury – a provincial fraud squad has a lasting memory of an octogenarian judge who directed a jury to find defendants not guilty of a conspiracy charge and some days later made the same direction on other charges based upon the same facts, saying they constituted a clear case of conspiracy, and two years' work went down the drain! Perhaps the Criminal Justice Act 1987 may facilitate the presentation of evidence and close loopholes currently legally exploited. It has now come into force (see p. 317) but it will be some time before its effects are seen in the courts.

Similarly, a private or in-company investigator must accept he has a time-consuming and possibly complex task, at the end of which he may find difficulty in getting companies to press charges, or Crown Prosecutors later to accept them. There is also a common preconception that this kind of offence only receives a nominal punishment, as it is considered crime against companies rather than people.

These problems have to be recognized by the investigator who must not let them affect the effort he puts in; even the police will not produce results if they do not commit themselves fully to the job. 'White-collar' crime it may be, but the offences are theft and fraud and the offenders simply criminals.

Computers may feature in an investigation and the fact should not be overlooked that, though they may be the source of problems, they may also be a useful aid where a quantity of material needs analysis, statements and information need to be recorded for immediate retrieval and comparison, etc., in fact for a variety of purposes according to the nature of what is being undertaken. There should be no reluctance or hesitation in seeking the assistance of experts; an investigator should not overrate his all-round competence'.

## *PRESERVATION OF DOCUMENTS*

Whether investigation is carried out internally or assistance of outside investigators is to be solicited, it is absolutely essential to seize all relevant documents at the very earliest opportunity. If not, 'accidents' will happen and material evidence will be destroyed or other steps taken to hamper the enquiry.

It may be that certain books must be retained in a department for use in the day-to-day work of the firm; if this is so, they should be photographed and inspected in the department itself where they would be available for reference if needed and such new entries as need to be made can be done under supervision. It is, however, preferable to remove everything of evidential value, and in the case of a police inquiry, they certainly will.

## *CAUSES OF IN-COMPANY INVESTIGATIONS*

Firms, at least responsible ones, will wish to firmly establish facts before calling on police assistance which, as previously mentioned, is unlikely to be forthcoming if suspicion only of an offence is referred to them. In addition, there are matters regarded as crucial by a firm which they will consider to be non-criminal until proved otherwise, or of an industrial relations nature in which they would not wish to be involved.

There are a variety of matters which a board of directors might well consider calls for a detailed and possibly costly investigation with attendant unavoidable interference with routine:

1 Unexplained shortfalls in cash-flow figures and profitability forecasts.
2 Stock checks throwing up unacceptable discrepancies in holdings/sales of materials and products.
3 Leakage of information about, for example:
   (a) commercial matters – pricing policies, tenders, new products, etc;
   (b) industrial relations – wage negotiations, policy decisions, confidential guidelines, personal records;
   (c) financial matters – budgetary, acquisitions or takeovers, forecasts, salaries;
   (d) planning – future intentions, patent applications, current and proposed research, new building and site acquisition;
4 Suspected corrupt practice in making purchases, awarding

contracts, or in other cases where undue preference has been shown and abnormal costs incurred.
5 Sabotage of machinery, computer interference, unexplained document loss.
6 Persistent and hurtful undetected theft with obvious employee involvement.
7 Any matters which have incurred the wrath of the managing director.

## *OFFENDER MOTIVATION*

Many of the factors that influence any criminal also operate on a dishonest employee who has the expertise required to profit from fraud. An investigator needs to keep these in mind and look for evidence of them as a means of pinpointing his target. Questions about the possibility of individuals having, say a grievance or known external pressures might be asked at the initial briefing – they may not have occurred to less suspicious minds and an enquiry might be thereby short-circuited. Such factors would include:

1 Plain dishonesty afforded an avenue by a vulnerable system.
2 Grievance against the employer rendering him susceptible to temptation to exploit his skills or to be receptive to bribery.
3 External pressures on financial resources resulting from:

   (a) financial over-commitment caused by adopting a lifestyle beyond his means, possibly by a change of home circumstances;
   (b) personal relationship problems;
   (c) gambling losses;
   (d) drink problem or drug addiction;
   (e) cost of illness at home.

4 Thwarted ambition and jealousy of fellow employees coupled with 'Walter Mitty' complex and desire to prove he can beat the system.

Note that points 2, 3 and 4 will be known or suspected by fellow employees and the indicator of the vulnerable point in a system should be shown up when discussing procedures.

## *PROBLEMS IN INVESTIGATION*

There is a very real handicap in the investigation of such crimes, often arising from the simple premise by companies of not throwing good (for investigation) money after bad (losses) especially where the cause of loss has been clearly established and its recurrence prevented – and possibly the person responsible removed for breach of company rules.

Most investigators are not ex-policemen, few of whom in any case have experience in this field, but they are persons with a specialized knowledge, such as internal auditors, accountants, computer experts, etc., or a competent chief security officer, utilizing their skills or expertise. To have the maximum expectancy of a successful outcome the investigators must be given:

1 a sense of importance and worth of their investigations;
2 the facility to discuss progress whenever necessary, and to make decisions;
3 the authority to request and acquire information and documents;
4 the ability and authority to contact other investigative agencies – police, customs, consumer officers, etc., for advice or special expertise;
5 adequate clerical confidential services;
6 authority to co-opt specialist staff to deal with aspects of the enquiry where they lack personal expertise.

The investigator must appreciate that he is unlikely to get full co-operation from everyone he interviews. When a fellow employee is the target, the prospective witness may be reluctant to be cast in the role of an accuser; on the other hand, personal dislike may emerge as exaggeration or innuendo. There will be evasion by those who feel they may be implicated or accused of inefficiency, and a disposition to point blame elsewhere. Persons in senior positions may attempt to 'pull rank' to prevent pertinent questioning; to counter this, a confidential directive from the managing director can be circulated to heads of departments likely to be visited, instructing that every assistance should be given to the investigator in any enquiries that he may deem necessary. One investigator, realizing obstruction would be met with, and aggressive obstruction at that, took the precaution of obtaining a written authority from the managing director that full co-operation had to be given and that 'all queries or dispute about the authority of Mr X in this enquiry should be referred to me *personally* with an explanation as to how and why these have arisen'.

In the course of the enquiry, anonymous information may be received and rumours heard and these should not be disregarded;

they are a means of passing information by persons who do not wish personal involvement to become known. Some will be malicious, some will be misleading, some will be of the 'crank' variety; nevertheless, note should be taken and quiet enquiries made to see whether they fit into the general picture.

Where there is an unexplained deficiency and no suspect, there are no short cuts, other than a stroke of luck or a moment of inspiration, to a long, hard slog of systematic checking, sifting, looking for procedural weaknesses, studying personalities, lifestyles, etc., and establishing as a priority that the loss is actually genuine and without any apparent reason other than theft or fraud. What to look for in those circumstances is anything unusual, unexpected, or irrational that is thrown up in record inspection or scrutiny of actions – anything which may point the enquiry in a certain direction. Rumours come into their own in this situation.

## *MANAGEMENT OF AN INVESTIGATION*

Before an investigator begins his work he must be clearly briefed by those authorizing the enquiry and given an accurate analysis of the facts already known. If there is a suspect he must be told all about him and the grounds for suspicion; he must know any limitations that are put on his investigations and to whom he must report his findings. An indication as to which personnel can be regarded as reliable will be helpful, but should not be taken as proof positive of integrity; a knowledge of those who should have known of malpractice and done nothing about it could be advantageous in evaluating any information they may subsequently give.

The investigator, if he does not already know, should familiarize himself with all the legal elements concerning the offence he is working on – the kinds of evidence he will need, the witnesses, statements, documents, etc. which he will require to substantiate it. He must also cast an eye on the snags that might be encountered and accumulate questions he will wish to be answered by potential suspects. He must appreciate his own limitations and know where to obtain reliable assistance in those fields; internal auditors are especially valuable in that they have both accountancy skills and a detailed knowledge of correct internal procedures, and their department is the one most unlikely to be involved in any defrauding of their employer. In preparing evidence against non-employees, the accountancy branch will be able to accumulate and analyse the paperwork better. Today the use of the computer is a clear example of a company tool used by both the criminal and the investigator, the one to commit the offence and the other to use it as an aid to establishing guilt.

If the investigator encounters opposition in the course of his enquiries, not only must he bring this to the notice of the person to whom he reports but he should ponder the real reason for that opposition and try to establish it. It may be something entirely divorced from his objective but equally culpable in a minor way – as in the bringing to light of the servicing of a car, belonging to a manager's wife, at company expense when the cashier's activities were the real target. Would that manager have been free to voice any suspicions he might have had about the cashier?

In a long-winded enquiry, progress reports will be expected and the frequency of these and manner of submission are matters to be clarified at the first briefing. One managing director was so doubtful about documentary security that he wished all reports to be verbal – the investigator wisely filed a precis of each at his home. Written reports should be terse and factual, rumour and opinion should be specified as such, the intended line of enquiry should be indicated – the managing director may have valuable comment to make on this, and care should be taken not to show undue optimism or pessimism. The penultimate one should be more voluminous with a summary and recommendations as to final action.

A most important point for external investigators: the conditions of remuneration and expenses allowed should be in writing and payment made at agreed intervals in the enquiry. A firm faced with a huge bill at the end of an abortive enquiry may prove argumentative!

## *CONDUCT OF AN INVESTIGATION*

In addition to being familiar with the law relating to the suspected offence, after making sure that all books, records, documents and papers likely to be needed have been safely impounded, the investigator before he starts interviewing and taking statements must make himself word-perfect on the authorized procedures on matters bearing on his enquiry. If he does not, his credibility with witnesses will be undermined and he may miss the importance of seemingly minor details. A thorough examination of the company books etc., with a competent and vouched for manager with the necessary qualifications, should be carried out. It is possible that certain facts will occur to a questioning mind which do not impinge on one which is concerned with figures. In a relatively trivial matter of petty cash, receipts to authenticate expense claims may balance on weekly submission and not be queried, but when receipts over a period are scrutinized it may be found that a string of consecutively numbered ones are being progressively tendered indicative of an 'acquired' book for personal use; or different names in identical writing are appearing showing blank receipts are being obtained possibly for a small bribe.

Armed with all the foreseeable necessary background knowledge, the investigator can move on to potential witnesses. If he can have sight of their personal records these may condition his approach to them, and may be indicative of where to direct his interest if there is no already established suspect. Interviewing and statement taking is dealt with elsewhere (Chapter 13) and note should be taken of the advice given. Patience must be exercised with fellow employees, let them gossip and assure them that any rumours will not be included in statements. Do not rush enquiries and continually reappraise the direction the investigation is going. It may be best to start with witnesses who have only peripheral knowledge or involvement, though a firm will probably tender a senior person to give a comprehensive picture at the outset.

It can happen that at some stage an accounting or other error will emerge to show the investigator is wasting his time, or the target is wrongly defined. He should pull out immediately and enhance his reputation by doing so, or, if he finds the second alternative, ask new instructions. A high-value loss of rare metal turned out to be not employee theft as suspected but fraud by a contractor not returning scrap after processing and a grossly inefficient system of record keeping.

In a complicated enquiry which has reached the stage of an interview with a suspect, it is highly advisable that the investigator should have a clear picture of the line of questioning he wishes to follow. By then he should be able to visualize the answers he may be given and thereby pose his questions in a sequence which becomes increasingly pertinent without alarming or antagonizing the interviewee. To do this successfully he should prepare a list which ensures nothing he wishes to raise is overlooked – it can be highly embarrassing to have to arrange interview after interview to deal with matters which have been omitted by oversight or by losing track of a line of thought after an unexpected answer or acrimony (see also p. 205). The questions can be set out in pro forma style with space for agreed answers, with a copy for the suspect and for his solicitor if he wishes to be represented at the interview. It is good practice not to hand over a sheaf of pro formas but to do so singly as the questions are reached so that the direction of the questioning is concealed and there is minimum prior warning of crucial ones. Throughout, of course, the investigator must have due regard for the relevant Code of Practice under the Police and Criminal Evidence Act (see Chapter 13).

### *Informants*

Malicious informants should be treated with special care, their nature may show in the manner in which they give their information. To follow it without maximum checking is to ask for trouble, but also it may legitimately stem from genuine grievance at having been defrauded personally, or innocently caught up in an embarrassing predicament. Internal matters rarely throw up an informant who wants money for what he knows, if it does and authorization is given, apply the old CID tenet and pay on results. Informants may be called upon to give court evidence and care should be taken to ensure their statements are entirely theirs with no question of pressure or inducement from the investigator. In only the most exceptional of circumstances, one is tempted to say never, should a paid informant be used as a witness.

### *External enquiries*

Circumstances could well arise where contact with customers or suppliers is needed, for example for invoice fiddling or corruption, or with former employees who may know pertinent facts, or for that matter firms in like business who themselves are thought to have been defrauded. If a personal interview/visit is needed, this is best done by appointment; the company secretary may act as an intermediary thus ensuring the highest level of contact. Contacting private individuals by telephone might avoid causing offence or inconvenience resulting in a hostile reception and no statement. If a circular letter-type enquiry is called for, care must be taken in its wording to avoid damage that might result from unfounded suspicion:

1 It must not contain any prejudgment or derogatory information about the company or individual under investigation.
2 Efforts should be made to minimize any harmful implications of the sending of the letter.
3 The letter should be worded to induce positive answers, possibly by using questionnaire tactics.
4 Photostats of any relevant documents, letters, etc. should be requested.
5 Co-operation and early response should be asked for in appropriate business terms.
6 In the case of a letter to a private individual, enclose a stamped and addressed reply envelope.

Obviously, such letters should be sent out with a company secretary/

managing director signature so their validity and importance are not queried. If confidentiality within the originating firm is important, pre-addressed envelopes marked 'confidential' could be sent out to all.

*Final report*

This should be a comprehensive one with copy statements attached, together with lists of exhibits: all originals and the exhibits themselves should be kept in a place of safety pending decisions on action. This report may be directed to a legally lay person so that an assessment of the evidential value of what has been discovered will be useful to him, as would opinion of the validity of witnesses statements. Any problems on action of any kind – disciplinary or court proceedings – should be emphasized.

*Contract investigators*

A case ending at Luton County Court on 11 February 1988, believed to be Britain's first prosecution for telephone 'bugging', will give cause for thought to those using external investigators; the Judge's comments will cause apprehension in a much wider field. The circumstances, as reported in *The Times*, whose Law Reports are legally quoted, on 12 February, are concerned with a failed take-over bid in which the losers thought leakage of sensitive information could be attributed to former senior executives who had joined the target firm. The losers had accordingly engaged a private detective agency to prepare dossiers on the private lives of those executives, over a considerable length of time.

Apart from other actions, a device intended to pick up conversations within a home by converting the vibrations of the windows was used. The Judge deemed that this constituted 'quite proper enquiries', but then a 'tap' and tape recorder were connected to the telephone lines which constituted an offence against the Interception of Communications Act, 1986. The owner of the detective agency received a prison sentence, part suspended, and the electronics experts acting under his direction were also prosecuted; it was said that there was no evidence as to whether the scheme had been suggested by the loser firm or the agency. Very large and reputable firms were invoved and the publicity could not be other than embarrassing.

The remarks attributed to the Judge are disquieting, namely that companies are entitled to engage in industrial espionage such as undercover surveillance work and electronic eavesdropping . . . that they had a legitimate right to protect their commercial interests so long as their agents did not resort to tapping telephone lines. A seal of approval seems implied to methods that many might think unethical; it is likely the dictum will be the subject of comment when it has been assimilated by interested parties.

## *COMMENT*

In addition to the legal and technical knowledge, above all a successful investigator needs patience and persistence. He must be methodical and meticulous in appraising, assembling, and presenting facts, and evidence. He must create confidence in those who employ him and at all times keep their interest paramount in his planning. It is most unlikely that he will be given authority to inform the police of a crime without prior discussion, that onus is vested in the firm that is paying him and it may decide on a civil remedy, though he will be required to act as intermediary if the decision is to refer. If he perceives industrial and/or union trouble arising from what he is doing, he should seek guidance before continuing – a lightning protest strike might cost more than the losses. He must have the techniques to elicit statements from unwilling persons without creating strained relations, either with himself or amongst the employees themselves.

As already mentioned, status must not influence judgement of a person's integrity or truthfulness, white-collar crime can be committed by the whole gamut of employees and it is generally regarded in this country as a 'management crime' with greater status equating to greater opportunity and greater potential loss. A director may have the same motivation of greed, jealousy, or resentment as the lowest clerk but with superior knowledge of the system to exploit loopholes.

Invariably, the investigation will spotlight such loopholes and it should be considered as part of the job to draw these to notice and recommend steps either to stop them by introducing preventive measures into the system, or to insert checks which will bring any discrepancies to early notice thereby minimizing temptation, cutting losses, and possibly enabling an offender to be trapped. There are few things better for this than an active internal audit department.

# 23

# *Shops and supermarkets*

Retail trading premises have all the potential security weaknesses of other business premises, and a few more besides. Open counter trading, customers *en masse* legitimately handling goods, massive and speedy turnover of stock, frequent changes in staff, pay scales that are not always an inducement to integrity, till and cash manipulation by junior staff, the constant presence of tempting articles of daily use – all these contribute to losses which are estimated to be far higher than those of any other industry. Many regard the self-service form of modern retailing as a major factor in the tendency of the public to regard theft as less reprehensible than formerly. Stealing has been made too easy; school children have come to think of shoplifting as a sport and an avenue for small luxuries that they would not otherwise have; organized gangs can make a good livelihood at it; and innumerable reasons have been thought of, and publicized, to excuse those unfortunates who have been caught. The clock cannot be turned back, and this form of selling is here to stay – but firms that do not take steps to stop the drain of theft will have their profitability eroded in a competitive environment where they cannot afford to compensate by price increases.

Estimates of loss are largely speculative, but the fact that in the past four years 'guesstimates' have risen from £500,000,000 to at least £1,000,000,000, speaks for itself. Accuracy is complicated by inadequate documentation, poor stock checks, clerical errors, and failure to account for genuine wastage by damage, and the simple fact that no comprehensive loss reporting system exists or is likely to be feasible.

Causation, though often wholly attributed to shoplifting, in the opinion of many senior security personnel is at least 50 per cent due to staff dishonesty and collusion with outsiders. This is not a purely British problem; it is worldwide, and if the percentage of visitors to this country who are caught is any indication, the UK is probably no worse than elsewhere.

## *PLANNING AND POLICY*

More than in any other industry, theft prevention must be a management policy dictated from board level. It is scant reward for effort if sales and profits both rise but the percentage profit falls because basic precautions are thought of as an encumbrance. A store manager criticizing a seminar speaker once said: 'I am in business to sell, not catch thieves.' The reply to this was: 'I am one of your shareholders. You are in business to make me as much profit on my money as you can, and stopping thieving, not catching thieves, is what I want you to do' – which just about sums up the objectives.

Security, as always, must be a planned and calculated effort whereby, for economic reasons, some remedial possibilities will have to be discarded as not commensurate with the possible loss. Customer resistance may also be a factor to induce a commercial decision to accept a risk; this has happened in some instances where closed-circuit television has been proposed in departments of expensive sales, though it is noticeable that this has received less publicity in recent years as if those who objected to surveillance now regard it philosophically as a necessity.

The first steps are to try to establish broadly what are the extent of losses, where they occur, and who is likely to be responsible. This can only be done adequately by a series of stock checks and audits which will be obvious to the staff and thus a deterrent in itself. Whether a problem is revealed or not, the general extent of retail losses is such that some preventive measures should be taken even where a reasonably acceptable situation exists, as it can very quickly deteriorate if action at other premises nearby drives the shoplifters away from them. The assessment will, however, indicate the degree of expenditure that need be incurred.

*Dismissal and prosecution*

A psychological step which costs virtually nothing is to agree at board or owner level a concrete policy towards detected stealing by staff, customers, or others and to publicize it so that no doubt is left of the consequences that will follow detection. A clear statement to this effect should be included in staff instructions or other forms of rules issued to employees. Dismissal should be an automatic consequence, and the phrasing 'will be dismissed' should be used, not 'liable to be dismissed' which has not found favour with industrial tribunals. As advocated in Chapter 2, a discretion should be kept *via-à-vis* prosecuting. For customers a printed notice near the entrance should be displayed, and in areas where foreign visitors are numerous this could be in several different languages – especially useful for the juvenile influx from the continent in the southern counties during summer months. The wording of the notice should make it clear that shoplifters *will* be prosecuted. In practice, the same considerations that apply in other industrial and commercial establishments should appertain; that is, the interests of the firm will take priority in deciding whether to call in the police. An area which also deserves such warnings is that of loading bays and the 'goods inwards' department where non-employees in the form of delivery drivers will be.

*Search*

A search clause in the conditions of employment is invaluable even if a decision is made to restrict it to handbags, shopping baskets, etc. However, it must be clearly spelled out who is empowered to do the search in the absence of established security staff on the premises; the manager and his assistant are obvious nominees. Spot checks should be intermittent but sufficiently frequent to make staff feel that they are likely to be searched at any time. To make the task easier a specific staff entrance should be designated with cloakroom facilities for outer clothing and handbags, etc. near the door. The size of many establishments will enable an office to be sited near this point which can be used for wages, personnel, or other purposes whilst also supervising the outer door. Instructions that staff will not use customer exits are desirable but much harder to enforce, particularly in supermarkets where assistance is given to customers in handling bulky purchases to cars.

A separate staff entrance may not be feasible for a smaller shop with limited workforce, and indeed the layout of larger premises might render it impractical. In such instances there should be a ruling that all staff leave together and by the same door, with the manager or a designated keyholding supervisor being last out; this is a responsibility best left to the manager unless there are cogent reasons to the contrary. The objective is to ensure that no employee leaves the premises without being scrutinized – or at least feeling that he is being so treated. Instructions should govern leaving during working hours or during lunch breaks, and there are obvious advantages in providing subsidized meals or good restroom facilities to lessen the inducement to go out.

Individual lockers should be provided for staff, and any form of carrier or container that they have brought into the premises must be left there; an absolute veto must be laid on taking them into the selling or stock areas. Many firms provide staff uniforms and smocks without pockets to hamper the removal of goods from counters.

### *Recruitment*

The fringe benefits of large retail organizations and stores, with prospects of continuity of employment and career development, lead to more stable staffing than at many supermarkets where there is rapid turnover of younger personnel. In either case as much care as time permits should be taken in the selection of employees. Too often a vacancy has to be filled quickly; and a manager, seeing an applicant who appears suitable and not wishing to risk losing him/her to another employer, offers the job forthwith without checking on references. It is much easier to acquire than get rid of a bad worker under current legislative conditions, and time spent in telephoning two (if possible) previous employers is time well spent as the best of interviewers can be deceived. If an experienced personnel officer is available, initial recruitment interviews should be his/her responsibility, with final selection in company with the manager of the department or shop if there are a number of acceptable candidates.

### *Staff purchases*

Allowing the buying of goods at preferential discounts has long been considered a means of reducing the temptation for staff to 'help themselves'. To what extent this is true is debatable, but it is most unlikely that the dishonest minority will in any way be influenced by the concession. What is not in dispute is the absolute necessity to ensure that no additional means for theft are created nor any cover for stolen goods is provided by the acquisition of receipts. The purchase procedure will depend on the nature of the business; but its general objectives will include preventing overpurchasing for resale, substitution of more valuable items for those paid for, use of the facility by non-employees, the inclusion of stolen articles amonst bona fide purchases taken away, and collusion between employees to defraud their employers.

The larger multi-department stores may find it necessary to give their employees means of identification by the issue of concessionary cards or even a photo-identity card. A variety of ways are in use for the granting of the discount. For accounting ease the full price is sometimes charged at the point of sale with a percentage refund when the receipt is subsequently presented at the cash office. Care should be taken that this is not subject to abuse, and the receipts should be marked at the time of sale (not by the buyer) to show that they are a staff purchase. Purchases exceeding a stated retail value, say £5, should be further validated by a voucher signed by the departmental manager showing purchase price, retail value, and discount, which should be handed in at the time of payment. A virtue of this system is that it enables records of staff buying to be kept centrally for a periodic inspection that might throw up instances of abuse.

Perhaps the soundest system is to have cash payment at one specific till or cash point near the exit after customers have left, the till being supervised by a manager and the staff leaving immediately after paying; payment by cash saves unnecessary paperwork and may preclude opportunities to falsify accounts. Receipts should be given; and if it is policy to record who is buying what and how much, these and duplicates or till roll could be marked to identify the purchaser. Care should be taken if receipt books are used with duplicate or triplicate copies and intervening carbons as these can be easily falsified in collusion between buyer and till operator. Independent supervision at the point of payment is a decided advantage. If purchases are allowed from any till, a staff purchase book should be kept at each and a sale record made immediately.

If a normally quiet period is allocated for staff sales, say 9 a.m. to 10 a.m., provision must be made for purchases to be taken away from the sales area and lodged in lockers until closure time.

## *TRAINING AND INDOCTRINATION*

There are fundamentally three different types of training required: for ordinary shopfloor staff; for management who will have to make decisions, control display layout, and perhaps exercise the role of store detectives in the absence of any such full-time operatives; and of course for store detectives themselves.

### *Shopfloor and warehouse staff*

The general instruction in their duties should be security biased, emphasizing with examples how they can be deceived by customers and suppliers. If this can be done in such a way as to convey a sense of personal affront as well as an attack on the employer, so much the better. Strictly on security, point out:

1. The effect of losses on profitability and thereby on the viability, amenities, and bonuses of the concern.
2. The inconvenience and misconceptions caused by documental and procedural errors.
3. The unnecessary wastage from damage and bad housekeeping.
4. The policies of the firm towards theft by staff and customers.
5. The forms of theft and fraud that are prevalent in their type of trading and the steps taken to counter them.
6. The procedures that should be followed if they observe or suspect dishonest behaviour.
7. Emergency procedures for fire, bomb threat evacuation, explosions, etc.

These are generalizations, and specific instructions will have to be given on normal routine matters which have security connotations – for example, till discipline, cheque acceptance, customer supervision, goods receival procedures, staff purchasing, and other internal instructions.

The action expected of sales staff observing shoplifting will have to be governed by commonsense and the availability of supervisory personnel. At least one large organization will not allow shopfloor staff to challenge suspected thieves whilst another has rewarded detections. 'One against one' situations must be avoided; it is essential that a manager or supervisor should be contacted immediately suspicion arises, and he/she takes charge of the action thereafter. A prearranged signalling code by lights, buzzer, or other means can be arranged to get this assistance without alerting the suspect.

### *Managerial and supervisory staff (see also p. 19)*

In addition to those matters mentioned for their subordinates, the higher grades of staff must have the knowledge and confidence to make decisions on security matters which will be well founded and not expose their firm to criticism by courts or actions for wrongful arrest. They have also a preventive role in limiting opportunities and temptation for theft.

Regrettably they must work on the assumption that their subordinates will steal and must watch their behaviour accordingly, making sure that laid-down procedures are adhered to. A point for particular notice, which may not be apparent on the first occasion that it is seen, is the occurrence of collusion between a till operator or cashier and a relative making a purchase. Agreed disciplinary rules and procedures should prohibit any employee from conducting a sale to a near relative as the temptation to grant a favour is too great to be an acceptable risk.

### *Display layout*

This is often made without thinking of security risk; for example, sales points of high value and small bulk items are too often sited near exits, thereby attracting the attention of the snatch-and-run thief. Ideally, a sales display should be arranged so as to provide maximum unimpeded view of customers' actions. If blind spots are unavoidable, mirrors can be advantageously placed to allow observation from a manned position, or actual or dummy closed-circuit television cameras be used. This question of visibility should be considered during the design of new premises, which might allow for the positioning of office accommodation on a mezzanine floor looking over the sales area or for a manager's office that is glass sided and elevated from the shopfloor – distinctly useful in a single-level supermarket. One of the objectives for management training is to induce consideration of preventive steps of this nature. As mentioned in Chapter 2, there is no better person than the manager, properly guided, to appreciate the immediate risk in his department and to develop workable economic solutions.

## *HANDLING OF SHOPLIFTERS*

Shoplifting is stealing, nothing more, nothing less; and all the guidance given in earlier chapters on the law related to stealing, powers of arrest, evidence, etc., is equally applicable. Frequency of arrests and court appearances, however, are much greater than in other spheres of business security, and it is the only one where deliberate attempts to induce false arrests may be encountered. A manager has to bear this in mind when deciding his actions and should not be oblivious to the possible defence and excuses that may be raised.

Legally there is no need to wait until a thief has passed the final checkout point that he could conceivably use in a shop or supermarket before a theft charge can be substantiated – *R.* v *McPherson and others* 1972 1973 CLR 191 – but in practically all cases it is desirable to allow this so as to confirm the evidence of dishonesty. Many firms have strict instructions that no arrests shall be made until the thief has left the premises, which in a very busy shopping thoroughfare is tantamount to allowing a chance of escape and possibly a struggle in the street. There seems little reason why the entrance to the shop should not be an acceptable venue, since the extra few yards are not going to give additional confirmation of lack of intent to pay when every possible cash point has been well passed.

If the suspect can be kept under observation continually from the time of taking the stolen property, so much the better since this will cancel out several possible defences, including that of genuine purchase. If other items have actually been bought and the till receipt is produced in proof of this, it should be retained and checked – it is a possible exhibit and may be required to refute allegations of having paid for the goods in question. For the same reason the audit roll from the till where payment was made should be taken and kept. Illness, drunkenness, and drugs are conventionally put forward as explanations/excuses; these may be genuine and have bearing on deciding whether to prosecute. The condition of the offender should be noted, questions asked sympathetically about illness, and the production of any drugs requested.

The training of managers and supervisors should therefore include instruction on:

1 The nature and meaning of 'theft' (Chapter 11).
2 The private individual's powers of arrest (Chapter 9).
3 The degree of force permissible in making an arrest (Chapter 9).
4 The manner in which an arrest should be made (Chapter 9).
5 The manner of speaking to a suspect and the Judges' Rules to be observed (Chapter 13).
6 The firm's policy towards persons found stealing and any internal procedural steps that have to be followed before calling the police (Chapter 2).
7 The police procedure on accepting a prisoner and the making of statements (Chapter 10).
8 The court procedure that will ensue (Chapter 14).

Of course, the various ways in which thefts can be carried out by both staff and outsiders should be outlined, and there are several excellent films available to demonstrate these. This is an idealized training content, and it is to be regretted that all too few firms even approximate to it.

*Shoplifter arrest procedure*

The routine that a firm might require to be followed before the police are called is worthy of mention. The ultimate discretion to prosecute is usually vested in the store manager. The arrest should be carried out by the supervisor, departmental manager, or store detective together with the witness of the theft in a quiet tactful manner, with minimum of fuss to attract bystanders' attention.

The employees should introduce themselves, tell the individual the reason for the arrest (if a store detective is involved, it is advisable that he/she should caution the prisoner in accordance with the Judges Rules), and ask that he/she should accompany them to the store manager's office. Care must be taken *en route* to make sure that no property which could have been stolen is dropped or otherwise got rid of; one walking with and one behind the suspect will achieve this.

At the store manager's office a full account of what has been seen and done, together with any explanation already given, should be related to the store manager in the presence and hearing of the accused, who should then be asked if he/she wishes to say anything or has any explanation to proffer. A request should then be made for the production of what is alleged to have been stolen. This is usually agreed; but if there is refusal, this is not a matter of great moment since the person must not be left unguarded before the police arrive. It is strongly suggested that, if the evidence indicates guilt, the police should always be called if the offence is denied. No discretion to only warn against repetition of the action should be exercised; otherwise the person might be tempted to think that a chance exists to get money for illegal arrest.

Two decided cases are of special interest in connection with this conventional procedure. In *R* v *Nichols* 1967 51 Cr. App. R. 233, a Court of Appeal case, the appeal was based upon the fact that the store manager, after the facts had been related in the accused's presence, asked her if what was said was true without the preamble of a formal caution. The judges held that the shop manager was not charged with the duty of investigating offences and therefore was not obliged under Rule 6 of the Judges Rules to administer a caution. The other case – *John Lewis & Co.* v *Tims* 1952 AC 676 – relates to an attempt to claim damages for false imprisonment based upon the delay involved between the arrest and the subsequent making of a decision to call the police. This was an instance where the charge against one of two persons arrested was withdrawn with permission of the court. It was held that the taking of the respondent to the shopowner's office to obtain authority to prosecute was not an unreasonable delay before handing over to the police as required by law. This of course would apply in other circumstances where an employee was detained pending a decision by higher management on prosecuting.

*Store detectives*

These may be specially trained or recruited employees of the firm or contract detectives from a commercial security firm. The pros and cons of using these alternatives are discussed in detail in Chapter 36.

It can be assumed that store detectives from a commercial firm will have been trained in the rudiments of their job; indeed some have experience over many years and are highly efficient. Unfortunately, however, relatively few of the security firms have such a service to offer. Usually their work is covered by an insurance policy against the consequences of wrongful arrest, which takes one source of worry away from their customers. No doubt similar policies can be negotiated for internally employed detectives. All that such outside assistance would need by way of guidance would be the procedure and policy that their customers wish them to follow by way of reference to the police and prosecuting.

In-company store detectives should undergo all the training given to other employees and to managers, but added to this must be consideration in depth of the ways in which shoplifting can be effected and the frauds that can be perpetrated by staff and outsiders. They should be encouraged to develop liaisons with colleagues employed at adjacent premises with like problems, with the objective of full and regular exchange of information to mutual advantage. In the centres of many of the larger towns such arrangements do exist in the form of 'early warning schemes' to report the movements of known shoplifters when seen in the premises of any of the participants, so that continuous observations can be kept on them. Such an operation concerns the larger city-centre stores primarily: experience has shown that liaison schemes can become unwieldy by sheer size and problems can arise from policy differences between the participants. Nevertheless, they have proved effective for both detection and deterrence; details can be obtained from local Police Crime Prevention Officers and in most areas the police do give considerable assistance.

This is an opportune point at which to mention 'APTS' – The Association for the Prevention of Theft in Shops, 6–7 Buckingham Street, London WC2N 6BU, the Director of which is Lady Phillips. It is now some years since this was established by a consortium of the larger retailing groups to co-ordinate and advise action to reduce stealing from shops and it has done excellent work, the value of which is not solely restricted to the retail industry. APTS operates an information desk (01 839 6614) between 8 a.m. and 4 p.m. on weekdays which gives a free advice service to its member companies and can furnish information about training courses for store detectives.

## *CASH POINT MALPRACTICES*

### *Test purchases*

Malpractices at cash register points are amongst the major risks, and when these are suspected checking is best allocated to an outside agency that specializes in test purchases. In-company detectives will be known to the till operators, and their value therefore is limited. The knowledge that a firm does use such agencies will in itself be a deterrent, and the dishonest employee can be further impeded by the use of automatic change dispensers and random checks on the audit roll of the till.

The layout of pay points at most supermarkets and large stores nowadays makes it much more difficult for an operator to gain advantage by 'under-ringing' and verbally asking more than is shown on the indicator. If the latter is seen to be obscured in any way to hamper the customer reading it, that is clear ground to suspect overcharging for the operator's illegal personal profit.

### *Coupons*

A till operator may have an opportunity to defraud when coupons are handed in. In practice the full amount should be rung up, the coupon accepted, and a cash refund made. However, if the operator rings up a reduced price for an item and tells the customer that she is doing so to compensate for the coupon, the customer may not object, and the operator can then withdraw the equivalent of the coupon from the cash drawer, substitute the coupon, and still have a balancing audit roll. Test purchases would immediately bring this to notice, as should spot checks on the till.

### *Till discipline*

The importance of this in the training of new staff cannot be overemphasized. Apart from security considerations, customer relations demand efficiency and accuracy, and only a fully-trained person should operate a cash register. The till gives proof to the customer that payment has been made and to the assistant that payment has been received, and it provides correct documentation of the transaction to the employer. Insistence should be placed on the handing of receipts to customers, and it should be stressed during instruction that the presence of apparently unwanted receipts at a cash point could cause the motives of the operator to be suspect.

To prevent opportunities for snatching quantities of notes from a till there should be regular collections of these during trading, either at specified times or when the operator considers that the cash drawer is becoming too full. Such abstractions must be properly documented and receipted in themselves to protect both operator and collector in any subsequent shortage investigations. The time of collection affords opportunity for a snatch, and the collector becomes an increasingly more attractive target as the monies collected from a series of tills accumulates. A male colleague solely there for protective purposes would be advisable where several tills are to be 'milked' of their excess notes before the bulk is handed in to the cashier's office. This early removal will assist cashing up at the end of trading and will have an additional advantage in that accumulated monies can be progressively banked during the day or at a time that can be varied to reduce the chance of a planned attack. It is strongly recommended that the services of a cash-carrying company be utilized where large sums of money are involved.

A fact worthy of note is the refusal of one of the largest insurers to accept a claim where a considerable sum of money was stolen from a till left open and unattended. They held that reasonable care had not been exercised and that this invalidated the policy.

Ultimate cashing-up is preferably done when customers have left and at the cash register, using a suitable form which identifies the register and the assistant in charge – also of course the total cash with an analysis showing composition by floats, cheques, Barclaycards, etc., and any other analysis that the employer feels desirable. After the total has been checked and recorded it should be compared with the audit roll – to do this first might lead to temptation to abstract cash if the total exceeded what was shown in the roll. Investigation should be made in the case of a substantial 'up' or 'down', and a record kept for each operator's discrepancies could be revealing. Experienced scrutiny of the audit roll may also disclose matters worth looking into – for instance, too frequent appearance of 'no sale' is a matter for query. 'Voids' – that is, mistakes on a till roll – must have been initialled by a supervisor on the back of the roll at the time; the operator, having made the mistake of say ringing up £50 instead of £5, should have called the supervisor at once to note the error. Floats should always be taken to the cashier's office and withdrawn again for the next day's trading, not left in a till drawer overnight.

As advanced types of cash register are introduced, including those used as point-of-sale terminals linked to an in-company mini-computer, the risk at this stage of the operation is likely to be progressively reduced.

*Cheque and credit card acceptance*

Apart from being a source of loss, worthless cheques lead to enquiries and correspondence which are time and profit wasting; and the acceptance of wrongly drawn cheques, whether tendered deliberately or accidentally, causes unwanted delays in payment. Proof of identity and guarantee of payment up to a specified amount is confirmed by a cheque card, but when a cheque book is stolen the card usually goes at the same time, as will other identifying documents in a personal wallet or handbag, for example, a driving licence.

On the reverse of a cheque card are conditions that must be followed for the honouring of cheques, and these must be the cornerstone of instructions for handling them:

1 The cheque must be signed in the presence of the payee.
2 The signature on the cheque must correspond with the specimen signature on this card.
3 The cheque must be drawn on a bank cheque form bearing the code number shown on this card.
4 The cheque must be drawn before the expiry date of this card.
5 The card number must be written on the reverse of the cheque by the payee.

Cheque cards must always be taken out of the wallet in which they are carried for examination.

Practically all retail businesses accepting cheques require the presenter to record his name and address on the reverse side of the cheque, usually using a stamped-on pro forma for the purpose that also requires the acceptor to note the cheque card number and particulars of any other identification presented; details to confirm the point of sale and acceptor are also needed. Any establishments not using these precautions should certainly do so. Discrepancies or inconsistencies between information on the reverse and what is on the front should be challenged. Some examples are:

1 name spelled differently;
2 initials wrong;
3 bank in different area from address;
4 address apparently fictional;
5 signature on front prefixed by Mr or Mrs;

6 on personalized cheques, initials or spelling of signature different from what is printed;
7 notable changes in signatures on front and back;
8 absurd care taken in forming the signatures.

Presigned cheques and those signed with felt pens should not be accepted. Other matters that should be observed include possible postdating, backdating, wrongly dating, not dating at all, entering wrong amounts, written amounts not agreeing with figures, and not signing. Any alterations must be initialled. All cheques over an agreed minimum sum must be 'sanctioned' by a supervisor before acceptance, thereby providing an additional check by a more experienced person who also signs the reverse of the cheque. Self-sanctioning by the employee accepting a cheque is not to be permitted in any circumstances.

Manuscript cheques of the type that can be purchased from stationers should be subjected to the closest check, and very many firms will not accept them except from a known customer.

Use of cheques opens an avenue for a till fraud by the operator. The person tendering a cheque for a single item may not notice that the operator has not rung up the value of the purchase; the cheque is then put on one side and subsequently substituted for cash from the till so that the audit roll remains correct. A preventive measure could be the enforced recording of all cheques as received.

*Cheque fraud prevention*

While carefully observed routine is a deterrent in itself, a more sophisticated measure is available in the form of a double-lensed camera which simultaneously photographs on a single print both the cheque and the presenter. This is an economically feasible process where cheques of relatively high value are permitted. A further development is a device which records the purchaser's thumb and forefinger prints. Such devices, though deterrents and means of convicting a fraudster, do not stop the commission of the offence, and on-the-spot commercial decisions must occasionally be taken as to whether to accept a risk or forgo a sale. If a purchase is to be delivered to a customer, cheque clearance can be effected before this is done. If the cheque is tendered during normal banking hours, immediate confirmation of validity can be obtained from the bank. But if banks are closed, the amount is substantial, and an unknown customer wants to take his purchase away at once, that is the time for a manager to exercise the judgement for which he is paid. A store dealing in high-value items should have clear rules for dealing with cheques over an agreed amount which may be quoted to a genuine customer to minimize irritation at delays.

*Credit cards*

Credit cards, as prophesied, have become a progressively greater source of loss; they are similarly stolen along with identifying documents which in any case are rarely requested when the card is tendered. Cards have been stolen in bulk, they have been reported lost when actually sold to someone who makes temporary use of them, they have been obtained by false pretences using temporary addresses, they have been switched during purchases – who checks that the identical card is returned when paying for a restaurant meal? They have been altered, forged, you name it, it has been done if there is possibility of gain. Credit card companies exist in an environment where margins are critical and continual efforts are made by them to produce cards which resist fraud – holograms are one example of many efforts. Practically all the frauds with these cards are carried out in a retail environment so staff training is of utmost importance with strict rules about card acceptance to be adhered to whatever pressures exist at the time. Apart from non-expiry and validity of signature checks, the blurred nature of a signature on a card, or it being in felt tip pen should be advised as suspicious – as should be any impatience or unease of the customer. When the salesperson has any doubts, the instruction should be to call for authorization, irrespective of whether the amount falls within the normal limits; if the card is being misused the customer will at that point disappear at speed. A system of rewards for detecting fraudulent use is operated by credit card companies, for the recovery of 'hot' cards. If an employer is prepared to supplement this by his own incentive scheme, it will stimulate staff scrutiny and help increase job satisfaction by catching an offender. Credit card companies are not unnaturally reticent about the extent of losses but they are substantial. This financing of buying is here to stay and the cost of losses ultimately devolves on the users of the cards; it would be taking an unrealistically optimistic view to foresee any diminution in the problem in the immediate future.

## *PHYSICAL PROTECTION OF SHOP PREMISES*

Relatively few retail premises will have round-the-clock protection by resident caretakers or security staff; and if their contents are insured, they are very likely to be required to have both burglar alarms and electronically-operated fire alarms. In the future, devices like 'Telemitter' (or 'Intermet'), which transmit both signals plus others from process sensors to a central station, will find increasing use in larger premises (see Chapter 34). Again there are no differences between shops and other business premises, and all comments relating to guarding, patrolling, and alarms are equally applicable. The real differences lie in the encouraged presence of large numbers of non-employees and location in the middle of busy shopping areas.

Design of the building should be such that it is easy to clearly segregate the sales departments from the storage and office areas. Thefts are not simply confined to the shopfloor; and if the opportunity is there to wander 'mistakenly' into offices, the 'sneak-in' type of stealing will be easy and profitable. Staff should be encouraged to challenge immediately any non-employee whom they encounter outside the normal selling areas. Wearing of a staff uniform, possibly with name tags, would be helpful for this – though if a determined thief could acquire an identical garment, he could probably walk about with a good degree of impunity.

Fire doors also are a heightened source of danger; and there is merit in having an alarm system in which all such doors are on a separate circuit that is permanently 'live', so that if any are opened the alarm is activated. The normal crash bar or panic bolt is best replaced by the variety which incorporates an audible alarm, which will at least limit what is taken out if not stopping the egress. Both electronic and audible alarms have the additional advantage of stopping staff from leaving the premises surreptiously via the fire door route for private reasons when they should be at work. The absolute minimum number of doors, other than customer entrances, should be left unlocked at any time.

### *Keyholding*

As customary with other premises, minimum keys should be kept, but because of the proximity to the public thoroughfare the keyholder should ensure that he is never alone when finally locking up, for the chances of being attacked and forced back into the building are too high. As mentioned earlier, the staff should all leave at once, the locking up being done as they go. Where cleaners work during the evening, the same principle should apply when they quit the premises.

### *Lighting*

Internal lighting on the ground floor will assist observation by patrolling police and attract attention from passers-by whilst deterring intruders. All too often the rear of buildings is badly neglected in this respect, which enables thieves to climb on to roofs or force doors and windows with a feeling of privacy.

## *DELIVERY AREAS AND LOADING BAYS*

Times of delivery are rarely predictable; the drivers concerned have limited loyalty to the recipients; the goods that they handle are likely to be of instant use and value; they have the means for removal immediately at their disposal (that is, their vehicle); they possibly have residual goods in their vehicle with which to merge what they have taken; and they have a credible excuse of mistake for deliberate shortages in deliveries. A pessimistic picture? Yes, but this is probably the greatest source of loss of all. There must be direct supervisory observation on to delivery bays and areas. At a single firm with seven loading bays and supervisors' offices out of sight behind racking, losses which totalled £500,000 in a three-year period stopped abruptly when those offices were moved to an elevated position overlooking the bays, and the throughput of goods speeded up noticeably. Closed-circuit television in larger units is again a sound proposition.

Planning of the premises must allow adequate space for reception and storage. If this is not done, checking of both deliveries and stock becomes haphazard in the cramped shambles that results. If available room permits, loads should be dropped into a cleared area in front of an office and checked there before removal into storage so that shortages come to light at the earliest possible time.

An adequate system of checking and recording deliveries is essential for keeping to desired stock levels and for accurate stock checks. Itemized comparison with goods received against invoices or delivery notes should be made as soon as possible so that claims for shortages are submitted to suppliers within the prescribed time limit. Shortages of any size should be advised by telephone so that enquiries can commence before everyone concerned has forgotten all but the vaguest details about the dispatch, carriage, and delivery – there is nothing more annoying to an investigator than a thoroughly cold trail.

In some instances, goods may be delivered directly on to the shelves in the sales area. The delivery staff should always be supervised during this operation, not just to ensure that the correct quantities have been accepted, but also to check that they do not take the firm's property away with them in the course of leaving. No exception should be made simply because the transport belongs to the firm itself, and supervisory staff should remember that the shorter the time for which the suppliers' staff are on the premises the less chance there is of relationships leading to collusion arising between them and the firm's employees – and less time wasted in gossiping too. When warehouse staff agree with a supplying driver to sign for a full delivery but part is deliberately left on the vehicle for disposal, the ploy may be carried out many times successfully before detection if supervision is inefficient or complacent. The goods-inward section is one where reliable older employees can be found posts with advantage.

### *Dispatch of goods*

(See also Chapter 28.) The added complication when goods are dispatched is that, when a firm uses its own transport, relationships between drivers and warehouse staff will already exist. Limited staffing may hamper implementation of desirable systems; but if the goods can be assembled by an order chaser, independently checked then for quantities, and checked off against a consignment sheet as loaded, at least the basic steps to prevent loss have been taken. Spot checks on loaded vehicles, if feasible, will add to the deterrent, but in many instances these cause so much time loss and annoyance that the additional risk involved in doing without them becomes acceptable.

### *Storerooms*

Move incoming stock from 'goods inwards' to storerooms as quickly as possible. Special arrangements should be made for high risk goods – for example, in the case of cigarettes in bulk, insurers may well insist on a steel mesh cage or separate strongroom with electronic alarms. Access to the stockrooms should be restricted, and sales floor staff must have clear instructions in connection with the removal of goods from them and the documentation to be implemented. Price ticket 'markup' of goods taken from the stock to sales, especially fashion goods, must be the responsibility of the departmental manager or section head.

A persistent snag in connection with dispatch and delivery bays is making sure that they are secured when staff are at lunch or engaged away from loading areas. If this is not done, a sneak thief can walk in, or his more enterprising colleagues can bring in a vehicle and load it. Their workrate on such occasions can be surprising – 1½ tons of cartons went in just under three minutes during a tea break at one London warehouse! Outer doors must be locked when the department is vacated, and this is a responsibility to be impressed on the supervisor.

## *THE SALES FLOOR*

### *Aids to security*

The best security measure is to have adequate trained and alert supervisory staff in the sales area who are not fully engaged in selling or at cash points and so can keep observation on counters and customers. This is especially true in conditions of self-selection.

When mirrors and closed-circuit television are installed, their prime objective is to deter, and notices should be exhibited drawing attention to their presence and purpose. Dummy cameras can be used, but the stupid error of neglecting to lead wiring to them must be avoided. The multi-lens hemispherical ceiling-mounted variety are especially valuable since the prospective thief cannot know whether he is being observed or not. This equipment may be purchased or leased and of course does not have the exorbitant cable-run installing costs sometimes encountered in industrial applications.

Portable radios and the like have long been attractive to shoplifters, and these can be protected by chains looped into an audible alarm to attract the attention of staff if they are removed. Special shelves have been devised which fulfil a similar purpose if articles are lifted off them; this is effective but hampers examination by customers. Price tags are available which will activate an alarm if they remain on a garment taken from the sales area without having been removed at the cash point on payment; these are called Electronic Article Surveillance (EAS) systems and may be in the form of flexible strips, adhesive labels or tickets, or a hard tag. Special means are needed to desensitize these efficiently and the system is not cheap. Its main use therefore is in connection with high-value articles and clothing. Suppliers should be asked to reveal current users that can be contacted to give their appreciation of the value of the system and drawbacks. Equipment may be leasable for a trial period.

### *Shoplifting*

The correct terminology for shoplifters is 'shop thieves'. The location of their activity is all that distinguishes them from other kinds of thief, and those who indulge in it are equally widely drawn from the general public – with, however, an unusually high female participation.

Stealing may be by one individual, two working in concert, small groups of youths, foreign visitors, relatives working in collusion with staff, organized professional gangs – the permutations are endless, and those involved often have respectable backgrounds and records. The means used are also legion and not always predictable; hence the necessity for regular interchange of information between those with a preventive role.

Organized gangs more frequently than not operate away from their town of origin, and in fact several have been known to work on an international basis as in the case of the notorious Australian groups in the UK. Their targets are high-value items, often fashion goods. There are planning visits to ascertain the best and safest ways to achieve their objectives; note is taken of tea and meal breaks when staffing is at a minimum, the presence of store detectives and supervisors, surveillance and alarm devices, location of the goods, sizes available, back stairs that may be used, and nearby parking spaces for cars and quick getaway. Suspicions vented by alert assistants about callers in risk departments who appear to have little intention to buy should not therefore be disregarded.

The attack itself is often made in two stages. The first to arrive are possibly two women who sort through clothing on counters and move the more valuable items that they want to the front for easy handling by the actual thieves, who follow afterwards and spend as little time as they can slipping the goods into bags or whatever they are using for quick removal. Thefts are not confined to the sales floors; furs and good quality clothing should not be left on rails unattended near lifts, or on landings when in transit between storerooms and sales. Whole rails of clothing are known to have been moved out to vans by thieves dressed in overalls or smocks similar to those used by the firm. The observant and keen shop assistant who challenges someone in her department, whose interest seems to be in the surroundings rather than the display, with a polite but pointed 'Can I help you?' or 'I'll be with you in a moment, madam', may well make a gang feel that too much attention has been paid to their scout and decide to try elsewhere.

Rails bearing clothing should always be placed where staff can see customers clearly so that nothing can be easily dropped unobserved into bags or shopping trolleys. Little groups of youths, some of whose members are playing the fool, should be carefully watched, the ones creating least commotion meriting by far the greater attention. Distracting the attention of an assistant is a usual technique; one partner enters into a discussion about whether or not to make a purchase from an isolated assistant and shields the colleague who effects the theft. Variations are the inebriated customer, the aggrieved one, and the indeterminate one who gets a choice of goods on the counter so that the assistant has to devote all her attention to seeing that some of those do not disappear.

Fitting rooms need a set routine so that all garments taken in to be tried on are accounted for. They should not be allowed to accumulate in the room as one after another is declined since each additional one makes it more difficult to be sure the departing customer is not wearing two garments – her own, and one underneath belonging to the firm.

Some common shoplifting gimmicks for removal of stolen goods are:

1 Inside rolled newspapers or folded umbrellas.
2 In false pockets in clothing and under clothing generally, pushed inside belts, up sleeves, in the folds of rolled-up sleeves, in gloves, down the tops of long boots, etc.
3 Inside hollow purchases (kitchen rolls etc.) or cartons unsealed and refastened.
4 In the false bottom of shopping bags.
5 By dropping a headscarf on to the counter and picking up articles in it.
6 In coiled-up female hair (watches, lighters, rings, etc.).
7 By swapping round a price ticket to pay a lesser price for a valuable article. (This has been confirmed as theft by a Court of Appeal decision in *R* v *Morris* 1983 3 All ER 288.)

These are only samples of the shoplifter's ingenuity.

One larger scale fraud in badly laid out or inefficient food supermarkets involves the use of two trolleys and identical shopping lists. The first person takes her purchases genuinely through a checkout and gets a receipt in the form of the audit roll duplicate, reloading her purchases into a trolley for carriage presumably to a waiting car. She then re-enters without the trolley, picks up the duplicate load from her colleague, takes an additional item, goes to the checkout (possibly even the one she has gone through before), says that she has gone back in for something that she had forgotten to buy, and produces the audit roll duplicate to show that she has paid for what is on the trolley.

## *ASSAULTS ON STAFF*

Perhaps the most disturbing manifestation during recent years has been the increased frequency of attacks on store detectives by thieves that they have challenged. Fists, feet and weapons have been used and occasionally an unnoticed accomplice has joined in making it obvious that the action has been premeditated. Even when no violence takes place, threats and the flashing of knives are hardly unrehearsed reactions to a challenge. The perpetrators have a good chance of escaping since the likelihood of members of the public intervening seems to lessen as time goes by. So far this problem is most serious in the bigger shops in city and larger town centres and shows signs of spreading rather than abating. In an area where it is prevalent and a store detective suspects the thief being watched may be violent, no approach should be made until male back-up is at hand – and this should be a matter of preplanning for the eventuality. Liaison between shops should include details of any attacks/attackers and close contact should be kept with the local police who should be notified at once when an incident happens – if their reaction time is unsatisfactory, senior management, not security, should raise this with their police peers to ascertain what improvements can be made. These are serious matters, the assaults are with the intent to resist arrest, not the run of the mill 'common assaults'. The target stores may well have 'bleeper' or radio paging systems for their staff, if so, and the facility is not already incorporated, the suppliers should be asked to incorporate panic-button facilities in their instruments and to discuss modifications which may be made to identify the location of a button being used. Where closed-circuit television is in use it may even be possible to activate a video recorder on a camera near the scene. Thought and money have to be applied to this problem, it is not going to go away unless there is a positive improvement in public morality and unless a store detective has the means of raising an alarm and securing immediate assistance.

## *COMMENT*

There is no easy solution to the problems of dishonesty in shops and supermarkets. Thefts will take place, and constant vigilance will continue to be needed to minimize them. Full use should be made of modern deterrents and of the psychological value of warning notices. The last thing that a retail establishment can afford to do is to allow either its staff or the public to think that it is indifferent to persistent petty losses whose cumulative effect could eventually put it out of business. Staff training is all-important and all too often neglected.

A checklist of important points is shown in Appendix 20.

# 24

# *Specialist security operations*

Certain types of enterprise have risks which are either peculiar to themselves or are afflicted by a cross-section of all the common problems. Predominant amongst the first variety are hotels and shopping precincts, both of which have specialized security needs not encountered elsewhere. On the other hand, hospitals by their very nature contain practically every variety of activity, complicated by a very large and mobile staff and almost unrestricted access at times to the public. Though not strictly 'commercial' or 'industrial' undertakings, security is of such increasing importance that further detailed consideration is merited.

There are also more conventional units whose nature necessitates work to which pressure groups may take exception and lunatic fringe elements to act unpredictably, sometimes without concern for property or life. Pharmaceutical and research laboratories in which live animals are used for drug testing or experimentation are proven targets for otherwise quite respectable persons declaring 'animal rights' justification. They have used excessive force to break in to 'liberate' the animals and search for any evidence which will give publicity to their cause; their enthusiasm has been shown in personal attacks on plant executives. To a lesser degree, meat packaging plants have been interfered with and even the odd butcher's shop window has been broken. These threats should be taken seriously; physical and electronic protection should be as sound as possible and the executives, although they cannot be expected to alter their lifestyles for a risk which, to be fair, is remote, should appreciate its possibility and be alerted by any suspicion of being followed or otherwise watched. 'Crank' letters should not be ignored unless there are sound reasons for doing so – they may be followed up by extremist action.

Of course, there are other numerically less important groupings of enterprises needing nuances in standard security procedures; some may be helped by bylaws or even legislation; others, like maritime fraud, may be almost purely investigative. It is proposed however to concentrate on the three already mentioned – hotels, shopping precincts and hospitals.

## *SHOPPING PRECINCTS*

If two circles were to be drawn, one for the police and the other for private security, and these were pushed together so that there was a part which was overlayed, an area would be seen where private security is working in the public sector, and this area is growing all the time.

The police role is at present contracting to maintain law and order, whilst private security is expanding to cover many aspects of public life, and events.

The Americans have definitely turned to the private sector for help, as the public demands better security at home and work, which the police can no longer provide. This is also happening in this country now, with a rising tide of crime, and more people becoming victims of the fear of crime.

Private security now represents a 'quiet revolution' in the way people 'police' each other with growing acceptance of changes in social control which can make it a major contributor to society's protection, through effective and efficient service.

The police are now continually trying to reduce their workload, and if constructive dialogue and creative planning between police and security can be achieved the public must gain in the wider sense of freedom to live and move about. Nowhere is this more highlighted than in the terms of 'policing' shopping precincts.

Precincts to be financially viable have to be large, with numerous shops catering for thousands of people at any one time. The security problem is sometimes as complex as that of a small town, because to 'police' such an area requires skills and knowledge far beyond the normal in-house security position.

Security personnel will be patrolling, acting as agents for the management company, but dealing with all the usual problems associated with a police force. It is quite usual for sudden deaths, missing children, lost and found property, disorder and crime to be dealt with by security officers, but with no other powers than that of a private citizen or authority granted by the company.

*Organization*

There is need for a full organization of security officers, working shifts, with supervisory and security management. The precinct owners have of course to produce and pay for this service, which means each shop within the precinct will have a 'security charge' included in its rent. If the security performance is poor, it is not surprising for shop managers to complain, and centre management will encounter difficulties if it has tried to save financially on its short-term arrangements.

Ranks in this structure are necessary and responsibilities should go with rank. There should be a centre security manager, answering direct to the centre director who is in charge of the whole precinct. With such a close liaison, problems can immediately be dealt with. There should be a senior supervisor on each shift, with minimum manning levels decided well in advance. Usually a two-shift system of 12 hours is worked – 7 a.m. to 7 p.m. and 7 p.m. to 7 a.m. At busy times and weekends there is a need for flexibility or overtime arrangements, and the use of part-time staff may be necessary to provide adequate cover. A very visible presence is required during sales times or when perhaps a local sporting event promises to bring in potential troublemakers in numbers. Unless a precinct is extremely large, daytime patrolling can be discretionary but subjected to supervision to prevent timewasting and image-damaging relaxation inside shops, not forgetting the attraction a uniform often has for gossiping members of the opposite sex.

Frequency and the nature of patrols after shopping hours and during the night will depend on several factors – whether the precinct is one of the smaller variety which can be locked off, the degree of coverage given by closed-circuit television, the universality of fire and burglar alarms, and the historical nature of crimes committed (which may show a propensity to night-time break-ins). Not all the points raised in security and fire patrolling (see pp. 105 and 503) will apply but an examination of individual premises as soon as possible after they have closed should be catered for, with no hesitation in calling out keyholders for 'insecures', suspicious lights, or anything else calling for immediate attention. Patrolling officers should have a remit to enquire whether 'everything is alright' if work is apparently going on in premises which at the time would be expected to be closed.

Moderation in attitudes is an essential instruction, inaction in security matters can be very damaging for management but overreaction can cause more lasting repercussions – as with disorderly behaviour, easily provokable into something much worse. The giving of facilities for adequate training is an essential ingredient in the organizational arrangements.

### *Uniforms*

There are differing opinions about the type of uniform precinct security staff should wear at work. Some precinct managers regard them in the role of public relations officers and want them to wear blazers and flannel trousers, with a blazer badge to denote status. Others want their staff to have more conventional security wear so they stand out amongst the shoppers who will be reassured by their presence and be more likely to regard them as representatives of management. Each precinct can dictate its own needs but most in city areas are aiming at the highest profile of security and safety for its customers.

### *Training*

The training of a precinct security officer must be extensive. He has to know everything he can about the shops and businesses in his area and special problems they may have. He must be competent in the use of all fire equipment, and know its whereabouts and the action he should take in the case of an outbreak. He must have first-aid qualifications to ensure that he can deal with emergencies – there is nothing worse in the public eye than a man in uniform standing helplessly by when someone has collapsed or been injured. He must be conversant with all the company instructions for dealing with incidents and disturbances, in particular for dealing with the necessity of clearing the area in the event of a bomb threat or other serious matter. He must be instructed in correct usage of his radio and be familiar with the operation of the control centre and the help he can expect from it. He certainly needs a good working knowledge of the law as many shopkeepers will hand over persons found stealing to him and expect that he will know what to do with them – the shopkeepers are after all paying for his services and deserve that return.

His training should include recognition of signs of drug abuse and what he should do if he suspects this is going on in toilets – glue sniffing, 'weed' smoking or heroin injection – or the smoking amongst groups of youths on or around benches in the precinct. The same would apply to homosexuality acts in the toilets which is not infrequent in some cities – many patrons will be alienated if this and drug taking is tolerated.

Shoppers will regard the security officer as an official of the precinct and he must realize that much of the respect they accord him will depend on his deportment, appearance and demeanour towards them.

Only on-the-spot experience in a precinct will bring all these requirements together which places a great deal of responsibility on supervisors to ensure their staff become competent and efficient.

*Administration*

The security department must have its own routine procedures of record keeping, dealing with lost and found property, lost children or missing persons, or those detained for theft. They should have all shop and business premises listed showing ownership, keyholders and managers for quick contact when needed. There should be recognized parameters of their authority to call the police with whom they should maintain a close liaison as a matter of paramount importance. Reporting to centre management should be a matter of daily routine.

*Communications*

A security control room should have been included in the initial planning of the precinct. It usually is, but if not it is a priority for centre management as the operational nerve centre of the precinct. As such it should have tannoy facilities for broadcasting announcements amongst the shoppers; personal radio control for the security guards who may need help, advice, or instructions, or simply to make intermittent contact; all fire alarm and burglary alarm systems should be routed through it. The latter should have been considered also during the initial planning so that they could be incorporated during construction and included in the standing charges made by the centre on occupiers.

If closed-circuit television is not already fitted to cover the footways in the precinct, with the increase of violence and disorder in public places, it certainly should be, with a video-recording facility in the control room so that the offenders can be identified and dealt with as the circumstances merit.

Full instructions should be provided to ‘controllers’ to deal with all foreseeable contingencies – including clearance of the premises for fire or bomb threat. These instructions should be in clearly printed form to prevent mistakes. It follows that the staff in the control room have considerable responsibility and this should be borne in mind during their recruitment.

## *Operations*

The considerable majority of precinct managements have opted for contract security cover for financial reasons; it is to be hoped that they have satisfied themselves with the answers to the questions posed on p. 582 before doing so. Centre Management will be criticized if the service it is getting falls short of what was expected and promised; it should periodically review it with the contracting firm. The problem with many such firms, even in these days of gross unemployment, is that they have difficulty in recruiting and retaining personnel of the calibre needed for the professional nature of precinct work. The pay offered is rarely attractive and it is not unusual for a company to have a manpower turnover of up to 150 per cent per year which does nothing towards supplying an efficient service either to the management or the public using the precinct.

All the problems of the public sector manifest themselves in a precinct and security standards need to be high; if, for instance, disorderly behaviour is allowed to persist, customers will go elsewhere and when that has happened they may not return. Those higher standards may call for extra expenditure but management has to face this. A major publicized incident that could have been prevented if the staff had been trained and efficient might cause indirect loss which is serious and lasting. That danger is increased if the turnover of staff is such that those on duty have only a sketchy idea of what the duties are and lack familiarity with their surroundings. The latter point is of particular importance when ensuring that specified access routes for fire authority vehicles are kept clear, to which could be coupled observance of actions by tenants, suppliers, and others which are not acceptable for safety reasons.

Management has the responsibility and the authority to create a security environment its tenants and their customers would appreciate; it should be regarded as a necessary and cost-effective measure, which might induce the setting up of an in-company force tailored to their own requirements, and performing a public relations function for management.

## *Comment*

Professionalism by security personnel is an achieveable aim for centre management, but they must be given both the tools for the job, and the considered authority to do it.

Many countries abroad give security officers the status of 'special constables' when working in precincts, and this is acceptable by a public who merely wants to visit and shop without hindrance, and management who need to protect its considerable investment and reputation. It is unlikely that such authority would be given in the UK, but it could be the answer to many problems. The Fortune 500 Company of America summed it up:

> the best security people in the world cannot be effective if they have to function in a climate where integrity and honesty are the exception rather than the rule. It's up to management to establish the highest ethical standards for business conduct, and to see that those standards are adopted throughout the company. You cannot establish such standards in proclamations only in practice. . . .

and never more so than in a precinct situation.

## *HOTELS*

Over the past few decades, the size of hotels has increased greatly, many of them expanding upwards in 15-storey buildings or even higher. Once again this creates a security situation dealing with numerous members of the public, but within a private area, the numbers limited only by the number of bedrooms or guest space.

The modern hotel is a complex organization, which has to accommodate the ever-changing needs of both guests and staff. These include a secure environment and the protection of a hotel's assets. Most hotels are totally self-contained, incorporating power plants, laundries, shops, clinics, and many recreation facilities such as swimming pools, saunas, gymnasiums and conference centres. Anyone involved in their security must be prepared to deal with all types of incidents, including major crime.

### *Personnel*

Careful selection of staff is important in most organizations, but in the hotel trade it is of paramount importance. The degree of access the staff have to guest rooms, often working alone, must be greater than in any other situation. Few hotels, however, make background checks on staff, but the responsibility for their actions must be taken by management. Pressure from lower managers to fill vacancies is high, and long hours and often poor pay do not help the situation.

### *General protection*

There are many public areas in a hotel, such as the lobby or reception area. Some have restaurants open to the public, all of which can attract undesirable people into the building. The essence of a security operation is to identify such people and to ensure they do not gain access to other parts of the hotel, and that they leave peacefully.

Many new hotels have taken security as a major factor in design and concept. Built-in security is far more cost effective than trying to improve an old building. The guest floors need security, not only in the form of a very good lock system for individual rooms, but also by a staff presence either in lifts or as floor security patrols. Fire exits must be maintained at all times, but those with vulnerable doors should have an alarm system fitted to deter persons from taking short cuts, or intruders. Staff entrances should be carefully monitored, and identity cards issued, especially where there is a high turnover of staff.

Deliveries must be dealt with at specific entrances at the rear of the premises and be subject to 'back door management'. Persons visiting from contract services, laundry, etc. must be made liable under the same terms as normal staff. The use of closed-circuit television to monitor these entrances can be highly effective, especially when supervised closely by security staff.

### *Guest rooms*

There is a historical common law liability that hotel owners have a 'special liability by virtue of the custom of the realm' not to refuse board and lodging to any respectable traveller seeking accommodation at any reasonable time of the day or night, provided they have rooms available, and the guest can pay the charges made. Hotel owners are also automatically liable for any loss or damage to their guests property irrespective of whether a member of staff is involved or not. Many hotels put up notices stating they will not be responsible for any goods lost or stolen, but there is no legal support to such notices. Some hotels invite their guests to put valuables in the hotel safe, and some authorities consider if a guest does not do so, he may be partially liable, or at least the hotel will be less liable.

The Hotel Proprietors Act 1956 allows the hotel to exempt guest's cars from their responsibility, although the common law remains intact. The hotel can also limit its responsibility to a certain amount for goods lost or stolen, but only if notices clearly show this restriction under this Act of Parliament, which clearly described the establishment as a place held out by the proprietors as offering food, drink and sleeping accommodation, without any special contract to any traveller presenting himself if he is willing and able to pay a reasonable sum, and is a person in a fit state to be received. The hotel therefore clearly has a legal responsibility for the guests well-being and safety.

The reputation of the hotel is at stake in the event of any untoward incident.

There has been a marked increase recently in security measures in locks on guest rooms, with new magnetic or electronic locks and keys being fitted. It has been a common occurrence for a thief to book in as a guest and make a copy of the room key, returning later to steal from other persons then using that room. Some hotels now offer a key card, which has to be produced at reception to obtain the key. In other hotels, guests are encouraged to retain the key in their possession until they leave. Other door hardware now includes a door chain, and a peephole to scan the corridor.

### *Theft*

It has been a matter of concern to hotel management that there is nothing a guest won't steal, or take as a souvenir. Some hotels capitalize on this fact and use the small items which are taken as advertisements for the hotel. This is no doubt costed into the guests account. Control of linen, however, by the staff is very important.

The primary responsibility for the prevention of thefts and losses must rest with the departmental managers, and ultimately senior management. All the hotel security staff have a role to play, but it is important that each area of management operates individually without undue interference from security.

Food thefts by staff is perhaps the major area of crime. It takes a skilled security officer to pinpoint where these happen – the way the food is purchased, stored, prepared and served are all where losses can take place. It is vital for security personnel in the hotel business to be familiar with all the practices of staff, even to the extent of working amongst them at times.

### *Security operations*

The security department must have a working relationship with all departments, especially the reception area. Good communication is essential with the telephone operators ensuring security is kept informed of all incidents. Fire alarms often terminate in the switchboard room and again immediate responses are required.

The cash flow of the hotel is often a sphere of responsibility for security staff. Many hotel staff are paid in cash, and the steps advocated in Chapter 15 should be observed for this.

Cash offices should be carefully designed for maximum security, and a location chosen away from the normal areas used by guests, for instance in the management suite. Several safes might be required, which can also be sited in this area. Access should be strictly limited, and electronic locking devices be considered. Cashier positions must be sited to give the impression of security and efficiency.

A separate room should be provided for guests safety deposit box systems. A screened area for guests to use the facility is a useful feature.

All reports of losses and theft must be reported to *and* recorded by the security department. If necessary reports should be passed to the police dependent on management's policies, or guests' wishes.

It must be borne in mind that there are a small number of people who pose as guests, invent a loss, and demand compensation from management, which is often given to protect the name of the hotel, although co-operation between hotel security managers can often identify such people and prove a system.

The hotel is of course open to the simple method of guests defrauding it by the use of dishonoured cheques or stolen credit cards. Many hotels are reluctant to accept cheques. Full training, dependent on management policies, must be given to staff dealing with guest payments. Again co-operation between hotel security staffs locally and with wider printed or telexed circulations can often prevent such crimes. Some guests will purport that accounts will be settled by a company. Again this can be prevented by good procedures of routine checking.

Counterfeit currency can also be a problem, but often this can be overcome with the use of ultraviolet lights. Many hotels of course offer currency exchange facilities, which need careful auditing.

### *Sudden death*

It is a reality that sudden deaths take place in hotels quite regularly. Security staff have the primary responsibility to ensure that the scene of any crime or suspected crime is well preserved. In cases of illness resulting in death, it is useful to call a Doctor who is retained by the hotel to certify death. If the guest is alone at the hotel, great care must be taken to make an inventory of all his possessions, and to seal them in a container for collection later by relatives. This should be done with a witness present, as relatives can sometimes make allegations which are hard to disprove. Clear procedures should be laid down by management, and instructions given to all staff to cover such an event.

*Prostitution*

This has always been a problem in the hotel industry, and again clear instructions must be given by management in dealing with it. There is often pressure on night managers from wealthy guests to 'bend the rules', and it is not for the security staff to judge the morality of such situations. If prostitutes start to use hotel bars, the reputation of the hotel itself can diminish, or often prostitutes will make a nuisance of themselves after leaving clients on upper floors by soliciting other guests in bars and reception areas.

These problems will affect the reputation of the hotel and can even put its liquor licence at risk. Strong action is required, although many 'contacts' are now made through telephone services, and it is difficult for security staff to respond. Male security staff should of course never be left in a room alone with a prostitute, and experience in London especially has shown a number of transvestites and transsexuals also working hotels.

*Legal powers*

The Theft Act 1978, s. 1 creates an offence of 'by any deception dishonestly obtaining services from another'. Falsely claiming that an account would be settled by the company would constitute an offence under this section. Presenting a postdated cheque which will knowingly be dishonoured is also an offence under this Act.

Section 3 is perhaps the most useful for security staff as it covers the person who fails to pay the bill and walks out of the restaurant – in fact dishonestly making off without payment.

The power of arrest exists for all citizens for all three sections and this is particularly important for security officers.

*Comment*

Hotels have a duty to provide a safe and pleasant environment for their guests, and especially in larger hotels, the security position offers a wide ranging and interesting challenge. The diversity of daily life raises all sorts of problems, and training in how to respond to incidents is essential or the security officer's actions can have serious repercussions on the hotel's business. An alert and capable person is required for such a position.

## *HOSPITALS*

'That a precaution is physically possible does not mean it is reasonably practicable' – the dictum of a judge on an issue of inadequate security measures (see p. 232). Nowhere is this more apposite than in planning measures to improve security within hospitals. The only comparable locales are those of universities and polytechnics where likewise risks are greatest when the sites lie in built-up urban areas.

Problems in hospitals primarily arise from the following:

1. Throughout the day there is constant movement and activity with numerous legitimate avenues into the building.
2. The public have virtually unrestricted access at patient visiting times in such numbers as to make the presence of intending thieves unnoticeable.
3. A high proportion of the ancillary staff perform relatively menial and low-paid jobs which do not attract the best quality of employee for an environment where there are ample opportunities for pilferage.
4. There are many jobs, porterage and the like, which provide an unquestioned right to go almost anywhere on the premises.
5. The buildings house a whole gamut of tempting targets – pharmacies and drugs, stores, valuable technical equipment, clothing, textiles, cash and wages, offices, canteens, shops, administration offices and their typewriters, calculators and desk-top computers, etc.
6. The presence of nurses and their residences attract the sexist-attacker type.
7. The casualty departments receive damaged and obstreperous drunks, the prevalence of whose attacks on staff is becoming a matter for real concern.
8. The potential havoc that could be caused by a major fire when many of those under threat may be virtually immobile.

Fire risk is of course the first recognized priority which will have been observed in planning and/or modification of buildings, access routes and equipment, in all of which the fire authority will have participated. Storm, flood and other natural or installation disaster potentials will have been considered and catered for and the local police should have emergency plans (see Chapter 32) for all such foreseeable eventualities. A security planner should note that these matters have been organized, familiarize himself with the procedures, and be prepared to notify any shortcomings and monitor implementation. His main task, however, will lie in the reducing of theft, fraud, damage and assault.

A fact of life which has to be accepted is that the overriding purpose, to which all else is subordinated, is that of the preservation of life, alleviation of suffering and care of patients; expertise, interest and expenditure are all targeted for this purpose. It would be a very brave administrator who would authorize costly security proposals in the face of demands for a body scanner or other essential equipment. It follows that recommendations have to create minimum interference with that main purpose; showing them, when possible, to be in furtherance of that purpose by preventing sources of impedence, and also to be financially acceptable.

The chances of any accurate figures of losses being available are negligible. The NHS have produced assessments which in the past have been so ludicrously low in comparison to its gross expenditure that they must be in question and relate only to its own property, in any case. Is there any trustworthy system of reporting? A conversation between nurses heard in a specialist's waiting room made the point – 'The box is empty and it was nearly full when I last used them, and it costs £140, I know because its happened before.' The reply was 'Well we can't help it, get another box out, he wants them right away.' Almost certainly those shortages were not reported.

*Preliminary action*

An early requirement is to inaugurate a system of reporting actual and suspected thefts by means of a simple pro forma which requires little time and effort. When collated, however, it should pinpoint where action is most needed, and may produce a blatant suspect who is always about when things disappear. A risk assessment questionnaire can be circulated to those with responsibility for the various functions in the hospital to establish where risk and consequential loss are the greatest. It would have to be detailed to be of value, it would need top level authorization to ensure completion, and would not be popular with busy executives. The accuracy of the answers might depend on whether any serious or controversial incident had recently focused attention on security, in which case they might provide managerial support for recommendations. Private discussion with individuals is likely to be more productive, even if more time consuming – personal acquaintanceship, indoctrination, 'brainpicking' and pooling of ideas are preferable to ticks on pro forma squares and familiarity with all aspects of the hospital work is expedited for the planner.

A 'plus' on the side of the planner is the general integrity of the medical and nursing staff, even amongst trainee nurses; it is rare to encounter a persistent thief. Those who enter the profession have its interest closely at heart and senior nurses in charge of wards and functions regard their responsibilities seriously and can be invaluable aids. Nursing staff have a long period of training and talks about security should be incorporated in this, remembering that they will be interested in things that they may encounter. Case studies and anecdotes will carry more weight than dry exhortation. Medical staff will probably have an induction period and/or actual training during which opportunity should be sought to impress the value and principles of good security upon them.

*Security manning*

The NHS Area Health Authorities are not yet uniform in their attitude to using security as a separate function, though incidents are causing a reappraisal of previous indifference. A policy of recruitment and training is spreading amongst the larger hospitals which for many years have had one security officer, almost invariably a retired police officer, going through the motions of theft investigation and prevention, whose results have depended very much on the internal rapport he has established and the esteem he has acquired with higher management. Occasionally a District Security Officer may be responsible for several hospitals within which he operates as a one-man band with no staff allocated to him; he will investigate and make recommendations but obviously he is limited in what he can achieve. Both he, and the single hospital security officer have to work hard on inducing the maximum co-operation at all levels of staff. As always, newsworthy or high loss occurrences have to be exploited to achieve improvements; at one of the largest training hospitals an extremely costly piece of equipment was stolen shortly after being acquired, with van transport and several persons being needed for its removal. This brought previous 'mislaid' equipment into focus and the net result was a security presence with drop-arm barriers at the traffic and main pedestrian entrance, a car identification scheme for employees, and strict parking enforcement – a start in improving security.

In many hospitals, other than the security officer, the only form of security may be a limited number of porters with other duties; they can only be indoctrinated individually during working hours. Nevertheless, they are the first line of defence and should be used and kept informed as much as possible. If staff are to be recruited specifically as 'security', care should be taken (see Chapter 5), and an additional appraisal of their likely reaction to constant proximity with attractive girls in large numbers – Don Juans are an unwanted liability! Their training should be that recommended for all security officers with added first-aid instruction to help credibility in their specialized working environment.

NHS unions are not the most placid and full liaison with the personnel department of the hospital must be kept to ensure acceptability of the proposed body. It may be that internal recruitment is a condition for peaceful acceptance. The intrusion of contract security is considerably more likely to meet objection. A search agreement in contracts of employment may be existant, if not, efforts should be made via the personnel department to instigate one.

*Physical protection*

Few hospitals have the luxury of sound perimeter protection of walls or fencing. The immediate surroundings of most are totally incompatible with either and where it is feasible to consider, financial restraints make it a waste of time recommending anything of that nature. Fuel tanks, gas storage, electricity sub-stations and the like, however, should be compounded against damage which could be excessive. Exterior lighting should be the best that can be provided subject again to budgetary limitations and especially concentrated on the walkways used by nurses going on and off duty. External doors offer a usage difficulty and it may be impossible to reduce them to the optimum number desirable; a survey of usage may help to convert some to fire doors. Expense may prevent fire doors themselves being fitted with the type of crashbar which sounds an alarm when operated, but this is a target to be aimed for. As soon as normal working hours have ended, all non-essential external doors should be locked off.

The main entrances should have a reception desk; those that are infrequently used should be in clear view of a manned office, and all should be well lit. Pharmacies, laboratories, computer suites and those offices holding important papers should have substantial doors and security locks with strictly imposed key control. The card-operated electronic locks have considerable advantages and are both an attraction and convenient to senior staff and doctors who may influence expenditure proposal approval. Such locks can be computer linked to give a printout showing all entries through those doors used by individually-coded cards (see p. 608).

Finally, clocking stations and cloakrooms for staff should be sited adjacent to reception desks to provide the scrutiny which in itself is a deterrent to theft. If outgoing pay telephones are similarly placed, this will reduce the damage by vandals and timewasting by staff.

## *Electronic protection*

Any major alarm company will survey and give estimates of the cost to cover the whole or part of the premises. The nature of hospital use limits what is practical to fully protect, but a system for high-security rooms to be linked back to a central control room or reception desk is feasible and this could include panic buttons to be fitted at points where personnel attacks might take place, either by outsiders or by patients in psychiatric wards. Some form of control room of course is essential for the bleeper systems and all fire and other alarms.

Closed-circuit television is becoming increasingly necessary in casualty reception where staff are subject to attack; its very presence being a deterrent – to stress the point, warning notices should be placed inside and outside the area. Union support is more likely than opposition to this, as in other areas where there is a risk to personnel, i.e. routes to nurses quarters, car parks and perimeters. Video-recording facilities should be part of the installation and when the basic requirements have been met and shown effective, efforts to extend the system to other risk areas, internal and external, will have a greater chance of success.

Attacks on staff can take place in wards or corridors; the risk during normal working hours is least, but at night the reduced lighting and fewer people about put nurses at risk. Many of the conventional bleeper systems can be modified or replaced to provide a panic button for emergency use – this would give greater confidence and peace of mind in addition to enhanced security.

## *Personal identification*

In any large hospital there are literally hundreds of people moving about at any given time, quite legitimately, but without means of identification, intruders could mix with them with impunity. There have been many American television films showing the use of a simple white coat to impersonate a doctor, a ploy which can be copied here. It is more significant that in more recent television series the doctors have been wearing identification badges. Periodically, there are scandals where individuals pretending to be doctors have examined and occasionally assaulted female patients – publicity that the NHS could do without. The white coat does not have the same symbolic value here; doctors dress much as they wish. Necessity may take them into wards where they are not known and the same could apply to porters, engineers or workmen. A large refrigerator was emptied of its contents and taken from the top floor of a multi-story hospital by three 'workmen' wearing hospital-type overalls.

Badges are already in use in many hospitals and should be implemented for all staff. To be effective they must be for all status and a donation-to-charity penal system, graded according to that status, has been found acceptable for not wearing the badge. The badges must be tamperproof; this means a sealed-in-plastic photograph with the name, signature and department, and status, with a perimeter or background colour-coded to indicate the nature of the job, and to confirm authority to be in certain areas.

For weekly-paid staff, these could be a necessary proof of identity to receive a wage packet. The scheme should be extended to regular contractors on site, though difficulties might be experienced with a photograph, and a special temporary card to be surrendered on leaving may be needed. Ward sisters and staff nurses will need little encouragement to challenge, but the right to do so should be impressed on all supervisory staff who can in turn brief their subordinates.

### *Car identification*

Unrestricted parking cannot be tolerated in hospital precincts, space may permit car parks for visitors and staff but congestion which will hamper emergency services and efficient working has to be prevented. This in itself may justify a security or a warden service for enforcement. Distinctive badges should be issued to authorized car users to be clearly displayed on the windscreens. Warning notices should be posted at entrances and wheel clamping is already in use at problem hospitals. A byproduct is the exclusion of thieves vehicles which reduces theft of bulky articles.

### *Theft*

The theft of anything which is disposable and accessible is possible. A senior nurse quoted the taking of stained pillows from a maternity ward as an example, the mind boggles at the reason for that. Linen is the conventional target from the wards, and this despite almost universal marking with the hospital name in large letters. Electronic tagging as in shops has been suggested; the cost of this would be fierce because of the vast quantity of linen, and would it stand up to laundering? Losses are such that the collection, laundering and return service may be the main source and should be subjected to regular checking. Nurses may cut down on the stealing by departing patients by helping them pack, or standing conversationally by as they do so. This would also cut down on souvenir taking.

Theft of a patient's property is continually alleged and the best prevention is a policy of insisting that relatives take everything away, other than a minimum of cash to buy papers etc., and what is retained should be recorded. The canteen staff will have more than ample opportunity to take away food unless supervision is strict and emphasized by regular bag checking where this can be done. The staff should have individually-keyed lockers with notices posted disclaiming employer liability for loss, as personal property is always at risk. Shops, stores, computer rooms, offices and the like are all subject to the same problems as comparable places elsewhere and are referred to in the relevant chapters.

The material purchases of the NHS create a continual inflow of stealable goods and if the checking and acceptance has loopholes these will be exploited. The availability of transport creates an avenue for organized and persistent theft if internal thieves get together with delivering drivers. Closed-circuit television installed over delivery bays can do much to cut down on losses – it will not stop acceptance of short deliveries but might deter other thefts. All the precautions for goods inward systems should be implemented as elsewhere.

*Comment*

The targets for the person in charge of hospital security are: indoctrination of personnel to check references and cut down on intake of potential thieves; indoctrination of all staff to be observant, challenge strangers, report losses and suspect losses; indoctrination of middle management to insist on inventory checks at regular and irregular intervals and to query discrepancies; indoctrination of higher management to accept security as a serious responsibility and to formulate and implement policies to deal with it.

Note must also be taken of things that happen in other hospitals; and in furtherance of this develop a liaison with colleagues with whom information should be given and received. Relating incidents to staff in similar work to that of the incident is one of the best ways to induce interest and co-operation. Typical items would be those of fake doctors, persons pretending acute symptoms to get overnight accommodation and steal, sex attackers operating at more than one site. The need for good liaison with the local police goes without saying, and an immediate response by them is essential for assaults in casualty reception. They should always be given every assistance in any enquiries they have to make.

Finally, the hospital service is one of the largest purchasing agencies in this country, it is too much to expect that there will not be instances of corrupt practices with suppliers. What the security officer may do in this connection may depend on his remit or the contacts he has made but he should always bear it in mind for it can induce greater losses than ordinary theft. An innocent enquiry of 'Do we get competitive tenders for these things?' may have unforeseen results.

*PART FOUR*

# *SECURITY OF BUILDINGS AND SITES*

# 25

# *Protection of buildings and factories*

It is wellnigh impossible to make a building, still less a complete site area, inaccessible to determined thieves. Only an optimist with no practical experience worth speaking of would assert otherwise. Nevertheless, by far the greater number of illegal entries are made by criminals who are able to do so simply because no thought has been given to the steps which would deter them.

To carry protection to extremes would be to make individual buildings into virtual 'Banks of England', and even then the subversion of an employee, as in the Heathrow bullion robbery, might circumvent all precautions. Risk evaluation and insurers' insistence might dictate maximum steps, such as reinforced concrete vaults with in-built alarm wiring in the walls or seismic devices – as could be needed to protect the holdings of fine art auctioneers or international jewellers. This would be prohibitively expensive in most circumstances, especially in conversion work needed to upgrade old buildings, and a compromise solution combining physical strength, electronic alarms and insistence on sound working practices might satisfy most demands. The objective is to achieve a level where prospective thieves would consider the risk too high to be acceptable or tactics to be too complicated to have any reasonable hope of success.

Apart from merely getting into premises, the thief has to get his proceeds out. This can be the more tricky and dangerous part of his operation, particularly when heavy or bulky goods are to be removed.

A second factor is that he wants to attract as little attention as possible. Not only does he wish to acquire a profit from his endeavours, he also wants to strictly limit the chances of arrest. In other words, he wants to get in and out as quietly, safely and quickly as he can, with a minimum of effort for the removal of property. He neither wants to be seen nor heard and will be looking for a clear and easy escape route in case of being disturbed. If he has to carry heavy and awkward burdens he will want it to be for a minimum distance before he can again be under cover or safely inside a vehicle.

If it is accepted that the criminal cannot be stopped from entering, it follows that the objectives of the defence are to make him as uncomfortable and apprehensive as possible when entering and leaving and to make the attractive goods on the premises as inaccessible to him as possible. This latter point is especially applicable to money and small valuable objects which should always be kept in the protection of a suitable modern safe.

## *ARCHITECTS AND NEW BUILDING*

Altering any building to increase its security is a costly matter and may be disruptive to the work going on inside. In new building work, this can be obviated by applying security principles at the design stage. These will not excessively hinder the architects' plans nor radically affect the final cost of the building.

The police have long recognized the importance and mutual advantage of a liaison with the building industry and architects and the HO Crime Prevention School at Stafford has taken the lead in training police officers to be competent in appreciating the problems of planners, constructors and users so as to give unbiased advice – which can be a thankless task as each of the other parties will have personal ideas of priorities. The school has gone further and provided a venue for security-based seminars and courses for architects and the building industry. A British Standard for the physical security of buildings, BS 8220 was introduced in 1986. Part 1 dealt with private dwellings and Part 2 with offices and shops. Despite a gestation period of nearly ten years and the background of similar long-standing US codes, there is little content which is not already known to anyone interested in the subject. It does, however, stress the need for senior management to be involved in security and accepts that planners have paid little regard to it in the past. All the practical points given are, and have been, made elsewhere in this text; the Standard does at least give official endorsement of them and this fact may carry weight in preliminary planning meetings.

Architects cannot be expected to be fully appreciative of the peculiar risks involved in every firm's business, nor of the importance that it attaches to the security of its property and products. However, once these matters have been fully discussed with them, their expert experience in materials and construction will be applied to meeting what they may consider to be an interesting challenge. Several draft plans are prepared before the final one is accepted by architects, local authorities, and the owners of the property; it is too late to consult security and fire specialists when this stage has been reached. There will then be a natural reluctance on the part of the other bodies to make amendments and it follows that the earlier consultation can take place the better for all concerned. It has been mentioned elsewhere that there may be conflict between security and fire prevention requirements; this can be resolved and accommodated early in the planning stage if a full exchange of points of view takes place.

What is applicable to shops and offices is also applicable to factories and to perimeter fencing – these should all be discussed at the design stage and the precise type of protection that is required should be established, having regard to the usage of the premises and the surroundings.

It is advisable that the design stage should be a formal process with meetings minuted to prevent misunderstandings, accidental or 'convenient', or matters being overlooked. Police Crime Prevention and fire brigade representation is highly desirable and, given adequate notice, those bodies are more than pleased to attend; failure to consult with the brigade cost one firm an extremely expensive and totally operationally unnecessary sprinkler system when constructing a big new warehouse in a district where a local regulation demanded an installation if the undivided capacity of an area exceeded a certain volume – the architects said the brigade was not involved in the early planning stages! Security cannot hope to be the overriding consideration on all points, as in an instance where local planning requirements demanded long narrow windows at ground level, which in the event gave a clear view of the valuable contents of a wholesale depot and provided an easy way to get in. The compromise was to fit translucent shatterproof glass with suitable electronic alarm protection, the cost of which was incorporated in the overall expenditure allocation.

Where does the security specialist employed by a parent firm fit into such discussions? First, he ensures that the discussions actually take place by stressing their cost-effective importance, if only to eliminate the possibility of additional expenditure and inconvenience after the premises have been taken into use. Then he acquaints himself with all there is to know about the intended purpose and usage of the premises so that he can present a balanced opinion, probably to a better degree than any of the other participants who are likely to be over-concerned with matters affecting their own disciplines. Finally, he must have a clear conception of what is necessary and feasible in the circumstances and be professionally competent to present a cogent case.

Petty-minded objections may be raised to such consultations by internal departments and architects, jealous of preserving their independence of action; conjoint representations from security, safety and fire interests in a firm should then be made if an impasse seems likely and the request for involvement made in such terms as to make it obvious that its refusal will be brought to notice subsequently if the necessity arises.

## *THE PERIMETER*

### *Perimeter protection*

The initial cost of putting any form of fence demarcation round a large site is a major item of expenditure which can reach prohibitive heights if the more effective varieties are required. Commonsense has to be used and constraints other than cash exist: opposition may be found from environmentalists and there may be prohibiting local bye-laws and planning restrictions. Rather surprisingly there is a British Standard (BS 1722) which has been progressively extended to refer to practically all types, ranging from wooden post and rail fences to steel palisade and security fences.

Obviously, the objectives to be achieved by the fencing are of prime importance: for instance, if exclusion of nuisance trespassers on grassland or other open areas is all that is required, agricultural-type fencing with warning notices might suffice round the boundaries with something more substantial where a real risk exists. At a factory where all sorts of objections were raised to prevent chain link fencing being put alongside a trunk road, thick hawthorn hedging was installed and blackberries encouraged to grow through it – it proved more effective than the coiled barbed wire at the rear of the building. Mediocre fencing can be improved by the incorporation of barbed wire which can also be used to increase its height; this can be obtained in several forms including barbed tape – this is more costly though. There are several types of this tape, one of which is of US origin. It is coiled and has very long barbs which are an effective visual deterrent. Electronic 'fences' can supplement others and can be used between double fences where the need is great enough and finance relatively unimportant; these are simply alarm systems by rays or microwaves to warn of the passage of intruders; the only kind of electrified fence allowed in the UK is the low-voltage type for the exclusion of cattle. Alarm devices can be incorporated into fencing to warn of interference with it and be adjusted to take care of minor disturbance by birds and small animals. Inertia switches are widely used and practically all the larger suppliers include suitable equipment in their range. Simple current-carrying wire has been interweaved with mesh and the use of optical glass fibre has been advocated with appropriate monitoring equipment. In cases of extreme risk, where expense is no object, the ordinary fence may encircle a parallel strip under which are laid devices to respond to pressure and vibration to initiate an alarm.

A jaundiced eye should be cast on the value of any natural obstacle to intruders – rivers or canals can give a false sense of security. In South Wales, intruders used empty barrels to float themselves across to premises they wished to attack and pulled a line of barrels back across containing their loot. At another factory in the north, a whole series of thefts were eventually traced to the use of barges on the adjoining canal, where the high wall of an office block abutted a river. Waterproof containers were dropped from a window at a set time each day to be picked up by accomplices downstream. There are many such examples of what might be thought a barrier proving to be an avenue for theft.

The comment on use of commonsense bears repetition. If 'overkill' recommendations are made, unless an 'open-ended' financial remit has been given, modern prices for perimeter protection may cause their rejection and arouse criticism of the recommender's assessment of requirements.

### *Perimeter walls*

Building brick walls purely for security purposes is a very expensive business and would rarely be contemplated on that ground alone if any alternative existed. An ordinary wall has little protective potential other than its height – it would not stop anyone getting in, but could be a formidable obstacle to getting anything sizeable or weighty out.

Nothing less than 8 feet (2.4 m) in height is worth considering as a security protection, and walls of any height should have further reinforcement by setting spikes or barbed wire on the top. Broken glass in concrete is only a deterrent to the unassisted climber and it is easy to negate its value by throwing sacking over it. An additional advantage of coiled barbed wire, particularly if arranged vertically, is that disturbance of it will give an indication to patrolling policemen or security officers that someone has climbed over the wall and may be inside the premises. This can easily and cheaply be applied to new and existing walls by setting mild steel bars in concrete or bolting them on to the wall to carry the wire. This must not be at a height where it could cause injury to an innocent passer-by.

In the older types of factory property the tendency was to build high walls which were excellent in themselves but remarkably prone to having obscure doors and gates set into them in the darkest corners, inevitably fastened by large rusty padlocks with hinge shackles, simply appealing for the insertion of a steel bar. Modern construction much prefers a perimeter fence and where an unbroken wall does exist, it will be for convenience and for a part of a building facing inwards.

### *Perimeter fencing*

Perimeter fencing is a much cheaper means of protection than brickwork or even concrete panels. The conventional type is that of wire mesh on concrete posts 6 to 8 feet (1.8 to 2.4 m) in height, with angled arms carrying several strands of barbed wire. The green or black plastic-covered wire mesh lasts longer and looks better but can be opened up something like a zipper and then easily folded back to allow entry.

It has to be appreciated that purely as protection against thieves, this form of mesh has only a nominal value – it is a delaying factor and minor deterrent only. It can easily be cut with wire-cutters and bent up at the base to allow crawling beneath. It may keep out marauding children and it demarcates the premises, but it does require additional measures if any confidence is to be placed in it. There are two ways in which the base can be protected: by setting in concrete, or by threading barbed wire through the bottom mesh.

Much more formidable, but much more expensive, is palisading of the 'Lochrin' type, although even this could be forced apart with a steel bar or cut with large bolt croppers. This is made from mild steel, 3 to 4 inches (7 to 10 cm) wide pressed into thin pieces normally with a maximum height of 10 feet (3 m) although any desired height could probably be produced. These are bolted on to a metal framework and the tops can be forked in several alternative ways to deter persons intending to climb over. The bolt heads should be burred over when installing this fencing, otherwise portions can be unbolted and removed. The manufacturers now produce this palisading in sections which can be transported between sites and are not unboltable.

Other variants are now available in the form of welded rod, similar in appearance to that used in concrete floor reinforcement; some, of tensile steel that is resistant to cutting, may be difficult to obtain and are expensive. At least one variety compensates for cost by dispensing with the necessity for barbed wire strands along the top by the rigidity of the rod ends.

It is useless to provide either a perimeter fence or perimeter wall and then allow bad housekeeping to negate much of its value. This could take the form of stacking cases, materials, and machinery against the fence, thereby providing both means of getting over and concealment for the intruder entering or thief going out. Frequently, because of the seeming unimportance of the task and lack of employees designated for that type of work, bushes and weeds are allowed to invade the fence. Apart from the concealment of both holes and intruders these are most undesirable in so far as they accelerate the wear and tear on the fence itself.

The obvious reinforcing factor for the mesh type of fencing is lighting, though coating with a non-stick paint can have advantages for trouble spots of limited area.

### *Security lighting*

It has been pointed out that no thief wants to feel that he is under observation at any stage of committing an offence, and this underlines the value of lighting. To go up to and cut a hole in a fence in the glare of lighting which makes the act clearly visible to all and sundry demands a risk acceptance beyond what may be thought reasonable by all but the crudest of criminals.

Like any other cost-effective security measure there should be planning to equate efficiency with minimum installation and running costs. It may also be possible to budget part of the expenditure to 'safety' or other sources, especially where a perimeter road follows the fence or wall or there are adjoining loading or storage areas, etc.

Ideally, the lighting should be such that it exposes the intruder and conceals potential observers inside the premises, that is, outward-facing lights into which the intruder has to look. A fact which must be taken into account is the potential hazard or nuisance to nearby road users or residents, which may result in a running battle of complaints and possible legal proceedings. Lighting angled along the fence may mitigate annoyance but would probably reduce the time of silhouetting an intruder. Tungsten-halogen lamps are those most conventionally used for the smaller lighting points, though mercury vapour or other high-intensity discharge lamps would be considered on larger installations. Specialist advice should be obtained prior to submitting an expenditure proposal, and this will be forthcoming on application to any of the regional Electricity Board headquarters which will also supply advisory notes.

Schemes can be tailored to the importance of the risk, ranging from simple fence-area illumination to the lighting of an intervening space between twin fences or of a single fence fitted with electronic warning devices. These may take the form of vibration detectors on the structure itself or the geophonic type buried in an adjoining gravel strip, both wired to automatically activate lighting if they themselves are activated. As mentioned in Chapter 35, closed-circuit television cameras can be added but tests should be done *in situ* to check that light intensity is adequate before a contract is entered into. If special lenses are required, these may prove expensive.

Having acquired a security lighting system, all practitioners know that one of the constant difficulties is that of ensuring that it is adequately maintained by immediate replacement of defunct bulbs. The best way to do this is to establish a reporting system in writing, not in oral form, preferably by a stereo-type pro forma – and to keep on repeatedly submitting reports until the work is done.

In designing a lighting system pay special attention to the access gates and adjoining areas so that there is maximum capability for the inspection of persons and vehicles. Try to avoid excess light in a security office in contrast to its surroundings – the requirement is to see, not to be seen. Keep the security officer out of the 'limelight' and his clients in it.

### *Perimeter gates*

From the point of view of supervision, and also economy, it is advisable that the entrances to any perimeter should be kept to the minimum to suit the needs of production. This will reduce the number of security officers who are needed and will ensure that all personnel and vehicles leaving are subjected to scrutiny. It must be anticipated that some union objection would be raised to the closing of existing gates where these are sited to the convenience of personnel, and rearrangement of clocking points might be necessitated. By their structure, gates are often easier to climb than the adjoining fence. They should be of the same height and style and hung close to the ground to prevent crawling under. Barbed wire on the top would prevent easy climbing up and over and this wiring could possibly be extended into the mesh or palisading of the gate itself.

Wooden and metal gates both need top protection as they are equally easy to climb from inside. An aperture to allow scrutiny of the interior area by patrolling policemen is a good idea – there is a sort of psychological urge which draws the eye where these exist. Where lightweight gates are in use the suspension must be arranged so that they cannot be lifted off bodily.

To carry out a major theft requires the assistance of transport as near as possible to the desired loading point so, from the security point of view, the further such vehicles can be kept away the better. Access for these will invariably be through the gates and it is therefore essential that the shackles and padlocks fastening them should be of good quality. A habit of thieves is to force off an existing padlock, open the gates, drive a vehicle through, and replace the padlock with one of their own so that patrolling police cannot see that the premises have been entered. This enables the thieves to load their vehicle at will and in peace and drive out, immediately locking the gate behind them so that they may have several hours of freedom before the theft is discovered. Thus a good-quality close-shackled padlock should be fitted, and the gate area especially well lit.

### *Traffic barriers*

Where the main requirement is the regular stopping of vehicles for checking or like purpose, barriers may be substituted for or used in conjunction with gates. They can now be of two types – the more usual with pivoted arms has a nominal stopping capability but the maximum security barrier is a recent innovation which should effectively stop the passage of anything less than a tank!

The former is in common use; it is balanced against a counterweight for easy operation but the factor rarely appreciated by users is that a safety device is usually incorporated to allow the arm(s) to shear off rather than damage the pedestal in the case of accidental collision – and will do precisely the same if someone decides to drive straight through it. It may be manually operated, or electrically by pushbutton from a gate office – which is decidedly preferable on grounds of efficiency and modernity. Collapsible 'skirts' may be fitted to the arm and both should be clearly painted in red and white; double-sided stop signs should be affixed to them so that there can be no question either of their purposes or of being visible to drivers.

### *Security barriers*

These function entirely differently; a hinged steel frame sunk into the ground supports a heavy hinged flap, heavy-duty air cylinders cause the flap to rise and present a vertical face to oncoming traffic in a maximum time of eight seconds. A third cylinder forces in locking bolts when the barrier is fully up. Safeguards are incorporated so that in the event of tampering or cutting the piping between the installation and the operating lever valve, the barrier will rise and lock automatically. There is little wear and tear in ordinary use and the compressor unit has sufficient reserve capacity to carry out one or two operations in the event of a power failure.

These are expensive but have an obvious application for roads and entrances with a high security risk.

## *EXTERNAL DOORS*

No matter how formidable these may appear, their strength will be no more and no less than the means of securing them. Solid wood, steel, or steel-faced doors offer the best protection. Extra strength to wide doors can be given by a swinging padlock bar across the full width of the door or fitting a substantial drop bar. Attention should also be paid to the jambs and surrounds of the door – the value of a substantial door is strictly limited if the jamb and surround are flimsy and would be forced away from the brickwork by putting pressure on the door itself.

The best locks for general use are the five-lever mortise deadlocks, and to carry these the door and jamb must be thick enough to accommodate them. Padlocks should be of the close-shackled variety to prevent them being forced off with a steel bar; they should also be used in conjunction with hardened steel locking bars – to make an expensive close-shackled lock dependent on a poor hasp and staples is a waste of money.

Where the exterior of the premises is subject to observation by patrolling police, there is no reason why the padlock should not be on the outside. In other circumstances, it would be best to have it on the inside where it could be difficult to get at. It should not be forgotten that a hacksaw blade can often be inserted between double doors to attack the bolt of the lock or the padlock bar. Mortise locks are now available with a hardened steel roller incorporated in the bolt: when these rollers are reached by hacksaw blade they revolve freely giving no bite to the saw edge. If cross-garnet or Scotch tie hinges are fitted they should be bolted through the door and the bolt ends riveted over.

A weak door can be strengthened by facing it with a steel plate; this should be bolted through the framework from the outside and the nuts riveted over the inside – there is no object in using screws for securing the plate to the outer door. Light doors consisting of timber frame and hardboard covering should never be used externally. Glass panels are not desirable, but there is no objection to the slit-type panel with wired glass. With existing glass-panelled doors, protection can be afforded by using grilles behind the doors – again providing the screw or bolt heads are burred over. Where two locks are fitted to any door – for example, a Yale lock for routine daytime use and a mortise lock for night protection – they should be sited as far apart as possible to spread the resistance to violence over the widest area of the door. Further strengthening can be added by nailing diagonal braces across the door.

*Double doors*

Where appearance is not an important consideration a substantial padlock and bar is the best method of securing double doors. The padlock should be close-shackled. The standing door of the pair should be bolted top and bottom preferably with riveted bolt fittings in the door end. A point to check on leaving premises is that the standing door is firmly fixed, otherwise even the best lock of the mortise type is no use.

*Fire-exit doors*

These can be a main source of conflict between fire and security requirements. They afford an immediate escape route for criminals and the crash-bar type can be opened from the outside by boring a hole in the woodwork and inserting an instrument, or occasionally by simply repeatedly shaking – if old and worn. This can be prevented by using the spring-loaded bolts of the Redlam Panic-bolt type which will not prevent a criminal leaving but cannot be got at from the outside. They have a further use in so far as an employee wanting to leave the premises before the end of his duty cannot do so without breaking the glass, thereby releasing the spring-loaded bolt.

A further precaution against such absentees, where the building is covered by an electronic alarm system, would be to incorporate all the fire doors into a permanently live circuit of the installation – this has been found very effective indeed. An alternative to the panic-bar type has a mechanical alarm which sounds when the operating arm is swung. The spring-loaded bolt variety should be checked periodically to see that the glass tube has not been chipped with a pair of pliers to create a hole on the rim which would allow it to be withdrawn without breaking: the tube should be rotated for examination. When one of these is found, it is suggested that observations are justified, the objective could well be theft rather than mere absenteeism. Some fire authorities may object to doors fitted with the normal crash bar being locked at night by means of a chain, padlock, and hasp when no staff are in the building – this should be raised with them before it is put into practice. A British Standard is now in use (BS 5725), dealing with emergency exit devices.

There is a new trick to watch for with this type of bar: an employee attaches a length of thin wire to it when the door is open and closes the door on the wire; later he can pull the wire from the outside to release the bar.

*Roller-shutter doors*

These can either be manually or electrically operated. Manually, the operating chain can be secured against movement by means of a steel pin and close-shackled padlock. Electrical varieties necessitate a cut-off of some kind of the electrical supply, which can easily be arranged. A further precaution consists of setting a ring into the concrete floor and welding a hasp to the bottom edge of the roller-shutter door, joining the two, with the door closed, by means of a close-shackled padlock. When this is done, care must be taken on opening up the premises: if an electrically-operated roller-shutter door is switched to 'open' with a padlock in place, the result can be a U-shaped door!

*Concertina-type or Bolton doors*

These are normally fitted with a specially designed type of claw lock, which in itself offers limited protection. Padlocks can again be used by welding on suitable lugs to the inside edges. A point to remember in connection with these doors is that they are often used with steel-framed sheet-walled buildings. It has been found on occasions that the attachment of the door edges to the H-type steel girder forming the door jamb, has been via a small number of bolts which it has been possible to get at and unscrew from the outside, thereafter simply pushing the door open from the edge. This will also avoid the normal electronic alarm fitting and care must be taken that where these bolts are accessible they are burred over to prevent them being turned.

*All-glass doors*

These are usually only encountered in shops and entrances to large office blocks. Whilst they would appear to be a weakness, they are of specially thick armour-plate glass and there have been few instances of entry being forced through them. The fear of attracting attention by the tremendous crash that would occur if anyone smashed a hole through them seems a very real deterrent. Where these are used, appearance is obviously a first consideration, and the best thing to do with them is to include them in a burglar-alarm system, with either a ray across them or suitable contact.

## *WINDOWS*

What can be done with windows depends on the importance that is attached to the appearance of the building. On the outside of a works, where security is paramount, a steel mesh can be bolted into the wall covering the windows, or bars can be fitted. In both instances, bolt-heads should be hammered over. If bars are fitted, spacing should not be more than 5 inches (125 mm) apart and steelwork used of minimum diameter of five-eighths of an inch (16 mm). The same type of bar can be fitted to fanlights where it should be in a cradle shape to allow opening for ventilation. If there is any question whatsoever of the window being used as an emergency escape route from fire, the fire authority should be consulted.

### *New construction*

Consideration could be given to using glass bricks or roof lights instead of windows where possible. The modern tendency in office blocks is to seek a maximum of light; in these circumstances, if sheer vandalism is not a danger, it is advisable to go the whole hog and use panes as large as possible and preferably of strengthened glass. A potential intruder is much more likely to be tempted to break a small pane, barely adequate to admit himself or to admit his hand to open a pivoted window, than to smash a large hole under circumstances which must attract attention.

## *PROTECTIVE GLASS*

### *Armour-plate glass*

The location and type of building may well militate against the use of grilles, bars, or shutters for the protection of windows, there is no gainsaying that these detract from appearance. Added security against both damage and theft can be gained at a comparatively moderate expense and without too much effect on the amenities by using a type of strengthened glass. All glass manufacturers produce some specially toughened kinds of the armour-plate type and others of laminated construction. These will withstand blows of varying strength according to their construction. Breaking such glass involves effort and creates noise, it also increases the time and difficulty in making a hole big enough to get through.

Two new British Standards deal with protective glazing: BS 5544 –

specification for anti-bandit glazing and BS 5357 – Code of Practice for the installation of security glazing.

*Wired glass*

For windows in the factory type of building and the smaller office building, particularly those which it is desired should be translucent rather than transparent, the Georgian-wired type of glass can be used. Both transparent and translucent varieties are ordinary glass, approximately a quarter of an inch (6 mm) thick, with an electrically-welded steel wire mesh reinforcement. The mesh usually ranges from half an inch (13 mm) square to approximately seven-eighths of an inch (22 mm).

This wired glass has an added advantage in respect of fire: whilst it will crack under heat, it will remain in position even when badly damaged, thus reducing draughts and retarding the spread of flames. For the same reasons it has a safety value too, since there is little chance of injury from falling glass when a window is broken.

*Locks*

Whatever else may be done with windows, the means of opening should be made as difficult as possible for someone trying to do so from the outside. All the better-known lock firms supply key-operated bolts, window locks, or window stops for every kind of window; ample literature is available on application to cover all contingencies. One kind of stop, fitted with a chain, allows ventilation whilst retaining the window secure against illegal entry; mortise bolts operating on a rack and pinion principle can be used for wooden casement windows and special locking handles can be fitted to metal windows. As in other cases, illumination is of deterrent value for window protection.

*Skylights*

These are in a class by themselves; they are difficult to cover and the only real protection is by means of substantial bars beneath them. It is little satisfaction to know that when these are seen by an intending thief he can very often remove tiles equally easily and climb through, forcing a way through the ceiling. Consolation can be found in that his egress might prove complicated and his carrying capacity be limited – he might also fall!

### *Bars*

Where the bars used are more than 1 metre in length, cross-ties, also of mild steel, should be welded to them to give strength. Cross-ties and bars should have ragged ends firmly cemented into the masonry to a depth of at least 3 inches (75 mm). Where bars have to be fitted to wooden window-frames, substantial crutch-head screws should be used for securing them or the slot should be drilled out.

## *GRILLES AND SHUTTERS*

As an alternative to bars and mesh, grilles and shutters offer protection without being permanently unsightly. There are many types of grilles on the market – collapsible, expanded metal, welded wire, etc. These must be adequately secured if they have to be fitted outside but should be fitted internally if possible. Collapsible grilles are manufactured for individual windows and tend to be expensive – these are most often used for the protection of small valuable items on display to the public, as in jewellers, camera shops and, more recently, the protection of licensed bars where there is a sizeable display of spirits almost within reach.

Shutters of any type provide excellent security for windows when fitted internally. Steel and wood are the materials which come most readily to mind and these are quite effective when fixed firmly in position by a drop bar across the back – this can be fitted with a padlock if deemed necessary. One point that should be borne in mind is that they equally effectively conceal the inside of the premises from the eye of a patrolling police officer, so a peep-hole is desirable to allow inspection of the interior, particularly if this contains a safe – it is a common practice to illuminate safes so that if there is any interference it can be readily seen.

## *CELLARS*

It is by no means unusual for cellars to be completely forgotten when the security of premises is being checked. The cellar-flaps, more often than not, are held in position purely and simply by their own weight. Gratings in footpaths giving access to unguarded cellar windows are apt to be treated with the same degree of neglect.

### *Cellar-flaps and gratings*

Both these should be held by substantial chains and padlocks to prevent them being lifted. Cellar-flaps can be further strengthened by means of bolts padlocked in position on the inside edges; flaps are normally of quite substantial construction but if strengthening is needed it can be given by putting cross-ties across the back of them. Where gratings are concerned, unless their removal for cleaning purposes is essential, they should be cemented into their surrounds.

### *Cellar-head door*

This should be kept properly secured with mortise bolts on the inside and the fixing of a steel plate if necessary to provide a higher degree of physical protection. In fact, where cellars are not being used it may be better to seal off entirely at the head of the cellar steps by bricking in.

## *FALLPIPES*

### *New buildings*

Access to a vulnerable roof and unguarded skylights can be prevented by making fallpipes so that they cannot be climbed. In new building-work it may be possible to encase them in the shell of the building or to shield them with plastic or asbestos sheeting. Plastic fallpipes and lightweight brackets are no good at all to a climber as they will break away from the wall easily; moreover they are exceedingly competitive in price with more traditional materials and need no maintenance by way of painting.

### *Existing pipes*

There is a variety of ways of protecting existing pipes:

1 Wrap with barbed wire starting about 3 metres from the ground.
2 Cement into the wall over the pipe a semi-circular set of downward pointing spikes, again about 3 metres from the ground.
3 Cover a section of the pipe, well above the ground, with one of the new non-drying paints which are decidedly effective.

If none of these suggestions are practicable, ensure that any windows that are accessible from the fallpipe are adequately guarded against entry.

## *CONCLUSION*

By all means upgrade whenever a source of security weakness comes to notice, that is stronger doors and locks, barred windows, grilles etc. but accept that a case will have to be made out to justify the expenditure. However, if substantial alterations or new building are proposed, take advantage of this to recommend security-desirable inclusions which are less likely to be opposed and the cost of which will be lost in the overall budget.

## *BUSINESS CLOSURES*

It is a matter for deep regret that the depression which commenced the 1980s caused the closure of many business premises which included very large sites as well as smaller factory units. The process has continued during the decade but at a progressively reducing rate which often takes the form of replacement of old and inefficient units by new construction housing modern machinery, facilities and technological improvements serviced by a slimmed-down workforce.

Closures are emotive occasions and it is understandable that security considerations receive meagre attention in the initial stages. A local security staff will be included in the general rundown and care will be needed to preserve morale and job interest; if there is a group or company head of security, he must impose himself at the local level and monitor that adequate action is being taken. If he is not asked to do this at the outset, he will be well advised to recommend he should be involved as soon as the decision is confirmed – this can save many problems later. Inability to foresee those problems will almost invariably add to the overall financial loss and that loss will not be confined to increased theft by employees.

This is a relatively new 'ball-game' for security and the comments and guidance are not just based upon unpleasant personal experience but are a summation compiled in discussions with senior colleagues from several major industries. It is remarkable that identical facts have been found in so many diverse sources and localities. The emphasis is on factory site closures but exactly the same principles have been found to apply in office blocks and the like.

Closure could represent the termination of the whole or part of an organization. If the former, there are likely to be only two possible intentions: to sell the premises and plant as a going concern, or a total closure with the sale of everything disposable followed by sale of the site, with, or without demolition. If it is a unit in a bigger company, the intention may be to 'mothball' on a care and maintenance basis with a view to recommencing activity on an upturn in trade; the transfer of machines, raw materials and perhaps even staff to other sites might also take place.

If there are prospects of a sale which will ensure the continuance of business, or mothballing with the possibility of production starting up again, there will be an incentive to those in charge to preserve their assets in the best possible condition and priority will be given to ensuring this. Workforce co-operation at all levels may be given in such circumstances. Unfortunately, much the more usual ones are those of complete closure and there, apathy, indifference, and indeed a non-cooperative degree of 'bloody mindedness' may show itself, even amongst senior members of management, and these are situations with which security is primarily concerned.

### *Phases of a closure*

Statutory notice has to be given to employees but management has to make a considered judgement as to whether this is adequate for the completion of outstanding contractual commitments, as well as carrying out the physical and documentary process of shutting down. Three separate phases emerge though action in implementing them will overlap:

1. Progressive fulfilment of outstanding orders and winding up of productive effort. Systematic disposal of plant and equipment.
2. Dispersal of workforce.
3. Disposal of site in condition to gain maximum financial return.

### *Risks associated with closure*

A closure, no matter how expected by employees, will cause shock, recrimination, and diminution of loyalty to the employer. Whilst there are hopes of the decision being rescinded behaviour is apt to be impeccable, but once it becomes irrevocable those standards may lapse abruptly – and standards of supervision are not excluded. Respect for company property will progressively diminish and a series of risks manifest themselves with limited assurance of workforce co-operation in containing them. Amongst these may be found, in sequence through the phases:

1. Interference, tantamount to criminal damage (sabotage), in productive work.
2. Vandalism by employees – window breaking, painting on walls, etc.
3. Theft by employees – minor items being regarded as 'perks' for those leaving.
4. Theft by demolition contractors employees, or others brought in to assist in the winding up.
5. Frauds by contractors on daywork repayments and in respect of materials taken away.
6. Corruption of employees in charge of the disposal of assets.
7. Acute vandalism by employees on last day of employment, prior to leaving premises. (Smashing wash basins and toilets, breaking mirrors, ripping fittings off walls, wholesale smashing of windows, upturning and smashing of lockers.)
8. Thefts of metal and fittings from unoccupied buildings by outsiders, resulting in rapid deterioration.
9. Senseless vandalism by outsiders, possibly resulting in arson.
10. Intrusion by squatters or gypsies.
11. Injuries to trespassing children or others resulting in legal action.

*Precautionary measures*

*Workforce* – These will be largely policy decisions to preserve goodwill whilst not relaxing discipline or security measures. Impending redundancy payments are a strong factor in favour of continuing employee honesty but may not have the same deterrent value against indiscriminate damage.

1. Consider discussions aimed at allowing workforce to leave at no financial loss, prior to announced date, on completion of scheduled production work and that needed to wind up the respective departments. (Objective – to get outstanding production done quickly and efficiently and to avoid workforce having time on their hands which might be wrongly applied.)
2. Prepare and publicize a scheme for selling non-essential or souvenir-type items to employees at nominal charges, and against proper sequentially-numbered receipts. Allow departmental head to fix prices, subject to approval of a designated manager to prevent abuse or lack of uniformity.
3. Simultaneously with 2, reiterate the firm's policy with regard to dismissal for theft by employees and stress that this will continue during the rundown period. (If such a policy has not been previously laid down – do so now.)

4 Where 'search clause' is in being for employees, check that it is being implemented during the rundown and not being allowed to lapse. Also, that contractors vehicles and personnel are being searched at discretion.
5 If feasible, minimize 'last day damage' by, without prior notice, allowing employees to leave immediately after they have drawn their terminal pay or, say, at lunch on closure day – any arrangement which precludes waiting about for the last few hours.
6 Do not allow re-entry after employment is terminated – except for valid reasons.

*Tools and plant* – Tradesmen are likely to seize any opportunity to supplement their personal stock of tools before leaving a firm, knowing that shortages are likely to arouse less concern than normal. Small articles of plant may be invaluable in setting up in business on one's own account and regular usage may give a sense of right of possession. The same would apply in offices where calculators will have a special attraction as will other portable pieces of office equipment. Preventing losses of this type is an essential security part of the closure action.

Maximum value will be looked for from the sale of plant and responsibility for doing this should not be left at too low a managerial level, nor be unsupervised – both temptation and opportunity for corrupt behaviour will exist. The same applies to the scrap materials, especially the metals, that may be generated during closure and subsequent demolition. If spare labour can be applied to the sorting of metal so that it does not have to be sold as 'mixed', it will prove a distinctly profitable use.

1 Cause departmental managers to check their inventory, or prepare one, showing what assets are available for disposal, listing those which may be sold to employees.
2 Nominate the person(s) to be responsible for the sale of removable assets, clearly define terms of reference, and procedure (i.e. tendering etc.) to be followed. Provide for monitoring by reporting and discussion.
3 Segregate classes of articles being offered for sale by tender, putting the more valuable portable items under lock and key, i.e. non-ferrous scrap, office machines, tools and portable equipment, etc.
4 Supervise removal of materials purchased and where weight is a factor, as in scrap disposal, ensure a responsible employee, well briefed, is in attendance at all weighing.
5 Keep a tight check on all requisitions for tools, equipment, etc.

that are made after the closure date is announced; restrict to a minimum the persons who can withdraw keys to stores.

6 Advise employees to remove personal property well in advance of closure date: where skilled trades' tool boxes are to be taken out have them checked and sealed in the department concerned before doing so.

*Contractors* – An accusation of prejudice might be made against the authors for their attitude towards contractors to whose control a separate chapter has been devoted (Chapter 26), but nowhere has suspicion been found more justified than during works closures and attendant demolition. A feature of many admirably-framed contracts spelling out what should be done, what should be taken out, what safety precautions have to be taken, etc. has been failure to also spell out the conditions which will enable adequate supervision to be given to ensure compliance with those requirements. One of the major shortcomings on a works site is the leaving of dangerous holes and structures for which the site owner remains liable.

1 Specify on all demolition and similar contracts, conditions imposed of a security and safety nature – recording in and out of employees, parking of vehicles, nature of materials to be removed, segregation of valuable materials recovered, acceptance of liability to search of vehicles and personnel, leaving of area of work in a safe condition, etc.
2 Nominate supervisory personnel to check individual contractors are doing work as per schedule and with due regard to safety requirements. Ensure conditions with respect to signing in etc. are being observed and that 'searching' is in fact being carried out.
3 If demolition work is to continue after closure and the materials taken out are to be scrutinized, limit entry and exit for their employees to a single route.

*Buildings* – Manpower to do any work of a security nature will become grossly curtailed when the main workforce leaves. It is therefore essential that as much boarding up of doors and windows or other precautionary measures as possible should be taken before this stage arises. If buildings are to be vacated but not demolished, intrusion by children has to be envisaged as a probability and steps taken primarily to exclude them, but also to prevent any conditions being left which could prove a source of physical injury to them.

1 Where demolition is intended of old buildings, check whether they are nominated as 'protected' because of age or uniqueness.

2 As rooms and buildings are vacated, lock them off after switching off non-essential gas, water and electrical services. Withdraw keys from those holding them. Maintain external lighting as long as possible. Board off accessible windows and doors, and complete protection of building before workforce is disbanded.
3 Protect buildings against vandalism whilst awaiting sale by:

   (a) board up windows, or shield them with corrugated iron sheeting;
   (b) seal doors off with planking nailed down with 6-in nails or bolted to masonry;
   (c) shroud fall pipes with barbed wire or paint thoroughly with non-stick paint;
   (d) shroud fire-escapes with barbed wire.

4 If an electronic alarm system is fitted to the building, consider whether it will serve a useful purpose in post-closure protection before having the alarm company terminate it.
5 If premises are covered by insurance for fire, damage or other purpose, check whether it is advisable/necessary to continue the cover. If so, ascertain whether insurers are in agreement with protective measures being taken.

*Post closure* – Whilst normal security measures should continue until the shutdown, the situation changes completely thereafter and ultimate responsibility for the premises may be vested in agents, a liquidator, or another branch or department of the owning group. With the departure of the bulk of the workforce, a skeleton staff is likely to remain to complete the disposal of removable assets and keep an eye on demolition contractors. An early decision has to be made as to whether these few should include persons purely for security purposes; there are alternatives to retaining the current security staff whose role would be merely that of watchmen. If use of contract security is contemplated, thought must be given as to how their work is to be checked. One of the alternatives is that of making the place as thief-proof as possible, removing all that is easily stealable, relying on physical obstacles to intruders and hoping for a divine dispensation against damage, remote though that chance may be! There are advantages in retaining selected members of the workforce whose integrity can be gauged from past performance; they will have intimate knowledge of the site, be likely to live in the near vicinity and prove ideal keyholders if full-time cover is not intended.

Any expenditure at this stage has to be fully justified and this will depend on what loss may reasonably be anticipated having regard to the locality and its record of theft/vandalism and the nature of the vacated premises. The nuisance and damage potential of squatters should not be disregarded – they are easier to get in than out. In the older type of building, the initial target will be roof lead and the deterioration that follows its removal quickly ruins the rest of the fabric.

If a watchman-type presence is decided upon, it is likely to be a one-person function which is lonely, infinitely boring and presents difficulties with supervision (the latter would be limited in effort if contract security were used). For the safety of the incumbent, apart from making sure that he is awake and on the job, some system of mutual-aid telephoning should be arranged to an adjacent firm with security manning, or the control room of a commercial security firm. He should also have detailed instructions in writing as to what is expected of him and who should be contacted in emergency. To summarize:

1 Make a considered judgement based upon nature and locality of site, potential damage loss, etc. as to whether the post-closure presence of a guard(s) is necessary, full or part-time, or in the form of intermittent visits; whether a portion of an existing security force should be kept in a watchman role or reliable redundant older employees used. If the services of commercial security firms are decided upon, obtain competitive quotations and give thought as to how their work will be monitored.
2 If a guard is to operate, provide a phone, full instructions, such keys as are necessary, and a list of persons for him to contact in an emergency. Notify the police in writing of what is being done.
3 If gypsy intrusion is likely, have a deep furrow ploughed round any part of the area to which they might get access; alternatively, raise a ridge of rubbish.
4 If feasible, make arrangements for premises to be periodically visited by a responsible person from agents selling, or other competent person, to ensure measures are effective.

A final thought, if a supervising security officer is redundant by virtue of the closure, it should be a matter of personal pride that he takes every possible step to maintain the value of his service to his employer, up to and beyond the closure day. Indeed, if the attitudes of members of senior management fall short of his own, he may find pleasure in expressing his personal views thereon with impunity!

# 26

# *Contractors*

Many firms have made increasing use of all kinds of contractors during recent years. This has brought in its wake proportionally increased problems both of a security and of an industrial relations nature. However, in the economic climate at the time of writing there is intense competition for such work as is available; conditions may therefore be included in contracts which would be refused in other circumstances.

Contractors' employees can hardly be expected to have the same respect for the property about them as members of the company's own workforce. In effect, so far as the host firm is concerned they are birds of passage with no particular loyalties and may be employed by the contractor only for the specific job in question; this would be particularly true of building operations. Much of their work will of necessity take place when works or offices are closed down. Consequently, there will be limited supervision on their activities, and this in itself creates greater temptation to misbehave. Where the contract labour is skilled it has much more to lose in lapses of integrity, and the risk is by no means as high; this would apply especially to such things as computer installations or working on highly-technical machinery. Unfortunately, though, much of the service customarily required is associated with the construction industry with its significant proportion of migratory labour.

Difficulties that may be encountered include petty or major theft, unauthorized use of the client's materials and equipment, submission of false records of hours and work done, accidental or even apparently deliberate damage, disregard of safety regulations, introduction of sub-contractors without prior notification, occasional drunkenness, inquisitive wandering into areas not concerned with their particular jobs, etc. Having said this, of course, it must be borne in mind that these are the acts of a distinct minority who are apt to be involved in the dirtier and more laborious type of work. With almost all, however, there is nearly always controversy in connection with vehicle parking, irregular times of arrival and departure, and variations in the number of employees brought in.

Where the opportunity arises, as when contractors throughout the country are 'hungry' for work, desirable controls can be implemented which in the future become routine without the past occasional acrimony. This is not a matter to be dictated by security: it primarily involves the members of management who are responsible for engaging contractors. Individually those persons' main interests lie in their own discipline, be it engineering, general site services, or specialist activity, and their personal objectives must be the early and satisfactory completion of the work in hand. As a first step it is therefore necessary to have a simple system of early notification by them of the contractors to be engaged, where they are to work, what they are to do, when they are expected to commence, and the duration of the job. This gives the person in charge of security a chance to consider what particular difficulties may be encountered and to arrange a discussion with the clerk of works, or whoever may be in charge of the incoming labour, to clarify respective viewpoints.

## *CONDITIONS OF CONTRACT*

All contractors are subject to 'conditions of contract' served upon them by their customer. Though there has been a general improvement in this field, these conditions may relate purely to the job to be done and not include clauses relating to safety and security. There is virtue in producing a booklet of site instructions which can be handed to every contractor against his signature to signify that he fully understands the conditions and restrictions under which his personnel will operate. These can be drawn up by a committee comprising company secretary, civil engineer, and safety and security advisers to spell out all the obligations that contractors are required to fulfil. Though these may be voluminous, most contractors will welcome them since they can be studied at leisure and any necessary prior arrangements and instructions given.

It is essential that the purely security requirements be clearly spelled out. They should include a requirement that any contractor's employee not conforming shall be removed from the site forthwith. Suggested clauses are:

1 Contractors, subcontractors, and their staff shall conform to the same restrictions accepted by the firm's own personnel with regard to security, safety, and works rules in general.
2 All contractors' personnel and vehicles shall be liable to discretionary search on entry and departure.
3 Contractors shall not use customer firm's equipment, facilities, and materials without specific, prior notification, and permission.
4 In relation to demolition work, no materials shall be removed from the site other than as expressly provided by the contract or in writing. (It has been custom and practice in the demolition industry, where part of the contract is the removal of the debris, to regard this as all inclusive and implying things like sheet lead and other metals, fittings, etc., of appreciable value.)
5 A specific agreement shall be reached on how contractors' employees shall be recorded, either by individual signing in and out or using some form of clocking to record their presence and the hours that they work. (This requirement in itself will deter the practice of 'overbooking' or using 'dead men' for work done on a time and materials basis. It also caters for checking in the event of any major disaster or evacuation.)
6 Any variation in the subcontractors being used shall be notified in advance.
7 Vehicle parking shall be agreed and in accordance with instructions given by the security staff. (The parking of vehicles near to the point of work allegedly for convenience in moving tools etc. should not be allowed as it creates an opportunity to facilitate theft.)
8 Any work done which may leave any form of potential risk shall be notified before the contractors leave, for example use of welding equipment.

These are of course only security requirements, and there should be detailed instructions on what is required with regard to legislation under the Health and Safety at Work Act so that responsibilities and liabilities are clearly defined for both parties.

Some firms have found it advantageous to extend their conditions to include the enforced submission at the end of each working period of a 'day sheet' to outline the number of personnel and hours worked. They have also issued identification passes for persons and vehicles, but this can be a mixed blessing unless there are very stringent arrangements for collection at the termination of validity.

On the industrial relations side it is similarly essential that the incoming labour should not be granted any privileges or immunities from the working conditions accepted by the firm's own employees as this could clearly lead to disputes. In a union-conscious environment a check should be made as to whether the contractors intend employing any non-union labour, the consequences of which are obvious.

Initial and final control is of course at the point of entry. Where there is a security staff this is not difficult to enforce, but otherwise inevitably there are more opportunities of loss in every way to the customer. Ideally, in all instances one of the firm's own employees with supervisory status, preferably in the trade in which the work is being done, should be given the task of holding a watching brief on the job. This does not imply that he should take over any responsibility from the contractor, but he should feel free to criticize the work being done. A watchdog of this type would be a restraining influence, particularly if he made it obvious that he had a suspicious mind. During shutdown periods in smaller premises there is likely to be a complete absence of own employees. This should be avoided if at all possible, even if it means specially detailing someone on overtime to be present, but it must be made clear to the main contractor that he himself will be responsible for preventing untoward incidents.

No particularly abnormal conditions, delaying, unreasonable, or oppressive on contractors, should be specified. The object is to limit the opportunity for theft and fraud on one's own employer, and the ready co-operation of the contractors should be sought.

## *CORRUPTION*

From the senior security officer's point of view, he must not lose sight of the possibility of corrupt practice on the part of those of his firm who are responsible for the hiring of contractors. Regrettably, experience has shown that this is not a remote contingency, though again we are speaking in terms of minorities. Offences like this are very difficult to prove, since both the recipient and giver of the favour are most unlikely to divulge the transaction and there will be little outward evidence of it. Some indicators which can be discreetly investigated include:

1 Failure to obtain competitive quotations.
2 The obtainer of the contract regularly being the last person to submit his quotation.
3 The person responsible for granting the contract repeatedly being taken out to lunch or granted other favours by the contractor.

4 Use of the firm's facilities and materials being allowed on oral authorization without this being incorporated in the contract.
5 Purchases or sales at prices distinctly favourable to the contractor.
6 Repeatedly drawing to attention of doubters that the contractor is giving the firm most favourable treatment in the urgency and attention that he accords to the work.
7 Invoices for fixed-price contracts continually endorsed with additional charges for work done 'as verbally agreed with Mr . . .'.
8 No delegation of authority to an assistant or deputy to sign invoices or work sheets for payment in the absence of the person responsible for contract acceptance – irrespective of payment delay (a bona fide contractor would complain bitterly if payment was held back, say, three weeks during the holidays of that person).
9 Excessive payments authorized for trivial casual jobs; orders been made out after work done. Work done by in-company personnel attributed to contractors.
10 By and large, overheartiness in the relationship between the contractor and his contact – and undue affluence on the part of the latter!

Unfortunately, this more often than not becomes a matter where there is suspicion rather than proof. This is not an indictment of either contractors or of management, but cases of this type are less isolated than they should be – a fact clearly evidenced by prosecutions, both recent and pending.

The situation is a most difficult one for the chief security officer whose suspicions may be falling on a manager of equal or higher status and even the one to whom he reports. Only a limited number of the ten points – 3, 4, 6 and 10 – will come immediately to his notice and if these satisfy him that deeper enquiry is necessary much will depend on his own remit, the relationships which he has established, and the organization of the firm. If there is an internal audit department, it will have the authority to inspect all records of contracts and payments from which verification of dates, times and personnel used could be checked with security records. A hint placed in the right place could start that quite normal process and the same might be possible in any other department, that is purchasing which has to process the paperwork associated with payment. However, not infrequently the investigation is triggered off by complaints from an outside source, a disgruntled competitor of the contractor alleging preferential treatment and possible malpractice, anonymous letters/

calls or police following up rumours that have come to their notice; these are all instances that have occurred.

Under *no* circumstances should security staff, on gate or other duty, accept presents of any kind, at any time of the year, from contractors who regularly visit the premises. These will be misinterpreted by observers and can lead to difficulties in relationships with the contractors, and to embarrassment if action does become necessary against them or their employees. It is good policy to make the relationship clear from the outset by a search on the first day that they leave, repeated regularly thereafter, and by equally frequent obvious-purpose visits to the area in which they are working. In searching it should be borne in mind that empty scrap drums and paint tins are excellent for the hiding of almost anything.

While contractor's employees and drivers can steal in the same ways as ordinary members of a workforce, a number of particular instances are worth noting which may account for unexplained shortages or losses.

1 Vending-machine mechanic stealing from the near vicinity of the machines on which he is working carrying out proceeds in tool kit.
2 Contract laundry collector stealing during collections and concealing property in and under bundles of dirty overalls (practically every driver of one particular firm was arrested for offences of this kind during one purge).
3 Collecter of canteen waste organized employees to put stolen parcels under swill.
4 Mechanical shovel used on site driven off with shovel lifted and containing stolen property.
5 Industrial window cleaners laid stolen property between ladders on top of their vehicle – stole during lunch break in absence of workers.
6 Industrial cleaners took out property under cleaning rags in bins and buckets.
7 Industrial painters dropped valuable metals into part-used tins of paint.
8 Travelling crane with shear legs – property secreted in recesses into which legs lowered for transit.
9 Concrete-mixing vehicles, stolen property placed inside drum for removal.

A final point where security vigilance may show results – if they have been adequately briefed – where a contractor has been given a fixed price and time and materials contracts to fulfil on the same premises, there is a great temptation to divert labour to the first and charge it to the second. Only by observing the number working on the second and checking later can this fraud be spotted.

# 27

# *Protection of building sites*

The building industry is one which is notorious for a floating population of workers; it is also one where property is dispersed over wide areas on occasions, under circumstances where it is difficult to give anything like the degree of supervision that would be desired. The workforce has a high percentage of semi-skilled and unskilled labourers; the conditions can be most unpleasant and it is not surprising that the percentage of these casual employees who have passed through police hands for dishonesty is as high, if not higher, than in any other industry. This makes security measures not only essential but most complicated to make effective.

## *NEED FOR SECURITY*

This need is pinpointed by the fact that it is customary to take into account probable losses due to theft in calculating costs – and the percentage that is allowed in some areas is high! It would be assumed by this acknowledged wastage of capital that, if reasonable steps could be suggested to managements, adequate finance would be made available to implement the recommendations. Too many firms hold to the attitude that they are in business solely to build houses to the price specified and if the profit margin is not too seriously reduced by stealing, the effort to curtail losses is not worth making. Once a suspect labour force gets the impression that a firm is indifferent to its losses a free-for-all ensues.

Insurance premiums for property on building sites are so prohibitive that many contractors philosophically accept that their losses will be the less expensive – the eventual customer is the sufferer in the higher prices which are charged. It is easy to blame theft for all that goes astray on sites, but this is by no means true and it is conjectural whether a greater amount is stolen than is wasted due to sheer carelessness, damage, and bad book-keeping. Good security measures are not solely aimed at theft – they should be directed to any sphere where the employer's assets are being needlessly dissipated. This conception will not be popular in some quarters as the building industry has many long-established 'perks' where interference can cause resentment – and not always at the lowest level of employee.

Plain wanton damage, by either children or youths, is a major factor on some sites; this is particularly true where there are limited recreational facilities in adjoining estates and the new building site becomes the accepted 'sports' area. On a compact site this can be restricted by fencing, but on a larger housing development the only real solution is preventive patrolling and for this the use of trained dogs is invaluable – their presence is worth that of several security officers.

## *SITE SECURITY: COMPACT SITES*

This type of site occurs, for example, in the construction of a block of flats or offices. A chief security officer or security adviser should be consulted in the planning of the site at the outset; not only could he make suggestions to improve the security of the final building but he should specify what is advisable in the way of fencing or hoardings, entrances, positioning of storage huts and areas inside, lighting, parking space, materials delivery procedures, and security manpower coverage during erection.

There is a somewhat better chance of insurance cover against theft being granted for this type of site, perhaps as part of an overall package; if so, approval must be sought for the measures proposed; when part of such a larger package, there is less likelihood of claims being disputed.

### *Fencing*

Consistent with the needs of the builders, entrances and exits should be as limited in number as possible; this at least ensures a better chance of checking vehicles on and off site. Where a perimeter fence or hoarding can be used, it should be. In congested town centres, hoardings may be an essential safety requirement for passers-by in any case. Construction sites exist for a limited duration and whatever is used in this manner would need to be transportable to the next job to add to the economic viability of the firm. Wooden hoardings are probably the most practicable solution, set with the planks vertically to make climbing difficult both internally and externally and to a height of 10 feet (3 m) or more. Gaps should be left to allow a patrolling policeman or security officer to have a good view of the interior but not wide enough to permit squeezing through. Sufficient lighting should be laid on inside the compound to make it obvious to an intending intruder that he would be able to be seen from outside.

All valuable materials should be put in storage in site huts or buildings under lock and key – and this does not just mean a first-class padlock put through two loops of wire or on a tinny hasp and staple screwed into the wood with the screwheads left intact. Where screws have to be used the heads should be burred or drilled out or a bolt put through the staple and the threads tapped over the nut inside. With mortise locks precautions of like nature should be taken to prevent a kick on the door breaking away half of a thin jamb and leaving the lock intact. There is a limit to what can be done with wooden site huts but these should be as strongly constructed as practicable and always kept locked when vacated.

Chain link fencing is the logical alternative to hoardings; this is usually set on scaffold poles to facilitate subsequent removal – or even on cranked concrete posts set into the ground less substantially than on a permanent site. This has the disadvantage of being easily cut by thieves and liable to damage by careless drivers. Metal palings of the Lochrin type or concrete panels are further alternatives, the former is probably best of all but is also more expensive than either of the more commonly used varieties, though the new concrete panels are easily erected, portable and durable.

### *Gates*

These should be fixed so that they cannot be lifted off and they should be locked with substantial chains and close-shackle hardened-steel padlocks – it is a favourite procedure for thieves to cut off the padlock and chain, substitute their own, and lock themselves in to accumulate what they want at leisure. As in more permanent structures, the gates should be of the same height as the adjoining fencing and should not be easy to climb.

### *Alarms*

In addition to attractive materials like copper tube, boilers, gas fires, and electrical equipment, which should be locked out of sight, other items which damage easily should be put where they cannot be damaged by stone throwing – a chipped lavatory suite is as much a loss to a firm as a stolen one and one ricocheting half brick could wreck several beyond further use. Where the value justifies it, temporary alarm systems can be installed by the burglar-alarm firms, either for operating bells or the 999 warning to the police. It is not beyond a firm's electricians to rig up some form of similar audible deterrent for nominal cost; alarms are not expected by the average building-site thief.

### *Security officers, guard dogs and security firms*

The question of live protection will to some degree depend on the locality – in a town centre with police patrols constantly passing, sheer physical protection of the nature outlined might be considered adequate. Additions can be either to use full-time protection in the form of the firm's own security officers, to tether a guard dog inside the compound, or to enlist the services of a commercial security firm to give periodic check visits.

If a dog is used it must be given as much latitude on a fixed lead as possible. It is not unknown for dogs to have been poisoned and indeed they have been caught with a noose round their necks and choked. The dog is really entitled to visits to check that it is safe; besides, if left to itself, it could be a nuisance to those living or passing nearby. If a commercial firm is to be used some check must be kept on their performance; the quality of these varies considerably and if a security adviser or chief security officer is employed it should be his responsibility first to obtain competitive quotations and then to ensure that his employer is getting the reliable and efficient service he is paying for. It could be found more economic to use one's own labour but in this instance some check must also be made by managerial personnel to see that the job is actually being done as required.

A further alternative, worthy of consideration, rises from the fact that some of the supervisory personnel may be living away from their home towns during the course of the contract and a comfortable caravan on the site and some financial remuneration might induce a reliable man to 'live in', perhaps with a dog provided; this would, with telephone communication to the police readily to hand, meet most contingencies.

At all times of day, private vehicles should be kept in a segregated parking area and commercial vehicles be got off site as soon as possible, once they have delivered.

It is a waste of time and money to employ watchmen who are physically and mentally inert. They are traditionally associated with coke fires, cabins, and sleepy old-age pensioners and they are dangerous in so far as an unthinking management believes it is doing all that is necessary when in fact the person's well-being may be in hazard and his protective potential is nil. Notices warning 'guard dogs' should be freely displayed with those offering rewards for persons apprehending or reporting thieves.

## *SITE SECURITY: DISPERSED SITES*

Except that compounds, similar to those on a confined site, can be established, these offer a more complex problem, if only because construction needs compel a virtual decentralization of much material liable to be stolen. For economic and storage reasons, scheduled deliveries are made to the site areas where the items are to be used, rather than to a compound, and this may involve them being left in uncompleted houses or, in the case of timber window-frames, stacked outside in the open. One manner in which this could be counteracted, without radical change to the building sequence, would be to complete outbuildings of the garage type first and then utilize them for lock-up storage. This would have the further advantage of reducing damage by vandalism and weathering; inadequate storage is the curse of building operations generally.

All that has been said concerning compounds on a restricted site is equally applicable to those on a housing estate; construction, lighting, and general precautions are the same. Wire-mesh fencing is much more likely to be used and, if this is so, materials must not be stacked against or near it on either side, as that would make climbing over easier and could conceal a hole from the eye of an observer or shield the actions of a thief. With adequate lighting being thrown on the fence and buildings which might contain a guard it would be risky to cut a way through.

### *Protection*

It is even more important to protect materials from damage on housing estates – stone throwing is more prevalent and the children have more latitude. The decrepit watchman, too, is even more of a risk – children are ruthless and may create damage to annoy and ridicule the unfortunate individual, who is physically incapable of catching them. There is more scope for the employment of the able-bodied person who will be needed to patrol away from the compound; this will relieve the tedium of long hours of comparative inaction. If the right people are selected, as in all other fields of security, management should be able to incorporate profitable tasks for them which are ancillary to, but do not impede, their main one of security. With several isolated compounds there is more scope for the use of dogs both on patrol and static – if the latter, they should be periodically visited. If no one is employed on a site after working hours, specifically to look after the constructors' property, no amount of mere structural protection will stop stealing. It is unrealistic to think that the police, with their multiple responsibilities, will be able to pay more than cursory attention where the road conditions may be such that they would be unable to gain quick access with their vehicles.

### *Machinery and plant*

There has been a progressive increase in thefts of all types of machinery used in building and roadmaking. Size in itself has proved no protection; tractors, JCBs, drotts, transformers, generators, compressors, diesel pumps, earth-moving machines, and even road rollers have been taken. It has been suggested with some measure of truth that disposal has been as far afield as Crete and the Middle East. This has been a matter of considerable concern to the building industry, and its advisory security arm. 'Consec', advocates the recording of full identification details to enable checks to be made at ports and help the police when querying ownership. Use of 'Camrex' paint which can be analysed is amongst the measures suggested. The calibre of the employees has bearing on some of the ploys of the thieves. It is difficult to believe that a low loader can be driven on to a site, the driver tells the workers that he has come for the compressor or JCB and asks their assistance to load it, and they do so without question; then off goes the thief, and no one takes a note of the number or can even remember the make of the vehicle. It does so happen and will no doubt do so again. The multiplicity of subcontractors admittedly helps open approaches like this.

Prevention is not the easiest of tasks due to traditionally casual workforce attitudes and the fact that employee involvement is strongly suspected in many cases. This makes the use of hidden switches to prevent ignition less effective than it might be; battery removal is an alternative found too much of a nuisance to enforce; moving into a locked compound is too much trouble at the end of a day, as is the locking of ancillary tools inside compressors. Until the actual handlers of the equipment have been indoctrinated the losses will continue.

### *Tidiness aids security*

An untidy estate will give an impression of lack of interest on the part of the contractors and there is a much greater temptation to take that which is scattered about and apparently unwanted than that which is reasonably and neatly stacked. Even commonplace items like bricks are much less liable to be commandeered if they are stacked and covered with polythene to guard against weather, than those in a shapeless heap. Similarly, when timber is stacked off the ground in racking and sheeted over, there can be no excuse that it was thought to have been thrown away, unwanted. If window-frames and door-casings are stacked against the side of a house and not for immediate use, they should have a wooden batten nailed inside to the top and bottom units to prevent them being thrown about or stolen; if unprimed and likely to stand for any time they, too, should be sheeted over against the weather. The enforcement of tidiness and of general security precautions for materials should be a matter of concern for the site agent or surveyor, and if he disregards elementary steps this should be brought to the notice of senior management, for it is neglect of his job.

Where materials and tools have to be left in unfinished houses they must be put out of sight and the doors made as secure as possible. If work can be so scheduled that heating pipes, boilers, fires, and so on, are virtually put in as they are delivered, so much the better. Things of value should not be left readily accessible – tradesmen take a degree of care over their tools that they do not exercise in respect of their employer's property.

It is even more important on dispersed than on compact sites for 'guard dog' notices to be displayed everywhere, together with warnings that all persons causing damage or removing property will be prosecuted and offers of rewards for the detection of offenders.

Finally, where unusually large amounts of metals are held at any time, the attention of the local police should be drawn to their whereabouts and nature. These constitute one of the major risks; copper and lead are easily defaceable and disposable and can be costly to replace – it is worth spending time and money to protect them.

## *PRECAUTIONS AGAINST THEFTS BY EMPLOYEES*

### *Casual labour*

In an industry where much casual labour is employed and personnel are thereby more likely to have records for theft (estimates place this as high as 20 per cent) it is absolutely imperative that no impression should be allowed to develop that management has a *laissez-faire* attitude towards both the theft and the handling of its property. Itinerant workers cannot be expected to have feelings of loyalty; those who are fundamentally dishonest will only restrain their natural instincts if they know that watch is being kept and prosecution will inevitably follow.

There is a great reluctance in the building industry to prosecute thefts unless they are of reasonable magnitude; this is particularly true where its own employees are involved. It is a bad attitude where petty theft is commonplace and simple dismissal is no punishment to the casual labourer who will automatically gravitate to the next building site – how often are references taken for this kind of employment? Prosecution will mean inconvenience in making statements to the police and perhaps attending court, but this will be more than repaid in the deterrent effect of a clear-cut policy on other similarly-minded individuals.

### *Security policy*

Where subcontractors are used, insistence should be placed upon their conforming to a common policy; trouble could be caused by different treatment being accorded to those committing identical offences. The same should apply to safeguarding property in their care, particularly where this belongs to the main contractor; any reports of theft from them should be carefully looked into to establish whether they are indeed genuine or are a cover for deficiencies arising from other reasons.

'Labour only' and 'piece work' are accepted components of the industry's manpower and wage structure; this does not help in checking who is actually employed on a site and when they come and go – it is quite impossible for a security officer to survey more than a percentage of workers when their work is finished. This gives ample opportunity for the removal of materials without being stopped and checked. The best protection during the day is that afforded by a reliable team of foremen and chargehands in constant contact with the labour force. There is an argument in favour of a periodic change round of foremen between sites, where the firm's size permits, to prevent familiarity or undue influence warping the foreman's judgement of what can or cannot be tolerated. The practice of a firm supplying transport to and from sites has an advantage over and above convenience and labour relations – it takes a certain amount of determination to carry stolen property in a lorry or bus load of fellow employers and foremen.

### *Tools*

Whether issued by the firm or owned by skilled tradesmen, tools are a persistent target for theft. By and large, despite the proportion of the unreliable, most workers respect other worker's tools on the same job; one tradesman rarely steals another's – perhaps because he appreciates the importance attached to them as virtually the person's means of livelihood and plying his trade. Where there are losses these should be enquired into fully and every effort made to trace the missing property; this is the ideal way to get the confidence of those on the site who are likely to be helpful and provide future co-operation. If a theft occurs during the course of a day, as many employees as possible should be checked before they leave the site; the assistance of union officials, if these are present, will be freely given.

The safekeeping of tools is of importance. There is an onus on the employer to provide a locked place where they can be kept – then at the owner's risk, except in case of fire where the liability reverts to the employer. This is a trade where tools are widely insured by their users and, when looking into thefts, it should be borne in mind that the complaint might be ill founded or exaggerated. A record of thefts proves valuable for reference purposes and could provide indication of previous suspect claims.

Spot checks on persons and vehicles leaving always have a salutary effect beyond those who are actually involved, and spasmodic visits to the site by those whose prime interest is security will prove a cautionary factor and be commented on.

## *CONTROL OF VEHICLES*

### *Employees' cars*

As in other branches of industry, employees' cars offer the best opportunity for the removal of property in bulk and the safest way to avoid the notice of security or managerial personnel. Under no circumstances should private cars, even those of senior staff, be allowed to be parked in compounded areas. There is no need for them to be there and, apart from any considerations of theft, they cause congestion and obstruction and their presence inevitably leads others to seek the same privilege – with the result that the refusal leads to grumbling and ill will. Cars can be a thorough nuisance on the roadways of the site itself; it is essential that heavy vehicles dropping materials shall do so at the nearest convenient point to where they will be used. This may be prevented by parked cars and damage to temporary kerbs can be extensive when lorries turn over them to get through to where they need to be. If possible a special hard-core stretch should be laid down near to the site office for all employees to put their cars; if this is kept in reasonable condition there will be no excuse for leaving them elsewhere and anyone who does so can be regarded with suspicion.

*Delivery vehicles*

The checking of materials on to a site is the responsibility of the site agent's clerk or checker, storeperson, where one is employed, or foreman/chargehand where sand, bricks, cement, and so on, are dropped out on a site. Whilst the latter are hardly of sufficient value to justify anyone coming out of hours to steal them in quantity, there is a ready market for them to be delivered by the carrier to places where small private building is taking place. This can easily be done by dropping off part of a load destined for a firm if it is known there is little or no checking or supervision of incoming materials.

No building site has its own weighbridge and loading notes have to be accepted for most bulk materials – a count could be feasible for the more expensive tiles, bricks, or bags of cement. The right should be reserved, and notified to suppliers, of occasional spot checking of loads by taking over a public weighbridge before and after delivery. When this has been done once or twice the word will spread and have the necessary effect.

Unless they are supervised, delivery drivers can clock in on a site, be directed to where they should drop their loads, and then only drop part if no one is about, or they can bribe someone to sign their consignment note that the full quantity has been received. Where there is any suspicion that this is taking place a watch should be kept on the driver in order to catch both him and the accomplice. It is a good principle to have a central reporting point from which drivers can be directed; they can also be clocked in at the time and asked to report back after delivery. Most suppliers would gladly accept this, for it gives them a check on the performance of their drivers; it also gives the site contractors the opportunity to make sure that they are not taking part of the load out again and, by comparison of times, fix a possible source of loss if items are found missing after the driver has left.

Deliveries to be made only during working hours must be specified; this will ensure responsibility is firmly established if a non-delivery insurance claim arises. A classic example of this was the dumping, or alleged dumping, of a complete load of plastic and copper tubes outside a builders yard after closure on a Friday evening, not to reopen until the Monday. It was over a month before the facts were established by which time the erring driver had left for pastures new and the dispatching supervisor was found to have a purely verbal arrangement for the delivery. The insurers were not sympathetic to the suppliers!

There must be a satisfactory system of recording receipt of materials. A responsible person must sign for them and accept the corresponding consignment note to pass to the site agent or his clerk. With both subcontractors and delivery firms, anything in the nature of security requirements – checking-in and parking – should be made in writing so there is no excuse of lack of knowledge, and objections can be discussed before they develop into incidents.

## *CONTROL OF PROCEDURES AND PREVENTION OF DAMAGE*

Prevention of damage has already been mentioned, with emphasis on tidiness on the site and stacking of bulk materials, but there are sources of wastage by rank carelessness during use, such as the following:

1 New sewer pipes left where they will be crushed.
2 Tie-wires in large quantities dropped in mud and used in hard core.
3 Floorboards damaged by plaster being mixed on them and soaked with water.
4 Cartons of nails broken open and left lying in uncompleted buildings.
5 Window- and door-frames left where they will be damaged by passing transport.
6 Cement bags left on damp surfaces, part-used bags left outside.
7 Sand tipped on soft ground with inevitable high proportion of wastage.
8 Tools, fittings, wiring, electrical gear, left in partly completed and insecure units.
9 Lavatories, wash-basins, and baths carelessly handled and damaged.

All these add up to an inaccurate assumption of theft when estimates of the quantities used are reviewed.

A reasonable system of control should be instigated so that damaged fittings may be utilized on sites in workmen's lavatories, canteens, site huts, etc. It is ridiculous to install new ones in these and either scrap or store damaged ones. All surplus stock should be returned to a central store and not retained where no one is really interested in it. Book-keeping should be tightened up to prevent over-ordering and to provide constant knowledge in a large firm of 'what is where' at any given time.

Tools issued to a site are rarely all returned at the completion of a contract; each contract seems to necessitate a fresh issue. This is a matter for the site agent – consideration might be given to providing him with an incentive in this direction by giving him a percentage bonus for all tools which are transferable to a new site.

Scaffolding poles are lent out, stolen and often little importance is attached to their collection after use. This practice was once tolerated to the extent that in 1967 a survey estimated that, even at the values then appertaining, not less than some £2,000,000 worth were unaccounted for. The shock of this certainly induced the larger firms to pay more attention to their security and to providing visible signs of ownership. The poles should be stacked tidily for easy appraisal of quantity and convenience of handling. Consideration should be given to marking to identify – different coloured paint splashes in a sequence could do this with minimum time and cost. If something of this nature is not done, any length or clip is indistinguishable from some other firm's property.

Timber can be stored under cover to prevent weathering, or it should be stacked and sheeted over. Employees should not have the privilege of taking short lengths, unless they are totally unusable.

Cement huts or garages should be used for cement storage and a watch should be kept on firms delivering ready-mixed concrete. The accuracy of the reputed load can only be estimated but an experienced foreman will soon have a good idea whether he is being defrauded – this is another instance where watching may have to be done from supplier to delivery point to substantiate suspicions. After working hours, a few bricks should be rotated in concrete mixers and the displaced hardening concrete swilled out; the practice of taking off starting handles should then be implemented. On a carelessly run site, mixers not standing on a wood base and regularly moved may become set in concrete which may necessitate breaking the wheels to move them.

Metals, in the form of unused ends and scrap, are the greatest 'perk' for employees in the industry. Some plumbers regard the disposal of any waste as their prerogative, which can amount to a major loss, with the present prices of copper and lead. A mandate to dispose of short lengths will inevitably lead to the deliberate creation of short lengths and scrap. This must be absolutely stamped out and, irrespective of protestations of 'we have always had it', exemplary prosecutions should follow. It would be interesting to know what savings would accrue if all short ends and scrap were saved for further utilization or sale. The same is applicable to surplus electrical fittings, conduit, and wire. If there is no return-to-stock procedure, this in itself is an incentive to over-order.

Perks of all kinds, including work done for senior staff, must be rigidly controlled – if they lay themselves open to criticism, how can they enforce their authority in cases of theft on those whose services they have used? As in more static industries, nothing should be taken off the site by way of tools on loan or materials on purchase or otherwise without a written permission from someone in authority. Damaged items of value should not simply be scrapped on the authority of anyone on the site, but should be inspected and 'written off' by managerial dispensation from the head office.

## *SITE LABOUR FRAUDS*

The nature of the building industry is such that a fluctuating labour force is needed and employment at any given site has a foreseeable duration. Avoidance of penalty clauses may dictate a sudden expansion of manpower, for example a small team of self-employed bricklayers – on the 'lump' – may be taken on for a short intensive spell to bring into line a phase of the operation which has fallen behind schedule – they will then leave. This creates a situation of casual itinerant labour being taken on at a moment's notice and departing at equal speed with accompanying difficulties of documentation – ideal for the dishonest site agent or foreman with a lax system which allows him to claim wages for non-existent or departed employees ('straw men' or 'dead men') or to exploit fictitious overtime claims or payments for alleged casual work done.

Paying out by persons independent of the site and against production of a clock card personally by the employee, unannounced 'head counts' against the means of recording labour, and rotation of site supervisory staff are feasible preventive measures. Collusion between supervision and employees on overtime or productivity bonuses is more difficult to combat, spot visits by senior staff asking pertinent questions and tight budgetary checking can help to keep these within limits. The more remote the site, and the less the interest displayed by the senior management, the more likely there will be fraud.

## *SUMMARY*

The precautionary steps that should be taken on the site are as follows:

1. Create as substantial a compound as possible for all materials which are easily stolen.
2. Ensure that the compound perimeter will be hard to climb from either side and any breaking into be immediately observable.
3. Establish a secure store within the compound for particularly valuable items: copper tube and fittings, fire-back boilers, gas fires, immersion heaters, etc.
4. Illuminate the compound, in particular the vicinity of the store for valuable materials.
5. Where the site is insufficiently large to justify the employment of a full-time security officer or capable watchman, consider using the services of a private security firm to give periodic visits.
6. On a large site, where security officers are employed, consider using guard dogs in each separate compound and for patrolling.
7. Segregate employees' cars to a given parking area which is easy to supervise.
8. Have a reporting point for incoming delivery vehicles and establish a practice that they 'clock out' at this point when leaving the site.
9. Thoroughly examine any accepted procedures whereby employees could remove any kind of materials from the site for personal retention.
10. Thoroughly check all means of storage of materials to reduce wastage due to weathering and carelessness.
11. Where materials of value have no means of identification, consider whether it is worthwhile to put some identifying mark upon them.
12. Post warning and reward notices to deter thieves and mischievous children.
13. Try to build up a security consciousness amongst all permanent staff, offer rewards for the detection of theft and for suggestions to minimize waste.
14. Do not leave any cash on site; arrange wage delivery to minimize risk of attack or snatch.
15. Keep liaison going with local police and invite resident beat officers to familiarize themselves with site.
16. Do not allow or accept out-of-hours deliveries of materials from suppliers.
17. Always support any prosecutions the police may wish to bring for damage, theft, or fraud.

18 Always claim compensation for losses and include mention of this intent on warning notices.

*PART FIVE*

# *SECURITY AND CONTROL OF ROAD TRANSPORT*

# 28

# *Security of vehicles and drivers*

Of all the offences coming under the heading of 'simple theft', that of property from unattended vehicles represents the biggest single problem to the police. Private cars are those mainly affected – this follows by virtue of their numerical preponderance, but individual incidents rarely represent loss to the user comparable with those where commercial vehicles are concerned.

## *COMMERCIAL TRAVELLERS*

Private cars fall into two categories: those that are used for strictly private purposes and those that are used as travellers' cars, either as ordinary saloon cars or estate cars. The casual thief is responsible for the greater part of the thefts from the ordinary private car user – stereo car radios are a particular attraction – but thieves attacking travellers' cars are a specialized species.

The very nature of a commercial traveller's work entails leaving his vehicle as near to his prospective customer as possible; this means that it will have to be left more often than not at the kerb-side rather than in a car park. The value of goods carried is often out of all proportion to their bulk and it is incumbent upon the traveller to take every possible precaution to protect his vehicle. It is absolutely essential that he should have a powerful and efficient alarm fitted and it is hardly less advisable that he should also have an immobiliser.

It is rare that the integrity of commercial travellers, in relation to their own goods, is ever in question, nor are their vehicles left at night other than in their own locked garages; with these exceptions, what is said in respect of commercial vehicles and their drivers is equally applicable to them. There is one special precaution that they must take: to glance round on each occasion that they stop near a customer's premises to see whether any vehicle, which they have seen at previous premises visited, is just parking or has just parked. If they do see this happen more than once they would be well advised to ring the police from the customer's office before returning to their car, giving the number of the vehicle and a description of any of its occupants. Another point is that of insurance; where high-risk goods are carried, special conditions and limitations are certain to be applied and the user must ensure that these are complied with to the letter or claims may be invalidated.

An unfair dismissal appeal hearing result is of importance in connection with commercial travellers, *Marshall* v *Alexander Sloan & Co. Ltd* 1981 IRLR 264. In this a commercial traveller's contract of employment contained an express condition that the company's merchandise should be taken out of the car at night and lodged safely; Marshall when sick, refused to do this claiming his contract was temporarily suspended – the dismissal was held to be fair. The condition would obviously be a useful one to consider including in such contracts; setting of alarms/immobilisers would be another.

## *INSURING COMMERCIAL VEHICLES*

Commercial vehicles represent a great temptation to any criminal and far too many firms fail to appreciate the likelihood of their being attacked, regarding insurance cover as being all that is required by way of precaution. Unfortunately for them insurers are tending to take a much stricter line with thefts of and from vehicles. Recent cases typify this new approach and it is obvious that where obligations are laid down upon a firm to ensure the locking of a vehicle and the taking of other precautions the insurance company will not be prepared to pay, and will be upheld in not doing so, if there is evidence that these precautions have not been taken.

One lawsuit worth bearing in mind is *Ingleton of Ilford Ltd* v *General Accident Fire and Life Assurance Corporation Ltd* 1967 2 Lloyd's Rep. 179. Here the policy excluded loss or damage by theft while the van was left unattended in a public place, unless securely locked. The van and contents were stolen whilst the driver was in a shop making delivery of goods; the van was unlocked and the ignition key left in the van. The defendants' insurance company were held not to be liable under their policy.

Only a relatively small proportion of all goods vehicles are actually attacked, but if thieves decide that a particular vehicle is a desirable objective its safety may well depend upon the protective factors that have been built into it, the intelligence, loyalty, and determination of its driver, and the planning of routes and procedures that are being carried out by supervision.

## *REQUIREMENTS OF THIEVES*

When considering how to combat a particular type of theft it can be advantageous to change places mentally with an intending thief. By doing this one can evaluate the things that the thief will most need to achieve his objective and, from the assessment of those, counter measures can be determined.

First, there are certain characteristics of all thieves which should be accepted: on the whole they dislike physical exertion immensely; they wish to get their theft over and done with as quickly and quietly as possible; they want to keep the stolen goods in their possession for as short a time as they can, converting them into hard cash at the first opportunity; and, above all, they want to incur a negligible risk of detection. If these characteristics are accepted, it is easy to see why theft of loaded vehicles is so popular with certain members of the criminal fraternity.

The essential requirements for a thief in respect of stealing loaded vehicles are listed below:

1 The capability of removing the vehicle and/or its load.
2 The opportunity to do so.
3 Sufficient time to carry out the theft without interference or attracting attention.
4 Adequate time to ensure being a safe distance from the scene before the theft comes to light.
5 An early market for the goods with reasonable reward.
6 Negligible risk of detection.

It can be taken for granted that any organized gang will have arranged a market for the goods before the theft has even been seriously considered.

## *METHODS OF STEALING*

Thieves have several means whereby they can achieve their objective; these are:

1. Stealing the vehicle or from its load, in the absence of the driver, whilst parked, either temporarily or overnight.
2. Taking possession of the vehicle by force by either stopping the driver by a trick or attacking him whilst his vehicle is stationary under normal circumstances.
3. Persuading the driver to join forces with them in simulating the circumstances of a theft.
4. Impersonating the driver, in his absence, to obtain possession of his loaded vehicle or, by using their own vehicle, persuading a firm to give them a load on the assumption that they are bona fide carriers.

There are many lines of diversification amongst these main avenues of attack and there is little doubt that others will occur to fertile minds with ample leisure for contemplation.

The more of these lines of attack and requirements that can be negated, the less likely it is that an attempt will ever be made, and if the steps can be offensive as well as defensive, so much the better. The greatest deterrent of all is the prospect of being caught – especially on the job.

## *VEHICLE PROTECTION*

Without the co-operation of the driver, the fitting of any device is useless and a waste of money, but by far the majority of long-distance drivers are honest people who will utilize any equipment placed at their disposal by their employer for their personal protection and that of their vehicles and loads. With no special equipment there is still much that they can do as individuals, but they should have more than the ordinary door locks to assist them.

### *Internal bolts*

Circumstances could arise where a driver might be attacked in his cab. This would be difficult in the first place since he may be two metres or more from the ground.

If the vehicle is required intact, breaking windows to get at him is out of the question, the thief therefore has to get into the cab. A very simple precaution is the provision of 3 inch (75 mm) tower bolts inside each door, to be shot by the driver as and when he requires; these will solve this problem almost entirely. If, added to this, the driver is furnished with a push-button in the cab to activate the alarm system and his own means of possible counter-attack – for example, poking a fire extinguisher through the window to discharge at his attackers – an attempt to get at him will probably soon be abandoned.

### *Vehicle alarm systems and immobilisers*

An increasing number of commercial companies produce a variety of both alarms and immobilisers. The electric circuitry of alarms, in particular, is not complex and, should they so desire, firms can construct their own variations that may be equally effective as the commercially-made ones and have the advantage of being unconventional. It is possible to incorporate both alarm and immobiliser in one unit but the disadvantages of this offset the advantages – if a means is found of bypassing the whole unit, both forms of protection will be eliminated together.

Both devices are really delaying ones and any skilled and determined criminal, with ample time at his disposal, could eventually overcome them. The presence of an immobiliser represents a challenge and an inconvenience, rarely a complete barrier; the alarm, a temporary embarrassment. It is no use at all fitting an alarm siren on a vehicle which is then left in a position where no one can possibly hear it; even if it is heard there is no guarantee that anyone will take a great deal of notice. The lack of attention paid to alarm bells sounding at jewellers' shops near opening time is proof of public indifference to such noises. Above all, no system is better than the operator – an uncooperative driver can nullify the effects of the best devices; an uninstructed or incompetent one can prevent them even commencing to fulfil their function.

### *Alarm systems*

Modern alarms are almost invariably electrical and either use the existing horn of the vehicle or are specially fitted with a siren of a distinctive note. There are several different means of activation, the most common is that of a key-switch turned after closing the doors and windows and with a separate circuit from the electrical equipment of the car. Contacts on doors, windows, boot, and bonnet, may be included as desired. Occasionally the alarm is wired into the ignition system or connected to the mechanism to operate when an attempt is made to start the engine. The switches themselves may be one of the following types: tumbler, combination cap or insert, relay, hand-operated, or key.

Other types of activation, which depend on the actual movement of the vehicle, are the pendulum type of contact, mercury switches, and vibrator contacts. These may have an additional value in being usable for protecting the load itself if they are sufficiently sensitive to react to movement on the vehicle or interference with either doors or sheeting. With this type of contact it must be borne in mind that the natural phenomena of wind or rain may occasionally be sufficient to cause false alarms. Few of the ways of totally protecting a load itself against interference have been entirely successful and a firm with a load of consistent nature and shape should be able to evolve its own alarm device connected to the locks, ties, or sheeting.

Under present-day traffic conditions it is absolutely essential that any alarm siren should be loud and distinctive, otherwise it will attract little or no attention. A high pitched or a warbling type of sound which causes irritation to the listeners is probably the best, but whatever is fitted should be tested under the conditions in which it will be required to operate.

If, on commercial vehicles, the power source for an alarm is in the form of externally-mounted batteries (see also p. 438), it is obvious that this can be interfered with and a separate internally-sited battery should be used.

*Immobilisers*

These are of two main types: mechanical and electrical. Consensus of opinion in the past has been that the mechanical types are more reliable and more resistant to interference.

*Electrical immobilisers* – These work on similar principles to the alarm systems, but cut off the ignition, starter motor, or the fuel supply. Those stopping the fuel supply are particularly suitable for diesel motors where they operate a plunger to lock the fuel-injection system. Operation is similar to alarms in the form of a key or combination lock mechanism, set after the driver leaves the cab. Variations have been introduced, dependent upon movement in the cab or again on an attempt to start the engine.

*Mechanical immobilisers* – The physical methods applied are, more often than not, in the form of locks upon the gear-shift, brakes, or parts of the engine and steering mechanism. More complex methods involve the transmission and the differential or the hydraulic braking system. There is ample scope for ingenuity in devising exclusive types – if these are of the lock and rod principle they should be in a position where bolt croppers cannot be applied to them and also be in a position where they are not obvious so the thief will have to spend valuable time locating them before determining how to counteract them. The unassisted driver quite often has simple and very effective means at his disposal – removal of rotors, leads, or parts of the mechanism, or slipping a piece of paper between the points and terminals can stop a thief, as effectively as any expensive mechanism.

A newcomer on the mechanical immobiliser scene is the ubiquitous wheel clamp which has proved its worth in deterring illegal parking in city streets. It is of course very effective, easy to fix, difficult to force off and a good visual deterrent for prospective thieves – drivers would probably have a certain amount of pleasure in using them!

Cost of equipment can vary immensely – the more complex the more secure, but also the more difficult to service and reset if accidently operated. Simplicity is a virtue for the driver – if he has to carry out a complicated and difficult procedure he will be tempted not to bother.

## *TRAILERS*

Trailers represent a problem since the normal immobilising devices are not functional when they are separated from their tractors as may happen at loading and unloading points. It is a practice with many firms to load the trailers overnight for picking up early next morning. Unless precautions are taken there is nothing to stop thieves backing their own tractor on to such a trailer and removing it immediately.

Chains around suspension and brake-locking devices are apt to be more dangerous than effective. A pivoted steel clamp with a hacksaw-proof padlock on the trailer pin seems to be the ideal fitting and a commercial appliance is available, although this is something which can easily be made. It is essential to remember to use a lock on the fitting which will be impervious to hacksaw blades and sufficiently close-shackled to prevent it being easily forced. Indeed, from time to time there have been spates of empty trailer thefts and there is a market for them after respraying.

If these trailer pin clamps are used they can form a line of defence for the tractor. If a driver can drop his trailer so as to leave his tractor between it and a wall with no room for manoeuvring, there is nothing a thief can do about it without forcing away the clamp.

## *EQUIPMENT*

The cost of accessories has reached a peak which makes the effort and risk of stealing them acceptable. Batteries, spare wheels, and radio equipment are first line targets from goods vehicles and even fuel syphoning is now worthwhile. Theft of the battery has a costly consequence in the delay of replacement which may put the lorry out of action for anything up to a day, and an interrupted attempt to steal an actual road wheel caused a front wheel to collapse in heavy traffic on the M6, nearly resulting in a major accident.

Batteries can be protected by fitting a cage of bars over them with a good padlock; locking wheel nuts are available that can be fitted to both spare and road wheels at no great cost. Radio equipment is a temptation to drivers as well as thieves; if the vehicle is fitted with an alarm system there is no difficulty in having a permanent contact fitted beneath the radio to sound off the siren if the equipment is lifted out. Conversely, one large firm plagued with such thefts and suspecting driver responsibility, simply had the sets and speakers removed and painted a revolting shade of yellow which rendered their sale value negligible! Where communication-type radio is installed, the loss–cost factor is considerably greater and even an unsuccessful attempt could cause gross damage; it is worth considering whether a form of obvious steel banding can form part of the installation to provide a visual deterrent and make the theft harder.

Fuel, likewise, may be stolen by the driver who, if his mpg is challenged can allege syphoning without his knowledge; locking caps will remove that excuse and thwart the outsider. Tarpaulins at one stage disappeared with an alarming frequency which fortunately seems to have lapsed somewhat, the remedy is simply to buy ones with the firm's name on them regarding it as an advertising as well as security measure.

## *LOAD SECURITY*

Where large vans are concerned there is no difficulty at all in wiring up the rear doors to the alarm system of the vehicle. In any case, these rear doors should be fitted with a sizeable close-shackled padlock and bar to discourage an attempt in the first instance. If such a fitting is to be made it must be substantial and not the flimsy hasps and cheap locks that are frequently encountered.

If the value of regular loads demanded it there would be no difficulty in protecting the walls of the vehicle as well with closed-circuit wiring of the type which is used with burglar alarms.

Open lorries carrying goods covered by sheeting represent a much more difficult problem. All loads must be sheeted down tightly to prevent casual petty theft during the course of deliveries, and also to conceal the nature of the load. A common practice is to steal from loads parked overnight at transport cafés by undoing the tie ropes on the sheeting, taking out property from the centre of the load, and resheeting down, so that the theft may not be discovered until the vehicle is several hundred miles away. More often than not this type of theft is almost certain to be the work of a dishonest lorry driver also parked at the café. Each driver has his own means of sheeting down and if he uses knots of a particular type or sequence, or in some way ensures that interference will be detected by him when he inspects his load before leaving, the chances of the police detecting the theft are infinitely increased.

With a little ingenuity, alarms can be attached to sheeting and for that matter, as mentioned earlier, switches functioning upon movement of or upon the vehicle can be installed. The easiest and cheapest way to stop thefts from loads is to put the vehicle on a guarded or well-lit car park.

Drivers making multiple deliveries to retail outlets of valuable commodities have to be aware that their greatest danger lies in opening up outside the customer's premises. A whole sequence of such attacks has been successfully made on cigarette carriers in the London area during the early months of 1987. These were the subject of an excellent television documentary reconstruction which, at least temporarily, seems to have either deterred the criminals or indoctrinated drivers to a realization of the risks.

A limited number of firms have experimented with carrying Alsatian dogs in the back of their vehicles – this is effective but has caused complications in some instances with members of the public!

### *Load packaging*

If articles are sent out loose, or in large numbers of relatively small cartons, from a central dispatch point this both makes undetected theft easy and provides a potential smokescreen for allegations of miscounting. It is also time wasting in loading and unloading and may result in unnecessary damage to the articles. If bulk packs can be made up in the dispatch warehouse, with adequate adhesive labelling to indicate any interference with the cartons, weight checking would be easy, speedy, and handling time grossly reduced.

Petty thefts by drivers from their loads would also become much less attractive. An exasperated security investigator, repeatedly frustrated by 'miscount' explanations, put up a case for a change of practice based upon security considerations, backed up by evaluated time savings – it was turned down as 'impractical'. Eighteen months later, to general acclaim (and a salary bonus) the manager who turned down the scheme introduced it!

*Tanker vehicles*

These fall into a special category; the sheer bulk of their contents seems to militate against the vehicle being hijacked. The main source of temptation is the possibility of repeated pilferage particularly of fuels, especially petrol, and wines and spirits. Therefore, preventive action is required primarily in respect of employees.

Where fuel is concerned, a driver's obvious ploy is to ensure that he has a quantity left after making all his authorized deliveries – any time before then, he is in danger of being detected in some form or another. To do this he has to ensure that he had the co-operation, either intentional or unintentional, of the person who is supervising or accepting delivery – sometimes this is unbelievably slackly carried out. If the form of measure is to use a dip-stick, this is frequently left to the driver. Despite the fact that he may have been visiting a particular site for years, an occasional check should be made to ensure that he is using it correctly and not just accept his measure. Drivers often have regular runs and it is easy to develop a sense of confidence and familiarity with them.

Where wine or spirit is concerned there are three main methods:

1 Short delivery (wine left in the tank).
2 Dipping through the top of the tank and transferring liquid to a container.
3 Removing customs seals to attach specially adapted couplings and pipe to the outlet valve.

Added to these possibilities there is the risk of dilution being used to cover volume deficiencies caused by the theft. Frauds can of course also occur during loading and offloading and the temptation is perhaps greater with these goods than most.

The vehicles should be surveyed to ensure that every possible means whereby liquid can be abstracted is either protected by padlocks of quality or is fitted with seals sufficiently reliable to give an immediate indication of interference. If the vehicles are outgoing through a security control, a practice of recorded inspection of seals and locks would have advantages. The Commissioners of Customs and Excise have produced a list of indicative and heavy duty seals which are acceptable to them for the movement of goods by vehicle or container of a nature which brings them within Customs and Excise jurisdiction. This list would merit consideration by any firm moving high-value material under circumstances where seals would be advantageous.

### *Seals*

Indicative seals do not provide additional security for the hasps, bolts or other means of fastening to which they are affixed. They are simply a means of indicating the possibility of tampering and theft and the necessity that a full check should be made.

For many years the conventional form was a wire with a lead seal which was crimped hard on to it and may or may not have carried an identifying number. With time and care these could be opened and refitted so as to avoid detection, but many new and improved varieties have come on to the market. Of these, those made of plastic seem to be advantageous both in price and performance; they have to be broken to be removed, and their nature makes it easy to carry a clear identifying number or marking so that they cannot be replaced by an identical seal. A system whereby the number of the seal is shown on a driver's consignment note to be checked at the delivery point before unloading takes place can easily be inaugurated with little additional clerical cost.

There are of course numerous other obvious applications – on cash bags, coin containers, and rope lashings on the motor vehicles. In any instance where it is necessary to recognize interference at a glance their use can be considered. A less obvious use is to carry a fire door key on a hook fixed on the wall or door in lieu of the conventional glass-fronted box, which has become absurdly expensive.

Heavy duty seals are a different proposition, they are usually made of case hardened steel and almost impossible to remove without totally destroying them. Sealable padlocks are also available and included in the Customs and Excise list.

### *Containers*

Theft from containers used for the transportation of goods is a universal problem and not restricted to motorized usage. Export shipments in all parts of the world suffer, and current thought is that the main interference takes place in dock areas, often as organized crime on a large scale. Unfortunately, almost invariably there is delay in learning of the loss. It is not until a customer notifies substituted cartons of rubbish or half-empty containers that any enquiries can commence, and by then the source of loss is hard to pinpoint.

Prevention commences right from the time of considering the desirable features of the containers to be used:

1 Doors must be substantial and hinged so that it is impossible to lift them off complete with padlocks etc. by overhead crane and then replace them apparently undisturbed.
2 Hinges preferably should be recessed to make it difficult to cut them off.
3 Padlocks and locking bars, or other means of protecting the doors, must be effective and attack resistant.
4 All locks used must be security type with strict control on key availability.

These factors are taken into consideration by manufacturers, and progressive design improvements no doubt will be made. The difficulties are emphasized by the fact that where concealment of the operation has been unimportant, cutting gear has been used simply to slice a way in.

Seals represent a valuable aid if there is an adequate system laid down of recording and checking. They must be individually identifiable, not duplicated, and a strict control should be placed upon their safekeeping and availability. After filling of the container is completed – and this must be done under proper supervision or rubbish-filled cartons may be put in at that stage – it must be locked, a seal put on, and the number of the seal recorded on the consignment notes or shipping documents. This in itself is useless unless practical steps are subsequently taken at stages during delivery to examine the seal properly. Practically all types are subject to opening in a manner which will be unobserved by casual inspection; for example, cutting carefully and resticking with special fixatives has proved effective with careless checking when plastic seals have been used. It should be remembered that those responsible are likely to have ample time and privacy to carry out their intentions, and it would be foolish to guarantee that any seal is tamperproof.

As if that possibility were not enough, the hasps through which the seals have been passed have been cut through with a fine hacksaw, the seal removed, the door opened, etc., the seal replaced, and metal filler used to conceal the cut – with the judicious application of a little dirt. Where seals have been used in conjunction with security locks, driver implication is indicated, but whether this is provable is another matter. Use of combination locks with the code independently notified to the customer may be another possibility if there are no obstructive safety-linked shipping considerations.

When all precautions have been taken, how often does the customer break the seal and throw it away without examining it! Seals *must* always be cut when removing and retained at least until the load has been checked and found correct.

### *Routes and schedules*

There is a limited amount that can be done in this respect. If long-distance trucking vehicles are to operate economically and to a schedule suited to the customer's needs they are virtually confined to the main route between producer and customer. Regular variations are ideal in theory only, and no route which involves passing through quiet country lanes, as opposed to main roads, should be acceptable. Using a trunk road with other drivers on it denies opportunity to the thief to carry out the theft in peace and quiet.

So far as variations in time are concerned, commercial considerations will be paramount and, in any case, except in the transport of wages, there is little likelihood that a regular time schedule will be adhered to.

### *Radio linkage*

A commercial security firm, Securicor, offers a system of radio contact with vehicles through the medium of its own chain of radio stations. A radio transmitter/receiver (transceiver) is installed on loan in each vehicle operating on wavelengths allocated to Securicor throughout the country. This enables a driver who is being attacked, or who has reason to fear that he may be attacked, to call the local control who will in turn notify the police. Other amenities that are offered by this service are those of passing messages, passing weather reports, and giving information on road diversions, etc.

The use of citizen band radio is now legally permitted in the UK, but its use has not grown in any way comparable to that in the USA where almost all the long-distance hauliers appear to have it installed. There is no doubt that the two-way facility gives confidence to drivers and is an ideal means of requesting assistance; it is already a standard fitting in the largest UK carrier and police monitor the channel – channel 19.

### *Overnight parking*

This is possibly the main danger area. There are two difficulties; first, the lack of adequate supervised parking facilities throughout the country, and secondly the difficulty in obtaining the co-operation of drivers. Unfortunately the impetus of a Home Office effort to provide a series of guarded parks in strategic parts of the country appears to have lapsed; few have been officially provided and those that are in existence are barely economic. At the moment they are limited in number outside London – the local police or road haulage associations will be pleased to advise on their location.

Well-lit parks are a second best but, where very valuable loads are concerned, it may be worthwhile to consider coming to an agreement with a local firm or garage which can provide enclosed accommodation. Every large group of firms should insist that vehicles belonging to any member of the group should be accepted overnight into any of their factories which is convenient; there is surprising parochialism to overcome in attitudes on this point. Where vehicles are to be left at a depot or warehouse with alarm facilities, it is possible to arrange external points where the vehicle can be joined to the circuitry so that any movement of it will trigger off the alarm.

Using supervised parks usually requires payment by the driver for which he will reclaim. As mentioned elsewhere there is a temptation for him to park in the open and claim the allowances. This should be clearly laid down as a disciplinary matter which will be regarded seriously. In no circumstances should lorries be left in back streets near lodgings for the personal convenience of drivers who are creatures of habit when it comes to having favourite stopping places. The regular presence of their vehicles there will eventually attract the attention of interested thieves in the area who can then conveniently pick their opportunity.

Police manpower resources are not, and never will be, entirely adequate for all calls, but if they are told of the presence of a high-value load in their area, they will try to periodically visit it and at least, if they see any interference taking place, they will know that action is required – even to checking the bona fides of the actual driver.

## *THE DRIVER*

The driver is all-important – he is not just another employee. His integrity is worth more than any amount of alarms and immobilisers which are useless without his co-operation. The best place for a suspect driver is in some aspect of transport where his potential to steal is strictly limited. There are numerous jobs of this type in the industry and to place temptation in such a person's way is foolish and unfair.

### *Selection*

Regrettably, integrity has acquired almost a cash value. The higher the risk the higher the standards to be set and the higher the rates of pay that should be offered to ensure a good choice of acceptable applicants. The idea that security is a vital matter for the employer is one which should be instilled in the driver right from the outset. This can be made clear by the application form to be completed and in subsequent interviews.

The Traders' Road Transport Association has designed a special form for the employment of drivers; a variation of this is suggested for those drivers who are intended to carry valuable loads. The object is to reduce the number of unsuitable applicants who cannot meet requirements and also to provide a ready avenue for dismissal, if it is subsequently found that they have deceived the interviewer. (See Appendix 21 for specimen application form.)

This form of application makes it incumbent upon the driver to declare any criminal convictions that he has. The object is not that these should be a bar but to allow the interviewer to use his discretion in full knowledge of the circumstances; some criminal convictions are so trivial or so long ago that they can be virtually disregarded. These are interviews that should not be hurried; qualities such as honesty, reliability and integrity are difficult to assess purely by interview but the effort must be made. A good and stable employment record is a first consideration; a person who has jumped from one firm to another at regular intervals for the hackneyed reason of 'advancement' is unlikely to have the steadiness of character required. Neither is a driver with a hard-luck story, or claim of victimization, or that he now wants to settle down. However sympathetic one might feel, the firm's interests are paramount and such a person is not a good risk.

Long service with previous employers, a settled background with domestic responsibilities, and personal recommendation by well-regarded long-serving present employees are all advantageous points. If there is any reluctance to answer questions it can be accepted that the matters are detrimental to the person concerned. It should be borne in mind that it never occurs to an honest person that he has to convince anyone that he is. A complete record of previous employers, as far back as possible, should be checked with the applicant; all gaps in the record should be queried and accounted for; references should be sought from employers – preferably the last three – and not from personal acquaintances furnished as referees. In one instance, those given were from a father, uncle, and brother-in-law and all three had long lists of criminal convictions.

Any interviewer, no matter how experienced, can misjudge a

convincing liar, so before any new driver is sent out with a load his references should have been verified and his driving licence, form P45 and national insurance card inspected. The time expended on these precautions will not be wasted – if it does nothing else it will impress upon the driver that security is the basic consideration of the post. A degree of casual supervision should be continued after a person's employment has been commenced. It must be repeated that the percentage of experienced lorry drivers, especially those in long-distance haulage, who are not entirely trustworthy, is remarkably small.

It is obvious that there should be a uniformity of performance on given runs, so if a new driver displays abnormalities in the time taken and the distances covered, note should be taken and a special watch kept. Even so, variations in mileage and times at stopping places are more usually for personal reasons rather than more doubtful motives. Breaches of discipline militate against a firm's interest and it may be equally necessary to dispense with the driver's services on these grounds as on those of dishonesty.

### *Driver collusion*

No one knows how many hijackings are faked, but they are far in excess of the genuine – these are estimates from persons in the transport industry and not those of sceptical police officers – and this is a major threat. The vast majority of drivers are strictly honest and it is a matter for conjecture how many casual approaches are made before a driver is found who is prepared to co-operate – at a price. Drivers appear to think that informing reflects upon themselves, and so hesitate to report these overtures; this is a pity, for they could be stamped out by police utilizing the common law offence of incitement to commit crime. It is immaterial that the solicitation or incitement of the driver has no effect upon him – if asked to help in stealing a load an offence is committed even though he refuses point blank. Proceedings would not normally be begun by the police unless there was a sequence of at least two such acts which could be substantiated to form corroboration.

A dishonest driver incurs infinitely more risk than the thieves; it is unlikely that they will let him know their true identity or even see him after the incident if this can be avoided. Such payment as he receives will be trivial in comparison with what he loses by way of trust and possible liberty. Having achieved their objectives it is more than likely that the thieves will simply ignore him as his usefulness is finished. He dare not talk and he has no redress of any kind. If they are caught they will have little hesitation in putting him forward as the instigator.

If the risk justifies carrying two people on the vehicle the danger of collusion is more than halved. Careful selection is the only countermeasure that can be taken, though it should be borne in mind that a driver who is contemplating participation in such a theft is apt to be transparently uneasy and worried before it takes place.

### *Observations on drivers*

With experienced drivers this is a most difficult thing to do. No driver of any calibre will fail to spot a following vehicle within a matter of miles and, if there is no justification for the suspicion, the practice will cause strained relations with the transport fleet drivers. If it is considered essential to 'tail', several cars of differing types should be used so that there is no continuity in the following vehicle. Police cooperation should always be sought in these cases – they have wireless facilities and inter-communication between vehicles will enable observations to be kept at a distance both from front and rear. Electronic devices of the radar type have been tried with some success, but this is a matter for the police.

## *PETTY FRAUDS BY DRIVERS*

It is somewhat surprising that a proportion of drivers who would not dream of stealing from their loads sometimes indulge in practices to obtain what they regard as 'perks,' which would however be regarded with equal severity by a court of law. The overnight allowance is a prime example and the consequential loss arising from the steps taken to justify this far outweigh the financial gain. A driver who could reach his home base within his permitted hours deliberately does not do so, but parks up near his home where he spends the night, then running in early the next morning purporting to have stopped some distance away. Perhaps £15.00 or £20.00 accrues to him but the employer loses a night in which to offload and reload and the whole transport schedule may be upset. A further security aspect is that the lorry will almost certainly be parked in the streets at risk.

In a similar manner, books of parking tickets can easily be purchased and submitted as receipts for stops on established parks when the vehicle has been left in streets near lodgings, or the driver may have slept in the cab. Again, whenever one of these is submitted it implies there has been an unnecessary risk to the load.

If the duration of a journey is such that refuelling away from base is needed, the employer can cater for this in a number of ways – none of them foolproof. A cash float may be allowed, agency cards supplied, a voucher system used, or purchases allowed only from accredited garages where a credit account scheme is maintained. Connivance between a driver and a dishonestly co-operative supplier can result in less fuel being put in the tanks than will be shown on the receipts, the difference being shared between the two individuals. Even at the accredited garages, there can be the same type of conspiracy between driver and pump hand.

Trucking drivers, who are most likely to indulge in these practices, are usually well paid and the first deterrent should be that of making it perfectly clear that the strictest disciplinary action will follow if anyone is found committing them. Drivers who are prone to overnight stopping near their base will show up repeatedly on their log sheets as being early morning arrivers into depots from certain runs. An enquiry to a driver as to why this is happening will bring an improvement at least for a period, the alternative is for visits to his alleged regular overnight stopping point and a tour of the area around his home. This is a matter which a firm may think fit to deal with internally, but it should also be borne in mind that inevitably a further offence of falsification of drivers' records is committed and drastic action is needed.

The difficulty with parking tickets lies in that they will normally be attached to expenses claims for auditors' purposes and will not be retained in the department. Where identical tickets are submitted, or those from a single numerical sequence, this may not be immediately obvious to supervision. Unfortunately, this is one offence which sometimes is connived at by supervision and it is therefore one which should be mentioned to and borne in mind by the cashier's department who pay the expenses or the auditor's department who check the receipts.

Whether the fuel receipts which have been falsified show up at an early stage depends upon the recordkeeping of the concern. If a mileage record is kept against the fuel usage, it will soon appear that some drivers are getting less mileage from their fuel than their contemporaries. Scrutiny of receipts may show up the regularity of a particular garage with perhaps identical writing but differing signatures, or even fuel purporting to be drawn from that garage when the vehicle is nowhere near that area which may happen if the driver has acquired a book of receipts.

### *Agency cards*

Agency cards are equally subject to abuse and it is not unknown for a large user to pay accounts rendered from petrol or diesel suppliers without any cross-reference to records submitted by drivers or other query. If any of these cards is lost or stolen it is imperative to inform the issuer at once and confirm by letter. Lists of lost cards are circulated to the garages of the fuel company concerned at monthly intervals and it would not be unreasonable for a user to refuse to pay if the card was accepted after such circulation has been made. Liability does rest on the user and if the drivers have regularly tendered the card at garages operated other than by the issuing firm a relationship by custom will have been established and the user will have to pay in respect of any fraudulent withdrawals by means of a stolen card despite the fact that they are not from garages of the issuing firm.

### *Own tanks*

If the company owning the vehicles has its own tanks of petrol or fuel oil and insufficient supervision of issues is exercised, discrepancies are likely to be made up from that source by the dishonest driver. Petrol and fuel oil pumps usually have meters registering issues and it is recommended that at least weekly checks of the figures and issues are made by someone independent of the person who is primarily concerned with the intake of the liquids and their issue.

### *Vouchers*

The system of vouchers used by some major carriers does reduce risk; they are usually in the form of a triplicate book and provide for work done or parts, in addition to fuel. Two copies go to the supplier who forwards one with his account to the user who already has the driver's copy for reference. There is less opportunity for fraud but it still exists.

Even when company's pumps are used, fuel may be supplied over recorded amounts or into vehicles for personal purposes. As has been mentioned this can be prevented by tight procedures and checks, or by fitting the pumps with a special device only responsive to a magnetically-taped token applicable to a particular vehicle. This causes the amount issued to be shown on a recording panel away from the pump or on printout in the device. It is costly to install but can show substantial savings in wages as well as cutting pilferage. Casual defalcations involving fuel will be hard to detect if done systematically but good supervision should bring them to light reasonably quickly.

## *PROCURING A LOAD BY FRAUD*

This is rarely attempted on a large firm with well organized traffic arrangements and insistence on precise documentation. The procedure is that the prospective thief with a large lorry or articulated vehicle, either genuine or stolen for the purpose, and in any case carrying false number plates, calls at a firm on the pretext of seeking a return load for a journey, usually to London. The vehicle may have a fictitious owner's name, address, and telephone number painted on the cab, and the driver will be in possession of copy delivery notes and other 'proof' that the offer is genuine and honest.

The field of operation is usually that of provincial cities well away from London and the driver will have documents showing a London telephone number which can be contacted if the firm so desires. This number may be that of a telephone kiosk or some other place where an accomplice is waiting to answer. By the time a check on non-delivery has been made the goods are beyond redemption, and identification of the culprit is remote by both time and by distance. If the thief is purporting to be employed by a well-known haulier he may even take the risk of contacting a local transport clearing house to be sent where transport facilities are needed at short notice.

That any firm should fall to a fraud of this nature seems ludicrous; nevertheless it is a regular occurrence – perhaps the convenience of a prompt dispatch at reduced rates is too tempting for the operator. The easy solution is not to use 'casuals' at all unless the carrier has been previously utilized, even those sent from a clearing house should be suspect until the driver's licences and his vehicle have been thoroughly checked and examined to ensure they are in order. A further check that could be made is to establish the identity of the customer to whom he purports to have delivered in that area and check with that person whether what has been said is true.

If it is thought that this could not happen refer to the case of *Garnham, Harris & Elton Ltd* v *Alfred W. Ellis (Transport) Ltd* 1967 2 All ER 940. This is a judgment given on the liability of a contractor who subcontracted the carriage of a valuable load of copper to what was subsequently found to be a non-existent firm. Damages were awarded against the contractor. Another large metals firm lost a complete load valued at tens of thousands of pounds in this way in July 1972.

## *DRIVER IMPERSONATION*

This is a rare form of theft, mainly confined to the London area. Loaded vehicles have been taken from guarded car parks by thieves posing as the legitimate drivers. Obviously, either a considerable amount of planning must go into an operation of this kind or there must be collusion with a driver to obtain suitable keys for the vehicle and knowledge of times and loads.

*Driver identification card* – This is the apparent answer. National Car Parks Ltd have circulated a request that the firms using their facilities should provide their drivers with a form of identification. A suitable card should carry a full or three-quarter face photograph of the driver, full name and address (and employer's address), date of birth, height, weight, and hair colour. His signature and that of someone of authority in the firm should be appended and the card stamped with an official stamp. This would not preclude this type of theft but would certainly make it considerably more difficult where guards are employed on the car park.

## *VEHICLE OBSERVER CORPS*

The menace of lorry thefts caused the Road Haulage Association to create this body in June 1962; it is now operating in several important centres in the country and is not restricted to members of the Association.

Fundamentally, the scheme is one of immediate search on notification of the theft of a lorry and load. In each locality a permanent area control is established which furnishes member firms with details of the stolen vehicle as soon as they are known. These firms in turn supply cars and crews to patrol small predetermined areas of an extent that can be covered in a matter of ten minutes. After an initial check they report back to their own base and then make a fresh search. With the co-operation of a large number of firms all these areas, which make up an entire district, can be searched simultaneously. No action is expected from the crew finding the vehicle other than to inform the police where it is and to follow it, if seen on the move. This is a form of self-help in the transport industry which has had success, particularly in London. It is still operating today, as lorry theft increases.

## *INTERNAL FRAUDS INVOLVING TRANSPORT*

Once a vehicle has left a firm's premises carrying goods in excess of what should have been loaded, detection of the theft becomes difficult. On many occasions it will be impossible to determine whether stock deficiency is due to bad documentation or theft from inside the premises, either by intruders or by progressive pilferage by dishonest employees. A driver and warehouseman acting in collusion represent a dangerous combination and could function for some time without detection. Any signs of undue affluence on the part of two friends in these categories should be viewed with suspicion.

There are only two physical ways of establishing whether a loaded vehicle carries what it should: by offloading and checking against the consignment notes or, where the load is composed of items of known weights, taring and grossing the vehicle over a weighbridge and comparing the net weight with the cumulative weight of the individual parts. The use of a weighbridge will have limited application with many products and especially with mixed loads but where bulk metals or raw materials are concerned it is always of value – if only because it will have a cautionary effect on those concerned.

Offloading is not a popular practice and could have repercussions in relationships if nothing were found; it should only be done where there is virtual certainty of fraud or where it is accepted that periodically such action will be taken. Spot checks can be made on odd items to establish that they are shown on the consignment note – in most cases a driver will want to get rid of unauthorized material as soon as possible and the most likely sources of disposal will be local, so this will be on or near the top of the load and readily accessible.

## *DOCUMENTATION*

Firms will have differing administrative requirements in this field, and no universal system can be recommended. Incidents have occurred, and no doubt will in the future, where goods in excess of an order have been deliberately labelled and consigned to a dishonest customer – this has even been known where no legitimate order of any kind has been in existence. Adequate documentation and good stock records which immediately show up deficiencies will effectively limit this type of fraud.

Some goods are inevitably lost in transit sooner or later, without indication of whether it is by theft, carelessness, or by misdirection; complete documentation will be necessary to support a claim on insurance and its absence may prejudice payment. Claims for reimbursement, particularly where outside carriers with trans-shipment points are used, may prove difficult in laying liability on a particular insurance company if the documentation has not been meticulous.

Each firm will have its own system of producing interlinked orders, job cards, advice notes, consignment notes, and invoices in sequence. The opportunity of a fraud really arises at the consignment note stage. If these can be created in a warehouse without the other notes coming into existence, it might be possible to conceal the method of getting goods out since the automatic procedural action, which would normally follow outside that department would never begin. The existence of an interlinked system where warehouse staff could only dispatch in accordance with notes sent to them would eliminate this possibility.

Normal distribution of consignment notes is from a set of three – in different colours for convenience and ready identification:

1 To be signed by the driver and retained in the warehouse.
2 To be handed over with the goods to the consignee.
3 Driver's copy for signature of the consignee on accepting delivery.

In a fraudulent arrangement where the third copy goes back to the warehouse originator and the second copy is destroyed, the whole could disappear leaving no trace of what has happened.

The third copy of this set is most important in another respect: claims of non-delivery of goods to a customer are commonplace and without this third copy, signed by the recipient, there is no basis on which to contest the claim. There may be no dishonesty on the part of the customer, whose own records may be faulty, but he will be convinced he is right in the absence of proof to the contrary. This third copy should therefore go back to someone other than the person responsible for making up the load and it should be filed and retained for a definite period, adequate to preclude possibilities of any claim being made – six months is a reasonable time.

## *FALSE SIGNATURES*

These come to light when a customer refuses to accept an invoice for goods thought to have been supplied to him. When proof, in the form of a signed consignment note, is produced, for the first time it is found either that the signature is indecipherable and bears no resemblance to that of any of his employees or the name shown does not belong to any of them.

To complicate enquiries, there will have been an inevitable delay between dispatch and the realization that something is wrong. Several alternatives exist: the driver may have forged a signature to cover his own theft or a genuine loss from his load – if he thought that might be held against him; a member of the customer's staff might be camouflaging a theft of goods inward; a total outsider may have posed as a customer's employee and deceived the driver; the goods may have been dropped with the wrong customer by genuine mistake and the latter has not brought the fact to notice.

There is a reluctance to accept that the loss is due to anything but mistake or bad records and the supplier, if the customer is a valued one, sooner or later has to come to a commercial decision and accept the loss despite annoyance and uncertainty. There will be little chance that his insurers will accept any liability in the circumstances.

If all firms would use an official receipt stamp as well as the initials or signature of the recipient, this increasingly frequent form of fraud would soon stop; unfortunately, this is most unlikely and suppliers should take such precautions as they can.

The possibility of fraudulent acceptance should be among the matters impressed on drivers. A driver should be instructed to ask the name of the signer when he cannot read what has been written; he should then print the name alongside the scribble. In the event of any complaint of non-delivery, the first step should be to forward the signed note to the customer with a request for immediate comment if not in order – not to go through several stages of correspondence before this is done. Finally, the driver should be asked whether he can throw any light on the discrepancy at the earliest opportunity. A check should of course be made with other customers.

If no logical explanation, other than theft or fraud, can be established, the customer and supplier must decide whether to inform the police; in the case of substantial loss, insurers will not be sympathetic unless this is done. The best protection again lies in a well-instructed and reliable driver with commonsense, in this instance backed up by a laid-down and speedy system of dealing with alleged shortages in deliveries.

## *TACHOGRAPH FRAUDS*

Regulations 3820 and 3821 of the EEC Regulations 1985 provide a requirement for all goods vehicles exceeding 3½ tonnes, and public service vehicles with more than 18 seats (including the driver), in the UK to be fitted with the tachograph meter.

Each day a new record disc is to be fitted into the tachograph, which will record the speed and distance travelled, time spent driving, time spent as rest time, and time spent on waiting, that is loading and unloading. Offences are committed if the driver and the company fail to use the tachograph properly or the records are falsified. It is possible to trace the exact route taken during a journey, the places stopped and the time spent driving or waiting, and, if the driver is suspected of making unauthorized deliveries, drop-off points can be checked using the disc and a road map. These can provide vital information to the police or the owner of the property investigating the losses.

The most common reason for tampering with the tachograph is simply to falsify the time spent driving, or rest periods, to comply with the legal requirements. The tachograph might not be used at all, or the top left open to prevent the chart being printed, although the speedometer mileage clock would still be working.

A new move to control the records on the chart is simply to fit a switch on one of the three wires which come from the sender box on the gearbox to the tachograph itself in the cab. With the switch off, rest periods are recorded, notwithstanding the vehicle is still being driven. With the switch 'on', normal driving time is recorded.

The Department of Transport makes regular checks on these vehicles, but with limited resources, there is ample scope for the dishonest driver to take chances, to avoid the regulations or to make unauthorized deliveries.

## *WEIGHBRIDGE*

The weighbridge, like any other mechanical device, is as good as its operator and no better. This is another post where honesty and reliability are of greater value than any other traits. If the operator is tempted to defraud his employer he has ample opportunity to do so and the chance of early detection, if he is not too greedy, is small. However, modern improvements in weighbridge construction and recording have provided safeguards which make the deliberate misreading of weights almost impossible. This is a good post for a reliable disabled person.

The potential danger to a firm's profitability in what would appear to be the simple operation of weighing a vehicle, loaded or empty, is rarely appreciated but the weighbridge offers perhaps the maximum source of risk that is likely to exist. In less than two years a single London factory, using one weighbridge, sustained losses approximating to £750,000 which was a factor influencing a closure decision; a northern steel and engineering complex estimated losses at £1,000,000 in a year; another, a known £20,000 in two weeks; another small factory using a public weighbridge was defrauded of £108,000 in a year – which effectively cancelled out its profitability. These are only specimen instances in which the weighbridge operators have been corrupted. More technically-minded criminals have realized the opportunities that can be exploited and specially altered vehicles have been encountered whereby a false recording of 2 tons either side of the true weight could be induced without the operator's knowledge – this will be referred to in detail later.

### *Types*

Improvements in weighbridge construction and recording mechanisms are continually being made, and there is progressive replacement of the older type of bridge. This is a costly procedure; to put in a 60-foot modern electronically-controlled weighbridge would probably cost for supply and constructional work not less than £60,000. Fortunately, for this sizeable item there are a limited number of suppliers, all extremely reputable and fully prepared to discuss ways in which fraud is prevented in their installations.

There are two older types still in common use which will no doubt continue so for a considerable time to come: the 'dial recorder' and 'steelyard' types. A high percentage of these still rely upon visual reading by the operator for accuracy, but they can be converted to an automatic printout device. The dial recorder, by virtue of its construction, is particularly adaptable for this and thereafter is less subject to interference than the steelyard. On the latter, readings are complicated by the progressive movement of the blade, and with many older installations it is possible for the weighman to stamp a card at a false reading. With dial-recorder precautionary devices it is impossible to mark the card when there is any movement of the indicator or vibration on the platform. A period of 1½ seconds or less standstill period is required, thereby reducing the possibility of accidental or deliberate error.

### *Faults*

It must be remembered that there are several ways in which a weighbridge can make a false recording without there being fraud. The worst offenders are weighbridges of inadequate length where double weighing has to take place. This is now technically illegal as errors of up to 200 kg (about 450 pounds) are common, even when both driver and weighman are expert in positioning the vehicle at the time of both taring and grossing. The absence of flat approaches to either end of a weighing platform will further complicate matters – there is also a legal requirement that these should be level but in the case of the older weighbridges this is rarely so – and this will render double weighing even more unreliable.

On an old weighbridge with mechanisms beginning to wear it is essential there should be even distribution of the weight – if excessive point loading at any part of the weighbridge exists this can lead to an excessive deflection on the recording mechanism and a false reading. If water and dirt are allowed to accumulate on the weighbridge, or a sudden shower soaks it, an error of considerable magnitude can arise unless the operator adjusts his zero accordingly. The steelyard type is particularly prone to error if the operator lacks experience and expertise – with too quick a movement of the blade, readings will inevitably be incorrect.

The new electronic load-cell type of weighbridge has brought into prominence a factor which has probably caused a great deal of excessive wear in the past and relates to the movement of the vehicle entering or leaving the weighbridge platform. Apparently, if this is done at an angle to the line of the platform, the sideways thrust accentuates a wear and tear on the knife edges which is even worse in the case of load cells. Similar effects may result from driving on at speed and braking abruptly. A design precaution which can be taken is that of constructing low parallel concrete kerbs at either side of the platform to ensure that entry and egress are made in a straight line. However, bear in mind the legal requirements of the operator being able to see both sides of the bridge. If the position of the office and recording mechanism renders this impossible with kerbs, cables may be used linked to an alarm bell – of the kind found on garage forecourts – so that any wheel overlapping the edges will be noticed. This susceptibility to load cell damage has been recognized and should not be experienced with electronically-operated bridges supplied after 1980.

### *Frauds*

There are two main legal aids to security at weighbridges: firstly, the weighman, from his position by the dial or steelyard, must be able to see both ends of the weighbridge clearly; secondly, the weighbridge office must be so constructed that the driver can see the readings being taken whilst his vehicle is being weighed, so that he can challenge them if he thinks fit. The virtue of these requirements lies in the facts that the weighman has the capability of seeing that the vehicle is correctly positioned which gives him the opportunity of questioning anything he thinks should not be on it, whilst the driver also has a potential check on the weight of his vehicle – this is of less value for there is little chance that the weighman would be involved with anyone else except in collusion with the driver.

Where stockpiling or selling large quantities of raw materials or scrap is being undertaken a fraud in weighing a loaded vehicle can lead to a loss which is difficult to detect or occasionally even to be aware of. This is a type of dishonesty which leads to over-confidence on the part of the perpetrators and will no doubt be practised regularly; the persistent drain should then eventually attract comment. One particular set of circumstances which are recurrent are those involving the purchase of scrap. Once the vehicle has loaded at the seller's premises the weight shown over a weighbridge will be that accepted for payment. If an agreement has been entered into between the buyer and the weighbridge operator it would be easy to earn quick money by under-recording the gross weight. The prevalence of this type of offence is demonstrated by the number of occasions these facts have been repeated in courts.

It is easy to establish that a load is under-weighed once suspicion has been aroused. Either the security staff or the police can stop a vehicle and send it back over the weighbridge or to another weighbridge. It is more difficult to prove that it is a deliberate act with fraudulent intent and not an accidental error that has taken place. If there are strong grounds for suspicion this is taking place it is better to enlist the services of the police and try to ascertain where the driver is dropping the excess material that he is carrying. A defence of negligence and inefficiency is a difficult one to rebut unless there is positive proof of illegal disposal of materials. It may be possible to gain additional proof by an examination of documents over a period showing regular deficiencies in the amounts that should have been carried. Where a publicly-owned, as opposed to a private, weighbridge is being used an independent person from the firm should always accompany the vehicle to the weighbridge, even if this causes inconvenience and loss of time. This precaution should apply to both tare and gross weighing, not just to the latter. A monumental loss was sustained where the unchecked weighing of a large empty skip was permitted, and this was then replaced by an identical skip constructed of lighter metal prior to loading and reweighing.

The opportunity that a driver himself has of perpetrating a fraud is mainly concerned with producing too high a reading for the tare weight of his vehicle. He could, of course, remove materials from a load in transit and substitute disposable matter to be jettisoned after gross weighing, but before offloading. However, the procedure would be complicated, dangerous, and unlikely to be carried out.

All types of industry use weighbridges for bulk weighing, and if their records can be falsified there are immediate and profitable opportunities for theft. These frauds are possible with or without operator co-operation, but regrettably the corrupted operator is more likely than the careless one.

*Frauds possible without operator collusion*

The following methods of inducing the recording of a load less than the true one have been encountered and should be brought to the notice of operators and security staff:

1 Person or animal remaining in the cab during the tare weighing operation but not stopping in when gross weighing.
2 Concealing heavy material – paving stone, scrap iron, sandbags – on the vehicle which is thrown off after tare weighing.
3 Discarding unserviceable old spare wheels, tarpaulins, batteries, etc. from the vehicle after tare weighing.
4 Carrying water containers – drums, duplicate petrol tanks, scrap containers – that are drained off after tare weighing.
5 Having a bowed tarpaulin covering the platform which is filled with water that is poured off before commencing loading.
6 Where tipper-type vehicles with welded bodies are concerned, likewise having water in the bottom which is tipped before loading.
7 Marginally overlapping the edge of the weighbridge during loaded weighing.
8 Driver standing on the weighbridge during tare weighing. Occasionally this has been found possible without notice by the operator – when the weighbridge is immediately beside the office window and the driver looks in during the weighing but his feet are actually on the weighbridge platform.

Parallel frauds with materials being supplied use the same tricks but in a reverse fashion. In this instance the objective is to increase the apparent weight being delivered by:

1 Delivering materials soaked with water, or carrying water containers which are drained prior to off loading.
2 Discarding the skids, iron bars, old spare wheels, etc. previously mentioned after gross weighing.
3 Person or animal staying in the cab during gross weighing and not being in at tare weighing.
4 Where there is a layout and lax system which permit it, substituting a lighter carrying vehicle with duplicate plates between gross weighing and tare weighing – for example, substituting four-wheel vehicle for six-wheel vehicle.

There are of course other frauds which take advantage of an inadequate system or an inefficient weighbridge. Specimens of these which have been encountered are:

1 Where skips are supplied for collecting purposes, painting false weights upon them and claiming mistake – if the deceit is discovered.
2 Supplying skips for collection which have been modified with removable lead sheets or iron bars – to be removed before check weighing.
3 Using duplicate vehicles one of which has a concealed concrete float under the platform, and working a substitution between weighings.
4 Where there is repeated collection or delivery during a day and for convenience only one tare weighing is carried out, dropping spare wheels, tarpaulins, etc. after that first weighing and reloading before any final check weighing.
5 Where the weighbridge is short and double weighing is necessary, persuading the operator to weigh without detaching the trailer, positioning the front wheels of the towing vehicle at the same point for both gross and tare weighing. This is almost invariably carried out with the horse box type of trailer; the whole load is thrown forward of the wheels so that an abnormal amount of it is carried on the tow ball of the tractor standing off the bridge. (The differences in weights have to be seen to be believed – less than one quarter in one instance.)
6 'Top dressing' goods of inferior quality beneath the genuine material being delivered.
7 Working a swap of carrying vehicles where two deliver loaded skips. Both gross weigh and one off loads; after taring the driver leaves the works then makes an excuse to re-enter; he swaps his skip with the one not unloaded and leaves again; then the second vehicle tares as it goes out as if its skip had been emptied.

The fraud involving a specially-adapted tractor-trailer combination deserves special mention and constitutes the greatest threat yet to weighbridge users. It is almost impossible to detect in action and will no doubt be met with again, though two separate offences have resulted in prison sentences. It can be used on weighbridges of a length which do not allow the full articulated vehicles on the platform (usually 40 feet or less) so that the trailer has to be dropped for separate weighing. The tractor unit is fitted with an extra tank of standard type – which is sometimes done to carry additional fuel – and the trailer with a flat one over the trailer pin, beneath the floorboards and concealed from underneath view by a plate over the trailer pin. The pin itself is drilled with a small hole to access the tank. A system of electric pumps and air pressure is controlled from the cab of the tractor to transfer mercury from one tank to the other as required when the trailer pin is brought into contact with a self-sealing valve in the 'third wheel' of the tractor as the two units are joined together.

When information about this device was first received, it was disbelieved to the point that there was reluctance to pass it to the police, especially as a weight transfer of 2 tons of mercury was quoted. It proved to be fact, the first vehicle traced being a gravity feed arrangement with outside air supply to pump the mercury back, but the second, some 18 months later, was the sophisticated and expertly constructed system described. It would be wishful thinking to suggest only one conversion has been done. The actual weight of mercury was found to be in excess of 2 tons and the profitability to the user of the ability to inflate or deflate his trailer weight by 2 tons at will by pressing buttons in the tractor cab needs no stressing.

*Operator collusion in theft*

Where the operator is party to the fraud, any of the above tricks become easy, but he can also assist other variants by:

1. Interfering with the mechanism or not stabilizing the bridge before recording.
2. Not zeroing his scale before carrying out a weighing.
3. Allowing rubbish with material to accumulate on the weighbridge and cleaning off between gross and tare weighings.
4. Deliberately making a false record where a printout device is not fitted.
5. Where this is possible, recording at the top of the swing when the vehicle is driven on to the weighbridge at speed and is abruptly stopped.
6. Where the weighbridge is known to be defective in a particular manner, perhaps by suspension wear, directing the positioning of the vehicle so as to get an advantageous reading.
7. Where the connecting beams between platform and recording mechanism of an older-type weighbridge are accessible by the removal of floor plates, putting weights upon those beams to alter readings.

There was an oversight in the earliest of the electronically-operated bridges which hopefully will have been removed by the manufacturers who realized it at once. It is worth checking as it provided an easier variation of the zero-adjustment of the older bridges. This provided for a 2 per cent adjustment either side of the mean zero and could be carried out manually – all new types have an automatic push button zeroing device which gives no such latitude to the operator.

Where public weighbridges are in use, any employee of the firm designated to accompany a vehicle to that weighbridge should be warned of the possibilities for deception. It is advisable that more than one person be used on a rota for this task, to prevent or at least make more difficult the corruption of that person.

There are advantages to firms in making security officers responsible for weighbridges. This is particularly so where the security office and weighbridge are both situated at the entrance to the factory, which of course is the best place for both. This enables the regular rotation of weighbridge operator and grossly increases the difficulty of any supplier or purchaser with thoughts of corruption. A list of general operating instructions which may be of value to security officers designated for this work is shown in Appendix 22.

*Precautions*

It is as well for an independent person to occasionally cast a paternal eye over what is happening at the weighbridge and ensure that the weighman is doing his job as he should.

A precaution that an honest weighman can always take, with regard to the tare weight, is that of asking the driver what his vehicle is marked as weighing and comparing what he gives verbally with what is recorded upon the dial. A further check is that of noting and comparing the tare weight printed on the side of the vehicle – though this is often inaccurate. Where regular records are kept of the same vehicle coming to the premises and it suddenly sustains a noticeable variation in weight, the reason for this should be challenged. Regrettably, in one instance this was noted after a lorry had called to collect scrap for the last time before being sold – it suddenly appeared to lose precisely 10 hundredweights (508 kg) after being constant for nearly two years. This was coincident with the police making enquiries at the weighbridge at the time of taring, the unresolved suspicion now remains that a profit of half a ton a trip had been enjoyed by the driver and his friend the weighman.

It is a good idea to keep a record of complaints of deficiencies so that if the same driver's name appears at regular intervals a more intensive investigation can be carried out. With a suspect driver the object should be to catch him, rather than induce him to mend his ways temporarily, or he will remain a potential menace. If it is thought that he is carrying an object into the premises on his vehicle to inflate the tare weight, observation should be kept upon him and if he is seen to throw material away prior to loading, after he is reweighed he can be stopped and he then has no defence at all to the charge of theft of the weight equivalent to what he has thrown off. Let him reload or at the best his offence will only be attempting to steal.

In conclusion, it must be remembered that wherever transport, either belonging to the firm, contractors, or carriers, is concerned, this is the avenue whereby a maximum loss can be inflicted in a minimum of time and with a minimum of risk to those concerned. Precautions should be taken accordingly.

## *ADVICE TO DRIVERS*

*If valuable loads are carried and lost by hare-brained and irresponsible drivers, those who selected and instructed them are at fault and not the individual who cannot help his nature* (from John Wilson's address to a security seminar at the Management Centre Europe, Brussels, in 1976).

If the correct care has been taken in selecting and interviewing drivers, it must be assumed that they are appreciative of advice and guidance on circumstances which they may encounter. However, it must at all times be remembered that the prime duty of the driver is that of driving and it would be both unfair and unwise to lay down precisely what he should do in given circumstances. Their reactions will be as varied as their physiques and temperaments and if there is an obvious threat of serious injury by firearms no driver should be criticized if he puts personal considerations above all others.

Most drivers will appreciate that if the obvious intention is to obtain possession of a load come what may, he will be well advised to try to get away because he is likely to be subject to injury in any case. This would particularly be applicable in the rare instances where an endeavour is made to stop the vehicle by force. Ramming or threatened ramming are the obvious ways, with a further possibility of obstruction by a barricade. Most drivers will automatically think of escaping and if the opposition get hurt in the process that would be their occupational hazard. A further point would be that a badly dented and undrivable lorry is better than no lorry and load at all, interest in both might have lapsed if the vehicle had finished in a ditch or jammed into a blocking car. These are rare incidents and it is much more likely that an attempt will be made to stop by trick.

### *Stopping by trick*

A driver cannot live in expectation of an impending attack at every routine happening. Nevertheless, there are several stereotyped approaches, of which impersonating the police is the current favourite, particularly in the Metropolitan Police area; here lies the advantage in internal bolts. A driver cannot with impunity disregard signals from a person in police uniform unless the circumstances make it obvious that no reasonable person would have complied. The police do accept that a commercial vehicle driver can insist on staying in his cab and driving to the nearest police station for examination of his documents. Commonsense and discretion are the guiding factors on the driver's actions. Police uniforms are easily obtainable and there is a wide variety in the types of car in use; black is no longer the predominant car colour and the presence of a 'police' sign can be meaningless. However, police vehicles can be expected to be reasonably clean with absence of damage and rust, fully roadworthy, and carrying intercommunicating radio, and few forces operate foreign cars or anything above the medium price range. As for the officers themselves: black shoes are always worn, with plain white or blue shirts and black ties; untidy appearance and very long hair are rare. In case of doubt the driver should speak through the cab windows with the door kept locked.

Fake accidents are difficult to cater for – especially where they purport to involve the carrying vehicle. The locality and circumstances could decide a driver whether he has or has not been in an accident or whether a trivial accident has been deliberately contrived. A collision with a car containing several men in a quiet area is very different to a bump in a traffic stream with a family car. In any case of doubt the driver should consider all possibilities before he leaves the safety of his cab. Where other vehicles are concerned in obviously genuine accidents, at the risk of seeming callous, a driver should again weigh the circumstances carefully before obeying his natural instincts to stop and help; under no circumstances should he stop out of sheer curiosity. Another method of stopping a driver, leading to the theft of a valuable load, was experienced in London. The rear registration number plate of a trailer was unclipped while it was stationary, either in a traffic queue or on a ferry. Later, at a point in the lorry's journey suitable to their purpose, the thieves overtook the vehicle in an old van and the passenger beside the driver waved the plate out of the window at the same time pointing to the rear of the lorry. The driver recognized the registration number as his and pulled up behind the van to recover it. The man in the van's passenger seat got out, ostensibly to adjust the nearside mirror, and when the driver leaned into the van to get the plate from the other person, he was pushed from behind into the interior and overpowered without attracting any attention from passers-by.

*Hitch-hikers*

There is a simple remedy for the dangers that can accrue from these: simply do not carry them. This should be made a specific instruction from a firm and should be the subject of severe disciplinary action if it is not obeyed. The danger is not always to the load: to a petty thief the contents of a driver's pockets are adequate – indeed it is a matter for speculation as to how often drivers have been threatened, blackmailed, or cajoled into parting with money and have never reported the matter for fear of ridicule. Female passengers are the greatest danger of all: the girl-tramp habituées of transport cafés can find it easy to persuade a weak driver to pull off into a lay-by where their accomplices could be lying in wait; alternatively, by virtue of their mode of life, they can offer a danger to health. Few respectable girls hitch singly and the female in distress is the most hackneyed way of inducing a stop.

*Other methods*

There is obviously considerable and continuous variety in the means that can be employed. Waving and flashing lights to indicate faults with the vehicle itself are common and diversionary messages have been left at cafés which are known to be regularly frequented by particular drivers, instructing them to divert to new premises. The latter should be queried if there is any doubt at all.

*Checklist of instructions to drivers:*

1 Do not park in quiet country lay-bys unless the presence of other similar vehicles gives a measure of protection by sheer weight of numbers.
2 Do not visit transport cafés where there are recurrent incidents or thefts.
3 Use guarded car parks, or well-lit ones when the former are not available. Under no circumstances leave vehicles in back streets near to lodgings for personal convenience.
4 When in doubt notify the police of the presence of your load.
5 Before leaving the vehicle for the night, check immobilisers, alarms, and all doors and locks, and sheet down loads tightly to conceal the contents and hinder removal; have a special arrange-

ment in the knotting to ensure that if there is interference it will be detected on sight.

6 If obliged to stop at an accident or by signals from police officers, stay in the cab with the door bolted if you are in any doubt.
7 If your vehicle is involved in an accident you will have to use your discretion as to whether it is a deliberate or an accidental occurrence when you can safely get out of your vehicle.
8 Casual passengers and hitch-hikers must not be carried.
9 Waving and flashing lights to indicate faults with a vehicle and even diversionary messages left with cafés should be viewed with suspicion.
10 Any approaches from strangers suggesting collaboration in stealing the load should be reported to the police or employer immediately.

### *Insurance*

Persons responsible for transport of valuable loads making overnight stops are recommended to look carefully at the small print of their insurance policies. Due to heavy losses sustained from vehicles parked unattended without satisfactory security precautions, insurance companies are insisting that the vehicles must be left in vehicle parks especially made for their security or, where one is not available, in others with at least first-class lighting. Where a security vehicle park is not used there is likely to be another requirement that the local police must be informed by the driver of the location of his vehicle and the nature of its load. Claims for reimbursement of losses are very likely to be disputed if reasonable precautions against theft had not been taken.

The degree to which, and the spheres within which, 'hired' labour is to be used should be carefully considered. Several organizations maintain pools from which their customers can draw temporary staff to substitute for permanent employees who are sick or on holiday, or to cater for a seasonal trade increase. The standards set by a company for its own drivers, and the checks it makes upon them, may not be met by these 'temporaries'. Fidelity bonds held by such organizations may not be considered by one's own insurers to be acceptable or an adequate basis for recovery in case of loss. If high-value loads are to be carried by such drivers, the insurers should be apprised and their prior approval obtained. They may well ask that a positive and satisfactory loss liability should be defined.

Hire transport is often verbally obtained without misgivings when it is known and locally based. On the vast majority of occasions this would be perfectly safe, but the precaution should be taken of ascertaining what insurance cover the carrier has. A case in point – a regular carrier for one unit of a large group solicited return loads from a second in the delivery area. The second, being of a cautious nature, and having materials averaging £1,200 per ton, asked for assurance of cover; after long delay, a reluctant admission arrived that the maximum was for £100 per ton. The group's insurers would have been decidedly annoyed had they been called upon to meet the difference for a loss and may have refused to do so.

# 29

# *Traffic control on factory roads and car parks*

The ability to travel in a private car and to have facilities to park safely close to the place of work is an amenity which modern employees have come to expect. If their comings and goings are to be inconvenienced because they cannot do this it will be an adverse factor in deciding whether to take up employment with a firm or, for that matter, whether they will settle down once in employment.

In many industries a high percentage of all personnel do travel to work in their own cars; this is particularly true of the more skilled, and accordingly more valued, type of worker. It is therefore in a firm's own interests to make car parking space available and ensure that users can leave their vehicles in safety and get away later without inconvenience. This has been recognized by most large firms who provide specific marked-out car parking areas. It is important that representatives of the employees should participate in discussions concerning parking, traffic flow and general safety of personnel on factory roads or parks. This leads to goodwill and better acceptance of restrictions.

## *CAR PARK SECURITY*

For all security purposes it is far better that employees' cars should be parked outside the factory perimeter fence so that they cannot go direct to their vehicles without passing a point of supervision. This segregation is virtually essential to ensure satisfactory control and direction of visitors. Moreover, if private cars are allowed to park inside the works and have to use the same entrances and exits as commercial vehicles there will be times of the day when searching or checking the documentation of the latter would be virtually impossible.

### *Internal parking*

If the available space and the location of the premises are such that internal parking is inevitable, specific areas should be designated for the purpose, and haphazard parking rigidly suppressed. This is not only in the interests of the motorists since production requirements demand the free movement of internal transport and this must have priority over personal convenience.

In selecting the areas, their proximity to easily stolen materials should be taken into account. Carrying a large carton a few yards to a car boot is a vastly different thing from carrying it the length of a well-lit roadway in full view. There are advantages in splitting up available space into blocks and allocating them to particular departments or to different managerial grades. Where this is done, without making the point obvious, it may be possible to place those least likely to steal in the positions where there is most temptation to do so.

Where partial internal parking is a necessity the only way to do it without causing continual complaint is to make it a privilege of the higher managerial grades. The important thing is that all employees must clearly know where they can or cannot park. No motorists will ever willingly leave their vehicles any further from where they want to be than is absolutely necessary, and they will find every imaginable reason or excuse why they should not. To be excluded from one's legitimate parking place by another, who has no right to be there, is most annoying. To prevent needless acrimony there must be some recognized disciplinary procedure against any individual who contravenes laid-down instructions designed for the communal good. These normally take the form of a written warning in the first instance, followed by possibly a final warning, then exclusion of the vehicle from the car park for a given period – if they refuse to comply with this, termination of their employment with the firm would inevitably have to be considered.

### *Wheel clamps*

Firms located in congested traffic areas can be plagued by outsiders taking up parking spaces needed for staff and official callers. 'Private parking' notices in themselves are no deterrent to drivers who are in a hurry, impatient, or just plain arrogant. If, however, those notices are reinforced by wheel clamping, and the threat of this is shown on the warning notices, the nuisance is likely to abruptly abate. Many firms include in the wording a financial penalty to be paid before removal – £25 seems the most frequent. Clamps are not ultra expensive and represent a permanent investment.

## *IDENTIFICATION OF VEHICLES*

In a small firm this offers little problem; the owners of vehicles are known to the security staff and can be quickly traced if necessary. This is not true of the larger concerns where the sheer weight of numbers makes it impossible; moreover there is a regular turnover and interchange of vehicles.

Space is a valuable commodity and the size of car parks will rarely be so greatly in excess of needs that obstructive parking will not occur. Where all employees leave at the same time this has a limited nuisance value, but where shift work is carried on it is not hard to visualize circumstances in which a complete line of cars may be blocked in by one left hurriedly by a latecomer who has just commenced work. If this individual cannot be immediately identified and the vehicle moved, not only will the irritation of those impeded be directed at the erring driver, the security staff and management will not escape criticism. Tannoy announcements may locate the driver, but in factory conditions this is by no means certain. It is far better to have some form of registration or identification whereby drivers can be traced to their department in the shortest possible time.

Such registration is also in the interests of the owners as a quick notification can be made in cases of damage, lights being left on, water or petrol leaking, etc. Nevertheless, reluctance to establish a system of this nature is occasionally met with from employees. It is hard to ascertain precise reasons; it would appear that the advantages to all concerned greatly outweigh the meagre inconvenience of notifying change of ownership and details so that a current card index can be kept. It may be that there is a basic thought that this is an extra piece of regimentation which should be resisted on principle.

If car parks are designated to particular departments or grades, quick identification is made easier by providing a colour sticker for the windscreen; the same colour can be used for the appropriate car park sign – this will at once show up use by unauthorized persons.

A card index system for vehicles can be very simply devised. Its sequence can be based either upon the numerals or the letters of the car's registration number; all that is required is that registration number, the name and department of the owner, and the telephone extension at which that person can be reached – and the make and colour of the car could be advantageous to prevent mistakes. A simple card of this nature could be made out every time a change of vehicle occurs, or a new employee enters the firm, to ensure the index is kept up to date.

## *CAR PARKS*

### *Design*

It would be the exception rather than the rule for a firm to have an ideal car park design and ample available space in which to put it. It is far more likely that an irregular area will be located for the purpose. When this is so, careful consideration must be given to making the best use of what is available.

Acceptable spacing to be allowed for each car has been found to be: 2½ yards (2.3 m) in width and 5½ yards (5 m) in length, with access lanes approximately 7 yards (6.4 m) wide.

No matter how good the plan looks on paper it is advisable to make the first markings so that they can be removed if necessary. Experience over a trial period can well lead to improvements in layout which will accommodate more cars and make entering and leaving the parking space easier.

Whether openings into the car parks should be marked specifically as exits or entrances will depend very much upon the manner of working of the firm in question. With most forms of day or shift working it is improbable that there will be incoming and outgoing traffic at the same car park at any one time. In these circumstances there is little point in designating entrances and exits as such, though it may be necessary to do so if there is any danger of collisions occurring inside the car park because of its shape. Directional arrows should then be used to indicate how the traffic should flow. When the final markings have been decided upon these should be made as permanent as possible and renovated when necessary.

If the parking area is well marked it is surprising how rarely bad parking and obstruction will occur. The offenders are apt to be persistent latecomers and it is reiterated that it is advisable that some agreed disciplinary procedure should be invoked against the regular offender. Where the parking is bad but the resulting obstruction limited a warning notice can be stuck on the windscreen. This effectively delays the departure of the offender and the result of his thoughtlessness is quite clearly visible to the other users of the car park.

*Illumination*

This is a matter for the management since obviously considerable expense can be incurred in its provision. It is possible that the proximity of the park to the lighting of internal and external roads may give an acceptable minimum to locate vehicles and prevent accidents. From security viewpoints, quite obviously, the more illumination the better, but the responsibility for protecting the car's contents must rest with the owner and not with the employer. Notices to this effect, disclaiming responsibility, should be clearly displayed at the car parks indicating that users use them at their own risk.

Points where there should be additional lighting in the interest of safety are the car park entrances and exits. Also for safety considerations, the car parks should be split off from footpaths running alongside by a barrier, either in the form of a fence or a low obstruction, to prevent motorists from driving straight out over the footpath and on to the main road.

*Supervision*

Though the responsibility for good parking and the locking of cars must rest with the car owners, who in the main conform, there is nothing more annoying after a long day's work than being held up by tedious and unnecessary delays caused by obstructing vehicles. For this reason the car parks should be patrolled, soon after each shift has arrived in the works, to ensure that all obstructing vehicles are removed before any general exodus begins – this only takes a matter of minutes. Similarly, when making normal patrols an eye should be kept on car parks to observe the presence of unauthorized persons who may be intending to steal cars or property in them.

Thefts from or damage to cars can cause extreme annoyance to their owners and can lead to attempts by them to save their 'no-claims' bonus by endeavouring to get their employer to accept liability when the incidents have happened at work and on the official car parks. Any inclination to 'welfare-mindedness' by *ex gratia* or other payment should be resisted since this would open the door to entirely spurious claims. 'Disclaimers', in the form mentioned in Chapter 17 under 'Security in Offices', should be displayed on car parks to make owner responsibility clear beyond doubt. Nevertheless, some areas of the country are plagued by persistent stealing from cars, and complaints from employees could be minimized by improvements to fencing, extra lighting, and perhaps continual surveillance by closed-circuit television. The latter would have the side effect of making the removal of company materials in car boots a much more hazardous proposition, and the cost might be treated as a 'site maintenance' rather than purely 'security' budget item.

Security staff should familiarize themselves with the changes in their powers which have resulted from the passage of the Criminal Attempts Act, 1981 (see p. 189), and bear in mind the possibility of charges being laid under the Health and Safety at Work Act 1974 against their employer if accidents on the site are caused by tolerated unroadworthy vehicles or uncontrolled reckless driving. Such charges were laid on two large northern sites.

## *INTERNAL TRAFFIC CONTROL*

### *Road markings and signs*

All road markings and signs used inside any works area should be identical with those used outside for everyday purposes. It must be remembered that production has paramount importance. Any scheme involving control of the direction of traffic flow or parking and loading areas must be discussed fully with production personnel to ensure that there is no conflict with their needs. Before any scheme of road marking is embarked upon, security, safety, and production representatives should meet to decide upon no-waiting areas, parking spaces reserved for works vehicles and equipment, where stacking may take place, the types of road markings which should be used, etc. Once decided upon, these regulations should be enforced by the security staff in respect of visiting commercial vehicles as well as the firm's own vehicles.

*Speed limits* – In some large works, the stretches of carriageway may be such that it is necessary to set speed limits. If this is so, these should be clearly shown in the conventional way and occasional checks made, either by stopwatch or other means, to make it obvious that these limits are going to be enforced. Offending private car drivers should be banned from the works, after a preliminary warning; the firm's own internal drivers can be dealt with departmentally and visiting commercial ones reported to their employers with a request that the works rules should be obeyed during future visits.

Do not put up mandatory signs and then not enforce them. The vast majority of employees are appreciative of the measures that have been taken for their safety and convenience and lack of implementation would cause resentment and criticism. Offenders will constitute a negligible minority who receive little sympathy from their fellows.

*One-way systems* – The introduction of such systems should always be considered. Even though they may increase the distances vehicles have to travel, their use is justified if a general improvement in the flow of traffic results; this, in turn, should lead to a higher efficiency in performance and a removal or, at least, reduction in road safety hazards.

*Traffic control*

It is again trite to say that the normal police signals for the control of traffic should be adhered to by security staff. These are those which are known and accepted and there is little doubt that, if requested to do so, the local police would be pleased to give guidance and instruction in their use.

White or yellow fluorescent gauntlets should be used when directing traffic under normal circumstances; in bad weather these should be reinforced by the wearing of mackintoshes of similar material. Where there is any danger to people performing traffic duty due to poor lighting, this must be rectified so that they stand out against their background.

Especially where the security officer is controlling traffic going out on to a main public road, the question of insurance must be carefully looked into. Not only in the person's own interests, to ensure that there is adequate cover, but also to ensure that the firm does not incur a gross liability should a signal be given which results in a major accident. One firm has been successfully sued for a very high amount indeed in respect of an incident of this nature.

Where an outgoing traffic flow crosses masses of pedestrians going to buses or other forms of transport, or walking home, serious consideration should be given to imposing a period of traffic standstill to enable them to leave in safety. This can be done by the use of stop-lights or even by electrically or manually operated traffic barriers controlled by security officers. If barriers are used they must be painted so as to be clearly visible, preferably with black and white bands along their length, and carry a red disc with the word stop on both sides clearly painted on it or outlined with glass reflectors.

All employees are most appreciative of efforts to expedite their comings and goings and anything that security can do in this connection will be amply repaid by appreciation and co-operation.

### *Vehicular accidents*

The reporting of these has been dealt with in detail in Chapter 8. It should be remembered that whatever the attitude of drivers might be at the time, subsequently insurance complications could give rise to ideas of suing the firm on whose property the accident has happened for some form of negligence. With the lapse of time, if there is no record for rebuttal, allegations which have little basis on fact find credence in legal proceedings.

A security officer on the scene of the accident should help employees in the exchange of the necessary particulars and get as many details as possible for future reference – especially if there is the slightest suspicion that negligence against the firm may be alleged or there is severe personal injury. A sketch plan and measurements are always useful; insurance particulars must always be obtained where the firm's own transport is involved or damage is caused to the firm's property.

## *CONTROL OF COMMERCIAL CAR PARKS*

The demand for off-street parking has led to an unprecedented growth in the construction of multi-storey car parks and the utilization of wasteground for supervised parking on prepayment.

Most have barriers and automatic ticket controls to prevent the owners being defrauded. This is not the case at many open car parks, be they operated by local authorities or by commercial interests. A vast amount of money changes hands at these but the attendant's job is one which is lowly paid and often allocated to disabled people. It seems to be accepted that the remuneration is likely to be augmented by 'perks'. Where simply using a ticket machine, the 'perk' takes the form of either the non-issue of a ticket, or picking up discarded tickets and reissuing them rather than using the machine. These practices are known but tolerated by users on a policy of non-involvement and the lack of action by employers can only be explained as indifference stemming from either inflated profits or the knowledge of the paucity of the wages paid.

The only certain way to counteract fraud by the attendant is to establish some means of monitoring the vehicles that enter a park. This can be done by a drop-arm barrier, electrically operated by the attendant and with pads coupled to an automatic recorder. An alternative is to dispense entirely with the services of the attendant substituting instead a self-contained unit at the point of entry with the drop arm functioning on payment of a fixed charge with an egress barrier pad operated. A yet more sophisticated unit issues a timed and dated ticket before the barrier is raised; this must be placed in the second control coupled to the egress barrier with the appropriate sum of money before that barrier is raised. Such installations which cater for a charge graded upon the length of time in the park are complicated and expensive, nevertheless they are worthy of consideration since they achieve both the saving of an attendant's wages and the less determinate sum of 'perks', thereby repaying their cost within a forecastable and limited period.

Card-operated barriers are increasing in number where parking facilities for senior staff are limited. Hospitals are a particular example where each authorized user can be given an identification card to place in the mechanism in front of the barrier causing it to lift and the card to be returned. Devices of this type save considerable controversy where precedence and dignity are symbolized by preferential parking.

*PART SIX*

# *FIRE, ACCIDENT AND EMERGENCIES*

# 30

# *Fire precautions*

Fire prevention is a basic responsibility of all security staff and management, irrespective of whether permanent full-time firemen are also employed. To get this risk into perspective: losses to the country resulting from fires average at least twice the rate of crime losses; there is no doubt the former can more easily be controlled than the latter.

The seven days a week, 24 hours a day coverage given by patrolling security officers makes them the most likely agents for the detection and consequently the prevention and extinction of fires. This fact is recognized in medium-sized and small firms by making fire prevention a dual responsibility for the chief security officer and the tendency to do so seems likely to become universal.

## *LEGAL OBLIGATIONS*

With increased immediate availability of units of the fire service, there is less need for firms to provide their own brigades and the trend, other than in large factories with high potential danger, is to rely on outside help rather than internal services. Despite this it is essential that all firms, be they large or small, whether they have a security force or not, should designate to a specific individual the responsibility for ensuring that legal obligations are observed, in respect of fire hazards and the safety of life from fire.

If there is any doubt about the effect of legislation on particular premises, immediate guidance can be obtained through the local fire authority which, under the Fire Services Act 1947, must maintain efficient arrangements for giving, on request, advice to firms in the area on fire prevention, means of escape, or restricting the spread of fires in respect of buildings and other property.

## *THE FIRE PRECAUTIONS ACT 1971*

This Act – as amended by s. 78 of the Health and Safety at Work Act 1974 and modified by the Fire Precautions Act 1971 (Modifications) Regulations 1976 – applies to factories, offices, shops, and railway premises, as do other relevant regulations including the Fire Precautions (Factories, Offices, Shops and Railway Premises) Order 1976. Enforcement of the provisions of the Act is vested in the fire authorities whose main weapon is the 'fire certificate' which will only be issued after the area authority is satisfied that the means of escape and other fire precautions are such as may be reasonably required. This certificate is mandatory for certain factories, offices, shops, and railway premises, but for similar premises not requiring one a number of fire safety requirements are imposed under the Fire Precautions (Non-Certificated Factory, Office, Shop and Railway Premises) Regulations 1976.

The Act is concerned only with the protection of life, and that of buildings is coincidental to this purpose. Where industrial processes are of such a nature or on such a scale as to have bearing on general fire precautions – as in nuclear installations, explosives factories, mines, and large fuel installations – enforcement comes under the Health and Safety Executive; the premises are exempted from controls under the 1971 Act and are specified as being within the provisions of the Fire Certificates (Special Premises) Regulations 1976. As will be seen, the obligation laid upon the fire authority to give advice upon legislation was a very necessary one!

### *Definitions*

The definitions used in the Factories Act 1961 and the Offices, Shops and Railway Premises Act 1963 remain fundamentally unchanged. In brief:

*Factory* – any structure in which anything is manufactured, or changed in shape or substance.

*Office* – a building (or part of a building) the sole or principal use of which is as an office or for office purposes; offices which are part of a building used for other purposes are included, as are ancillary rooms, staircases, storerooms, etc.

*Shop premises*:

1. *Retail* – shops in the everyday sense of the word, such as butchers', grocers', tailors', etc; it includes buildings or parts of buildings where the sole or principal use is for retailing.
2. *Wholesale or warehouse premises* – building, or part thereof, occupied by a wholesale dealer or merchant where goods are kept for wholesale selling (parts of factories are excluded, also those premises associated with docks or wharves).
3. *Catering establishments open to the public* – everyday interpretation.
4. *Fuel storage premises* – used for storage of solid fuel and for the purpose of a trade which consists of, or includes, the sale of such fuel.

*Railway premises* – building or part of one occupied for the purpose of the railway and situated in the immediate vicinity of the permanent way.

*Multi-occupied premises* – where any of the aforementioned premises are:

1. held under a lease, or agreement for a lease, or under a licence, and consist of part of a building all parts of which are in the same ownership; or
2. consist of part of a building in which different parts are owned by different persons.

The owner, in (1) premises, or the owners, in (2), are responsible to the fire authority for legal obligations of fire precautions.

### *Fire certificates*

These are required under the 1971 Act for any premises (except 'special premises' mentioned earlier):

1. in which more than 20 persons are employed to work at any one time; or
2. in which more than 10 persons are employed to work at any one time elsewhere than on the ground floor; or
3. which are in the same building as other factory, office, shop, or railway premises where the sum total of all employees in all the premises exceeds 20, or 10 other than on the ground floor.
4. if explosives or highly flammable materials are used or stored on the premises (unless the authority considers the risk acceptable).

Applications for certificates must be made to the area fire authority outlining the use or uses of the premises and such other information as may be required. Further information and plans will be asked for, and the premises (all the building if in multi-occupancy) will be inspected. The person doing so will prove his identity on request; he will usually be a member of the fire brigade (s. 5). If he is satisfied with the means of escape and precautions, a certificate will be issued, and this will specify (s. 6):

1 The particular use or uses of the premises that it covers.
2 The means of escape in case of fire (a plan will be used).
3 The means of ensuring that the escape route can be effectively used at all times (this would include measures to restrict the spread of fire, smoke, and fumes and the provision of emergency lighting and direction signs).
4 The means of fighting fire for use by persons in the building.
5 The means of giving warning in the case of fire.
6 Details of explosive or highly-flammable materials stored (applies to factories only).

Additional requirements may also be included:

7 The maintenance of the means of escape and keeping them free from obstruction.
8 The maintenance of other fire precautions specified in the certificate.
9 The training of those employed in the premises in the action to be taken in the event of fire and the keeping of appropriate records.
10 Limitations on the number of persons to be in the premises at any one time.
11 Any other relevant fire precautions.

These requirements may apply to the building as a whole or to separate parts of it (s. 6). If the authority is not satisfied about requirements 2 to 5, it must serve a notice on the owner to improve them and withhold the certificate if these are not done within a specified time (s. 5). A power of appeal is given for refusal (s. 9), but the obvious course is to take advantage of the expert advice given and comply with it unless there are extremely good reasons to the contrary.

A fire certificate must be kept in the premises to which it refers (s. 6), and in the case of multi-occupied premises a copy must be kept in those premises to which it relates.

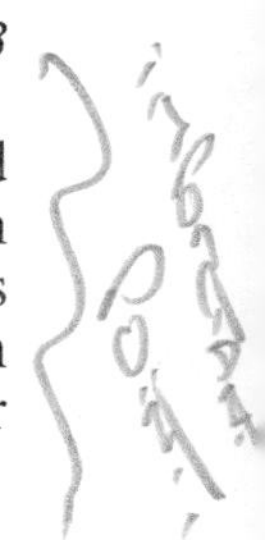

Whether or not premises within the scope of the Act are required to have a certificate, the fire authority may apply to the court for an appropriate order if satisfied that the risk to persons in any premises in case of fire is so serious that the use of the premises in question ought to be prohibited or restricted. This is an emergency power for unacceptable fire risks (s. 10).

*Certificates issued under the 1961 and 1963 Acts*

These, if still valid as at 1 January 1977, will continue to apply as if issued under the 1971 Act, though they will be subject to being replaced, modified, or revoked in accordance with the latter Act. The requirements under those certificates include:

1 Specified means of escape to be properly maintained and kept free of obstruction.
2 Notice to be given of any proposed material increase in the number of people to be employed in the premises or factory (or in any specified part of a factory).
3 Doors not to be locked or fastened so that they cannot be readily opened.
4 Exits to be conspicuously marked by notices.
5 Contents of rooms not to impede free passage to means of escape.
6 Effective means of giving fire warning to be provided, maintained, and tested.
7 The ensuring that employees are familiar with means of escape and fire routines.
8 Appropriate means and equipment for fighting fires to be provided and maintained.

At factories the alarm tests should be entered in the general register, doors should open outwards, and hoists and liftways should be enclosed with fire-resistant material.

*Alterations to premises (s. 8)*

The fire authority must have advance notice where it is proposed:

1 To make a material extension of, or a material structural alteration to, the premises.
2 To make a material alteration in the internal arrangements of the premises, or in the equipment or furnishing.

'Material' is interpreted as being such as to render the means of escape and related fire precautions inadequate for the normal use of the premises. Alterations apart, the fire authority has power to review requirements from time to time to check that precautions remain adequate, and it must be notified in advance of an intention to store or use explosive or highly-flammable materials.

### *Non-certificated premises*

The Fire Precautions (Non-Certificated Factory, Office, Shop and Railway Premises) Regulations 1976 provide for certain fire precautions to be taken in those premises, within the scope of the definitions, which do not require a fire certificate nor are specified as 'special premises'. In the case of factories these are:

1. Doors from any room in which more than 10 people are employed which open on to a staircase or corridor, and all other doors affording means of exit or access to the premises, shall open outwards, except sliding doors (reg. 4).
2. All windows, doors, or other exits affording means of escape from factory premises in case of fire or giving access thereto, other than the normal means of exit, shall be distinctively and conspicuously marked by notices (reg. 4).
3. Hoists and lifts must be enclosed by construction of not less than 30 minutes' fire resistance, and access doors to them will have similar resistance. If not vented at the top, the top enclosure material must be easily broken by fire (reg. 4).
4. While any employee is at work in the premises or having a meal, none of the doors leading out shall be locked or fastened in such a way that he cannot easily and immediately open them (reg. 5).
5. Contents of workrooms must not obstruct free passageway to a means of escape (reg. 5).
6. Means of fighting fire shall be provided and kept readily available (reg. 6).

In some circumstances, exemption can be granted from precautions 1 and 3. In the case of offices etc. only precautions 4, 5, and 6 of the above are required. There is a very obvious similarity of requirement whether premises are certificated or not.

### *Powers of inspectors and enforcement*

The enforcing authorities for premises covered by the 1971 Act are the fire authorities, who are empowered to appoint inspectors for the purpose. These will take such action as is necessary for giving effect to the Act, or Regulations made under it, with powers:

1. To enter at any reasonable time premises to which the Act applies, or seems to apply, as well as the rest of the building containing the premises.
2. To make such enquiries as may be necessary to find out if the Act and Regulations are being complied with.
3. To request facilities and assistance to be given them in the exercise of their duties.

### *Identification and responsibility*

The inspector will be in possession of a duly authenticated document to prove his identity and authority which he must produce on request (s. 19). He is prohibited from disclosing any information obtained by him in any premises entered in the course of his duties unless it is necessary for the performance of those duties or for legal proceedings and reporting (s. 21).

### *Offences*

On summary conviction, a fine up to £400, or on indictment, a fine and imprisonment up to two years, may be imposed for the following:

1. Using premises for a designated purpose without a fire certificate (s. 7).
2. Contravening a requirement of a fire certificate (s. 7).
3. Carrying out changes to premises or beginning to store explosives or highly-flammable materials or materially increasing the quantities without giving notice to the authority (s. 8).
4. Contravening a direction given by the fire authority in connection with proposed alterations (s. 8).
5. Contravening any specific regulation under the Act (s. 12).

On summary conviction only, a fine up to £400 may be imposed for (s. 22):

6. Recklessly giving false information in purported compliance with an obligation under the Act.
7. Forging a fire certificate with intent to deceive.
8. Recklessly giving false information to procure a fire certificate.
9. Making a false entry in a book or document required to be kept under the Act, knowing it to be false.

On summary conviction only, a fine up to £100 may be imposed for:

10 Pretending to be an inspector under the Act (s. 40).

On summary conviction only, a fine up to £50 may be imposed for:

11 Failing to keep a fire certificate, or where applicable a copy of it, in the premises to which it applies (s. 7).
12 Obstructing an inspector in the course of his duty or failing to give him any necessary facilities and assistance (s. 19).

Where an offence is due to the act of someone other than the person liable to the penalty for the contravention, that other person may be charged whether or not the person liable is proceeded against. It is a defence to charges that all reasonable precautions and diligence have been exercised to avoid committing the offence (s. 25).

## *GENERAL OBSERVATIONS*

These legal requirements are designed to limit loss of life, and to ensure that means are available for the extinction of fire. However, a large majority of fires are caused by the unpredictable careless acts of individuals for which no legislation can provide. It is therefore essential that precautions must be taken on an assumption that inevitably some fires will be caused by carelessness which may eventually produce a major conflagration.

The ultimate responsibility for extinguishing any blaze of consequence lies with the local fire authority and therefore the higher the potential risk the closer liaison that should be maintained between the chief fire officer, the fire insurance surveyor, and the fire prevention officer. The latter should be encouraged to visit as regularly as possible so that the maximum advice can be obtained and the local fire brigade can be familiarized with the buildings and with the materials in use. Recommendations by the fire prevention officer as to the siting of appliances and the marking of hydrants should be followed meticulously. A plan of the building or factory area showing the location of all these and alternative sources of water supply should be kept, preferably by the main entrance to assist the brigade.

There can be no single procedure recommended for action which is suitable for adoption at all premises in the event of a fire. This is inevitable owing to the divergent number of storeys, structural standards, processes, and materials, and the number of people employed, as well as possible means of escape. One thing is common to all:

comprehensive instructions must be available to all employees, clearly displayed, and easily understandable; no room for doubt must be left by them as to what action should be taken in the event of fire.

## *IMMEDIATE ACTION ON FINDING FIRE*

1 Raise an alarm and make sure that the public fire brigade is informed.
2 If there is reasonable hope of extinguishing the blaze, attack the site immediately.
3 Put into operation a prearranged plan for evacuation of personnel, notifying the management and the trained staff required for action in case of fire (see Appendix 23).

The magnitude of an outbreak must dictate whether attacking should take a priority over reporting – it is obviously foolish to allow a small fire to spread by spending time on reporting; it is equally foolish, in any case where there is doubt, to delay reporting whilst making an abortive attempt to put the fire out. Normally there will be other persons at hand to assist and to notify the switchboard, but a patrolling security officer at night would be well advised not to overestimate his capabilities.

Providing there is no danger to the persons concerned, every effort should be made to contain the blaze pending the arrival of the fire brigade. All staff other than those actually engaged in fighting the fire should vacate the surrounding area. Onlookers should keep out of the way and doors and windows as far as possible should be kept closed to prevent a quick spread of the fire.

It would be unrealistic to stop all work because of a fire in one isolated department and only the alarms in those most likely to be affected should be sounded. These can take the form of:

1 Manually-operated bells – these are only suitable for small single-storeyed buildings with a low fire risk.
2 Electrically-operated sirens, actuated either from points throughout the factory or from a switchboard.
3 Automatic electrically-operated alarm systems activated by heat or smoke. Such systems would normally incorporate manually operated points as well.

Where a mass evacuation is necessary it should be done with a minimum of fuss and panic and the employees assembled at a prearranged point to check that all have left. A search for stragglers should be made before the building is vacated, if this can be done with limited risk to the searchers. The head of each department should be responsible for checking that all his personnel are out.

It is essential that the fire brigade should have been directed in the first instance to the entrance nearest to the fire. They should be met and the person meeting them should be fully aware of the location, the best route of approach, whereabouts of water supplies, special hazards, and any factor which might need priority action, for example trapped persons.

## *REPORTING AND INVESTIGATING*

All too often these are matters which are neglected, with the inevitable result that lessons are not learnt and used equipment is not replaced or refilled before it is needed again. There are really two separate groups into which industrial fires will fall for reporting purposes: the first, minor fires where damage is limited and there is no personal injury (these will include the recurrent trivial ones which may be almost inevitable, perhaps because of the nature of a manufacturing process being used); the second, larger fires where insurance claims are likely to follow.

The first group must be reported and recorded if only to show up any pattern whereby preventive measures can be strengthened, also to ensure that extinguishers which have been used are replenished. Experience shows that at least as many extinguishers are used without notification to the fire officer as are reported to him. A simple card system could be used for this type of incident giving date, time, department, installation involved, apparatus used for extinction, nature of fire, and signature of supervisor.

With a more extensive fire, every effort should be made to establish the cause, irrespective of whether this might cause embarrassment to particular individuals or departments. This will necessitate a comprehensive report compiled as soon as possible, whilst memories are fresh, preferably on a printed form to ensure the details are not overlooked. A suggested list of items for inclusion is:

1 Time, date, general area of outbreak and person finding.
2 Time, date last seen in order, and by whom.
3 Precise location of fire.
4 Times of notification, arrival, and departure of the fire brigade.
5 Identity of officer in charge and means of contacting him.
6 Cause of fire, if given by him.
7 Senior personnel notified of outbreak and times of doing so.

8 Appliances used that need replenishing.
9 General description of fire, including structures and equipment damaged, factory services affected – electricity, gas, etc.

The foregoing details can be filled in by the senior security or fire officer attending the fire but further space should be then available for the chief fire officer to complete with:

1 Time and date of notification to fire assessor.
2 Appreciation of effect on production.
3 Comments, and observations to prevent a recurrence.
4 Possible causation.

The completed form should be circulated to the manager and heads of department concerned. Fire assessors do not take kindly to being informed of a fire after everything has been tidied up before they have an opportunity of inspecting the site – hence inclusion of that item. (Specimen report form, Appendix 24.)

Even in the case of a minor fire, a note should be made in the security department's occurrence book so that security personnel subsequently coming on duty can pay especial attention in case smouldering material has evaded detection.

Where the fire is of such magnitude that the management decides that an official inquiry should be held upon it to establish the cause and to consider future prevention, and desirable variations in procedures, it is advisable that the composition of the panel should be independent of the departments concerned and the services affected – if there is blame to be ascribed it is unlikely they would be unbiased.

*Arson*

This offence, the wilful setting fire to buildings, is a crime under the provisions of the Criminal Damage Act 1971 (see p. 175) which also invariably leads to imprisonment for adult culprits and comparable sentences for juveniles. It is an offence which is not always recognised as such, being attributed to accidental causes for lack of proof otherwise. Nevertheless, in 1986 there were no fewer than 19,240 recorded cases with only 4,219 detected – an unsatisfactory 22 per cent at a cost which has been estimated at between one third and one half of the total loss due to fire. Consequential loss when business premises are the target is very difficult to evaluate – an important customer may go permanently elsewhere after a crucial delivery date is not met due to disruption of production. Jobs may be lost and firms that have been operating on a financial knife edge may go out of existence altogether. Insurers by habit regard arson claims with a jaundiced eye and settlements may be delayed.

There is no single 'arsonist type' to be looked for after an outbreak, though on the non-industrial scene the numerous incidents involving schools are almost invariably the work of juveniles. An occasional thief will set fire to a building he has broken into to destroy any evidence of his presence, another will do the same maliciously because he has not found what he wanted to steal, or because he was drunk – they will not be found amongst the spectators afterwards. Pyromaniacs may have a sense of physical or mental inferiority which they relieve by exploiting the power of fire to destroy and it is suggested they may achieve something in the nature of sexual gratification by watching the blaze. They are much more likely to be lurking back on the outskirts of spectators, watching their handiwork. Disgruntled employees or former employees may give an indication of identity by starting a fire where it will do most harm in the shortest time; a classic case was that of a man who set fire to a series of firms where he had been refused employment – each in the office of refusal! Where a business is concerned which is on the point of bankruptcy, or has large stocks of non-saleable goods accumulated, the insurance connotations at once will point to a suspect. Office fires that destroy records which might prove fraud – employees in cashiers, accounts, purchasing, etc. then have a potential causation reason that demands enquiries. Alternatively, the whole conflagration may be the totally unintentional act of a vagrant who nearly burns himself to death and promptly puts as much distance as possible between himself and the fire. There are ample avenues for investigation but the acquiring of adequate evidence is another matter.

There are a number of indicators at the scene of a fire which can point to arson.

1 Fire breaking out at several unconnected sites.
2 Simultaneous outbreaks at several points, unexpected speed of spread.
3 Smells or other indications of inflammable liquids which should not have been there.
4 Signs that the premises have been broken into, doors or windows broken, or left open by an escaping intruder.
5 After the fire, burnt out tins or other containers found near the point of origin, or traces of flammable materials that should not be there; apparent piling up of materials where fire started.
6 Absence of remains of valuable machinery or office equipment that would be expected to be there.
7 Interference with fire alarms or equipment.

There will be police and senior fire brigade attendance at any outbreak of consequence and both have at their disposal experts to examine scenes for confirmatory evidence. What they do not have is an intimate knowledge of the buildings before anything happened and this can be provided by the security staff plus the actual occupiers of the parts affected who should be summoned as soon as possible – a telephone call for this purpose effectively rebutted the alibi of a cashier who did not get home in time to substantiate a claim to have spent the evening reading. If the security officer at the scene senses something is not quite right about the fire – something 'does not fit' – he should at once tell the fire officer in charge.

In the event of experts being called in to make an examination, this will probably be made known by the fire/police before they leave the scene, if indeed they do before these arrive. Members of the firm must be kept out of the area before the examination is made; it is not unknown for equipment and pieces of wiring to have been removed prematurely by electricians either mistakenly or in apprehension of that being a negligent cause of the outbreak. The experts will appreciate any additional information that might have bearing – smoke colour and smell, colour of flames, apparent point of origin or greatest intensity, direction and speed of spread, contents of area, anything amongst the debris that should not be there, weather conditions – in fact anything that may be of value to them including any previous suspicious incidents whether resulting in fires or not.

An additional point should never be forgotten. It is not unknown for fires to be started deliberately by a person responsible for finding or extinguishing them to impress his employer with his vigilance and efficiency. These will rarely be of any size and if a security guard is the person in question he will almost certainly be a non-performer and a non-entity. The thoughts of a chief security officer who had so suffered are apposite: 'The first time one of my men finds a fire, I congratulate him, the second time I again thank him, but keep an eye on him, the third time he really has to be gone into.'

Ordinary security and fire precautions properly applied are the appropriate deterrents to arson; the former must be adequate to exclude intruders or at least give early warning of their presence. The consequences are such that any threats, warnings, or information must be treated seriously and notified to the police. Retail premises in large cities in particular must do so since they have been the target of an addition to the routine run of arsonists – the terrorist (see pp. 528–9).

*MEANS OF ESCAPE*

The requirements entailed for the granting of a fire certificate ensure the availability of adequate means of escape. The fire authority will dictate what is acceptable, but an indication of its probable recommendations can be obtained from a very detailed section dealing with means of escape found in two excellent HMSO booklets produced under the common title of *Guides to the Fire Precautions Act 1971: Factories and Offices, Shops and Railway Premises*.

The general premises are:

1 No one should have to go towards any fire in order to escape.
2 Each route should be as short as possible and of adequate capacity to allow the speedy passage of the number of persons who may have to use it.
3 Each route should lead to the open air at ground level either directly or through a fire-resisting barrier.
4 Enclosed parts of escape routes should be protected against penetration by smoke or fire.

These principles can be met by:

1 Each occupant having a choice of escape routes.
2 No occupant should have to go more than about 100 feet (30 m) to reach open air or a smoke-free fire-resistant stairway, corridor, or lobby (lifts should not be considered as escape routes).
3 All corridors, stairways, and exits should be adequately wide to allow occupants to leave quickly without panic.
4 The walls, floors, and ceilings of any passageway forming part of an escape route should have a fire resistance of at least half an hour with self-closing fire-resisting doors at every entrance to the enclosed area.
5 Exits from escape routes should be sited so that people can disperse from them in safety without difficulty and without being confined in yards near the building.
6 External escape staircases should be provided where there is no possibility of using enclosed fireproof stairways.
7 All apertures, stairways, lift shafts, and hoists, whereby smoke and fire could spread rapidly through a building, should have been surveyed and enclosed as far as possible.
8 All escape routes should have been clearly marked and all the doors on them should open outwards to avoid any confusion.

## *FIRE-FIGHTING EQUIPMENT*

It is mandatory that premises for which a fire certificate has been issued under either the 1961, 1963, or 1971 Act, or those governed by the Fire Precautions (Non-Certificated Factory, Office, Shop and Railway Premises) Regulations 1976, should have appropriate means of fighting fire readily available for use.

Precisely what is used depends on the type of risk but portable appliances should be placed at conspicuous, properly maintained fire points which are readily accessible, similarly located on all floors of a multi-storey building if possible, and include a fire alarm contact for manual operation with instructions clearly displayed. While proximity to ignition sources is important, the potential user must have a safe route of escape. Where different types of extinguisher are sited together, they must be clearly coloured or labelled to prevent use on sources for which a particular one might be dangerous.

Fire needs fuel, oxygen and heat; starving – cutting off the fuel – is rarely possible; smothering – cutting off the air, and cooling – removing the heat are the main methods of extinction, the choice being dictated by the 'class' of risk.

1 *Class A*: Fires involving solid, normally carbonaceous, materials which form glowing embers, i.e. paper, wood and their derivatives. Most effective extinguishing agent is water, in the form of a jet for deep-seated fire, and a spray for surface; a wetting agent is sometimes added to reduce surface tension of water in a container.
2 *Class B*: Fires involving liquids or liquifiable solids, i.e. oils, fats, petroleum jellies, petrol, paraffin, etc. Use (a) dry powders – sodium bicarbonate, (b) inert heavier than air gas – carbon dioxide, (c) foam, (d) non-toxic vapourizing liquid or (e) in some cases an incombustible sheet. Water should never be used on Class B substances except in the very rare instances where the substance is known to be miscible with water which can then be applied in spray form. The general rule should be – do not attack with water.
3 *Class C*: Fires involving gases or liquified gases in jet or spray form – propane, butane, methane, etc. Containers are cooled by water spray; foam should be used to control resultant fires which include spilled liquid.
4 *Class D*: Fires involving metals – water may be dangerous on these and should not be used. Dry powders, carbon dioxide or dry sand are advised.

### *Electrical fires*

These are not now treated as a class since, if electricity is the causation, one of the other classes results; also many of these other types will include electrical wiring and appliances in their spread. If the electricity supply can be cut off, this must be done and the appropriate extinguishing agent for the class of fire then used. If this cannot be done, a non-conducting agent must be used which will not damage equipment present, i.e. dry powders, inert gas or vapourizing liquid. Water again is dangerous because of possible shock risk to the fire fighter.

### *Colour identification of appliances*

British Standards recommend that appliances be painted with specific colours for quick recognition; regrettably some manufacturers do not comply and adhere to the traditional red. Red signifies water; cream, foam; blue, dry powders; black, carbon dioxide; green, vapour forming.

### *Equipment using water*

#### *Static equipment*

1 Automatic sprinkler systems. These are brought into operation by an automatic fire detector which gives an alarm and delivers water in spray form to the seat of the fire.
2 Hose reels. These are adequate lengths of non-kinking rubber tubing wound on a reel and permanently connected to a mains water supply. (There are recommendations for the length and diameter of these tubes and for the water pressure, nozzle bore, and siting. Advice should be sought from the fire prevention officer on these.)
3 Wet or dry hydrant systems or rising mains (in buildings). These are to provide a substantial supply of water close to where it may be needed by the fire brigade. In the 'wet' systems the main is charged with water permanently up to each outlet valve; in the 'dry', connection is made by the fire brigade on arrival.
4 Special water spray systems. These are fitted for the protection of specialized plant and each system is designed to meet individual requirements.

*Portable equipment*

1 Extinguishers. To simply supply water in either jet or spray form by expelling water from a container by virtue of pressure, either generated by gas pressure, soda acid reaction, or by pumped pressure.
2 Buckets. These should be kept filled and lids fitted to avoid wastage and contamination.

*Equipment using foam*

Fixed systems are usually individually designed to protect large flammable liquids such as oil storage tanks, they may be automatically or manually operated. They can be from self-contained systems in which the foam supply is part of the installation or from systems of pipework to which a fire brigade may connect foam-making equipment.

Portable equipment is of two types, both extinguishers: in one the foam is generated by chemical means, in the other by mechanical. These are specially suitable for use on small fires – deep fat frying ranges in kitchens or oil fired boilers.

*Carbon dioxide equipment*

Fixed equipment is individually designed and may be automatic or manually operated. There is danger to employees of suffocation from carbon dioxide if they are trapped in a confined area.

Portable equipment is for use on small fires of flammable liquids, electrical equipment which may be live and where it is necessary to avoid damage or contamination by powder.

*Dry powder equipment*

Fixed systems can be automatic or manually operated. The powder is blown out in a cloud from outlets in a system of piping by pressurized gas, usually carbon dioxide. Installations are individually tailored for dealing with fires in flammable liquids and electrical equipment.

Portable equipment is in the form of extinguishers. These are most useful for all-round purposes in extinguishing electrical fires and fires of flammable liquid which have spread over a large area.

*Equipment using vaporizing liquids*

These are portable and mainly used to apply a non-conducive extinguishing agent to fires in live electrical equipment. They can be used for other types of small fires but have no advantages then over the conventional means. The toxicity of their vapours, so far as early types were concerned, restricted their application in confined spaces and indeed all CTC (carbon tetrachloride) cylinders should now have been phased out. Halon is generally regarded as the form giving satisfactory efficiency coupled with acceptable toxicity but care should be exercised if used indoors. Practically all cylinders have a controllable discharge. Apart from their suitability for fires coupled with electrical danger they are valuable for dealing with small flammable liquid fires and vehicle engine fires in particular.

## *ESSENTIAL PRECAUTIONS*

1. Water jets must not be used on electrical fires unless it is certain that the current has been cut off – otherwise the jet could conduct electricity to the person holding the hose.
2. Water jets should not be used on oil fires unless they are fitted with a foam attachment – otherwise the oil would float on the water and continue to burn.
3. CTC cylinders should be treated as obsolete; they may be dangerous in confined space and care must be exercised when using any vaporizing liquid extinguisher in such circumstances.
4. All extinguishers must have the last date of recharging and the last date of pressure testing recorded by fitting tags or by painting on them – an office record should also be kept of these.

With regard to the testing of extinguishers, full information about these and inspection can usually be obtained from the manufacturers and from the Fire Service drill book; the latter also gives instructions concerning maintenance of fire hose and hydrants.

## *PREVENTION OF FIRES*

Apart from any requirements of legislation, there is no reason why action to prevent fire in new buildings should not commence before a stone is laid. If the elimination of fire risk is laid down by prospective owners as a basic consideration to be taken into account by their architects and planners at the design stage, the latter will have to consult with the local fire authorities from the outset to exclude unsuitable materials and potential fire hazards. Access for mobile appliances can be assured, adequate hydrant points dispersed to give full coverage, and optimum means of escape incorporated. The siting of storage areas for high-risk materials and fluids can be discussed, as well as their isolation in the event of outbreaks. Economies in installation cost and improved appearance can be achieved by incorporating permanent installations of the sprinkler type during the actual construction work – it is considerably more expensive to add these subsequently.

A wide range of British Standards deals with fire equipment of all kinds. Of special interest are: BS 5306 Code of Practice for fire extinguishing installations and equipment of premises; BS 5423 Specification for portable fire extinguishers; BS 5445 Components of automatic fire alarm systems; BS 5588 Code of Practice for fire precautions in the design of buildings.

### *Sprinkler system*

Sprinkler systems are very effective in operation and rarely subject to fault or malfunction; they immediately cover areas which may be difficult to reach with hose jets and to a degree that is not possible with portable equipment. They are efficient in extinguishing fire in all normal combustibles including those used in the construction of the building. The fear of accidental water damage, or unnecessary damage in the case of fire is grossly exaggerated – only those heads in the region of the fire operate. It is advisable to protect all the building when fitting an installation otherwise an outbreak in an unprotected part might interrupt the water supply to the other or become intense beyond extinction.

Insurers will insist on sprinklers where their potential loss is high but it should not be forgotten that reductions in premiums ranging from 50 to 70 per cent may be made where an installation is put in, also that automatic sprinklers are classified as plant and may qualify for allowances against taxable profit. To obtain premium discounts, the systems must conform to rules made by insurance companies which govern density of 'heads' type of water supply, rate of discharge, etc.; it follows that no action should be taken without prior consultation with the insurers.

### *Automatic detection devices*

Damage will have been done by the time sprinklers operate and there is good reason to incorporate electronic warning equipment which is much more sensitive and of course can be fitted in premises by itself. Where the main purpose of having security staff on premises is the fire risk, the job might more efficiently be done by a warning system with or without sprinklers; the cost will soon be written off against that of manpower.

Operation is similar to that of burglar alarms; heat or smoke sensitive detectors are incorporated in an electric circuit with audible and visual warning devices. It is important that the circuit is 'closed' so that it is broken when a detector operates – this enables it to fail safe and operate the alarm if a wire is cut or a detector broken.

### *Considerations of new construction*

In connection with materials and construction, this is a specialist subject and the Fire Protection Association has produced a large number of technical information sheets which are invaluable; a full list of all publications can be obtained on application to the Association's office at Aldermary House, Queen Street, London EC4N 1TJ, and will be found to give expert detailed guidance on every aspect of fire prevention.

It is at this planning stage that the fire and security officers should get together to ensure there is no clash of interest – it is the former's responsibility to enable people to get out of buildings in emergency, the latter's to prevent other unauthorized persons getting in. A satisfactory compromise is not impossible provided discussions start early enough.

With regard to the equipment required to be installed and its siting, again the fire authority will make detailed recommendations which should be followed – the only one which may be in dispute is that of sprinkler systems which are an expensive item. In the event of opposing opinions on the necessity of the installation being irreconcilable, the matter may have to be resolved before a court if the certificate is withheld.

For the maintenance of equipment, contracts with manufacturers offer the best guarantee of skilled checking, but security staff, in addition to receiving instruction in the use, should have at least an idea of the rudiments of servicing and refilling. One precaution which must be taken in respect of water-filled extinguishers is that of protection from frost for obvious reasons. The necessity of replacing expended extinguishers must be re-emphasized; any user is likely to put the empty one back from where it was taken and if the reporting system has not functioned properly this may escape notice; if this is a recurrent feature the solution will be to devise a system of storage or suspension to which the extinguisher cannot be returned after withdrawal. This again can be expensive and it is far better to educate staff to a consciousness of fire prevention's importance, so that they will meticulously follow a laid-down system to ensure that all equipment at their disposal is always serviceable.

When all the precautions of a permanent nature by way of construction, design, and provision of equipment have been met, there still remain the dangers which accrue from human fallibility and unpredictable breakdowns in services. These can only be countered by: enforcing good housekeeping standards to prevent the accumulation of flammable material to form fire hazards; instituting a routine of checking when premises are vacated and systematic patrolling thereafter; and, perhaps most important, the instruction and training of new employees.

*Fire instruction*

A fire certificate may impose a requirement for 'securing that persons employed to work in the premises receive appropriate instruction or training in what to do in case of fire, and that records are kept of instruction or training given for that purpose' (s. 6(2)(c)).

All employees in premises covered by a fire certificate should be trained to understand fire precautions, the action to be taken in the event of fire, and their own responsibilities. Training should be based upon written instructions, and all employees should have a session on this given by a competent person at least once, and preferably twice, every twelve months.

The general content of the training would provide for:

1 Action to take on finding a fire.
2 Action to take on hearing the fire alarm.
3 How to raise an alarm – location of alarm points, internal alarm telephones, alarm indicator panels, telephone-operator call procedure.
4 How to call the fire brigade.
5 Location and use of fire-fighting equipment.
6 Knowing of escape routes.
7 Appreciation of the value of fire doors, coupled with the need to close all doors at time of fire or at sounding of fire alarm.
8 Stopping of machines and processes and isolating power supplies in case of alarm.
9 Manner of evacuating premises and, in the case of shops, assistance and guidance to be given to customers.

Employees with supervisory or operational fire emergency duties should be specially instructed in the roles that they have to fulfil. Engineering and maintenance staff, telephonists, security staff, departmental heads or floor supervisors (shops), and safety officers will come into this category, and, of course, a co-ordinator or fire officer should be nominated.

The necessity to keep records requires the following to be noted:

1 date of instruction or exercise
2 duration
3 name of instructor
4 name of trainees
5 nature of training.

## *FIRE DRILLS*

The question of fire drills is a difficult one. The object is praiseworthy and advocated by all responsible bodies. It ensures that everyone knows: how to leave in an orderly fashion without fuss and commotion which could engender panic; the alternative means of escape in the event of the obvious being blocked; and where to go to assemble for a roll call to check that no one has been trapped. But, and this is a major 'but', in many large industrial concerns such a drill would entail the closing down of machinery and processes which would have an effect on production perhaps equivalent in financial loss to that caused by a sizeable fire – the views of management in such instances are obvious! Where this applies, efforts should be made to carry out drills piecemeal by departments; the instructions on what should be done in an emergency should be reiterated at every opportunity and the printed directions prominently displayed (see Appendix 23).

A form of 'dry' or 'static' fire drill has been evolved and practised in the past by a number of firms to avoid the expense of a full-scale exercise. This is really no substitute for a genuine rehearsal of what would be needed to be done in an actual emergency; it is no longer to be recommended as a satisfactory alternative.

When a full-scale drill is to be carried out, the local fire brigade could, with advantage, be informed and their comments on the manner in which it is done would be valuable.

### *Fire wardens*

An adequate number of willing and responsible employees should be designated and trained as wardens so that there is representation in every department. The names of these should be shown on fire notices so that everyone is aware of their identity, can approach them with queries, and will respect their orders in the event of an emergency.

Notices marked 'fire warden' are recommended outside applicable offices.

Written instructions should be held by the wardens showing precisely their duties and areas of responsibility. Dry drill is useful for ensuring that these have been given and comprehended. The wardens should endeavour to keep an up-to-date list of the employees for whom they are responsible.

Checks should be made at frequent intervals on the movements of personnel so that any necessary replacement wardens can be appointed and trained.

## *CAUSES OF OUTBREAKS*

Knowing the main causes of fires is the first step in knowing how to prevent them and what to look for if there is suspicion that a fire is smouldering in a given area. No effort has been made in the list hereunder to place these in priority, obviously they will vary from industry to industry.

### *Electrical faults*

These can be of a variety of causes:

1 Overloading of circuits beyond capacity.
2 Short circuits due to wear or damage to insulation.
3 Careless maintenance.
4 Electrical equipment overheated due to ventilation failure or overloaded mechanically.

### *Heating appliances*

Portable heaters, gas and electric fires, stoves, open fires, oil burners, steam and hot water pipes, etc. All harmless if correctly installed and maintained but dangerous by proximity to flammable material, or by generation of sparks or breakdowns. Fixed fires and stoves should be on non-flammable bases.

### *Process dangers*

Usually accidental or due to carelessness in operation as safety from fire risk is taken into account in designing any process. Typical dangers:

1 Accidental ignition of liquids or flammable gases (including spillings of liquids).
2 Accidental overheating of substances under processing.
3 Flame failure in heating equipment causing explosion and fire.
4 Frictional heat and emission of sparks.
5 Breakdown in ventilation or cooling devices.
6 Chemical reaction getting out of hand causing explosion.

*Static electricity*

Particularly dangerous where solvent vapours are in use, all fixed machinery has to be bonded to earth and increased humidity will reduce the chance of an unexpected spark. Instruments are made to detect and measure the presence of static charges.

*Flammable dusts*

These are both an immediate explosive risk and a long-term one. In their finely divided form, and adequate concentration, certain metallic and carbonaceous dusts can form what is in effect an explosive cloud. The remedy is regular overhaul of dust extraction installations. Deposits of dust in roof voids can smoulder unobserved for lengthy periods before breaking into flame. In an enclosed area this can lead to a progressive build-up of heat which will cause a sudden flare-up over the entire area.

*Spontaneous combustion*

Certain substances, by decomposition or chemical reaction, gradually heat up to ignition point without any outside agent other than the presence of oxygen. Oily rags, sacks, and oil seeds are examples. These should be cleared regularly and, if they have to be retained, kept in fireproof containers.

*Rank carelessness*

Smoking heads the list here. Works regulations should be strictly enforced in prohibited areas. Wastepaper baskets are the most prolific source of fire in offices. Repair work with blow lamps and welding equipment can also leave smouldering material if care is not taken. Means of extinction should be to hand while the work is in progress and security staff should always be notified for special attention to be paid subsequently.

## *FIRE PATROLS*

A fire during working hours is unlikely to escape detection for more than a limited time and consequently has restricted opportunity to do real damage, as opposed to those which break out in the absence of the main body of staff. Even those in daytime are concentrated in the areas least frequently visited. The moral is obvious: fire patrols must be worked to a specific system which ensures that all parts are visited at regular intervals. This does not conflict with the principles of security patrolling and there is no reason why the two functions should not be combined with correctly trained personnel.

Again, the practice of employing aged watchmen is to be deplored; to be of any value at all for fire prevention, the individual must be in full possession of all his faculties and of sufficient intelligence to use his own initiative in deciding whether immediate action by him can prevent an outbreak developing or whether he should seek outside help. An incompetent person is more dangerous than no person at all and by his own carelessness he might even start fires.

Familiarity with the premises and with the processes carried on therein are essential assets of the fire patrol. For this reason, staff specially employed by the firm are preferable to engaging the services of an outside professional agency; these do compile a very detailed list of instructions for their employees to follow but there are unavoidable changes in those detailed to supervise various premises which must reduce the efficiency of their patrols. Nevertheless, where a firm is small, it may well be economic to engage the services of an agency to give periodic visits; the cost of these will vary appreciably with the area and the commitments of the particular agency in the near vicinity – competitive quotes should be obtained. It should be possible for several firms to reach agreement to jointly employ staff to patrol their combined premises – this would certainly reduce costs and improve the quality of the patrols – but agreement rarely seems to be reached in this manner.

During daytime the necessity for patrols should be a strictly limited one, other than to those areas where either there is a regular incidence of fires or where few people are employed. After working hours, however, a full inspection of the whole premises should be made as soon as possible. The object of this is not just to detect any incipient fires but also to eliminate factors which may lead to an outbreak. A full inspection should imply actually entering every room – a good sense of smell is as valuable as any in the detection of smouldering material. Clocking points can be incorporated and in cases of high risk, an insurance firm may insist on this, though a degree of flexibility should be insisted on in the interests of security. Where they are used, they should be sited at the far extremity of any area to be visited to make certain the whole is seen. Too often the point is placed beside the entrance door and the lazy or hurried patrol may go no further in.

A form of checklist to be borne in mind by a fire patrol should include:

1. All electric fires or heaters left on, other than those essentially needed, should be switched off.
2. Gas and electric cooking facilities should be checked off.
3. Plant running but not in use should be switched off, and checked if cooling down.
4. Heaters obstructed by overalls or other flammable materials left on them should be cleared, missing fireguards should be replaced.
5. Doors and windows, internal and external, should be closed – the external locked against intruders and the internal closed to prevent possible fire spread, and also to give indication of the presence of intruders if subsequently found open.
6. Flammable materials left near any source of heat should be moved to a safe position.
7. Any leakage of oil or other possibly flammable liquid should be investigated immediately.
8. Check that all fire-fighting equipment is present, serviceable, and unobstructed; that access is available to all hydrants; and that fire alarm points are intact.
9. Smouldering fires due to electrical shorts or cigarette ends are not existent.
10. Any naked flames should be extinguished and soldering irons checked that they are disconnected.
11. Recognized avenues of access for the fire brigade must be unimpeded.
12. Sprinkler heads are not obstructed by piled goods.

Inspection during a patrol should not be confined to floor level – ceilings and roofs are equally important as much electrical cabling and ventilation conduits are sited in the roof and the first indication of trouble can be when fire breaks through the exterior of the roof.

Where automatic processes continue to function during non-working hours, or heating devices have to be switched on at given times, details of dial readings that have to be checked and other actions that have to be carried out should be in printed form to prevent mistakes. Though not precisely a fire problem, where a computer installation has to be visited the patrol should be conversant with the room temperature that is specified – overheating, without any suggestion of fire, can cause massive damage to tapes if the ventilation and air temperature control fails.

Finally the exterior of premises should not be neglected, particularly where timber for case-making is stacked or tarpaulins or sacks are stored or waste material has accumulated. In cold weather there is a temptation for workers to use open braziers for warmth; these are doubly dangerous if left burning when work has finished – a sudden gust of wind can generate a swirl of sparks. There is the ever-present danger from irresponsible children or vandals and perimeter fencing inspection has a fire as well as security purpose. Flammable goods should be stacked a safe distance from fences to prevent accidental or deliberate fires being started by outsiders.

It is a waste of time simply correcting risks which result from carelessness; a report should be made of each of these so that they can be drawn to the notice of those responsible to prevent recurrence.

## *GENERAL FIRE PRECAUTIONS*

In conclusion, a checklist of general precautions, other than those concerning equipment and patrolling, and perhaps under a form of good housekeeping heading:

1 Clearly demarcate areas where smoking is not permitted.
2 Provide adequate non-combustible ash trays where smoking is permitted.
3 Enforce a period, say half an hour, of non-smoking before the end of each working day everywhere in buildings.
4 Where cleaners are employed in offices each evening, make the emptying of wastepaper baskets a priority. Use metal bins instead of the conventional baskets.
5 Provide metal lockers for clothing and overalls to counteract the cigarette or pipe left in a pocket.
6 Have a regular and frequent system of floor sweeping and waste collection and removal.
7 Store collected waste at a safe distance from buildings and either dispose of it in an incinerator or have it removed by outside contractors.
8 Oily rags and items that might be the subject of spontaneous combustion should be collected separately.
9 Try to convince the head of each department that he should be the last to leave and to give a quick look round to see that all is in order before he does.

It might well be thought that the most difficult matter has been left to the last item of these precautions and indeed, this is so. The whole essence of fire prevention is that of educating staff at all levels to the understanding that fire can kill, can deprive of livelihood, and, whilst it is a good servant, it is an unreliable and treacherous one which needs constant supervision. Once this is instilled in everyone its danger is reduced to a minimum.

Appendix 25 lists a synopsis of the main Acts and Regulations relating to fire.

## *ADDITIONAL READING*

Any reader who requires more 'in depth' information on fire precautions is referred to *Practical Fire Precautions* by G. W. Underdown (Gower) and all publications of the Fire Protection Association, Aldermary House, Queen Street, London EC4N 1TJ.

# 31

# *Accident prevention*

Employers and occupiers of factories and premises, where any kind of work is carried out, have an obligation not to create unnecessary risks or to expose employees and others to danger of injury or damage to their health. Society through various Acts of Parliament lays down minimum standards of safety which must be attained by industry.

It is not unusual for the chief security officer of a small or even a medium-sized factory to combine a safety function with that of security, but for rank and file members of staff it is a purely ancillary matter in which the 24-hour daily cover that they give provides a potential source of surveillance of value to their employers.

Normal risks and incidents will be dealt with by the safety officer, it will only be those rising outside normal working hours which should receive attention as soon as possible to prevent injury or damage which affect the security officer. There must be no suggestion of interference in what is a responsibility of specially-trained staff – security's role is a purely helpful one, consistent with the overall mandate to protect the interest of employers and employees whenever possible. However, it should be realized that industrial fatalities have approximated to 500 per annum over a period of years and any action that can reduce that figure should be taken even if it causes embarrassment to a specialist who should have foreseen the particular danger.

## HOW THE SECURITY OFFICER CAN HELP

It would be unrealistic to suggest that security staff should be given special training in this subject except in very exceptional circumstances where recurrent risks may arise. Nevertheless, there are a number of things which may be temporarily overlooked by personnel intent on production where commonsense observation by a patrolling officer may be utilized to advantage. These, when observed, should be drawn to the attention of the supervisor within whose jurisdiction they occur; where repairs are necessary, it is his responsibility to requisition the work and then ensure that it is done as expeditiously as possible.

If a security officer sees what he considers to be unsafe practices, again he should draw them to the attention of the supervisor for his action – he should not approach the worker. It might be that a procedure that appears unsafe to the uninitiated is one which is acceptable. In all cases, a note should be left for the safety officer and a record made in the occurrence book for the information of other staff, to prevent duplication of reports and to ensure that attention is in fact given. No doubt items will be reported that are already the subject of action, or are known and catered for, but the residual balance will justify the practice.

Items which should attract the attention of a patrolling security officer are perhaps best given in the form of a checklist:

1. Patches of oil or grease which constitute a danger on roads or footways.
2. Protruding slabs or cavities in footways, broken tiles or holes in tiled floors, loose floorboards.
3. Defective lighting over staircases or any place where people may have access after dark.
4. Defective stair-treads or where the stair edges are badly worn or broken.
5. Broken or defective handrails.
6. Damaged ladders.
7. Obstruction of gangways, fire points, and exits.
8. Dangerous stacking of materials.
9. Leaking valves, joints, etc.
10. Repeated dangerous parking of vehicles.
11. Reckless or dangerous driving of vehicles.
12. Unauthorized riding on fork-lift trucks etc.
13. Failure to use protective equipment in dangerous areas, where specific requirements are laid down in factory rules.
14. Leaving unattended loads suspended on overhead cranes.

15 Horseplay by employees, anywhere on the premises but especially amongst machinery.
16 Deliberate interference with anything provided for first aid, welfare, fire or safety purposes.
17 Unauthorized personnel interfering with electric services and switch-gear.
18 Contraventions of 'no smoking' regulations.

The foregoing is only a sample list, it should be reiterated that anything which might adversely affect the well-being of employees should be commented on, if observed by the security officer in the course of his patrol. It must be stressed that these are matters in which he himself should not take immediate action except in emergency but should inform those with direct responsibility to do so.

## *LABORATORY TESTS AND EXPERIMENTS*

In many industries, and frequently under laboratory testing conditions, machinery or equipment is left running and unattended. This places the patrolling security officer in something of a quandary if he has received no prior notification. If the running is unintentional he has a potential safety/fire hazard coupled with power and other wastage. If intentional and there is a breakdown he would be bereft of instructions; if he uses his discretion and switches off that which should continue to operate he could cause delay, spoil a longstanding experiment or interfere with the production programme.

All such usages, of course, should be notified and logged but if this is asking too much the use of a standard informatory notice attached to the equipment or machinery should be introduced. A recommended printed form which covers all the contingencies is shown in Figure 30.1.

## *HEALTH AND SAFETY AT WORK ACT 1974*

The chief security officer, irrespective of whether he has a specific responsibility for safety, should have some knowledge of this Act. Indeed he should make sure that his own staff are conversant with the more important of its provisions.

FRONT

Operating overnight
and
outside normal working hours

| | |
|---|---|
| Apparatus | |
| Running conditions | |
| Instructions and/ or special hazards | |

Action to be taken in event of supply failure

| | |
|---|---|
| Water | |
| Gas cylinder | |
| Electricity | |
| Air | |

Contact (in emergency only)

| | | | | | |
|---|---|---|---|---|---|
| Name | | Telephone | | Date | |
| Address | | | | Signed | |

BACK

Instructions

1. Essential instructions must be clearly written, in ink, on the reverse side of this card
2. This card must be displayed in a prominent position on or near the equipment left running
3. The location of the equipment left running must be entered in the Safe Running Permit book held by the Laboratory Safety Officer. One entry in this book will cover a maximum period of one month
4. Equipment found running which is not recorded will be promptly shut down
5. Please remove this card when apparatus is not running

*Fig. 30.1*

### *The Health and Safety Commission, and the Health and Safety Executive*

The Act brings responsibility for all the main safety and health statutes under a single authority, and there is a complete reorganization of the existing structure of enforcement. The Commission is essentially a policy body setting the pattern of legislation, approving codes of practice, preparing regulations and advising generally. The Executive is the operational and enforcing arm and provides a technical and advisory service to both sides of industry and commerce.

## *LEGISLATION*

The relevant Acts will continue in force until replaced by 'agreed' codes of practice and regulations. Codes may be prepared by government departments, employers' organizations, and trade unions and more probably a combination of all, including independent professional bodies. The code must be approved by the Commission and then passed to the Secretary of State before it is accepted as a standard under the Act. It does not then impose a general legal obligation, but it is admissible in evidence in any prosecution for failing to comply with the statutory requirement. The offender has to prove that he has used a better or equally effective method than that mentioned in the code (s. 17).

The main objectives of the Act (s. 1) are:

1 Securing the health and safety and welfare of persons at work.
2 Protecting the general public against risks to health or safety arising out of or in connection with the activities of persons at work.
3 Exercising control of the acquisition, keeping, and use of explosives and highly flammable or otherwise dangerous substances.
4 Controlling the discharge into the atmosphere of noxious or offensive substances.

This is an Act with penal sections and duties laid on both employers and employees. Fortunately, it contains a saving phase repeatedly employed – 'so far as is reasonably practicable'.

*So far as it is reasonably practicable* it shall be the duty of every employer to ensure the health, safety, and welfare at work of all his employees – a duty which in particular includes (s. 2):

1 Provision and maintenance of risk-free plant and systems of work.

2 Safety and absence of risk to health in use, handling, storage, and transport of materials.
3 Provision of information, instruction, training, and supervision to realize the objectives of the Act.
4 Provision and maintenance of a place of work with means of access and egress all of which are safe and without health risks.
5 Provision of a working environment without risk to health and adequate as regards facilities and welfare arrangements.

The words 'reasonably' and 'practicable' are defined so as to provide a defence that nothing more could be done to satisfy the duty or requirement or that there was no better way – the onus of proving this is on the accused (s. 40).

A Court of Appeal case is important to employers – *White* v *Holbrook Precision Castings Ltd* 1985 IRLR 215 – which held that where work involves risks which are not commonly known, other than to the employer, he must make these known to an applicant so that person can make his own decision as to whether he is willing to accept them. This opens an avenue for injury compensation claims which, though not firmly based, may be exploited successfully on the grounds that the employer has not carried out this obligation.

The employer's duty is extended beyond his own employees to members of the public (s. 3), and the same phrase 'reasonably practicable' is also applied:

> The undertaking has to be conducted in such a way that those who may be affected thereby are not exposed to risks to their health and safety.

Though not particularly of interest to security personnel, the same obligation is placed upon self-employed persons in the conduct of their business, for example those working in the construction industry.

From a security point of view perhaps the main emphasis will lie on the fact that the employer will be responsible for ensuring that contractors and visitors, either legal or illegal, are not exposed to risks arising out of or in connection with the work carried on by the employer. This has been the basis of some civil actions in the past, for example entering by children and their subsequent injury by some existent dangerous circumstance.

### *General duties of employees*

There is a duty (s. 7) laid upon all employees whilst at work to take reasonable care of themselves and other persons who may be affected by their acts or omissions, plus a specific obligation to co-operate so far as is necessary to enable the employer or others to comply with any duty or requirement laid upon them by or under the statute. A previous piece of legislation is re-enacted in s. 8:

> No person shall intentionally or recklessly interfere with or misuse anything provided in the interests of health, safety and welfare . . .

The security staff may well find themselves reporting misbehaviour by employees which infringes those duties.

*Enforcement*

The inspectorate now comprises a number of experts in different fields whose work is co-ordinated through the Executive, so that factories and other premises may have visits by several different technically-qualified officials.

*Powers of inspectors*

These are given under s. 20 of the Act and include:

1 Power to enter at any reasonable time any premises which he has reason to believe that it is necessary to enter for the purpose of carrying into effect any of his legal responsibilities.
2 If he has reasonable cause to think that he might be obstructed, he can ask assistance from the police.
3 He can take with him any authorized expert assistance that he requires, together with equipment and materials for measurements, photographs, recordings, etc., that are necessary for any purpose connected with the reason for the power of entry being exercised.
4 He is not restricted in what he examines inside the premises and can investigate any activity going on.
5 He can order that any of the premises he enters, any part of them, any machinery, etc., shall be left undisturbed for as long as is reasonably necessary for the carrying out of his purposes.
6 He can take samples of articles and substances present on the premises, or make atmospheric tests therein or in the vicinity. (If he does take such samples he must leave a proportion of each with a responsible person.)
7 If he thinks an article or substance could cause danger to health or safety, he can have it dismantled or tested. He can take possession of it for as long as is necessary to examine it and to ensure that it is not tampered with – also to ensure that it is available for use as evidence if necessary.
8 He can require any person to give information relevant to an investigation or examination and to answer any questions that

the inspector thinks fit – also to sign a declaration of the truth of what he says.

9 He can require any person to afford him facilities and assistance within that person's control or responsibilities that are necessary to enable the inspector to exercise his powers.

10 He can demand the production of books or documents required by statute or any others which it is necessary for him to see for examination or investigation.

These are very extensive powers. Apart from the legal requirement and moral necessity to co-operate it would be most unwise to do anything that might obstruct in any way. However, an inspector is appointed in writing by the enforcing authority; so if there is the slightest doubt of identity he can be required by security to produce a copy of his instrument of appointment.

*'Improvement' and 'prohibition' notices*

If an inspector is of the opinion that a relevant statutory provision is being contravened, he may serve an 'improvement notice' requiring remedy of that contravention in a stated period of time. If this is not done, prosecution can follow with the accompanying compulsory stoppage of the activity concerned with the contravention.

If an activity involves the risk of serious personal injury, the inspector may serve a 'prohibition notice' forthwith. This may direct the activity to stop at once or be effective as from a time specified in the notice.

It is thought in police circles that these powers may be invoked when a firm resolutely refuses to accept police advice where there is a total lack of protection afforded to employees when dispensing wages. In such a case the police would inform the inspector who, if he agreed with their appreciation of the circumstances, could probably serve an improvement notice (see also p. 232).

Any appeals against either of these notices must be to an industrial tribunal, not a Magistrates Court. Failure to comply with either notice makes the person upon whom it has been served liable to prosecution; indeed failure to comply with the prohibition notice can lead to imprisonment.

*Penalties*

Prosecution can be by summons in a Magistrates Court, or a Sheriff's Court in Scotland, or may be on an indictment. The maximum fine on a summary conviction for most offences is £2000; but there is no limit to a fine on indictment, and imprisonment for up to two years can be imposed in certain cases.

The main penal section of the Act is s. 33, which caters for failure to discharge duties, misuse or interference by employees, contraventions of requirements, failure to comply with notices, falsification of records, etc.

*Obstruction*

Under the Criminal Law Act 1977 penalties for some of the offences deemed most culpable are raised to £1,000. They are:

1 Intentionally obstructing any person in the exercise of his powers under the Act.
2 Preventing any person from appearing before an inspector or from answering questions to which the inspector requires answers.
3 Intentionally obstructing an inspector in his duties.
4 Falsely pretending to be an inspector.

It should be noted that a person to whom duties *have been delegated* may be liable to prosecution if he fails to comply, thereby causing an offence to be committed (s. 36).

*General comment*

This is a most far-reaching Act. In addition to the matters mentioned it imposes duties on manufacturers and upon firms to produce statements of policy, and also provides for the setting up of safety committees, safety representatives, etc. and for wide consultation and co-operation. A senior security officer should take an intelligent interest in any codes of practice which affect the premises in which he works and should take note of any policy statements which might affect the working of his department.

## *NOTIFICATION AND RECORDING OF ACCIDENTS*

Certain accidents in factories have by law to be reported to the factory inspector forthwith. These are those which:

1 Cause loss of the life of a person employed in the factory.
2 Disable any such person for more than three days from earning full wages at the work at which he was employed.

The notification has to be in writing on, and in accordance with, a prescribed form (Form 43). Those under (2) are referred to as 'lost time' or 'reportable' accidents and particulars of these are amongst entries which have to be made in the general register for the factory.

### *Dangerous occurrences*

Whether injury has occurred or not, the inspector also requires notification of certain dangerous occurrences as specified in various Acts and Regulations; they include explosions of boilers; fire and explosion involving petroleum spirit; explosions of explosive materials or fires in premises where they are kept; bursting of revolving wheels; collapse or failure of various lifting appliances; explosion or fire due to ignition of dust, gas, vapour or celluloid, or from electrical failure; explosion of compressed gas containers, etc.

### *'Frequency rate'*

This term is used in conjunction with 'lost time' accidents in the compilation of statistics: simply, it is the number of lost time accidents per 100 000 man hours worked in the department or works being considered. Monthly, quarterly and annual figures are usually produced; a monthly frequency rate calculation, for example, would be: number of lost time accidents × 100 000 divided by the man hours worked in the month. An 'all injury incidence rate' is referred to – this is an inclusive figure of reportable and non-reportable injuries shown in statistical returns; also 'severity rate' – indicative of the average number of hours lost per injury. In some industries an 'accident incidence rate' is used calculated upon the number of injuries per 1000 employees.

These records facilitate performance comparisons between similar factories, departments and indeed industries. If prepared in graph form for monthly periods, they will show at a glance the adverse or beneficial effect of new safety techniques or of the introduction of new machinery or processes. Comprehensive records will also focus attention upon areas where insufficient attention is being paid to safety of employees.

## *ACTION BY SECURITY STAFF AT ACCIDENTS*

Security attendance at the scene of an accident will almost invariably be to render first aid. Whilst this is the first essential, the cause of the accident should be ascertained from the injured person, if possible, whilst treatment is going on. The knowledge of this is desirable from a medical point of view, and it has two further advantages; first, it may prevent an immediate recurrence to someone else, second, it might obviate a subsequent unfounded claim for compensation on the grounds of negligence by the employer.

The practice of many firms in having a form which must be completed when treatment is given, or when an accident, not apparently requiring treatment, is reported, is worthy of being copied. This enables information to be passed to the safety section immediately, facilitates their records and, if the form is correctly made out, gives a basis for action to remedy possible defects immediately. It also compels the supervisory staff to take cognizance of the accident and consider preventive action. A sample form is shown in Appendix 26.

Security officers should familiarize themselves with the manner of reporting and recording in use in the relevant department of their particular firm as part of their general knowledge of works techniques which they themselves might have to use. They should also familiarize themselves with regulations which have a safety application. One of their duties is reporting contraventions of the regulations; in doing so they are not only upholding the authority of the management but are doing a service to the individual concerned.

## *ACCIDENT INVESTIGATION*

Firms will almost always set up their own internal investigation of serious accidents irrespective of what may be intended by the factory inspectorate or other bodies. A chief security officer may find himself co-opted on to such a committee of enquiry, especially if his own staff are involved or there has been a fire. He must remember that the purpose is primarily to establish causation and decide future preventive action, not to establish who is to blame – that will be done elsewhere – and pitch the questions he wishes to ask accordingly. It would not be unusual for a competent officer to be asked to make out a draft report for the chairman.

## *FIRST AID*

Even if management has not incorporated first aid as a part of its security officer's job description, everyone who undertakes that type of employment should consider he has an obligation to acquire proficiency in it. The wearer of a uniform which is associated with both authority and the giving of assistance will be expected to act with those characteristics in an emergency, and injury is an emergency. The efficient handling of casualties can lead to improved goodwill and co-operation from the general workforce. Superficial knowledge can be dangerous; so far as legislation is concerned, and for all practical purposes, anyone who is termed a person with first-aid responsibilities must be a holder of a current certificate from a recognized organization.

## *LEGAL REQUIREMENTS FOR FIRST AID*

The Health and Safety (First Aid) Regulations 1981 which came into operation on 1 July 1982 define 'first aid' as follows:

1 In cases where a person will need help from a medical practitioner or nurse, treatment for the purpose of preserving life and minimizing the consequences of injury and illness until such help is obtained.
2 Treatment of minor injuries which would otherwise receive no treatment or which do not need treatment by a medical practitioner or nurse.

The previous legislation, Factories Act 1961, s. 62, and Offices, Shops and Railway Premises Act, s. 24, is repealed and specific obligations are laid upon employers. They must inform employees of the arrangements that have been made in connection with the provision of first aid, including location of equipment, facilities and personnel; they are to provide, or ensure that there are provided, adequate and appropriate facilities and personnel who have training and qualifications to the requirements of the Health and Safety at Work Executive who also have the power to grant exemptions. Facilities have to be such as to provide 'on the spot' treatment within three minutes of its being needed.

### *First-aid boxes*

Adequate first-aid facilities in the form of first-aid boxes must be provided on the scale of one for every 150 persons who are employed at any one time and an additional box or cupboard for every additional 150 persons working or a fraction of 150. This means that if 160 persons are employed on a shift, two first-aid boxes must be available. Where there is a surgery or first-aid room where people can be treated immediately the requirements with regard to first-aid boxes may be waived and the factory be given a certificate of exemption by the factory inspector.

Minimum contents for a first-aid box, where more than 50 persons are employed at any one time are laid down by the First Aid Boxes in Factories Order 1959; these include small, medium and large sterilized dressings, adhesive wool dressings, triangular bandages, eye pads, rubber bandages or pressure bandages, adhesive plaster, cotton wool, etc. These are minimum contents; the mandatory contents should be studied and added to as advised by a firm's medical officer, having regard to the special requirements of the industry or offices.

*First aiders*

On all occasions where more than 50 persons are employed in a factory, the first-aid box or cupboard must be under the charge of a responsible person trained in giving first-aid treatment. This person must always be readily available during working hours and a notice must be displayed for the information of employees, quoting the name of the trained person or persons available. In the case of fuel storage premises in the open, with a greater element of danger, a notice bearing the same information must be given to each employee. Where office buildings form part of a factory or an engineering site of any kind they will be considered as part of a factory for numerical purposes in respect of requirements.

It is the practice in many factories to nominate shop floor or clerical employees as first aiders. Indeed many firms have built up highly competent and competitive first-aid teams which have entered national competitions. Where shift work is a regular feature ensuring that an adequate number of trained employees are always available can cause administrative headaches. If it is a requirement that all security staff are trained to acceptable standards they can take over the responsibility and their duties make them always available. A security department which is always manned and has telephone communication with the outside will always be a place to which persons who are hurt or in need of assistance will gravitate. For this reason alone there should be a fully-equipped first-aid box in the security department and staff trained to use its contents.

*Training*

Approved training is done by three recognized societies: the St John Ambulance, the St Andrew's Ambulance Association, and the British Red Cross Society – all three now use the same *First Aid Manual*. Certificates granted by them are for a duration of three years, after which they lapse if the holder does not undergo a course of revisionary training and a further examination. Where security personnel are concerned they are likely to be more regular practitioners than the average first aider and a revisionary course every two years should be regarded as a minimum requirement. First aid is a practical subject and it requires practical training and instruction – not just the reading of the *First Aid Manual*. It must be borne in mind that first aiders no matter how well they are qualified, must never take upon themselves the responsibilities or duties of a doctor.

### *The* First Aid Manual

This is issued by the St John Ambulance Association and is regularly revised and very comprehensive. Slight variations in techniques are progressively introduced with experience and an up-to-date copy should always be held available for reference by the security staff. In view of the availability of such an authority there is no object in giving other than essential points for the treatment of injured employees.

## *FUNDAMENTAL PRINCIPLES OF FIRST AID*

1 *Prevent further injury* – This can be done by either removing the source of danger, or removing the casualty, for example:

   (a) switching off electrical current;
   (b) stopping machinery in motion;
   (c) moving casualty away from fire or falling debris.

2 *Restrict movement* – Do not move a badly injured person unless it is absolutely necessary to prevent further injury. If he has internal injuries or a broken spine, unnecessary movement could cripple him permanently or kill him; do not move him before the arrival of fully-trained and fully-equipped personnel, nor allow anyone else to do so. Do not give him anything to drink if there is any suspicion of internal injuries, even if he is fully conscious.
3 *Artificial respiration* – If there are no signs that the casualty is breathing but if there is any possibility at all that he is alive artificial respiration must be attempted. Mouth to mouth respiration is now accepted as the easiest and most effective method to apply. Unless given artificial respiration a person who has stopped breathing may die within three minutes of doing so.
4 *Stop severe bleeding* – If the rate of loss of blood is obviously causing danger to life apply a clean dressing pad or anything sterile and press firmly on the wound. Do not waste time cutting clothing

unless this is essential and, in the absence of sterilized materials, with blood gushing, stop the bleeding with anything which is available.

5 *Send for assistance* – Get assistance by any means possible which does not involve your having to leave the injured person to his detriment; get a reliable person to telephone for help. The message he sends should mention the precise position, some idea of the extent of the injuries, and an indication as to what has happened.
6 *Alleviate pain* – With the minimum of movement, manipulate the casualty into a comfortable resting position (see paragraph 2 above). Talk quietly to him in a confident and reassuring manner until skilled help arrives. Warn onlookers to keep well away, leaving a passage for a doctor or ambulance. Do not discuss the accident with the injured person other than to ascertain how it happened, and do not discuss it with other people in his hearing.
7 *Minimize effects of shock* – If there is to be appreciable delay in the arrival of skilled help, do not forget that shock can have a serious delayed effect. This can be recognized in a casualty by a pale, cold, clammy skin, and with a rapid pulse and breathing rate. The treatment: keep the casualty warm but without overheating. The object is to conserve his body heat. Loosen his clothing and reassure him.
8 *Assist skilled help* – Give those taking the casualty away as much information as you can about how he received his injuries. If he is unconscious tell them anything of value he may have said before he lost consciousness.

### *Movement of a severely injured person*

Unless movement is necessary to prevent further injuries do not change the position of the casualty until the nature of his injuries are known. If he has to be moved do so by moving his body lengthwise, preferably by supporting and pulling from the shoulders – not sideways. Slip a blanket or other material under him so that he can ride upon it without direct force being applied to his body. If he has to be lifted do not 'jack-knife' him by lifting heels and head only. Obtain assistance and support each part of his body by persons kneeling alternately at either side and lifting his body in a straight line. If he then has to be transported, for any reason at all, endeavour to do so on as rigid a stretcher as can be improvised, fastening him to it. Several forms of improvisation are laid down in the St John Ambulance *First Aid Manual*.

## *CRIMINAL INJURIES*

In most cases the facts will be obvious but if there is an injury for which there is no apparent logical explanation, the possibility of it being the result of violence from another should not be overlooked. Facial or head injuries may leave the sufferer unable to tell how he got them so look around for the presence of a possible weapon which has been used. A mental note of persons in the vicinity who may have seen something should be made. In all instances of serious injury, whether criminal or accidental, it is advisable to seal off the immediate area to prevent interference by unauthorized persons pending examination by those appropriately qualified.

# 32

# *Emergency plans*

## *EMERGENCY PLANS FOR MAJOR INCIDENTS*

During the past few years there has been a drastic revision of attitudes towards planning in anticipation of accidents occurring on industrial or commercial premises of such a magnitude as to be termed 'disasters'. The Flixborough and subsequent incidents have had the effect of inducing many reluctant managements to draft contingency schemes and provide necessary equipment to implement them. Pressure from public opinion, the Health and Safety at Work Act, and fire and police sources have expedited action where risks are obvious and the consequences likely to extend beyond the perimeter of premises.

The main industries at risk are those which use explosives, chemicals, and high concentrations of fuel, but even the steel industry has been a disaster sufferer. It is incumbent upon any firm to review the potential threat that its operations hold for employees, customers, and people living in surrounding areas. 'Negligence' is actionable; and whilst the prime responsibility rests with senior management, if security has responsibilities in preplanning, as it should, and does not bring hazards to notice, criticism will quickly work downwards.

Only a small minority of concerns will carry potential dangers that can give rise to a 'major disaster', which is defined as follows:

> A serious disruption to life arising with little or no warning, causing or threatening death or serious injury to such numbers of persons in excess of those which can be dealt with by the public services

operating under normal conditions at that time, which calls therefore for special mobilization and organization of those services.

Major fire, explosion, poison gas release, and abrupt flooding are the types of things envisaged. There can of course be the totally unforeseen emergency caused by a crashed aircraft.

### *Emergency controller*

The views and experience of a number of persons must be taken into account during preplanning, but it is essential that when an emergency occurs there should be no split authority and the controller is clearly designated. This will not necessarily be the senior executive on the site as his abilities may lie in an entirely different field from the practical co-ordinating role required. A works manager, engineering manager, or engineering director is the type of person best nominated, but the choice will depend upon the individual and his personality. Protocol or status-consciousness must not dictate in this post. In turn there must be nominated deputies and an accepted department from which a temporary controller will assume authority for out-of-hours incidents pending the controller's arrival.

### *Planning*

Ideally this should implicate as many departments as reasonable to ensure that no factors are overlooked. Obviously the security, fire, and safety departments must be involved, together with the service departments controlling gas, electricity, and power installations. Police and fire brigade authorities will be only too pleased to assist and join in any conferences that are necessary.

The first step is to establish precisely what emergencies can occur. This requires methodical evaluation of activity, environment, and equipment.

Having drawn up a list of possibilities it is next necessary to consider likely consequences of the more serious possible incidents, reviewing potential damage and casualties both internally and externally. Following this come determining the radius likely to be affected, the consequences on access for relief services, and the possibilities of further snowballing incidents. Apart from generating a warning to alert employees and limit avoidable risks, two primary communications are essential: (a) to warn outside authorities that assistance is required, and (b) to expedite the utilization of resources that are immediately available.

Many firms will have willing manpower to deploy, site equipment that can be used, and skilled staff in the form of electricians, plumbers, engineers, and labour gangs who can do much to save life and avert damage pending the arrival of outside assistance, which will inevitably be subject to some delay in these days of traffic congestion. Moreover, these people's intimate knowledge of the premises and processes make them invaluable to the incoming help. It follows that there must be a plan to concentrate skilled internal personnel without panic and as quickly as possible. This may be done by a prearranged system of tannoy, siren sounding, light flashing, signals, etc., to cause automatic ringing-in to an information point for instructions. Many firms will already have such arrangements in anticipation of fire, and little variation may be needed to extend them. It is easy to say that expert help with specially trained personnel will soon be on the spot, but it should be remembered that many lives are saved by those who have basic first-aid knowledge and are prepared to use it promptly, or those who have fire training and can stop a small blaze from becoming a conflagration.

A point which is not always appreciated but must be understood from the outset is that the co-ordinating authority for outside assistance is the police. This has been agreed by all the other services that will be incoming, and many senior police officers have been specially trained for the purpose. Indeed specialized services of the type provided by mines rescue teams will only respond to a police notification. The first information to the police should be as concise and accurate a statement of what has occurred as can be given at that moment. It should specify the location, installations concerned, probable casualties, best routes of access, and services likely to be required. If there is any chance of special hazard for surrounding areas, this should be emphasized.

### *Control point*

The necessity for communication with outside resources presupposes that the means of doing so remain intact. It follows that a control point should be specified which has telephonic communication and is unlikely to be put out of action by any conceivable emergency – that is, in a safe vicinity and preferably on a direct access road from the surrounding area. In many instances where there is a security lodge or office, this is likely to meet the requirements, being on the exterior of premises or areas and having both internal and external telephone capability. If the site size merits it, serious consideration should be given to installing radio equipment.

The manning of the control point is a matter for the controller. He should ensure that he has no shortage of persons for the tasks that he can envisage, and they should be of a temperament not given to panic or excess emotion.

*Control point information*

Duplicated detailed plans of the premises must be held in the control room for emergency purposes and for the information of incoming services. They should clearly show access routes, installations with special hazards, fire hydrant points, and indeed anything which might help services with only limited knowledge of the area.

There should be action checklists for the various types of emergency envisaged, giving concise and clear instructions on what is to be done so that no essential steps are overlooked and errors and delay are minimized.

Lists of key internal personnel and VIPs who should be informed should be kept in conjunction with the action checklists and regularly updated. It is not unusual in these days of periodic reorganization and restructuring to find that responsibilities and people are regularly shuffled, and those who do keep their posts change addresses and telephone numbers. The lists should show both home and internal telephone numbers, with those of deputies designated in case of unavailability.

*Action on incident*

The police should be informed as soon as possible and internal warnings broadcast. The controller should go to the control point as quickly as possible with such staff as he has decided he needs. A log of the incident should be commenced; and if the security lodge is used, then provided the right calibre of person has been employed the keeping of this could be made the responsibility of the officer in charge. The controller should dispatch a deputy to the scene of the incident with instructions to keep in constant touch. The controller must not get himself engaged personally in the actual work going on at the incident, whatever may be the temptation to do so. He should control the flow of internal expertise and assistance to the point where they are best deployed by his designated deputies, and he should acquire as much information as possible from them to facilitate the work of the incoming services.

Prearranged messengers should assemble at the control point to cater for internal breakdown of communications and to provide guides for incoming services. Amongst the matters to be ascertained are the routes whereby the rescue service may safely reach the disaster scene, and the plans available should clearly show where sources of water supply are established. Another consideration for the controller is where incoming vehicles can be parked for maximum convenience. Apart from the immediate vicinity of the incident, parking areas will be required for fire service, ambulance, heavy rescue, and police vehicles that are in reserve or waiting to be called forward.

At the earliest opportunity a senior police officer will arrive to act as controller to co-ordinate all action. The internally designated controller should stay with this officer and give him every assistance possible.

Any major incident is going to cause casualties, possibly in large numbers, and priority will be given to the treatment of the injured, not to the removal of obviously dead bodies. It is therefore advisable to select a building which can be used as a temporary mortuary, unpleasant though this may be. The police will be able to contact a firm which has specially refrigerated vehicles which can come to an area of devastation at the shortest possible notice, and they will do this if necessary.

In the early stages of discussing the likely consequences of disasters there is merit in a managerial exercise to determine what the effect upon production will be and what alternatives can be used to offset the consequences of specific incidents. This will go beyond the immediate effect to consider subsequent complications like the layoff of manpower and the possible mobility of labour to allow for essential work to go on – a matter for the personnel or industrial relations department in conjunction with other departments.

### *External services*

The police service has prearranged links with the following services to assist:

1. Fire service
2. Ambulance service
3. Hospitals
4. WRVS
5. Local authority and public authorities
6. National Coal Board rescue teams
7. Woodland and Highland rescue teams
8. Ministry of Defence and armed forces rescue teams
9. Mine disposal units.

### *Evacuation*

Whilst stress has been laid on using the expertise of internal personnel, it must be realized that the majority of employees are likely to be an encumbrance rather than a help. For their own safety it is necessary to get them off the premises as quickly as possible. Evacuation schemes should already exist for fire and bomb threats. It is likely that these plans can be applied here to ensure that all personnel not specifically designated for rescue work and the like are got to a place of safety and recorded, so that there is no possibility of victims being left unsearched for.

### *Conclusion*

The foregoing can only be general guidelines having regard to the infinite variety of premises and circumstances that may be applicable. Like every other matter with which security principles are associated, often the greatest difficulty is in persuading senior management to turn its mind to the possibility that this is something which could happen to its unique firm. The saving grace in this instance is that the sanctions that can be applied if something does go wrong are nowadays obvious to all. To a senior security officer with misgivings – make sure you are not being unnecessarily alarmist, then commit your comments to writing so that the queries are ventilated and responsibility for not taking precautionary action is established.

The ideal result of detailed and sensible planning is that the emergency never happens – the logical consequence of evaluating and minimizing all the foreseeable risks.

## *LETTER BOMBS AND EXPLOSIVE OR INCENDIARY DEVICES*

The terrorist activity which began to spread in Northern Ireland nearly two decades ago has continued spasmodically to spill over into mainland Britain with bombs and incendiary devices, mainly in shopping areas, and attacks on selected individuals or targets. Palestinian, Iranian, Libyan and other bodies have been responsible for incidents in the UK but the terrorist disease is international and there is evidence of liaison and mutual support between some factions. It is not a situation which is showing any signs of improvement despite increasing co-operation between governments and it would be highly optimistic to expect any. The mere fact of doing trade with a particular country may be considered adequate by extremists to justify 'reprisals'. It is significant that of nearly 500 killings attributed to terrorists in 1984, almost one third of the victims belonged to the business community and if the news media accurately depict the current state of affairs there has been no diminution since then.

The past pattern of attack in the UK has been senseless and indiscriminate; immunity is a privilege that no employer can assume. It therefore remains essential to know the form that the threat to employees and the business itself may take and the countermeasures and precautions that are advisable.

For the terrorist, letter bombs represent the safest mode of attack as there is limited chance of their being traced back to the originator, who may even be in a different country. They are a danger both to the addressee and to the mailroom staff of any firm receiving them.

### *Letter bombs*

These have been mainly in the form of substantial envelopes or parcels containing paperbacked books delivered through normal postal services. They are likely to be in the form of flat letters weighing up to 4 ounces (120 g), or in packages the size of a conventional library book. The degree of caution advised in handling depends on the cumulative effect of points which give rise to suspicion.

Points which may make unfamiliar material received suspect:

1. The postmark – if foreign and not familiar.
2. The writing – which may have an 'un-English' appearance, lack literacy, or be crudely printed.
3. Name and address of sender (if shown) – if address differs from area of postmark.
4. 'Personal' or 'Private' letters addressed to senior management under the job title, for example managing director.
5. Weight – if excessive for size and apparent contents.
6. Weight distribution – if uneven, may indicate presence of batteries inside.
7. Grease marks – showing on the exterior of the wrapping from inside may indicate 'sweating' explosive.
8. Smell – some explosives have a smell of marzipan or almonds.
9. Abnormal fastening – sealing excessive for the type of package; if such an outer contains a similar inner wrapping, this may be a form of booby trap.
10. General – damaged envelopes which give sight of wire, batteries or fluid-filled plastic sachets should be left strictly alone; those that rattle or feel springy should be treated with caution; and, naturally, any ticking noise should be treated as a 'red' alert.

Where a conventional paperback book has been used in the making of a letter bomb, it is likely to be noticeably softer in the centre than at the edges.

If suspicions cannot be alleviated:

1 Do not try to open the letter/parcel or tamper with it. It has been made to withstand postal handling and is designed to function during a normal sequence of opening.
2 Do not put it in water or put anything on top of it.
3 Isolate it where it can do no harm, that is enclose it in a nest of sand bags but ensure it is in a position for easy visual inspection.
4 Open any windows and doors in the vicinity. Keep people away from it.
5 Inform the police and seek their guidance, give them full details of the letter/parcel, its markings and peculiarities which have led to suspicion.

*Points of note*

1 So far as is known, no letter/parcel bomb has been received which has borne a franking mark on the envelope or wrapping, *but* a franked stuck-on address label has been used on letter bombs originating on the Continent.
2 Spraying with an aerosol of the Boots painkilling type as used for sporting purposes or Holts 'Cold Start' (the first is the better) will make a manilla envelope or wrapping transparent to a degree which may allow identification of the contents sufficient to alleviate suspicion, if not to establish dangerous nature.
3 Conventional components include detonators, connecting wire and minute batteries; no device *yet* used has been known to have been activated when tested with a low-power metal detector or X-ray (a wide variety of hand-held metal detectors are commercially available). The military advise that if X-ray equipment is used, it should be operated by remote control.
4 The police have already dealt with innumerable suspect but innocuous letters/parcels and may be able to give immediate clearance, if provided with pertinent details, that is town of origin, size and shape, franking or other distinctive stamping.

*Explosive incendiary devices*

These can be encountered in almost any form. In retail premises small incendiary packets put in clothing pockets in cloakrooms or amongst inflammable materials have had the objective of starting fires which would obliterate their cause. Explosive types have been placed in dustbins, suitcases, biscuit tins or lodged inside cars; the weights used varying from trivial amounts up to several hundred kilograms.

The means of detonation seem to have been mainly battery operated after a clockwork timing delay – but electric detonators have been fired from the safety of a nearby vantage point and there is strong suspicion of radio control being used. As a precautionary measure, all personal radios and 'bleepers' should be switched off in the near vicinity of a suspect device.

If there is a suspicious object *do not* shield it off from view by sand bags etc. The bomb disposal people will want to give it a very full visual inspection before deciding on their course of action. (NB: in cases of doubt they are likely to blow it up wherever it is sited.)

*Hoaxers*

The bomb hoaxer is still with us, but at least there is now a specific offence (Criminal Law Act 1977, s. 51) with which he can be charged. Under s. 51(1):

> A person who:
> (a) places any article in any place whatsoever; or
> (b) dispatches any article by post, rail, or any other means whatsoever of sending articles from one place to another,

with the intention (in either case) of inducing in some other person a belief that it is likely to explode or ignite and thereby cause personal injury or damage to property or under s. 51(2):

> A person who communicates any information which he knows or believes to be false to another person with the intention of inducing in him or any other person a false belief that a bomb or other thing likely to explode or ignite is present in any place or location . . .

is liable to a penalty, on summary conviction, of imprisonment for up to three months and/or a fine of £1,000 and, on indictment, of imprisonment for up to five years. The possibility of claiming compensation for loss due to such calls should not be overlooked, though the persons responsible are likely to be 'men of straw'.

## *BOMB THREATS*

These can be made by direct telephone call, anonymous letter or notification by some means to police or newspaper offices to be passed on to the intended recipient. Those by telephone are by far the most numerous and lend themselves readily to the 'hoaxer'; they are sufficiently numerous that every firm should decide a policy and procedure to be followed.

*Instructions to telephone operators*

The receipt of a threat message by an operator who has not been instructed in how to treat such a contingency, could cause considerable alarm with subsequent confusion about the substance of the message.

Specific instructions to operators should be given – it is suggested that a pro forma questionnaire could be used (see Appendix 27).

*Policy*

Fundamentally, there are three possible alternatives:

1 to evacuate, and search before re-entry;
2 to search without evacuation
3 to ignore the message.

The decision must rest with the senior person available; amongst points to be considered are:

1 Nature of the call – apparent age of the caller, speech, attitude, general approach, etc.
2 Recent history of such threats, genuine or otherwise, locally and nationally.
3 Prevailing conditions of industrial tension, strikes and political unrest in the neighbourhood and particularly at the recipient's premises.
4 The implications and dangers of an evacuation.

In all instances, police and fire authorities should be informed immediately, whether an evacuation is to be ordered or not. As a good neighbourly gesture, adjoining firms should be told what is happening.

*Evacuation*

Communications must be such as to ensure that all personnel can be warned speedily and with a minimum of alarm without affecting areas not intended to be evacuated.

Based on American experience, a clear radius of 100 metres should be allowed from the threatened area, 200 metres if a car bomb is suspected. Assembly points for evacuation will not of necessity coincide with those used for fire drills. Car parks are definitely not acceptable for obvious reasons, and there are advantages in housing evacuated staff in substantial buildings – if of suitable size and outside the prescribed distance.

*Search*

1 Responsibility for search of premises lies with the occupiers. The police cannot be expected to accept this: though they may assist, they will be unfamiliar with either buildings or likely contents.
2 Bearing in mind the multiplicity of forms a bomb may take, it is the unusual object, not normally in the particular environment which is suspect. Again, the occupants are best qualified to identify.
3 Any search made must be methodical with areas designated to individuals to ensure the whole is covered and with a co-ordinator to ensure that this is done.
4 Unaccountable or suspect objects should not be interfered with – when such are found the police should be advised and they will then instigate any necessary action.

*In case of evacuation*

1 Persons who are instructed to get out must take their personal parcels, bags and other belongings to avoid complications during searching.
2 If the time limit given by the warning permits, there should be a quick search by supervisory staff and/or designated employees before the premises are vacated. A system should be devised to ensure everyone is out.
3 After the time limit of the threat has elapsed, a reasonable margin should be allowed before a search by security/supervisory personnel takes place and employees are allowed to re-enter.

*No evacuation*

A search should be made by security/supervisory personnel of likely 'planting' areas, that is entrances to buildings, cloakrooms and toilets, and perimeter of buildings with special attention to parked cars.

## *GENERAL PRECAUTIONS*

A general tightening of precautions in or around plant and offices can reduce opportunity for causing incidents and also have a by-product in minimizing other sources of loss. Some suggestions are shown in Appendix 28.

# 33

# *Extortion by kidnapping*

Extortion of cash or some other consideration has continued spasmodically during recent years. The incidents in the Middle East, particularly Lebanon, have attracted media coverage but only those which fall in the category of *cause célèbre* come to notice elsewhere. Add to that the fact that police in all countries, in the interest of kidnap victims and relatives, attempt to agree a news blackout with the Press, and the difficulty of presenting an accurate assessment of frequency becomes understandable. In the UK the first public intimation may be the appearance of an accused person before a court, but there is no reason to believe that the offence is yet a major risk in this country though the very rich who have high ransom potential may justifiably think it wise to take precautions. Possibly the quickest way of inducing a firm or private individual to part with money is to threaten the well-being of an essential member of that firm or a valued relative of the individual. However, 'kidnapping' of goods has emerged as an alternative – paintings have been stolen, not for their intrinsic merit, but for what can be extorted from the owner; the taking of the racehorse *Shergar* has every appearance of such an intention which went wrong.

The reaction to events in the UK has been creating the Taking of Hostages Act 1982 which provides for a maximum sentence of life imprisonment for anyone convicted who 'detains any person, and, in order to compel a State, International Government organization, or person to do or abstain from doing any act, threatens to kill, injure, or continue the detention . . .'. This chapter is set in the context of the UK conditions but the principles are of universal application and have been discussed at length with both victims and internationally-known security practitioners, experienced both in precautions and actual incidents.

To stress the degree of apprehension that can be caused by this threat – in the US some high-powered firms when recruiting senior staff offer as an 'executive perk' protection for the individual and his family.

Unfortunately, criminals in the UK tend to mimic their opposite numbers in the USA, where kidnapping is an accepted means of financial gain in criminal circles. Whilst the same is undoubtedly true in other countries, political gain is the objective of terrorist groups in Germany, Italy, the Netherlands, Argentina and the Middle East. The borderline between the two is distinctly blurred, criminals use political motivation as an excuse and cover, terrorists need cash as well as freed colleagues and publicity, and of course there is also the lunatic fringe. Future trends are difficult to forecast, but this is a risk where loss can be of very sizeable proportions, and it cannot now be entirely disregarded in any serious assessment of potential risks, particularly for overseas representatives in 'danger' countries. Whilst it is quite wrong to sensationalize hazards, as is fashionable in some service-supplying quarters, nevertheless any reasonable firm should take cognizance of its existence. A senior personnel executive receiving notes upon this subject said: 'Thanks very much. I will put them on file in case this sort of thing develops.' Though in many ways his attitude is understandable, nevertheless those notes should have been made available for his overseas-travelling colleagues to consider and adopt at their discretion.

## *ASSESSMENT OF EXECUTIVE KIDNAPPING RISK*

### *Motivation of kidnappers*

1 *Political* – wants publicity, wants concessions (for example, prisoner release), but may accept cash or equivalent, justifying demand by stating that proceeds will be given to poor; IRA type is to coerce relative to plant a vehicle containing bomb; may be used as cover for a group 2 objective.
2 *Criminal* – encouraged by success of group 1, demands cash; intention may be to acquire keys to premises or to coerce friends or relatives to carry out thefts; occasionally used as means to endeavour to escape arrest.

3 *Mentally ill* – minute group inspired by publicity given to other groups; often demands impossible conditions.

*Acceptance of the risk*

In the USA a firm which offers protection for fees reaching $1,500 a day for a single client promises only to minimize risks, not to eliminate them.

It is quite impossible to devise reasonable and satisfactory precautions to protect a person who is not known as a specific target and leads a normal routine life. Where definite threats are made against individuals they might accept restrictions on activity, but where simply travelling in a risk area they are apt to take the attitude of 'why should it be me' and behave accordingly.

A solution would appear to lie with insurers; firms who do not accept that there are danger areas will have to bear the consequences; insurance cover is available, but may have stringent conditions attached to it. Lloyds are the only recognized risk acceptors in the UK, and whilst they play the subject in 'low key' they will probably evaluate the special circumstances of each application. The IPSA have a direct interest in the problem and their insurance brokers, Messrs Darwin Clayton will assist on request.

At least one of the leading security companies will give advice to companies on precautionary measures that should be taken for the protection of executives travelling in recognized danger areas and governmental guidance notes have been issued; another, in addition to this, offers its services as a negotiating intermediary in an advisory capacity after a kidnapping has taken place. If enlisting any source of guidance on this subject, the question of competence yet again raises its head. To pay a 'consultant' who is reading from a standard questionnaire or mouthing official guidelines may be wasting money.

Representatives travelling in 'danger' areas must maintain a low profile. It is reported that the kidnapped contain a preponderance of persons who have featured in the public eye or have engaged in the normal sightseeing and social activities. If a competent local representative is available, consider the feasibility of briefing him fully and allowing him to do what is required with ready communication to headquarters for advice and instructions. In the late 1980s, a prime example of where that should be done is in Beirut.

## *GUIDELINES FOR FIRMS WHOSE SENIOR STAFF VISIT RISK AREAS*

1. A standard 'precautions' briefing should be prepared and given as routine to those visiting for business purposes. Key personnel should have the sense not to do so for pleasure.
2. Insurance cover should feature as a part of the planning of visits.
3. Hotel and travel bookings should be made in the name of the firm, not the individual. Arrangements should be made for the individual to be met at the airport on arrival and to be taken there on departure.
4. Select accommodation so that the need to travel to appointments is as small as possible – hotels preferably in a city centre.
5. Contact the Foreign Office for guidance in cases of doubt or where it is felt that there is real risk to the representative.
6. Instructions to representatives should include the observing of sensible precautions, for example:

   (a) Do not use casual taxis but engage cars from reputable agencies. If a series of visits to the same destination is needed, vary timing and route.
   (b) Keep to a minimum number of appointments and minimize time spent in transit between appointments.
   (c) Do not give press, radio, or television interviews unless these are essential for business purposes.
   (d) Do not take any steps which would attract attention to yourself.
   (e) Unless adequately accompanied and conveyed, do not attend social functions or go sightseeing in areas where there have been previous kidnappings.
   (f) Do not answer your hotel door to strangers, nor discuss your business or movements with hotel staff or drivers.
   (g) Do not form casual friendships with persons, either male or female, whom you do not know, or accept invitations from other than your hosts.
   (h) Avoid travelling during darkness hours and in remote areas either rural or urban.
   (i) Do not be a sightseer at accidents, disturbances, or demonstrations.
   (j) If you are of high status and an obvious potential target because of your firm's trading relationship with particular countries, establish contact with the local UK representative.
   (k) In short, realize that you are a potential target, use your eyes

accordingly, do not court publicity, and take precautions whenever you have the slightest doubt.

7 Have available at headquarters a photograph, a description, and personal details of the individual going to the target area in case of emergency.
8 Have in being a plan for dealing with a purported kidnap demand. This should include the questions to be asked of the 'contact' to establish whether the kidnapping is genuine and the prisoner alive. Nominate provisionally the panel of senior management to be convened to discuss and decide action and policy in respect of notification to family and to police authority.
9 Nominate a spokesman to deal with inevitable publicity and define precisely the discretion that will be allowed him.
10 For a kidnapping verified as genuine, have an identifying code to agree with those responsible, to negative spurious callers who (experience has shown) will endeavour to capitalize upon the situation.

## *PRECAUTIONARY MEASURES WHERE A RISK IS ASSESSED AS REAL*

These recommendations have been made for persons resident in the USA or temporarily resident in South American risk areas. Even if acceptable to the person under threat, it is difficult to visualize them being effective for more than a token period under UK conditions, even if a bodyguard were provided. It is conceivable, however, that a person with good reason to feel under direct IRA or Arab terrorist threat might find it advisable to conform.

The undermentioned are really consolidated abstracts of guidance primarily for whoever has been given responsibility for the safety of the VIP:

1 Compile descriptive file with photographs of executive and his family showing activities engaged in – for example, church attended, golf club, schools attended. Nicknames used and telephone numbers of possible usefulness should be included.
2 Make a complete survey of the executive's daily routine at home and work. Note anything that might constitute a health hazard if kidnapped (tablets etc.).
3 List travelling undertaken with normal routes – to work, to other sites, to conferences, etc. – and consider what variations will reduce risk opportunities. Note details of vehicles normally used by him and his family.

4 Try to dissuade him from regular activity where he is obviously open to the possibility of successful attack.
5 Survey the access available to the executive at work, and introduce any necessary filters against unwanted visitors. Check on general security of the area round his office, and consider installing an alarm button at his desk. Preplan external visits so that he is always accompanied to functions and uses the safest means of transport.
6 Apply extra screening to the recruitment of any of his immediate staff.
7 Apply the same measure of screening to any domestic staff employed at his home.
8 Check the physical security of his home, and apply conventional crime prevention principles to the security of the house. Install a burglar alarm with personal attack buttons if necessary; suggest a viewing lens in each door as an advantage to identify callers; fit external lighting; try to persuade him to keep a guard dog.
9 Brief those who work with him at his home or office on the suspected threat, and discuss desired precautions with them. Do not cushion family members but include them in the discussion.
10 Formulate a policy of regular contact to secretary and home if there are changes in routine movements.
11 Brief the executive on action to be taken if suspicious circumstances arise – for example, prowlers round house, following by suspect car, involvement in minor accident, doubtful telephone messages, unexpected invitations to functions.
12 Cater for family members, in particular children travelling to and from school, who should be escorted or travel in a regular taxi.
13 Advise general precautions such as avoiding unnecessary night travel or functions at night where there is risk which cannot be mitigated, keeping clear of demonstrations and accidents, and staying in known company socially as much as possible.
14 Brief the executive on what he should do if indeed picked up, and try to agree a predetermined code to be used in messages in this eventuality.
15 Above all, by means fair or foul convince him that he is a target and not to ask for trouble. The suggestion that he should take out a personal insurance in the interests of his family may prove the most cogent argument!

'Defensive' driving training is offered in courses run by a few commercial firms – this is targeted on frustrating attempts at in-transit kidnapping.

## *PRECAUTIONS WHEN USING MOTOR TRANSPORT*

Practically all advice given makes reference to precautions to be taken in connection with vehicles used. Many kidnappings have been transit matters where danger appreciation may have prevented the opportunity from arising. It is also indicated that booby trapping with explosives should be guarded against.

### *Choice and inspection of vehicle*

1 Abroad, use vehicles that do not attract attention and are locally a common type.
2 If a pool of cars is available, work the changes on those you drive.
3 Keep the car locked at all times when not in use – in a locked garage or supervised area if possible. Outside, put it under a street lamp.
4 If there is any reason to think that you may be the object of an explosive attack, check under car, under chassis, under bonnet and in boot before switching on ignition. Bonnet and boot locks will save time and reduce 'planting' places. If any object or interference causes suspicion, leave things alone, do not attempt to move car but call police.
5 Where a regular car is used, fit a vehicle alarm with siren – the latter with an internal operating switch like attack button.

### *When in transit*

6 Before leaving the safety of a building, look up and down the street to see that there is nothing out of the ordinary.
7 Keep doors and windows locked.
8 Vary times and routes of journeys, but keep to busy streets as much as possible.
9 Do not get too near the vehicle in front in case you want to take quick evasive action.
10 Keep your petrol tank full.
11 Have good mirrors for rear view.
12 Travel in company with another car if possible. Carry a reliable passenger, but under no circumstances pick up a stranger.
13 If you think that you are being followed, drive to the nearest police station.
14 Keep your eyes open, and accept that it *could* happen to you!

## *INSTRUCTIONS FOR CONTINGENCY OF ACTUAL KIDNAPPING*

Obviously, circumstances will dictate what may or may not be done by the person kidnapped to alleviate the situation, but the following guidelines are worth consideration:

1 Treatment and survival will depend as much on personal behaviour as on efforts made by the police and others to free.
2 In the face of weapons or physical disadvantage and without hope of assistance, do not struggle as kidnappers are under stress and their reaction is unpredictable.
3 Try to stay calm and appear phlegmatic about the occurrence. Work on establishing an atmosphere of friendship with your captors, with as much good nature and humour on your part as can be summoned. Certainly do not insult or provoke them.
4 Play down your importance, and disclaim knowledge of the action likely to be taken on your behalf.
5 Try to remember times, distances, and directions travelled together with any identifying noises or other factors to establish location.
6 Memorize descriptions and peculiarities of the captors, with any special knowledge that they display about your business and movements which may account for your abduction.
7 You may be asked to speak or record messages to prove that you are alive and endangered. If you can safely use a predetermined code, do so. If not, consider the probability that you will be asked to speak and think what you may say, again safely, which may assist the searchers.

### *Recording messages*

A pro forma of the type used conventionally for bomb threat calls (see Appendix 27) is recommended to be completed for each message. It should show all details and characteristics which might identify the caller and the location from which he is speaking, plus any indication of specialized knowledge which might be informative.

## *CHECKLIST FOR KIDNAP-TYPE TELEPHONE CALLS*

### *Objective*

Keep the caller on the telephone as long as possible. Ask as many questions as the caller will answer, bearing in mind that the safety of the hostage is all-important. Note everything that might help to identify the caller and to establish that what is said about the kidnapping is true.

*Questions to ask if possible (abbreviated abstracts)*

1. Who is calling, and whom do you represent?
2. How do I know that what you say is true?
3. Can I speak to him?
4. If I cannot, how can you prove that you have him? What is he wearing? (Also questions of a personal nature that can only be answered if the person genuinely has been kidnapped.)
5. Is he alive and well? Has he any messages to give us?
6. What do you want from us, or what do you want us to do?
7. If money is wanted, how much? When and how do you want it? How do we know that you will keep your side of the bargain?
8. Where do you want it? How will we know that we are giving it to the right people? Will you let him go at the same time?
9. (Stressing that you cannot give an immediate answer) Will you give us time to see what can be arranged and ring back to us? If other people get to know what has happened, they may try to hoax us, so will you agree a code so that we know we are talking to the right people?

*After assessing that kidnapping is genuine*

10. If there is delay in getting the money or something unforeseen crops up, how do we contact you?
11. Where, when, and how do you want the money handed over? How are you going to release our person so that we know that you are playing straight?

## *NEGOTIATIONS*

The safety of the kidnapped person must be preserved as best possible, and nothing therefore must be done in the course of negotiations which would cause precipitate action by the captors. A balanced team should be assembled to ensure this.

It must be anticipated that persuasion and pressure will be introduced with the object of obtaining conformity with official political policy towards kidnapping demands.

*Suggested composition of negotiating group*

1 *Co-ordinator* – to see that a complete log of action and information is kept, and to maintain constant contact with members to ensure that the group can be convened immediately and with full attendance.
2 *Psychologist* – to guide the group in understanding the probable thought processes of the kidnappers.
3 *Management representative* – board-level man with authority, to make decisions regarding monetary or other demands in accordance with company policy.
4 *Negotiator* – preferably a legal man with experience in negotiating techniques, to be contact man with the kidnappers on behalf of the group.
5 *Police representative* – to co-ordinate police action and form a liaison between them and the group.
6 *Insurance representative* – for obvious reasons, and to link with the management representative.
7 *Press liaison man* – to be a liaison point for the media, thereby providing a 'buffer' for other members of the group; to issue media statements and answers to questions in accordance with the group's wishes.

*PART SEVEN*

# *AIDS TO SECURITY*

# 34

# *Alarm systems*

## *ELECTRONIC ALARM SYSTEMS*

If, as is said, the security industry has the greatest postwar growth rate of any, the spearhead of that growth has been the burglar alarm sector. Apart from the longstanding suppliers, several of the giant electronic industry groups have acquired interests by takeovers or by starting their own alarm divisions.

There is little variation in the types of equipment in use as comparatively few firms manufacture their own. Certain devices are subject to allergies or preferences – for example, infra-red is anathema to some installers, others refuse to use ultrasonics.

Customers should realize that, no matter what the size and reputation of the supplier, there are variations in expertise and workmanship between branches of the same firms, as well as between the firms themselves; efficiency in one area does not of necessity mean efficiency everywhere. For this reason, when customers are contemplating the installation of alarm systems their first step should be to contact the police Crime Prevention Officer. They should ask advice on what is required and for a list of those suppliers who have a branch office and engineers locally. The police cannot be expected to recommend one particular firm, but they may be prepared to give an indication as to which has the greatest nuisance value in false alarms in that particular area. This will automatically give an indication of workmanship and of the service provided. It is hardly likely that the list will include any small firm of whose integrity the police hold any doubts

– several instances have been known where alarm installation contractors have been controlled by persons who have been convicted of dishonesty.

If an installation is being made at the insistence of insurers, the proposed level of protection should be checked for their approval. It is unlikely that they will clash seriously with what is recommended by the police. An individual insurance surveyor may suggest the use of a particular installer; but provided he has agreed the specification of the system there is no reason to accept this recommendation as binding, and quotations should be obtained from three separate firms – experience shows that they may vary widely. If insurers are involved for burglary loss with premises not fitted with an alarm, it may be worth checking with them to ascertain whether use of a system will result in a reduced premium – they normally display little enthusiasm for this!

A well-established firm in the area must be used. This is a popular field for small electrical contractors to start up business in hopefully, then disappear from, leaving unfortunate customers with the prospect of approaching those whom they should have utilized in the first place for them to take over the maintenance. If they do this, there will be a large bill for conversion of those parts which they will undoubtedly find unacceptable.

The point about the availability of local engineers is most important. Invoices for work done subsequent to installation will include travelling time charges, and these may be very substantial – as in the case of a Brighton installation where a London-based supplier did not implement his assurance that a Brighton office was to open. Before signing the contract it is wise to find out how additional work will be charged – for example, one supplier makes a minimum charge of one day's work in all cases, irrespective of how long the engineer is engaged.

It is advisable to look closely at the small print on agreements before completing – some endeavour to exclude every conceivable form of responsibility and may make the user pay for equipment replacement and services that he believes would be included in rental/servicing. When in doubt about the equipment which is being suggested, ask to see a similar installation that is already functioning – then ask that user what snags, if any, have been encountered. A highly expensive linked system of rays to cover a very long factory perimeter was 'sold' and on the point of signature by the local manager when a gentle request 'to see one that's working' produced a reluctant 'well there isn't one yet' and a further voluble attempt to persuade 'a firm with so much at risk to experiment' – at the user's expense!

### *British Standards*

BS 4737 has been introduced to provide a 'Specification for intruder alarm systems in buildings'; it is in three parts:

Part 1 Systems with audible signalling only.
Part 2 Systems with remote signalling (to signal an alarm condition to a remote point).
Part 3 Detection devices.

BS 5979 relates to specification for direct line signalling systems and for remote centres for intruder alarm systems.

### *Alarm activation*

Installations are almost invariably 'closed circuit' which when interrupted by, say, the breaking of a wire, parting of contacts, obstruction of a beam, etc. causes an alarm condition which continues until reset at the control panel. Police policy is that all such resetting should be done by engineers so that the cause of the alarm can receive competent attention thereby reducing false alarms.

### *Message transmission*

Apart from the simple installation with warning bells only, which is not recommended for any commercial premises with any risk of consequence, all systems transmit messages to alert the police; this may be done in a variety of ways, via:

1 Telecom '999' emergency service;
2 direct specially-laid line to the nearest police station;
3 direct line, or shared line from a satellite, to a central station manned by a commercial security firm;
4 electronic 'digital dialler' normally to an alarm company central station (an agreed telephone number is automatically dialled and the message transmitted to it);
5 British Telecom ABC system (alarm by carrier). This uses existing lines to the nearest exchange from which the message is redirected to the emergency service. Transmission is at a frequency which does not affect the line's use for speech purposes. The system has been on trial for several years;
6 radio transmission of the warning message to the control room of the installing company whose own patrol staff attend.

The use of wireless signal in lieu of telephone link has been under development for many years after a pilot system in the Southampton area did not find favour. It seems likely with increasing pressures on the police that the attending of alarm calls by security personnel will spread and radio should prove an economic and efficient link with limited susceptibility to interference.

It is worthy of mention that one of the largest retailing groups has utilized the digital dialler concept so as to route alarm calls from all its premises in the UK to its own central control room.

The police have at their disposal means which can be installed on a temporary basis to provide for a positive risk; triggering off this alarm superimposes a coded bleep on a wavelength used by them. The number of such devices that can be used in a given area is limited and their loan will only be made where substantial grounds to anticipate attack have come to light.

Another scheme which is prospering is that of satellite automatic stations. These are of the direct line type and radically reduce Post Office charges on their customers. The satellite unit, which is unmanned and comparatively small, is strategically placed amongst a number of installations which are connected directly to it. It is then linked to a centrally manned station; Telecom charges are on a per mile basis and could be prohibitive when the distance to the station is great; by this system the cost is split between the users by virtue of the single line needed. This is ideal for a small industrial estate and enables the best form of protection to be provided at reasonable cost.

Central stations are likely to continue their rapid increase in number. There are indications of growing co-operation between alarm companies so that one who has such a station locally may, with suitable financial arrangements, allow others to terminate lines from their installations in it. Unpredictable changes of police policy have dictated this increase. Discretion is given to police chief officers as to whether they allow or continue to allow direct lines to their stations; with certain praiseworthy exceptions the result has been a mandatory withdrawal of such lines with inevitable cost to customer, reduced efficiency of his alarm protection, and certainly no enhancing of his opinion of the police regard for either crime prevention or their public image.

### *'Telegrouper'*

A variation of the direct line principle has been found acceptable to some police forces. This is the Telegrouper where direct lines go into a special electronic device which provides for continual line monitoring. In the police office there is simply a printout device, no phalanx of Home Office-approved panels. There is no communal police policy towards such systems, indeed in one police area direct lines accepted by them were vetoed in favour of compulsory use of Telegrouper only to see a further change of mind almost before the ink on amended agreements was dry. As this necessitated use of '999' with all its defects, or (in the absence of digital dialling at that time) long-distance direct line links to commercial stations, user reaction to the cost involved need hardly be commented on.

*'Telemitter'*

The Telemitter can be operated via digital dialling or direct line and has been found a valuable device to incorporate a number of warning circuits at a customer's premises, for example theft, fire, computer humidity, automatic machinery malfunction, and so on. On activation the device dials the number of a central station and continues to do so until the connection is made. The receiver accepts and prints out a precoded message identifying the source and nature of the alarm. Distance is no object, the cost of direct lines is saved, and some interesting variants may be developed. Alternative names for similar apparatus are already in use.

*Electronic message-transmitting systems*

These all work on the principle of interference with a warning circuit which activates a tape or disc mechanism, automatically dialling 999 and conveying a pre-recorded message which indicates the location of the premises and the presence of intruders; or, in the case of a direct line, causes the illumination of a light and the ringing of a bell on a panel at the terminal station. This alerts the police to put a prearranged plan into operation to assemble an adequate number of police at a fixed point near the premises, but out of sight and hearing; the area will have been previously surveyed and when sufficient number are assembled they move simultaneously to selected places to entirely surround the premises; keyholders from the firm are notified in accordance with the list supplied to the police – but, to save time, firms are often agreeable that keys shall be lodged with the police. Where this is done the keys are retained in a sealed envelope at the police station and are only withdrawn on the sounding of an alarm. Where a central station is in use the operator notifies the police, either through a direct connection or by dialling 999; apart from this brief delay such listening posts have advantages in picking out false alarms due to faulty opening and closing down procedures, thereby saving police time. On the adverse side, the calibre of the employees in experience and perhaps in resistance to external pressures is scarcely likely to be comparable, and the regular change round of police staffing restricts in comparison the possibility of conspiracy with thieves.

### *Audible bells*

Many electronic alarms are fitted with audible bells functioning on a delay mechanism at a fixed time after the initial telephone message has been transmitted. This is a precautionary measure, often insisted on by insurance companies, but disliked by the police, who thereby have a limited time in which to carry out their plans before the criminals are alerted. The insurance view is that the alarm will make the intruders decamp before they can steal anything, if the arrival of the police should be delayed for any reason. However, if the buildings are at such a distance from normal police coverage that it is not feasible to assemble the required number of personnel in less than ten minutes but the bells work six minutes after the alarm is activated, the police cannot be blamed for having a jaundiced view since they will obviously arrive when the premises have been vacated.

Once premises are known to be protected by alarms, as they must be if the bell sounds, thieves can either leave them strictly alone or make efforts to see how this can be circumvented. There is a growing tendency to try to defeat alarm systems and the course that is subsequently followed will depend purely upon the determination of the thieves.

Audible bells have outstanding disadvantages – their only virtue is that of cheapness. Experience has shown that people other than police either take no notice or do nothing about it on hearing one. Frequently such bells are fitted where no one can hear them in any case. They do not catch criminals: they may chase them away – or they may induce them to take a look and see how they can stop the bell next time and work in peace. Such is the oldest type of alarm, cheap and widely used on small shops and premises where expense is a major consideration. Unfortunately, most thieves will have little trouble in putting the bell out of action – this can easily be done by either cutting the wire to it, filling it with cotton wool, or even discharging a fire extinguisher into it. A bell is only effective when it causes thieves to leave the premises immediately.

## *SURVEYING PREMISES FOR ALARM INSTALLATION*

All commercial firms supplying alarm systems will survey the premises of prospective customers without any charge or obligation. They will specify the most suitable type of protection which could be installed and submit estimates of overall cost. Insurance companies employ surveyors who work in a similar fashion at the request of applicants for insurance cover, and in this instance there is no charge for the survey. Police crime prevention departments also furnish free and detailed advice on a firm's problems; it must be appreciated that they will not be able to detach a man for a prolonged survey extending over a period of days – this is the task for a professional firm. Cost considerations apart, the practice of requesting quotations from several alarm firms has an additional advantage in that the coverage proposed by each firm can be compared and the best system selected to conform with the firm's usage of the premises.

Where experts are called in to advise and make surveys they must be informed, in confidence, of every risk that may be encountered so that they can include it in their considerations. Justifiable standards of protection will vary according to the risk attached to the particular premises: a jeweller's shop would require much more detailed protection and consequent cost than an ordinary grocer's shop. Similarly, a tobacco warehouse would justify provision far in excess of a furniture equivalent. The location of the buildings is a further factor; if these are sited within a works, the perimeter of which is well lit and patrolled by security staff, there is little object in having detailed protection, other than at an essential point. If an office block in the dimly-lit back street is concerned, it will obviously need far greater coverage.

Overall protection of premises may not be required; at some points nothing of value will be stored; other areas may contain safes, valuable metal stores, or office equipment, all of which merit security. A single contact at one particular point may obviate the need for six at other locations; certain parts of the building could be strengthened by physical means to make an intruder's task almost impossible. These are all matters for the surveyor, who must be fully qualified in every way, not only in connection with equipment but in knowledge of the probable modes of operation of prospective thieves.

Two separate main schemes exist for the protection of premises: the first is that of protecting the perimeter so that no one can get access to a vulnerable interior; the second is that of 'trap' protection, which, in effect, allows the intruder in but ensures that he subsequently trips an alarm in endeavouring to reach his objective. The perimeter variety can be avoided by attack over roofs and dropping through; this of course limits the quantities of material which can be taken out, but, where values are high, this can still be productive for the thief. The ideal is a combination of 'trap' and 'perimeter'; this allows the police a maximum warning before the thief can get to his main objective and also ensures that should he get past the first line of defence he is caught with a subsequent line which he may not expect.

It is not always easy to disguise perimeter protection since this involves windows, where tubing and wiring will be visible, as could be vibratory contacts. Not only that, but should a thief carefully examine the outer doors he will probably notice the various contacts which may be fitted there. Trap protection usually takes the form of invisible rays, pressure pads, pressurized systems, etc., which on the whole are less obvious than the perimeter protection.

Where a number of surveys of the same place are made, it should be borne in mind that it is likely that the insurance surveyors will specify a much higher degree of protection than either the police or the alarm company (because their own finances are directly involved); the police will advocate the next; and the alarm company, not wishing to price themselves out of work, will hit a medium which will preserve their reputation and supply an acceptable degree of security.

## *MEANS OF DETECTION*

There are a variety of devices; some of them have a multi-purpose nature which may be utilized by installation companies. The main ones are described below in non-technical terms.

### *Door contacts*

These are the simplest yet most effective and troublefree devices. There are two types: the mechanically plunger-operated type and the magnetic type which are much more widely used.

*Mechanical* – The first, sometimes called the switch type, is normally fitted in the door jamb. The plunger is depressed when the door is closed and when it is opened about 4 inches (10 cm) the plunger, which is spring loaded, operates and the alarm system is completed. The mechanism is micro-switched and the plunger should be adjustable to take into account variations in the fittings of different doors.

*Magnetic reed switch* – The second type consists of a combination of a magnet and a glass capsule containing two contacts which are closed by bringing the magnet into close proximity to them. The magnet and the switch are housed separately, one in the door surround and the other in the door itself. New varieties of this magnetic contact have been devised which are resistant to any attempts at interference by other magnets.

This type has advantages over the plunger type as it is easily rendered invisible in wooden doors, there is no corrosion effect upon it, and it is perhaps more difficult to counteract.

*Heavy-duty contacts*

These are used for larger doors and again may be either of the mechanical or magnetic type. They are used on garage and factory doors of the large roller-shutter and concertina type. They are basically the same as the smaller contacts but more robust. The disadvantage of the nonmagnetic variety is that they are clearly visible and, once seen, it may be possible to circumvent them.

*Closed-circuit wiring*

This again can be in two types: tubular wiring and simple wiring.

*Tubular wiring* – This is widely used for the protection of external windows and consists of light steel tubing mounted on a wooden frame, shaped to fit the whole vulnerable space of the window. This of course gives a bar effect to the windows which cannot be concealed; it is obvious what they are to anyone who knows anything about burglar alarms.

The tubes themselves each contain a fine non-stretch wire throughout their full length, and this is connected into the circuit wiring. After breaking the window any further entry must be made by forcing the tubing; cutting the tubing breaks the wire and interrupts the circuit, activating the alarm. Stretching the tube, forcing the tubing apart, will similarly break the wire. This is very effective but also very obvious.

*Simple wiring* – Not contained in tubes, this is used for protecting doors, vulnerable walls, ceilings, and forming protective enclosures around safes. Wire of the same type is stretched over the area to be protected in the form of a network and similarly linked into the circuit wiring. The whole is then covered over by hardboard or some similar material to blend into the surroundings. Any strain upon the wire caused by physically breaking through will break the wire and activate the circuit.

This form of protection is ideal for lightweight doors, in combination with an ordinary plunger or magnetic contact. The object of this is to defeat the criminal who has some knowledge of alarm systems and will break a hole in the door, rather than open it, or force a hole in a plasterboard wall. This is a comparatively cheap and certain means of protection, unless there is expert interference with the circuitry.

*Window foil*

An alternative to the tube and wiring of external windows is the use of foil, which is by no means as obvious to the intruder. Very thin lead or aluminium foil is stuck firmly to the inner surface of the window, completely around its perimeter, and indeed, if required, a form of pattern could be included. The foil is connected to the circuit wiring in the same way as the previous methods, and, if the window is broken, the foil likewise will break.

The foil is normally given a coat of clear varnish to protect it and while this is a good method of window protection it requires expertise in application. It can also suffer from the efforts of enthusiastic window cleaners!

*Floor pads (pressure mats)*

These consist of two layers of conducting material separated by an insulator, all of which is enclosed in a waterproof and non-conducting cover. The idea is to form a very thin sheet which can be put underneath a carpet or other floor covering so that pressure upon it will cause the closing of the circuit and activate the alarm.

These can be used as a form of trap protection on stairs or landings, in front of safes, behind doors or beneath windows. Care must be taken on their installation so that they do not show their presence after a time by an outline on the surface above them – this may involve the removal of carpet underfelt, or the placing of some suitable material around the perimeter of the pad and under the floor covering so that the outline of the pad is lost. It is advisable not to use these on concrete floors or under circumstances where they may be subjected to chafing or the effects of condensation.

*Ray-operated devices*

These operate on the principle that anyone or anything passing through a beam of the equipment will break the circuit and thereby activate the alarm. Infra-red light, invisible to the human eye, is used; and though by careful inspection a dull red glow may be seen from the transmitting bulb, this is of little value to an intruder since he will have to break the beam to see it.

The ray has to be modulated; otherwise it would be easy to overcome by shining a torch into the receiver. In effect the light is coded by the transmitter to an intermittent pulse to which the receiver is designed to respond. The older types of rays had a fan-shaped rotating disc to do this, but it is now done by electronic means. Different types are available to cover distances up to about 300 feet (91 m), and the prices may vary from company to company. A power source must be installed adjacent to the rays, and provision must be made for regularly cleaning dirt and dust from the equipment. Mirrors can be used to take rays round corners, but this is a complicated matter, entails thorough cleaning of the mirror surfaces, and multiplies the chance of misalignment.

Single rays can be a nuisance, since they are extremely sensitive and the passage of a bird can activate the alarm. This is obviated by twin rays adequately spaced, both of which have to be broken. A check should be made with the installer to ensure that if one of these becomes defective the twin system shows up as a fault; otherwise the system reverts to being a single ray. An alternative is to site the ray so closely to fixtures as to make it unlikely that birds will cross it. Where there are very high risks, multiple rays can be provided to form an invisible fence, that is 'barrier rays'. Needless to say this is a very effective but very costly system.

In the older installations infra-red bulbs needed regular replacement, but fortunately longer-lasting alternatives have now been invented. Newer renovations should also take care of a fault which passed unobserved for a long time, namely, that of wiring the power source provided into a section of the supply which was liable to be turned off for lengthy periods for servicing. This brought into use the standby batteries which would be effectively flattened, causing an alarm condition in a prescribed number of hours. There has been a vast expenditure on replacement batteries because this factor was not visualized.

Where rays are to be installed in new buildings and a specification quotation is to be made 'on plans', one drawback may be that of ensuring that they have an uninterrupted line. If the area is intended for vehicle or materials storage, their presence may hamper the effective utilization of costly space. The customer must clarify, to the best of his knowledge, how the space will be employed.

### *Vibro detectors*

These are mainly used in connection with windows and fences, though efforts have been made to extend them to wall protection with a certain amount of success. Basically, they are devices which, when subjected to vibration or shock wave, cause a breakage in the circuit running through them. Most types are adjustable for local conditions to prevent false alarms, and variations of such devices have been used in the protection of vehicles. It is this need of adjustment which makes them unreliable: if too finely adjusted they might react to children banging against windows; if too coarsely adjusted, it might even be possible to cut a complete square out of the window with a diamond without upsetting the vibrator contact.

These can be used for the individual protection of filing cabinets and safes to which they can be clamped by means of powerful magnets. Any effort to remove them will of course set them off as will any attack upon the container to which they are attached. An obvious disadvantage is that they are immediately visible and any expert thief will endeavour to find other ways of counteracting them by interference with the circuitry. The varieties used on safes can contain a heat-sensitive device to ensure that the back is not carefully cut away with oxy-acetylene or other cutting gear. These could have a fire-prevention value too.

A variation which has had a military application is that of utilizing vibro/sound detectors for outdoor purposes. It is unlikely to be practicable for ordinary commercial or industrial purposes but the principle is to set a chain of weatherproofed units in gravel or other suitable media round the area to be protected so the sounds made by the passage of an intruder will be picked up – very expensive, and avoidable if the presence is known.

### *Sound detectors*

These can only be used satisfactorily where there is either a complete absence or an almost constant level of sound to which the apparatus can be adjusted. Essentially, a sound detector is a microphone and amplifier system connected into the circuitry; where a sudden increase in sound is detected this circuit is broken.

These are primarily for special risks and may be combined with a heat detector which could have a dual purpose as a fire alarm and also a means of detecting the use of a thermal lance or other heat-using cutting device.

Sonic detectors are of a special value for retail premises with their plate-glass windows and display showcases in that they react to the sound of breaking glass before the actual theft takes place. Indeed, sensitivity can be adjusted to pick up the first attempts to break in. There are good grounds in this application for coupling to instantaneous bells since the intruder is likely to be on the scene in any case for the least possible time.

### *Pressure differential systems*

These are further specialized systems which can only be used in comparatively small self-contained areas. The principle is that a fan is installed in the wall of the compartment which can blow air either into or out of the area to be protected. A diaphragm constantly monitors the pressure difference between inside and outside; this is connected to a pendulum-type contact which in turn is included in the alarm circuit. Any additional aperture made into the protected area causes a variation in pressure which is registered by the diaphragm and transmitted to the pendulum contact.

The use of this device is almost entirely restricted to strong rooms and the like. It is an expensive and sensitive installation which, by virtue of its sensitivity, can be subjected to false alarms. It has a drawback in that no warning is given until the protected area has been violated and the intruder may have time to escape profitably if he has planned carefully and has knowledge of the system in use.

A pressure differential system of a different kind has been used in the same way as geophones for creating a sensitive zone inside perimeters. In this, pressurized liquid filled tubes are buried to a safe depth and pick up vibrations caused by surface movement – costly to a degree that is prohibitive for normal commercial purposes.

### *Ultrasonic systems*

The general principle is that high-frequency pulses are transmitted by the apparatus which are inaudible to the human ear. These are reflected by objects in the surrounding area and are picked up by a receiving unit which adjusts to a pattern when switched on. A change in the pattern will cause an alarm condition – any movement in the area covered by the apparatus will do this but sensitivity can be controlled.

There have been considerable improvements in the efficiency of this type of equipment during recent years. It has benefited by 'spin-off' from the American space industry; the field each 'head' can protect is limited but it can be adjusted to exclude areas likely to cause faults, some types will accept regular movement such as rotating fans or constantly working machinery, without going into an alarm state.

These are still expensive forms of equipment, but can be very effective. They need careful and expert installation, the optimum positions being found to some degree by trial and error.

*Radar detectors*

These too can be extremely efficient but are also expensive. The principle is widely known nowadays; a radar pulse is sent out from a combined receiver/transmitter which is reflected by all stationary objects in the immediate area at the same frequency, movement will trigger off the alarm by causing a variation in the reflected frequency. The sensitivity of the equipment can be adjusted to exclude the effects resulting from the presence of birds or rats, etc., and portable forms are available.

The apparatus can be used directionally and miniaturization has reduced the size of the equipment needed to the extent that it is possible to build it into the base of a telephone handset. Like the ultrasonic system, the area each unit will cover is limited and the optimum placements would have to be ascertained during installation. The development of this type of detection is by no means complete, experiments have actually taken place with rotating beams for long-range outdoor coverage where there is obvious scope if faults can be eradicated; under some conditions the very sensitivity of this type of equipment can negative its possible application.

*Proximity detectors*

The full potential of these may not yet have been fully utilized. The principle is that when an object is brought near to an area connected with a proximity detector, or touches it, a capacitance change takes place which operates the alarm on the circuit of the detector. It is essential that the proximity detector be linked with a metal surface so the use is restricted to safes, filing cabinets, and articles which, somehow or other, are connected to metallic surfaces. This is done in the case of paintings by placing aluminium foil over the rear of the frames and similar steps can be taken in connection with showcases by placing strips of foil over or under the articles to be protected.

The disadvantage of this system is that its limited detection range restricts its application. Efforts to utilize the principle on a wider scale have so far been impeded by the liability to false alarms.

### *Heat sensors (passive infra-red)*

These work by monitoring the rate of change of temperature within their range and are adjusted to activate the alarm when this exceeds a predetermined level. They have an excellent secondary purpose in the early detection of incipient fires. They are now in general use, compact and unobtrusive in design, and deemed sufficiently efficient and reliable by some surveyors for them to dispense with conventional door contacts where they can be installed. There are many manufacturers, some of whose research and development has been done under wartime conditions.

Many of the original objections no longer apply, though care must be taken that they are not specified for an unsuitable environment.

### *Trip wires*

These are the most elementary types of warning devices but they can be effective because they are unexpected in modern society. The drawback is that an intruder will almost certainly become aware that he has been detected and will make his escape. Not only that, if he wishes to renew his attempt at a later date he will be aware of the barrier and can easily avoid it.

Nylon trip lines are probably the most effective and least likely to be noticed. Where used externally, false alarms must be expected from animals and possibly children. They could be of considerable value inside compounds and building sites when connected to instantaneous bells.

A variation of these could be used for the protection of windows and roofs by arranging that they could be under constant tension which if varied in any way would set off an electronic alarm. This would enable almost invisible steel wires to be stretched behind windows as an alternative to closed-circuit wiring, thereby facilitating the use of the windows for ventilation. A similar arrangement, with suitable intervals between wires could be put on the underside of roofs. A period of settling down would have to be allowed before these could be wired into the system to avoid false alarms caused by initial stretching or straightening of the wire.

### *Developments*

Whilst these are the more commonly-encountered devices, there is a constant development being carried on by the larger companies to improve and miniaturize their equipment. Much of this is in the ultra high frequency field and will be most difficult to counteract since the presence of the intruder will be noted before he can possibly interfere with any part of the system. This, coupled with a radio link, could well produce the ultimate in protection. It will not be cheap and users might even consider dispensing with insurance coverage if no substantial reductions in premium ensue.

## *PRECAUTIONS AGAINST INTERFERENCE AND LIMITATION OF FAULTS*

Expert thieves are fully aware of the potential danger to them from burglar-alarm systems. They also realize that premises equipped with burglar alarms must also contain property of attractive value. It can be anticipated therefore that these will offer a challenge to a certain class of criminal who will apply expert knowledge to overcoming them. It is not proposed to discourse on how this can be attempted, but only on precautions that can be taken to prevent any success being achieved.

The normal telephone '999' line is most vulnerable. The following steps can be taken which will enhance its value:

1 Have the transmitter coupled to an ex-directory line which can be made 'outgoing only' on application to British Telecom. It is possible for the technicians to arrange that this shall be a normal two-way line for certain periods of the day and outgoing only after a fixed time. This will involve an extra payment.
2 Where an ex-directory line is used, remove the disc from the centre of the telephone showing the number.
3 Ensure that the line connected to the telephone exchange is in the form of an underground cable.
4 Site the transmitting apparatus in such a position that any intruder, even if he knows its whereabouts, must pass through a form of trap protection and be unable to reach the transmitter before the message has been sent out.
5 Get the alarm company to fit a 'line sense unit' which, should the Telecom linkage be non-operational, automatically brings into action external alarm bells which are usually self-contained with their own battery supply and encased against interference.

### *Power supplies*

Most systems are mains operated, but technical developments are in hand which may make them self-contained. Rectifiers and transformers are included in the circuitry to produce a direct current output of 6 or 12 volts. It follows that a mains failure means that the system becomes non-operational unless there is an alternative source of supply. Alarm companies therefore install wet or dry batteries and the testing of these is part of the periodic servicing carried out by them. These offer a standby of adequate duration in case of failure of the power supply.

### *Control panels*

These are a necessity to test the state of the wiring and the state of the power supply. Multiple control panels can be installed whereby prearranged sections of the premises may be made subject to the alarm circuits, allowing others to continue to be used for production purposes. More specialized panels can combine this facility with one of fault location whereby, should the control panel show a fault on test, it is possible to identify at once the offending contact. All control panels now incorporate locking devices to prevent the system being switched on unless all the triggering devices are set correctly.

### *Operating procedure*

The normal procedure is that the operator tests the circuit and switches on when he has found that it is in order. He then goes to the main exit door possibly via other contacted doors and either the last of these or the main exit door itself will be fitted with a shunt lock which he will turn to incorporate those doors into the alarm circuit, including the one he is locking. These locks are usually of multiple lever type, the keys to which cannot be duplicated other than by the alarm company who can be anticipated to be very circumspect about to whom they issue them. Maintaining the security of the keys is obviously of vital importance and strict control must be exercised on the number of persons who have access to them.

An alternative to the use of shunt locks is to incorporate a time delay into the control panel which allows a fixed time for leaving the premises and closing exit doors before the alarm becomes operative after switching on at the panel.

### *Bells*

If premises are insured, the insurers will have to approve the location, number and operation of any bells incorporated in the alarm system. Their main insistence is likely to be on a minimum time delay between the alarm being triggered off and their operation; some agreement will be necessary with the police to achieve a compromise. Police practices vary but they do not normally want external bells openly displayed on premises where there are direct line facilities into a police station.

Where the major reason in installing a system is that of damage prevention there is a case for immediate bells to disturb the intruders. This is currently particularly true of the lowlands of Scotland where damage by children and youths has become excessive.

## *FALSE ALARMS*

The police will discontinue attending installations where repeated malfunctions cause excessive waste of time for their limited personnel. The user must do everything in his power to prevent these, including analysing their causation, not just leaving an alarm engineer to record the reason – it is just possible that he might be biased! The user should keep an alarm log for that purpose.

### *Alarm log*

This should record everything which happens in connection with the system, listing every servicing visit. Not all companies are meticulous in keeping to their maintenance schedules, and the log provides a ready check to ensure that what is being paid for has been done.

All faults should be shown – time, date, what happened and why, time service engineer attended, his opinion of cause – in fact all the detail that might be relevant to pinpoint recurrent faults either in equipment or in careless usage. It will provide a picture of the service given by the alarm company in speed of attendance and elimination of persistent malfunctions. Most important, it will show the police that the customer takes his responsibilities seriously, and this will weigh in a decision to withdraw response at troublesome premises.

### *Alarm system supervision*

The alarm industry has fully appreciated the dangers of persistent alarms and of inefficient installers, and in fairness to them it must be said that they have taken firm steps to put their communal house in order by establishing the National Supervisory Council for Intruder Alarms (NSCIA) in 1971. This is an independent body with a register of approved installers who agree that their work will comply with British Standards. It operates an inspectorate who will check new systems and investigate complaints. Installers certainly have accepted their recommendations and criticisms, but doubts expressed in Home

Office Working Party reports of the impact of the scheme on the false alarm rate are yet to be proved unfounded. The NSCIA headquarters is St Ives House, St Ives Road, Maidenhead, Berks SL6 1RD Tel: Maidenhead 37512, from which a new-user advisory booklet may be obtained on request.

It says little for co-operation within the industry that since the inception of the NSCIA other bodies have been set up with identical objectives: the Independent Associated Alarm Installers (IAAI) representing mainly smaller firms; and, from January 1987, the ECA Security Group formed by members of the Electrical Contractors Association in recognition of the fact that many are involved in alarm installations. To a customer the end effect of membership of any will be an indication of reliability; if membership of none – *caveat emptor*!

### *Keyholder failings*

A high percentage of false calls are caused by the keyholder when switching on or switching off the system. They are usually due to rank carelessness, but they may also be due to the fact that the person doing so has not been adequately instructed and is acting as replacement to the regular person. A detailed instruction card approved by the alarm company should be provided, showing exactly what is required to be done to test the circuit, how to check when the test indicates faults, and whom to contact if these faults cannot be rectified. If there is timed entry or exit delay to enable the operator to activate the system and then leave the building before the alarm becomes live, this should be shown, with the egress and access route which must be followed.

All keyholders should be on the telephone, be reasonably convenient to the premises, and have their own transport, and at least two persons should be nominated. If they are varied, the variation should be immediately reported to the police.

A requirement, which it is hoped will become mandatory – and probably will – is that a keyholder should be helped to avoid faults by a shunt lock on the final exit door with an internal warning, audible from the time he switches on the control panel, till he leaves the building and operates the shunt lock. The inbuilt precaution is that if the warning buzzer does not go off when he uses the shunt lock, he knows that there is a fault on the exit route which he must check. No signal is of course transmitted whilst the buzzer is operative. It is essential that the exit/entry door cannot be opened without prior operation of the shunt lock.

A keyholder who accidentally causes a fault must notify this to the alarm company for their record purposes in case a site meeting is later called by the police to discuss possible withdrawal of their attendance.

This notification may be obviated in the future by police insistence that systems can only be reset after functioning by an alarm engineer. This is usually a simple variation in the control panel.

## *FAULTS ASSOCIATED WITH APPARATUS*

### *Magnetic and mechanical contacts*

1 Warping and distortion of contact-bearing surfaces.
2 Inadequate door-closure devices to hold firm against draughts.
3 Damage to flexible leads fitted to closed-circuit wiring on doors.
4 Damage to wiring, pressure pads, or foil due to normal wear and tear.
5 Jamming of plunger-operated contacts by paint.
6 Damp and dirt in shunt locks.

*Defects in electronic devices* – Microwaves sited in an area where there are large reflecting surfaces, for example roller shutters, metal containers, skips, mirrors, or in circumstances where rainwater can cause reaction to their movement or on the surfaces upon which they impinge. Microwaves might also react outside the area for which they are intended if the walls and partitions are of a type through which they can pass; this should have been discovered on test.

### *Infra-red rays*

1 Damage to projector or movement of fixture to which it is attached.
2 Impeding dirt upon them.
3 Receiver sited where it might be affected by another infra-red source, for example strong sunlight.
4 Standby battery failure in instances of prolonged mains failure or disconnection.

### *Ultrasonics*

1 Siting in an area where there is possibility of legitimate ultrasonic noise, for example bells.
2 Siting in an area of substantial air movement – for example, space-heating devices automatically switching on, large fans coming into operation.

*Passive infra-red* – Sudden injection into the area of an infra-red source, for example the electric heaters automatically switching on.

*Space detection devices in general*

1 Fitting detectors on surfaces which are subject to damage or vibration.
2 Siting in an area where there may be a legitimate movement of wildlife – for example, birds, rats, animals.
3 Siting in an area where there are possibilities of movement – for example, badly-fitting roller shutter blinds, loose asbestos sheeting, wind-operated fans, draught-movable hanging signs and lights.

*Vibration contacts* – The main difficulties with these will be of bad siting in the first place on loose surfaces affected by external factors of passing traffic, strong winds, or the effects of children playing against walls and windows. Similarly, if breaking glass detectors are fitted, they may be affected by sounds of similar resonance – for example, telephone bells ringing, milk bottles breaking when blown over by wind. (Equipment of this type, however, has been rapidly developed to considerably greater efficiency.)

## *LEGISLATION*

Some 98 per cent of alarm calls received by the police are classified as 'false', that is there has been no intruder found. These false alarms can cause a great deal of nuisance to the general public and to neighbours in particular. For this reason there is a Code of Practice on Noises from Audible Intruder Alarms issued in 1982 under the Control of Pollution Act 1974, which required all intruder alarms to be installed and maintained in such a way that no unnecessary disturbance is caused to the public. It was suggested that compliance with BS Standard 4377 was the right way to achieve this. It also suggests that a 'cut off' should operate after some 20 minutes, although a 'flashing light' could continue after this time to show an alarm condition.

It must be noted that the occupier of the premises must provide the police with the names and addresses of two keyholders who should know how to operate the alarm and silence it. Any changes of keyholder must also be reported.

The local environmental health authority must be informed when alarms are installed or taken over from a previous owner.

The Code of Practice suggests that where a keyholder fails to silence an alarm within 20 to 45 minutes (dependant on the circumstances) consideration should be made by the local authority to demand a 'cut off' be fitted, using powers under s. 58 of the Control of Pollution Act 1974. This is an area which local authorities are now taking more seriously in an attempt to stop some of the nuisances caused by faulty alarm systems.

# 35

# *Radio and television*

The value of these aids to efficient security still does not appear to be appreciated by management. The use of radio, in circumstances where there is a static control point with patrolling officers, would appear to be self-evident, but instances are still encountered where the facility is rejected as unnecessary. It is possible that support from a safety officer or safety committee might be enlisted to reverse such an antiquated attitude. Closed-circuit television (CCTV) has so many cost-effective cogent arguments in favour of it that it is easy to see why its use was increased with the advent of the general depression. Where it has not been given adequate consideration, the fault may lie in the security manager's failure to appreciate its potential, or ability to present a properly financially-orientated case for it.

## *RADIO*

Many large firms with numerous security officers and large areas to protect have already appreciated the enhanced value of their patrol staff when they have means of instantaneous communication with a base station. Observation on intruders is made easy, when there is no need to break off and perhaps lose contact with them to make a telephone call for assistance. A most satisfying experience for one security officer was to sit on the top of a tower watching thieves struggle with heavy slabs of metal while the police assembled in accordance with his commentary. There was, of course, no hurry and the exhausted thieves had not the energy to run when the police eventually closed in.

Despite the fact that he is guarding premises a security officer may be overpowered and left out of action for a considerable time before his colleagues are aware that there is anything wrong. Where there are no regulations concerning 'ringing in' procedures from various points on his round this period could be almost the full duration of the round, leaving the thieves with ample opportunity to do as they wish. There is less chance of this happening where a portable set is carried; at least one modern version is equipped with an 'attack' button to give an automatic alarm if the carrier is suddenly assaulted. This modification, at some considerable expense, can be accompanied by a computerized addition to the base station, which will indicate exactly where the incident has taken place.

The fact that there is radio contact can also be a protection for the base station, which will almost certainly be sited in the office at the entrance to the premises. Numerous instances are on record of attacks being made on the gate office before anything is attempted elsewhere on the premises, as by eliminating the person in that office the main means of contact with the outside world can be cut off. Hence the recommendations to keep the office door locked or bolted at all times after dark, to have an inspection hatch through which visitors can be seen before letting them have access, and to have an alarm siren, which is preferably linked to the factory siren, and can quickly be operated by a push button inside the office.

### *Licensing – a necessity*

The Ministry of Posts and Telecommunications is the sole authority in the UK for regulating the use of radio and broadcasting equipment, and for the granting of licences for their use. The Ministry is also responsible for the allocation of radio frequencies and for honouring obligations binding on Great Britain as a member of the International Telecommunication Union. The Ministry has a special responsibility for tracing and measuring interference to radio reception, especially in suggesting methods of suppressing interference, and for encouraging the responsible use of equipment – whether for business or pleasure.

Radio frequencies available for use in this country are overcrowded and must be shared between many services, such as aeronautical, defence, maritime and land mobile services, radio-navigation, radio-location and communications satellites.

To ensure that maximum benefit from the frequencies is obtained and that they are operated in an orderly manner with the minimum interference, any use of radio in the UK must be licensed.

There are certain kinds of transmitters, however, for which the Ministry cannot issue licences because they are liable to cause interference to authorized radio services.

These include 'walkie-talkies' of foreign manufacture designed to operate on frequencies around 27 MHz. The use of these without authority may result in prosecution.

Would-be purchasers of radio transmitting equipment are advised to confirm that the equipment has been approved by the appropriate authority of the country in which it will operate.

### *Call-signs*

Specified call-signs must be rigidly adhered to. In any individual case the operative name will be the same for both base and portable sets; for example, base station will be 'Yorkmet Base', the first portable 'Yorkmet Alpha', the second 'Yorkmet Bravo', and so on in sequence, using the phonetic alphabet.

Only spoken messages are to be used and only authorized persons may operate the stations. An onus is laid upon the licensee to ensure the latter provision is strictly carried out and that those engaged in operating the stations will observe the terms, provisions, and limitations of the licence at all times. The apparatus must be designed, constructed, and used so as to avoid causing interference with any other wireless telegraphy. The call-signs must precede every message and be repeated at the end, though the two identifications are not mandatory if the transmission is less than one minute in duration. Messages should be brief, clearly phrased and spoken, and those making them should never forget that freak conditions may lead to their words being heard over a wide area. A senior police officer was once very embarrassed to find that an installation which went wrong at the crucial stage of a ticklish operation, and which he was addressing in terms which were certainly not of endearment, was transmitting all he was saying but was not receiving the frantic appeals addressed to him by the base station.

### *Limitations*

Both stations and licence must be available at all reasonable times for inspection by duly authorized officers of the Post Office, at whose request the station must close down. The licence is not transferable and can be revoked or varied as desired by the Minister of Posts and Telecommunications, who must be notified of any proposed changes to be made. Amongst a series of conditions and limitations are those that stations shall not be linked to the public telephone exchange system; near airfields and power lines restrictions have to be complied with concerning the transmitting aerial; and transmission of certain misleading messages is forbidden.

*Uses*

Modern portable sets are small and efficient. They are being progressively improved and no longer impede those who have to carry them. The range that can be covered will naturally be governed by the terrain and buildings, but intercommunication between portables should be possible up to about 4 kilometres and considerably further with the base station. In fact, the range will be in excess of requirements in most cases, thus care must be taken with matter that is transmitted. The installing firm will advise on the siting and type of aerial to be used; remote-control facilities can be provided when it is inconvenient to operate the fixed station itself, or, for instance, to reduce the length of down lead from aerial to set in the case of high buildings.

Full use must be made of such equipment where it is available. For security purposes this will mainly be after normal working hours, but there is no reason at all why, in a large factory, it should not be used to maintain contact with important managerial personnel or visitors touring the site.

Regular contact with base station should be kept during patrols by positional reports made every fifteen minutes or so. This enables the base station to know almost exactly where their person is at any given time, and if no message is forthcoming his whereabouts can be pinpointed with a reasonable accuracy. Where vehicles are in use, or dog patrols maintained, the additional facility of wireless inter communication enables greater use to be made of them. From all points of view this is an expenditure which can well be justified and there are few running or maintenance difficulties to be encountered.

## *VHF 'POCKET PAGING' SYSTEMS*

These are better known as 'bleeper' systems, the name being derived from the warning sound which is put out by the pocket receiver carried by the person who is wanted. They provide a means of locating key members of staff almost instantly, wherever they may happen to be. The person it is wished to contact is called individually, without disturbing the other carriers of these pocket receivers, by a transmission to which only his will respond. Dependent upon the system that is employed, or the type of equipment that he is carrying, the person will either ring the telephone switchboard to receive a message or listen to the message direct from his base station on his personal set. A 'talk-back' version may also be used which will give certain key personnel two-way speech facilities anywhere on the premises. The more sophisticated the installation, naturally, the greater the cost. Though normally used inside very large buildings this type of communication is equally effective externally.

Equipment can be designed to cater for a large number of 'bleepers' to be covered by the system; modifications can provide for units of two, six, or twelve individuals to be called simultaneously as teams. Receivers are very small and easily carried in a pocket quite unobtrusively. The system is particularly useful for office blocks, large shops, and hospitals.

### *Comment*

Manufacturers are not as numerous as they were. The smaller and less reliable ones can be forced out of the market and several security departments have found themselves in the embarrassing predicament of having unserviceable equipment with no spares, no willing repairers, and the prospect of having to explain the need for complete replacement. Uncharitable though it may be to newcomers, it is suggested that choice be restricted to established suppliers who have a local servicing/replacement facility.

The initial budgetary proposal should include allowance for a percentage of standby sets – they are apt to receive ham-fisted handling and durability is important. When they are vital for efficient performance, non-availability can be disruptive.

Selection should involve on-site testing for freak conditions limiting potential usage which should be known before recommendations are made. Beware esoteric capabilities pushed by pressure sales personnel which, when thought through, are superfluous to present or envisaged needs.

## *CLOSED-CIRCUIT TELEVISION*

This equipment may be obtained by direct purchase, or on a leasing rental basis. It is already in use in many placces where high security risks are incurred, but its cost and a reluctance on the part of employees to undergo continual surveillance has possibly limited its development for all the situations where it could be of value. In some instances where it is in use there is a tacit agreement between the management and unions that matters concerning minor breaches of factory regulations observed by this method will not be acted upon. The attitude of some large retail shops may be governed by the thought that a percentage of their wealthier customers might have objections to being 'televised' in this way.

The basic installation is that of a fixed television camera linked to a single television screen or monitor. This can be indefinitely extended by the use of additional accessories and equipment, the purpose-designed monitor gives better results than the ordinary television set. Fixed cameras are of main value in watching a single vulnerable point with no necessity to obtain closer viewing. A wages office, a corridor, and a small compound would be suitable objectives. By linking several fixed cameras and using a simple switching arrangement a single viewer can inspect a whole series of vulnerable points from a remote position; or automatic switching from camera to camera at fixed intervals can be incorporated. Alternatively, a battery of monitor screens could be used so they could be seen simultaneously.

The practice of linking a fixed camera to a video recorder has become almost standard practice in banks and building societies where a massive increase has taken place in the scale of armed robberies. These have been so successful, as evidenced by the BBC *Crimewatch* programmes, that the time is surely coming when the first act of an intending robber will be to put a bullet through the camera. For this reason, thought should be given to camouflaging the camera if the environment permits. It is an excellent deterrent to such attacks.

### *Accessories*

The value of individual cameras can be very much increased by accessories to control the rotation and up and down movements of the camera – 'pan' and 'tilt'. In addition, protective housings can be fitted with screen wipers to offset the effects of condensation; cameras can have thermostatically-controlled demisters for operational use out of doors in all conditions; 'zoom' lenses can be used under remote control. However, it must be appreciated that the cost of these accessories, plus their means of activation, may be individually in excess of the cost of the original installation.

The quality of picture received depends on light and shade in the 'target area' which will of course vary at different times of the day and season. Automatic lens adjustment can be provided on the cameras to ensure an even brightness of pictures under varying conditions of scene illumination. Artificial lighting can be quite satisfactory internally and existing lighting on roads may provide usable pictures. Recent use has been made of special types of camera tube which are sensitive to infra-red radiation and have proved effective in conditions of apparent total darkness where the area under inspection has been lit by infra-red lamps mounted on top of the cameras. These vidicon tubes are about three times the price of the ordinary kind but have a similar life of some 2000 hours. In all cases where television cameras are to be used for security purposes it must be made certain that there is always adequate lighting in the areas to be surveyed for reasonable pictures to be obtained.

A further development has been the use of video-type recorders in conjunction with closed-circuit television; these record pictures for future reference which have been held as acceptable evidence (*R* v *Fowden and White* 1982 CLR 588). Specialist advice should be sought by interested parties; foreign equipment, of relatively low cost, can be quite satisfactory.

The continual watching of monitors can be a strain on the observer, in addition to which his attention may be distracted from the screen at times by other duties. A 'movement detector' is now available which will actuate a buzzer if there is movement on a previously static scene, and activate a video recorder.

There is progressive improvement in equipment, especially in connection with lenses of the 'low light' type which render vision possible in conditions which would previously have made it out of the question. The correct lens is essential to the effectiveness of the installation and it is better to request the best quality available for the particular application.

### *Installation*

Both time and expense may be saved if the supplier can carry out demonstrations or tests at the exact point where the camera is needed. Lighting may be inadequate, the field of view not that which was envisaged, there may be problems of mounting and cable run or distorting electrical emissions. At least one manufacturer has a landrover vehicle specially adapted to carry an elevating beam so that the camera can be placed in the exact operating position intended. This makes an excellent selling point when the expenditure proposal for the project needs support – get interested management to attend the demonstration. The presence of union representatives has also been found beneficial so they have advance notice of what is envisaged and can raise such comments as they wish. In one instance this led to an immediate request for all the employee car parks to be simultaneously covered!

A large item of cost is the cable run between camera and monitor. This is normally coaxial cable which has to be laid manually; with long runs, line amplifiers and other equipment might add to the cost, but it may be possible to utilize internal telephone lines or other developing techniques amongst which the use of fibre optics is predicted for areas of strong electrical interference. The availability of internal lines would be extremely useful for separate buildings divided by a public area which made use of coaxial cable impracticable. 'Slow Scan' equipment can be used for long distance surveillance via the ordinary telephone network. This is a relatively recent development in which the signal is transmitted as audio tones; definition is less precise than with ordinary equipment but this may only be a temporary shortfall and the system is an ideal one for linking with intruder alarms. It can be engineered to switch on simultaneously with the creation of the alarm state and may prove very helpful in monitoring false calls from a complex fault-prone system. It has been mentioned earlier, in connection with perimeter protection that closed-circuit television can be combined with other facilities for that purpose; fence interference detected by inertia switches and activating both cameras and video recorders as well as warning devices would be a case in point.

Technology in this field is developing at such a rate that what seems cost-prohibitive or environmentally impossible at first sight may prove at least reasonably feasible after discussion with specialist suppliers. Again, competitive quotations are advised, much of the standard equipment is of common origin and a deciding factor could be the availability of service though this equipment is normally less prone to fault than its detection counterpart. A substantial number of firms that have set up in this business have disappeared from the scene and it is recommended that firms of proven reliability be given preference despite any adversely-priced quotations.

### *Remote control of gates*

Economies in manpower, or constant supervision of an uncontrolled access can be effected by a combination of closed-circuit television and an electrically-operated gate release mechanism. Loudspeaker arrangements can be incorporated to challenge individuals, pass instructions, or answer enquiries. The cameras used are normally the fixed variety, though there is no reason why the pan and tilt facilities should not be fitted – other than price. At dispersed sites the extent of cable runs for the more expensive variety will probably cost them out of acceptance.

Optimum circumstances are where it is not economic to constantly man a gate or turnstile which is needed for a matter of minutes at shift changes only and used by a limited number of personnel. The system does not provide perfect security, except where the only property to be stolen is so bulky that it cannot avoid being seen, spot checks are really a necessary adjunct. Even where grounds of suspicion do exist, the egress of the persons concerned may be difficult to prevent until enquiries have been made. The siting of a fixed camera so as to observe both persons entering and leaving sufficiently to be able to identify is an added complication.

The same type of installation of course can be used at the entrance doors to high security risk buildings and to control the entry of vehicles into a restricted area. A successful television procedure in respect of these has been to allow access on visual inspection and verbal identification into an outer compound, itself sealed off, to await supervision of unloading or collection in the restricted risk area.

The combination of outer and inner gates remotely-controlled eliminates constant manning and ensures the driver cannot do anything overt until it is possible to directly supervise his actions.

Apart from the pure security applications, sight should not be lost of the potential of closed-circuit television for the remote observation and inspection of dangerous processes, where it also could have a very important safety application.

### *Retail store supervision*

This is an additional application in which television not only provides a means of catching shoplifters but also constitutes a monumental deterrent where it is known to be in use in a store. Cameras have to be sighted above normal head height which allows a much wider field of view than would be available to the normal store detective. Observers therefore have a much better chance of seeing shoplifters in action without bringing to notice their presence. Multiple viewing points can be installed in shop ceilings which will give a very wide field of coverage and these can be fitted to use sequential photography for a permanent record. These viewing points are hemispherical and have five 'eyes' giving an all-round field of vision; as these become better known the detrimental value of 'dummies' for future consideration needs no comment. They can be fitted with indicator lights, audio equipment or other variations that may fit the environment and purpose of use; they constitute a formidable deterrent to stealing by staff when the store is not in public use. Elsewhere in retail premises the loading and delivery bays, cash points and cash office are examples where cameras could be sited with advantage.

### *Surveillance by miniature television and automatic cameras*

Frequently there is shrewd suspicion as to who is responsible for internal theft, interference with papers, and petty malicious acts but an absence of sufficient evidence to justify action against the culprit. Preventive measures may put an end to the problem at least temporarily but will leave the suspect free to recommence operations as and when he pleases. In one instance an employee evolved an almost foolproof way of getting valuable material off the premises which was found out by accident and could have resulted in an easy and early capture but for stupidly premature stopping up of the avenue of removal. He quickly established a viable alternative which functioned for months before again being terminated by accidental discovery followed again by managerial and security action which can only be described as injudicious. Two years after the first method was ascertained, he was caught by the CID in the course of arresting a receiver of stolen property – he refused to say how his latest ploy was worked. In those two years he had gravitated from riding a bicycle to work to driving a 3-litre saloon! It is obvious from an example like this that there is virtue in catching a persistent thief rather than just making his task more difficult.

Advancing technology has produced miniature cameras and sensing devices to activate them which can easily be concealed in mock books, files, air grilles, etc. without being conspicuous. A passive infra-red sensor may be used with built-in battery supply to start up an automatic camera using 8mm ultra fast cine film coupled with wide angle lens. The camera may be programmed to take single exposures on cassette film at set intervals, be tripped mechanically, or started by remote control to produce a photograph of evidential quality if the lighting is adequate. Small television equipment with video recording can be used but this may be complicated by wiring and power supply difficulties.

Practical applications would be pilferage from canteens, drinks cabinets, and directors who rightly or wrongly suspect their offices have been tampered with. It should be noted that patrolling security staff have been responsible for these in the past and knowledge of the use of this special equipment should be restricted to the absolute minimum of persons who have to be informed. The cost is not excessive but cameras are available on hire.

## *VIDEO RECORDINGS*

Video-recording equipment is becoming increasingly used with success in most aspects of security work and will obviously progress further with its relatively low cost and the acceptability of its results by courts. Evidence from it is the basis of the massive prosecutions pending (in 1987) for the deaths arising from the Belgium soccer stadium tragedy. It will be noted that it has been advocated for the protection of practically all premises mentioned in the preceding chapters. Its evidential value is now fully recognized both in statute and case law (see p. 222); warning notices of its presence in conjunction with closed-circuit television has had a marked effect on the level of theft in some large retail shops – a good deterrent as well as a good detector.

A measure of court acceptability is reflected by a court of appeal finding in which the contents of a video recording were held to be admissible when, in the absence of the recording, they were related by the persons who had observed it. A submission that evidence of what witnesses saw on a video recording was not different in principle from their own evidence of what they saw in direct vision was accepted. Two comments on this case. First, the time may come in the none too distant future when video recordings will be preferred to conflicting evidence of live witnesses. Secondly, what happened to the missing recording? The reason is not reported, but was it that the recommendations for safekeeping of exhibits were not followed? (*Taylor* v *Chief Constable of Cheshire*, 1986).

# 36

# *Services of security companies*

The commercial service-supplying security industry, with the possible exception of computers, is the only one whose growth rate in recent years has equated with that of crime in the UK. This has opened the door to the springing up of small firms without expertise, finance, equipment, adequate manpower, and even, occasionally, integrity. Unfortunately, as the regular police have found, the misdemeanours of a distinct minority are more newsworthy than the behaviour and deportment of the vast majority. It is a matter of user regret that the Home Office has not seen fit to bring forward legal controls to exclude undesirables, both as individuals and firms, but has advocated regulation by the professional bodies involved. Whether this approach will survive political change is a matter for conjecture.

As far back as 1966 the British Security Industry Association (BSIA) was formed by the principal UK companies then providing security services and security equipment. The membership is divided into four sections, alarms, guard and patrol, safe and lock, and transport, each represented on the BSIA Council. BSIA does not represent firms providing private detective or private enquiry services nor consultancy. Its rules lay down clear parameters within which its members are expected to conduct their business and an inspectorate was set up in December 1982 in conformity with the Home Office conception of self-regulation. BSIA headquarters is at Scorpio House, 102 Sydney Street, London SW3 6NL, Tel: 01 352 8219.

Many of the smaller commercial companies are members of the International Professional Security Association (IPSA) who are likewise requested to regulate standards and integrity; the IPSA for many years have implemented a strict vetting procedure for membership and expelled several firms whose conduct has fallen below what customers are entitled to expect. Their Code of Ethics and Conduct (Appendix 29) show clearly their conception of what is essential.

## *SERVICES OFFERED*

Safes and locks with one or two notable exceptions are manufactured, supplied, and installed by different firms to those concerned with electronic alarm equipment, systems and maintenance. Other chapters of this book deal in detail with those subjects but there are two other important types of service available; transport of cash and valuables, and uniformed patrol and guard.

### *Transport of cash and valuables*

Initially this was confined to collecting and delivering cash for wages but has spread rightly to the collection of money from large shops and supermarkets for transfer to banks. Facilities have been created by all the main carriers to take over the make-up of wage packets for their clients prior to the payroll delivery and actually do the paying out if contracted to do so – which can be of advantage to clients operating a series of dispersed sites as in constructional work. This merits cost-effective appraisal as against use of own staff.

Inter-place movement of computer data is now widespread and the transport of valuable property will also be undertaken by special arrangement and contract. Firms with national coverage have a flexibility which enables them to give an overnight service for urgent or confidential information that might not be available from normal postal services.

### *Uniformed patrol and guard*

The conventional ways in which these are supplied are:

1 To use in lieu of a company's own personnel for complete cover on a 24-hour 7-day-a-week basis and subjected to periodic supervisory visits from senior personnel of the supplying company.
2 For complete weekend coverage when premises are vacated by customer's staff. This is more or less in the form of a fire-watching/ watchman commitment. Most service firms are unwilling to take

this work because of manning difficulties unless it is guaranteed to continue over a substantial period.

3 By pre-agreed routine visits to premises during shutdown periods. This takes the form of spot visiting to ensure that all is in order. A snag that might arise is that such visits may have to be done at predictable times because of other visits that the patrolling officer has to carry out. Also, delay at any point will wreck his routine.
4 To supplement a company's own security staff in emergencies – for example, sickness, holiday commitments, special occasions. This will be charged at a much higher rate, possibly even double the normal.
5 To provide security coverage in uniform or plain clothes at special functions or exhibitions, or to guard valuable property on display.

Service companies on request will supply and install clocking points free of charge for either their static or visiting patrolling officers. Some system for reporting incidents and other activity should be devised which can be scrutinized and filed. A spot visit by a senior member of the customer's staff at an unlikely hour might give some indication as to the efficiency of the service. The authors hesitate to suggest that an item meriting action and report should be deliberately left to test the service given – but it has been done.

### *Mutual aid schemes*

On repayment many security companies are prepared to accept calls from watchmen or isolated company employees at their own switchboards at prescribed times. If such a call is not received within the agreed period, the security company checks, first by telephone, then by visit of its own patrolling officers with notification to the police if this is not feasible. If a prescribed form of calling is agreed, any variation from it – for example, by impersonating the security officer or compelling him to make the call – will immediately cause suspicion and action.

## *PRECAUTIONS WHEN ENTERING INTO A CONTRACT WITH A COMMERCIAL SECURITY FIRM*

It has been mentioned before that standards vary considerably between security firms; the lowest quotation is no guarantee of the most cost-effective service – most probably the reverse. The more glowing the description of the facilities and capabilities of personnel on offer, the more reason to raise queries. The security 'consultant' is essentially a salesperson in a highly-competitive market whose own well-being is tied up with getting contracts; he is therefore unlikely to bring to notice local deficiencies of his firm unless he is directly questioned. There is no common form of contract and the wording of these should be carefully scrutinized – they are designed to protect the supplier, not the customer.

A most important House of Lords ruling is worth bearing in mind – *Photo Production* v *Securicor Transport* 1980 1 All ER 556. A night patrol employee of the security firm deliberately lit a small fire in the premises he was supervising, it got out of hand and caused damage to the value of £615,000, the firm was subsequently sued on the grounds of its liability for the acts of its employee. There was a liability exclusion clause in the contract between the two parties; the Court of Appeal to which the case was eventually taken held that there was a fundamental breach of contract by the defendants which precluded them relying on the exemption clause. However, on further appeal to the House of Lords, this ruling was reversed on the grounds that there was no rule of law whereby the clause could be over-ridden. Parties were free to agree to whatever exclusions or modifications of their obligations as they choose – this is now binding. The moral is obvious!

The first precaution to be taken lies in the selection of firms from whom to request quotations; the local police crime prevention officer will supply these and it can be taken for granted he will not nominate one about which he has reservations. An office in the near vicinity would be an advantage. Before arranging interviews with representatives, a clear idea of requirements should be decided and a list of points to be covered drawn up. Ample time should be allowed for discussion as each representative will have points on which he will need clarification, and will want to make a thorough inspection of the premises – though he may defer making up an actual instruction sheet for his guards until the contract is agreed.

It is suggested that a customer should ask at least the following questions and have the answers he deems important included in the contract or in other written form:

1. What assurances are given that guards will be persons of integrity?
2. Will they be fidelity-bonded? (See p. 40.)
3. Has the customer any recourse for non-performance, or negligent performance of contract?
4. What insurance cover does the supplier have?
5. Will regular people be used for the assignment, or will there be a succession who may be unfamiliar with the premises?

6 If the contract is one of complete cover, are the suppliers agreeable to new people being interviewed for acceptability by the customer?
7 Will casual or part time staff be used, if so when and why?
8 What training will the staff supplied have had, who has given it and what does it cover; are part-time staff equally instructed?
9 Has the customer discretion in deciding a guard is unsuitable and asking for his removal from the premises?
10 What are the termination conditions for the contract?
11 What will be the age limits for the guards to be supplied and are there any physical standards they must meet? What working hours are expected of them?
12 How will supervision be given to the guards? How will the customer know this is carried out? What checks will be made to ensure the job is being done?
13 What means of communication will guards have whereby they can request advice or assistance, or make periodic contact with their superiors?
14 What is the proposed system of reporting routinely to customers?
15 Will the guards be competent to make written reports or records?
16 Will they speak fluent English? (important in some areas)
17 What uniform will the guards wear and how will they be equipped?

Each representative will naturally submit a specification and quotation for what is required – the manner in which this is done may be revealing as to the efficiency of the firm. It should be made clear to each from the outset that competitive tenders are being sought. By comparison of the specifications an optimum way of doing the job will emerge and if a decision is made to use a particular firm, a check could advantageously be made to see whether their quotation would be affected by implementing that way. The customer should have the deciding word on where clocking points should be placed – at the places of highest risk and at the extremities of the premises to make sure the whole area is covered.

Adequate accommodation should be made available together with such stationery as may be needed; a book suitable for use as an occurrence book is essential. Also, a detailed plan of the premises showing the whereabouts of all fire hydrants and equipment. Instructions for all special tasks – switching on and off of machinery and heating etc. should be in written form and a list should be provided of all the employer's personnel who may be needed in an emergency.

Finally, before coming to a decision on the supplier, ascertain which other companies in the area are using their services and check whether they are satisfied customers – and *check the small print before signing*.

## *SPECIALIST SERVICES*

### *Consultants*

The comments made in connection with risk appraisal (see p. 6) merit repetition in the present context. When it is decided that an assessment of the state of security in a concern should be obtained this can be done by employing a security consultant. He can be an individual who is self-employed or a member of a security services company. Before employing the latter it should be borne in mind that he has a vested interest to encourage the engagement of the services of his company. Therefore an opinion based on good professional experience by someone independent of such companies has certain advantages. There are no criteria of the capability required to warrant the description of 'consultant' and personal knowledge or recommendation is desirable.

### *Investigators and store detectives*

The employment of such persons can be for a specific inquiry to find out the causes of losses of goods or cash in circumstances outside the responsibilities of the police. Store detectives can be hired hourly, weekly, or for longer periods for duty in one or more premises belonging to the hirer. The employer of the security company personnel must satisfy himself he is indemnified in writing by the company against a claim of civil damages which may result from their action.

Private investigators frequently offer their services for detailed vetting of prospective employees where it is thought necessary to go beyond the normal references to more personal details which may render him/her less acceptable for a particular post. Some specialize in patent and trademark infringement investigations, a lucrative market with few alternatives to their use. Insurers and firms may engage them to obtain evidence to discredit the alleged degree of injury in industrial accident claims – the use of own employees for this (they would have to be at least of supervisory level) is timewasting, embarrassing and can lead to strained industrial relations.

*Keyholders*

To reduce the managerial inconvenience of attending premises at night and weekends in consequence of something occurring which requires attention, usually brought to notice by police, security services companies will hold the keys and become registered with the police. Should their visit to premises result in no further action being necessary, the instance is subsequently reported to the customer. If the incidence is such that the authority of a representative of the owners of the premises is immediately necessary a previously nominated person is informed at once.

*Undercover investigations*

Member companies of the BSIA have an undertaking that they will not deploy undercover agents and perhaps a majority of non-member commercial security firms take the same view. Private detective agencies, however, are much more flexible and many will supply such a service but it must be anticipated that the exercise will be expensive to the client, and a conclusive result cannot be guaranteed.

Continuing frauds and thefts with certain employee involvement are the main instances where the use of an agent might produce results. There are two ways of looking at the ethical aspect of doing so; it is a desperation measure when all else has failed and the continuance of the situation is unacceptable, or it is a method of combatting illegality which is not in itself illegal and the end justifies the means. These may be over-simplifications but the policy decision is not an easy one and the negative factors must be fully evaluated against the possibility of a positive conclusion.

The decision to use the agent must be kept secret and confined to the absolute minimum of senior people; his identity must be restricted to the knowledge of one only, if this is feasible; his recruitment must be such that it attracts no attention; his presence in the particular environment must be acceptable and he must be competent to carry out the required tasks in a manner which again attracts no attention; he must be placed sufficiently near to the trouble source to have a reasonable chance of acquiring the desired information without his special interest being noticeable; he must have a safe means of communication if speed in doing so is crucial; the evidence expected from his presence should be exploited without his identity and purpose becoming known. These considerations alone may make the proposal a ‘non-starter’, but there are others. If the agent is not sufficiently experienced the suspects may sense his purpose, he may then be at personal risk, his position could be made untenable in a variety of ways, or he would be made known to the workforce in general – which would generate all sorts of industrial relation problems, or the dishonest operation might be closed down until the cost of his continued retention with no end-product causes the firm to terminate the contract.

Further problems could arise from the way the agent achieves his results; he may embellish facts to make his performance seem more impressive resulting in a misleading picture of what is going on – even leading his clients to take misguided action with painful consequences; he may even provoke the commission of crime, *agent provocateur*, to expedite his work; or be stupid enough to join in to the extent of being criminally responsible. The only criteria for judging these possibilities are the reputability of the supplying firm and the experience of other users.

To achieve success in minimum time an agent has to be very thoroughly briefed about the problem and the most likely suspects. He must be warned about any sensitive aspects of the operation and told the type of evidence that is being looked for from him; also, the way in which such evidence will be followed up. All personal contacts should be made well away from the premises in which he will be working. 'He' has been used but more frequently than not in retail business, the agent may be a woman.

The decision to use undercover agents should be confirmed at the highest level likely to suffer embarrassment if things go wrong; there is apparent reluctance in the UK to use the practice but this is a matter for the person who has responsibility to make the decision in the light of all the circumstances and after weighing the good or bad consequences that might ensue.

## *USE OF SECURITY COMPANY AS AGAINST OWN STAFF*

*Cash carrying* – Where the risk is adequate, a cash-carrying company should always be used in preference to employer's own staff.

*Alarm installation* – It is always advisable to use a professional alarm installation company, as internal electrical engineers will not have the necessary specialized experience and it is better that knowledge of the system be kept from employees so far as possible. Specification of what is required should not be left entirely to the installers; it should be checked on every occasion with the police, and their views should be given priority.

*Part-time cover* – In part-time guarding or patrolling it will probably be impractical to utilize company personnel for weekend coverage only. This also will almost certainly apply to spot visiting during shutdown periods.

*Stores* – Not all service companies have a store detective section, but where they do the staff are almost invariably adequately trained and reliable. A small proportion actually specialize in this work and in the supplying of 'customers' to make test purchases.

A large stores group may have adequate need to run its own section and supplement this only at peak periods from security companies. Smaller units will find advantages that accrue from periodic use of the service if a full-time staff is not economically justified by the known or suspected losses. In a single large store the ideal could be a resident officer in charge of security, supplemented by visiting contract detectives.

*Full-time cover* – Serious thought and costing must precede a decision to utilize a force of a firm's own choosing for full-time cover as opposed to employing a security company. A wide range of factors have to be taken into consideration.

1 There is a rapid turnover of personnel amongst a service firm's employees; over 100 per cent per annum is not unknown. However, where their staff are assigned to a regular commitment this change is very much reduced. Nevertheless, it does create a situation where there will be a number of persons, unconnected with the firm or its current guards, who have intimate knowledge of the premises.
2 The customer has to accept the standards of training, physique, integrity, etc. deemed by the service company as acceptable.
3 Many in-company forces can be trained in ancillary jobs which effect economic savings to their employers. This is infinitely more difficult with the staff of the service firm, and impossible where they are regularly changed.
4 Specialist training in first aid, fire, weighbridge, and other duties, easily done with employees, is barely feasible with contract guards.
5 In favour of a service firm: the customer is completely relieved of the administration and the overheads normally incurred – also staffing problems which would accompany holidays, sickness, etc.
6 Guards employed by security firms are generally paid less than their in-company equivalents. It is therefore logical that the types of person most suitable for the job will gravitate towards in-company status.

*Costs* – In estimating the expense it must be anticipated that there will be an on-cost applied to the basic wages outlay of the security firm, which will take care of their own administrative charges and give them a reasonable profit.

It is difficult to avoid prejudice in making recommendations as to which is the more desirable. Obviously, if a firm's selection and training is what it should be, a more efficient service can be provided by its own personnel. Goodwill can be built up, continuity provided, and detailed knowledge of premises, processes, and persons acquired. Perhaps most important, the staff will have no split loyalties and can identify their interests with the firm's. On the other hand, if the need is a temporary one, perhaps because of intended replacement by electronics, there is no object in recruiting a force which is to be made redundant within a foreseeable period. A service company can best meet the requirement in this case.

## *VICARIOUS LIABILITY*

Contract security firms and customers should take note of the ruling given by the Court of Appeal when hearing three similar cases consecutively. The relevant one was *Heasmans* v *Clarity Cleaning Co.* 1987 IRLR 286 in which a contract cleaner used the customer's phone to make a series of overseas calls costing £1400 – which the customer tried to recover from the cleaner's employer. The common decision, confirming previous rulings of inferior courts, was 'An employer is not vicariously liable for the act of an employee outside the scope of the employee's employment'.

A firm suspecting that this is happening would be well advised to cut their losses and insist on a replacement, or they could seek BT's advice and assistance to confirm their suspicions and then get rid of the offender.

# 37

# *Guard dogs*

It is only in recent years that the recognized qualities of guard dogs have been applied to police and security purposes. Even now industrial firms do not use their full potential, and the Guard Dog Act 1975 has deterred commercial security firms from expanding their 'dog handler' sections.

The Alsatian is generally recognized as one of the most intelligent breeds of dog in the world. This is the essential quality to ensure that, in addition to being fearless without being aggressive, they can be eventually trained to act as companion dogs, guide dogs, guard dogs, or working police and security dogs. They have an appearance which is consistent with the general conception of what a guard dog should look like: sturdy build, combined with a physical ability and strength to work hard in all types of weather. Other types of dog have been used for the same purpose but these have become increasingly restricted in numbers; the only other one which now is in general use is the Dobermann which is widely used on the Continent but has found little favour in this country; breeding is less likely to guarantee planned development of required characteristics and it is accepted to be less predictable in behaviour. For these reasons it should be assumed that when guard dogs are referred to in this book they are Alsatians.

## *ALSATIANS' INSTINCTS*

There are characteristics of the Alsatian which render them to be of particular value to users; these are of a hereditary nature arising from their origin in pack life – hence the alternative name, wolf-hound. They have a real sense of self-preservation, quick reaction to danger with an alertness to every sight and sound; loyalty to those with whom they are in regular contact coupled with a suspicion of strangers, and a quickness of supple movement. They are comparatively short-sighted but their range of hearing covers seven octaves above the human ear and is infinitely more acute. Their power of scent is even more greatly developed in comparison – it can warn them of the presence of intruders or strangers even when they are motionless and cannot be seen or heard. Having an ancestry associated with cattle and sheep herding, under all conditions, they can pass instantly from sleep to alertness. It is no hardship for them to work up to 14 hours a day if this should be necessary, although for humane reasons they should not be worked for periods of this duration. While they may be used every day, they will reflect the benefit if one clear night's rest each week can be allowed. There is no difference whatsoever between dogs and bitches in security usage – they are equally efficient.

## *ACQUISITION OF GUARD DOGS*

Even police forces and the commercial security firms with their large numerical requirements rarely find it either advisable or economical to breed their own dogs. This is a specialized field and there are kennels owned and manned by persons who have spent a lifetime in this work and can supply dogs to meet all needs. Over a period of years these breed to eliminate undesirable characteristics and to select those which are of value for particular purposes. Dogs, like human beings, vary considerably in their mental attitude to training and in their ability to assimilate it; the more experienced their handler the earlier any shortcomings in the dogs will be observed and the less time invested in training a dog which lacks essential qualities. To acquire a dog of unknown background is to run a risk of complications and unforeseen behaviour which is best avoided.

## *SELECTION OF DOGS*

No breeder of repute will risk selling a dog which may reflect unfavourably upon his standards; the more reputable the establishment the more certain it is that any dog purchased from it will conform to requirements.

An expert will have detailed and minute points for a dog to match up to, but the main general characteristics that are desirable are:

1. An appearance of alert watchfulness to both sight and to sound, an interest in everything around him, and an immediate and unafraid suspicion in the presence of strangers.
2. Well proportioned, strongly boned, muscular without being too heavy, smooth and supple in movement.
3. A long lean head in proportion to the body size, black nose, and a long strong muzzle.
4. Ears rather large, pointed, carried erect giving an alert expression; eyes normal – not bulging.
5. Muscular broad-backed body with straight legs viewed from either front or back.
6. Feet should be round with strong toes, firm pads, and short strong claws.
7. The outer coat should be straight, hard, and lying flat, the inner coat woolly in texture, thick, and close. The tail should normally hang down in a slight curve.

These are general points from which a would-be purchaser may gauge whether he is being correctly informed by the breeder.

## *USES OF DOGS*

The value of a dog lies in the full utilization of its basic instincts. However intelligent they cannot replace the presence of humans, but correctly trained and handled there are many instances in which a dog and handler are more valuable than two people. In any case, the view has been expressed by a handler, accustomed to work in a dangerous area, that whatever else his dog did it gave him the confidence to go anywhere at any time. The main uses of dogs are:

1. As an aid to more thorough and effective patrolling by using their organs of sense, smell, and hearing to detect the presence of strangers.
2. As a means of apprehending intruders. Few will run in the presence of a dog and they will have little chance of escape if they do.
3. As a protection to the handler. Their appearance and, perhaps undeserved, reputation makes them a formal deterrent to the use of force against their handler.

4 To search premises in darkness or with an involved layout and find intruders quickly. This is of particular value in complicated premises where a person can easily hide himself from the security officer but be detected by the dog's sense of smell.
5 Utilizing their sense of smell to track intruders from the scene of an incident.
6 Searching for small articles lost over a wide area.
7 As a deterrent where there are signs that a situation of potential violence is building up.

So much importance is attached to the presence of dogs that merely the display of a notice warning 'danger, guard dogs' has a marked effect in keeping intruders away from premises.

To these industrial and commercial uses have been added more specialized roles to which they can be trained by virtue mainly of their excellent olfactory capabilities. The most important of these is that of searching for drugs where they have proved extremely effective, as they have with explosives, and the Armed Forces have tried them for mine detection. A public use to which they have been put is as an aid to mountain rescue and the like for the tracing of persons lost or injured.

## *THE LAW AND GUARD DOGS*

The most relevant law is the Guard Dog Act 1975. Brought in as a Private Member's Bill and given a hurried passage through Parliament, it resulted in the withdrawal and death of many dogs and the consequential removal of their protection from small premises. Nevertheless, it is now the governing legislation though prosecutions based upon it are extremely rare. It does not apply to Northern Ireland. Its predecessor, the Animals Act 1971, has not been repealed, but there is no evidence that it has been used to any extent; it may be one of those items of English law which nearly everyone prefers to forget.

Under the provisions of the Guard Dog Act 'guard dog' means a dog which is being used to protect:

1 premises; or
2 property kept on the premises; or
3 a person guarding the premises or such property.

The section of the Act of prime concern in security work is s. 1:

(1) A person shall not use or permit the use of a guard dog at any premises unless a person ('the handler') who is capable of controlling the dog is present on the premises and the dog is under the control of the handler at all times while it is being so used except while it is secured so that it is not at liberty to go freely about the premises.
(2) The handler of a guard dog shall keep the dog under his control at all times while it is being used as a guard dog at any premises except—
(a) While another handler has control over the dog; or
(b) While the dog is secured so that it is not at liberty to go freely about the premises.
(3) A person shall not use or permit the use of a guard dog at any premises unless a notice containing a warning that a guard dog is present is clearly exhibited at each entrance to the premises.

The subsequent sections, which relate to licensing of kennels, have not yet been brought into effect, and there is no indication as to when they will. In any case the section dealing with interpretation excludes a dog 'which is used as a guard dog only at premises belonging to its owner', thereby also exempting in-company use from kennelling provisions. If the dog owner keeps dogs for use elsewhere than on his own premises, his kennels will in due course have to be licensed.

Section 1 quoted above is self-explanatory, and the only real query was resolved by a case before a Divisional Court on 12 October 1977 – *Hobson* v *Gledhill* 1978 1 All ER 945. The point at issue was whether handlers should always be on the premises while a guard dog was being used, even though the dog was secured so that it was not at liberty to go freely about the premises. Home Office guidance notes indicated that the handler should always be present; the Divisional Court decided that this was not necessary but did state that interpretation depended upon the circumstances of individual cases. In the instance in question the dogs were on chains which restricted their area of movement and left space in which any intruder was safe. In the authors' opinion, in the absence of a test case, it would both be unwise and illegal to use a dog on a running line, that is, a lead attached to a ring on a taut wire between two points.

The essence of the Animals Act 1971 is that, if a dog is known to be vicious and does inflict a severe bite, the keeper is liable unless the injury is wholly the fault of the person suffering it, or the person voluntarily incurred the risk, or the injury is to a trespasser and caused by an animal kept on any premises or a structure in circumstances which are not unreasonable, for example guard dogs. It is specifically stated that an employee bitten in the course of his employment will not be regarded as accepting the risk voluntarily. This does not appear to be a piece of legislation which should give undue concern to security management.

## *HANDLERS*

It must be clearly understood that where a dog is used with more than one handler its efficiency will not be that of a dog which has only one master. This is inevitable since no two persons can have the same attitude towards a dog which is not their sole property, nor will they place the same emphasis on their words of command. The main use of guard dogs is, of course, at night and it therefore follows that should a decision be reached to utilize one dog with one handler, that person is committed to permanent night duty unless the staff is big enough to employ sufficient dogs and handlers to have round-the-clock cover. Though special financial allowances will have to be made for this unappealing perpetual tour of duty, experience has shown that firms with smaller security staffs have difficulty in keeping a dog handler for this task for any length of time.

It is possibly this limitation, which has been widely publicized, that prevents more firms making use of guard dogs. Whilst it must be reiterated that a multi-handler dog cannot be as efficient as one with one permanent handler it still has a massive value, sufficient to justify its keeping. War-time experience showed that a dog need not be considered expendable purely because its handler was killed or wounded; it could be readily transferred to another handler and it is a short step from that practice to accustoming a dog to a series of handlers – multi-handling.

## *TRAINING: DOG AND HANDLER*

Serious training of dogs starts at the age of adolescence. This can vary between different dogs from the age of seven months to fourteen months, according to whether the dog is a slow developer or not. It should already have been given some elementary obedience training by a professional. This is the make-or-break stage with a dog; if it is not trained properly then, bad tendencies are likely to persist later. Similarly, there is no specific age at which a dog becomes suitable for practical usage. This will depend upon the expertise of the trainer, in addition to the age of adolescence which is indicated by the dog suddenly becoming aggressive towards other dogs in its presence. Generally speaking, a dog which is purchased from a training school will have been given basic obedience training and will have had its most undesirable traits corrected. Most important, it will have become accustomed to receiving instructions.

The training of dogs requires a clear mind and the ability to observe and study each dog individually to diagnose the degrees of loyalty, tenacity, cunning, stubbornness, wilfulness, ability to absorb instructions, and other desirable qualities. This is where the sheer professionalism of dog training shows its real value, for unconventional methods may be needed with a particular dog which would be beyond the capabilities of a novice handler. Transformation of raw material into a trained dog cannot be achieved overnight. There must be systematic and planned periods allocated to each day and patience must be exercised by the trainer who, by his experience, will know when a dog has reached the point of frustration. He will then divert it to an exercise with which it is fully conversant, then flatter it and return it to its kennel.

Even when a dog will have to be multi-handled, when it is transferred from the kennels to its permanent home, it is advisable to associate it with one particular handler. Training men for handling is at least as vital as the dog's own training – or even more so. If one particular individual can be designated he should spend preferably two weeks at the kennels with the dog before it is brought back to the firm. The first week would be spent mainly in developing an association between the dog and the handler and giving the handler instruction; the second week would be basic training with dog and handler working together. During this period the handler should obtain expert advice from the kennel supervisor, not only on the future training, but on health, housing, and any particular problems or tasks which the dog will subsequently be asked to perform.

If a person is to be trained solely for duties as a dog handler the first essential is that he must be already a sound and experienced security officer, even-tempered, patient, and mentally alert. If he is a cheerful individual this will be reflected in the behaviour of the dog, and if he has a strong character and persistence to achieve his objectives the training of the dog will be much more thorough. An irritable and bad-tempered handler will soon ruin a good dog.

It is those latter characteristics of handlers which have to be watched for where a dog is being multi-handled. It is easy to ensure that a standardization is achieved in words of command that are used; it is much more difficult to see that the dog is treated equitably and reasonably by all who have its handling. A dog, like a child, will soon realize and react to variations in the degree of control which is imposed upon it, particularly if one individual acts maliciously or sadistically towards it.

Where a dog is used by a number of handlers it seems even more essential to display appreciation of good behaviour and good work by making a fuss of it. It is in the human sense 'deprived' by not having a single person to look up to and, in this sense, the wearing of a security uniform can be an advantage as it can tie the dog's affection to a particular body of people more easily. One of the virtues of Alsatians is that they are a suspicious breed, which makes it all the more essential to assure them that they are regarded with affection and to keep them amply fed and warm.

Before a dog is taken into full use at a firm it must be given a period of familiarization with all the sounds, sights, and smells of the place where it is to operate. This time should also be utilized for training in the particular tasks which are to be required of it. The individual who has already had experience in its handling can, during this period, introduce to the dog those of his colleagues who are to assist in its handling and ensure that they are fully conversant with the correct words of command, in effect passing on the knowledge which has been imparted to him by the professional who has supplied the dog.

The training must not stop when the dog is taken into use for the first time; a specific period should be allocated each day during which the dog should be given obedience training and progressively more advanced training. Since this will probably be done by persons who are not specifically professional trainers not only will the process be lengthy but certain faults may creep in. For this reason it is advisable, after the dog has become thoroughly acclimatized, for it to be returned for a period, say a week, to the kennels from which it came for a form of professional check-over, comment on how the training has been carried out, and the formulation of a further programme.

### *Obedience*

An excellent handbook, *Police Dogs, Training and Care*, is published by HM Stationery Office, and deals at length with obedience training; reference should be made to this book for the detailed manner in which it should be carried out.

Standardized words of command are used throughout and obedience is brought about by serious habit-forming exercise to specific orders. Patience is required, for it must not be expected that a dog, no matter how intelligent, will become immediately aware of the meaning of a new command. It must be shown what is needed and there must be repetition, until it automatically associates the command with a required action.

The training must be progressive and care must be taken to avoid boring both dog and handler. Single words of command must be used wherever possible and in the early stages of training it is better that the dog should be on a lead so that it can be guided into the action that is required. Disobedience to or disregard of commands must not be allowed, but physical punishment should not be used; use of the choke chain and a reprimanding tone of voice show disapproval. Praise should always be given for good performance. At the conclusion of any exercise period one of the simpler procedures should be carried out so that the dog can be praised and encouraged before being put away.

In multi-handling, each person who has the dog should be taken through the full sequence of obedience training to ensure that he is familiar with it, and the dog accepts him as its master. Every time it is taken out it should first have a ten-minute spell of obedience training to accustom it to accepting commands from the current handler.

*Tracking and searching*

There is wide variation in the abilities of different dogs in both these fields. The sense used is that of scent; an intruder may leave a trace on either the ground or in the air of his presence or passage.

*Ground scent* – This is caused by disturbance of the ground – movement of the soil or crushing of vegetable life – causing the emission of a scent which to a degree adheres to the outerwear of the intruder.

*Wind-borne scent* – This is the scent given off by the person of a particular individual and his clothing. A dog can make use of either of these in his efforts to track.

Climatic conditions have a great effect upon the work of a dog for these purposes, thus a scent is persistent:

1 in dull mild weather with little wind;
2 in low-lying areas where there is little air movement;
3 at night time when the temperature of the ground is higher than that of the air.

Conversely, scent is soon dispersed in:

1 hot sunshine;
2 winds;
3 heavy rainfall.

There is a limit to ability to track on footpaths and road, pedestrian and motor traffic quickly break up any scent and ground disturbance is negligible. The age of a track is a major factor and too much must not be expected of a dog which has not been thoroughly trained specifically for tracking.

A dog's obedience training includes teaching him to 'speak', that is to bark on particular occasions. This is of importance in searching for intruders – there is no desire that they should be bitten and it is adequate for a dog to stop their escape and indicate their presence by barking. This, in itself, will have the desired effect in rendering the criminal submissive enough to surrender readily.

### *Attack*

A dog must be thoroughly advanced in other aspects of its training before this is taught, otherwise serious injury may be caused out of all proportion to the offence which is being committed. Moreover, the dog must not anticipate that every time it sees a person running he is doing something wrong – this could have complications at any factory when employees are leaving. With the use of a protective sleeve a dog should be taught how to attack the forearm only, and to let go immediately on the word of command 'finish'.

*Police Dogs, Training and Care* deals with all matters of training at considerable length, with suitable diagrams.

## *KENNELLING*

The best position for a kennel is near a works entrance where it is in full view of everyone and within full sight of the gate office so that employees have no opportunity to bait the dog and supervision can be given to ensure that no poisonous material is thrown into the run. Whilst warning notices have got a preventive value, the obvious physical presence of a dog is even more beneficial.

The kennel itself should either be a permanent or semi-permanent structure. Whichever it is it should have certain characteristics. It must be:

1 Waterproof and windproof.
2 Correctly ventilated so that there are no draughts but yet a reasonable admission of fresh air.
3 Of sufficient size not only to allow the dog to lie down in comfort but also to allow the entry of a handler to clean out the bedding and interior.

4 Fitted with a removable floor, a few inches off the ground for warmth and easy cleaning.
5 Connected with an outer run for exercise purposes.
6 So placed that rain water and moisture will run away from underneath the kennel and the enclosed run.

For both prestige and psychological reasons the kennel and run should be well made and substantial in appearance if they are in the public view.

Either sawdust or straw should be used for bedding purposes, inspected and shaken-up regularly and swept out for cleanliness as and when required.

The dog should be thoroughly groomed each day with a brush and comb. The coat should first be brushed against and then with the direction of the hair growth, then combed to remove dead and tangled hair. Apart from keeping the coat and skin healthy and clean this helps to keep up the appearance of the dog which, like the dress of a security officer, reflects upon the efficiency of the department. Bathing is seldom necessary and may remove the natural grease in skin and coat, thereby being detrimental to health and appearance. If it is necessary to wash the dog, soap must not be allowed to enter the eyes and ears and it must be thoroughly rinsed off before the dog is dried with towels.

## *FEEDING*

Adult dogs need to be fed only once a day. Opinions differ somewhat as to the best time to do this but once a time is fixed it should be made a routine to avoid disturbance of the dog's digestive processes. Where a dog is used overnight it seems best to feed it in the morning immediately before putting it back in its kennel to sleep. An adequate supply of clean water is essential for a dog and this should be available at all times.

A full-grown working Alsatian needs about 3 pounds (1.4 kg) of mixed meat and biscuits, or their equivalent, to supply the energy it requires and the necessary protein, vitamins, and minerals. Meat and/or fish should supply about half the total, either raw or cooked. If bones are given they must be raw and only the larger type of beef bones that will not splinter. Food should be prepared as near the time it is to be eaten as possible. The dog should be left to consume it in peace, without being diverted by grooming or anything else. Frozen meat must be thoroughly thawed before being mixed.

## *HEALTH*

A sick dog suffering discomfort or pain may act totally out of character, which, in an industrial environment, could lead to considerable trouble with personnel. Handlers must recognize the symptoms of illness since a dog, naturally, has no means of telling anyone about them. If there is any suspicion of it being ill, rest, quiet, and warmth will go a long way towards remedying the condition; if any illness appears to be serious or there is doubt as to what it is, a veterinary surgeon should be called in without delay.

Usual minor illnesses are those of constipation and diarrhoea; both are due to an improper diet, possibly coupled with a chill of the stomach. If these are recurrent, checks should be made that the kennel is not draughty and causing a chill to develop. Worms are also a common cause of complaint; there are on the market proprietary drugs to deal with all these conditions and instructions are clearly shown on the packets. Regular inspection should be made of the dog's ears for excessive wax or irritation due to foreign bodies; great care should be used in dealing with the ears and again veterinary treatment may be advisable. Canker is a fairly common ear complaint in some breeds but is fortunately rare amongst Alsatians; this has a foul smell and causes the dog to scratch and shake its head – it definitely needs the early attention of the veterinary surgeon. Feet and eyes are other points which require close periodic inspection.

## *POINTS IN USAGE*

Different environments will require different techniques in the general use of dogs. Whenever there are employees about the dog should be kept on a lead, unless it is one hundred per cent trustworthy. There should be a gradual build-up for a new dog in the peculiarities of its new environment. This includes getting used to employees in mass without desplaying signs of nervousness or aggression by 'giving voice'. If the dog does misbehave it must not be struck under any circumstances, except in an absolute emergency. A well-trained dog will probably attack if it sees a raised foot. Adequate reprimand is by scolding and using the choke chain. The handler should be on the alert at all times and ready with the command 'finish' to prevent unnecessary action by the dog.

On normal patrol the dog should be kept on a lead to ensure that it remains close to the handler and also to prevent incidents with employees. It should be trained to walk quietly to heel on or off the lead, unless it is ordered to do otherwise. Its natural reaction, if the handler is attacked, is to counter-attack and it will evade kicks and blows by natural instinct.

The dog can be left in a vehicle or chained to something that it is required to protect; it must be remembered that doing so restricts the dog's ability to defend itself and as much latitude to move as possible should be allowed. It is possible for a dog to protect a considerable area inside premises by tethering it by a chain to a fixed point where it can cover important stocks or access to buildings, but due regard must be observed to the provisions of the Guard Dog Act 1975. However, this also gives more opportunity to attack the dog by throwing poison or missiles at it; if it is used in this fashion it should be visited at regular intervals. The law no longer permits guard dogs to be turned loose in buildings or compounds during the night; it is an offence to do so.

Extra care should be observed where the dog has to be used in an environment where there are acids or metal which could cause severe injury to its pads. Similarly, where (as in many industrial premises) rat poison is put out the dog should be kept on tight leash otherwise it might die in agony. Under no circumstances should it be turned loose for guarding purposes in any premises where any of these dangers exist.

## *CONCLUSION*

To get the maximum value from a dog it is necessary to use one handler only if this is an economical proposition. Training must be thorough initially and continual thereafter to maintain the sense of obedience in the dog and progressively extend its intelligence. If it is engaged in incidents which cause complaint these must be thoroughly investigated to ascertain the cause and prevent recurrences. If the dog is at fault this fact must be admitted and no blame laid on the person who is objecting to its behaviour – but it must be borne in mind that the allegations may be malicious or made for an ulterior motive. Above all, where a dog is multi-handled, there must be absolute uniformity in orders and treatment by all concerned. Firmness and affection to the dog will be rewarded in like measure.

# 38

# *Locks, keys, safes, strongrooms*

Despite sad experience, there still exists a tendency to regard anything given the magic name of 'lock' as adequate protection against any form of intrusion. As in everything else, the true value of what the buyer gets is reflected by what he has to pay. The price range of locks and safes in general is wide, but there is little variation between the cost of those of equivalent strength produced by reputable manufacturers.

## *LOCKS*

In effect, locks can be regarded as falling into two broad categories; those that provide privacy and those that provide security. It is incidental that the latter also give privacy, but the former only give security in the imagination of the user or the advertisement of the maker. Needless to say, locks of the first type are the simpler and also the cheaper.

### *Night latch or pin-tumbler rim lock*

These locks are fastened to the face or 'rim' of the door and the locking device is a spring-loaded bolt, which engages in the staple or bolt-housing when the door is closed. The principle is that the correct key will align a series of pins and tumblers, revolve the cylinder, and withdraw the bolt. These locks have the advantages of being low in cost, easy to fit by any amateur handyman, unobtrusive in appearance, and usually trouble free in use. Moreover, their keys are small and can have an immense number of variations or 'differs'. These are quite easily the most commonly fitted locks in Britain for internal and external house doors.

Unfortunately, such locks rarely give any trouble to a determined thief and their limitations are no secret. If a partial list of the means whereby they can be defeated is given, it is to be hoped that this may deter persons from placing too much trust in them.

1 The door or an adjacent window will almost invariably provide a breakable pane of glass from which the operating knob inside can be reached and turned.
2 Both lock and staple are surface mounted and the staple is held by only two or three screws. A firm kick will remove these with the minimum of effort or expertise. The body of the lock itself is probably only secured by four similar screws.
3 The bolt is spring loaded and, unless the locking catch is down, if a thin sheet of mica, or other flexible strip, can be worked between the door edge and the staple, sufficient pressure can often be exerted on the head of the bolt to force it back against the spring. The more forcible intrusion of a screwdriver at the same point can give identical results.
4 By clamping a pipe wrench on the outside edge of the key cylinder it can be turned bodily – thereby opening the door.
5 The key mechanism can be drilled out or the entire cylinder pulled out by force.
6 Duplicate keys are readily available to anyone who knows the number and they can also be easily made if possession of the key is obtained for even a limited time.
7 Most of these locks can be picked with patience and some experience – though this is rarely necessary in the light of easier alternatives.

Manufacturers have overcome some of these weaknesses by various steps, all of which, of course, increase the price of the lock. The main improvement is that of dead-locking the latch and securing the operating handle on the inside, by either an extra turn of the key or other means according to the pattern of the lock. Nevertheless, it is not liked by insurance companies and, while it gives a degree of privacy against the casual caller who hopes to find an unlocked door, it gives little real protection.

*Box locks*

These too are fastened to the face of the door and the bolt housing to the face of the door jamb. Again, these easily give way under bodily pressure since the usual means of fixing is by screws. Moreover, they usually operate on the 'ward' principle; obstructing plates are placed between the keyway and the operating cam, and the key is cut to miss the obstructions. The result is that any key which is cut away simply to leave the operating part will provide an effective 'skeleton' key. Extra strength can be given by putting strapping over the lock and the housing on the door jamb. With the older types of this lock it is even possible to get hold of the end of the key with a sharp pair of pliers and turn it from the outside.

The two above-mentioned forms of lock offer a low security value but even with the better types of lock which are available, it is essential that the door itself should be of robust construction. It is useless putting a most expensive mortise lock on a flimsy door with a weak and narrow jamb. If a door is faulty, in that the edge is not close to the locking plate or staple, it will be possible to spring out the bolt by means of a jemmy.

*Mortise locks*

When installed in a suitable door a good-quality mortise lock eliminates many defects of those previously mentioned. Both lock and the housing for the bolt are fitted inside the woodwork of the door and jamb and will therefore stand considerably greater pressures. These locks operate on the 'lever' principle: they contain thin strips of notched metal held in position by springs and the correctly cut key brings all the levers into alignment and allows the bolt to be moved. It follows that any increase in the number of levers brings additional security out of proportion to the multiplicity of key variations that become possible. If these locks are made to operate from both sides of the door – 'double sided' – this reduces the maximum number of basically different keys; in the case of the four-lever mortise lock this is only nine for a particular type, though additional key variations may be obtained by the use of wards on the inside of the lock case. Nothing less than a five-lever lock is normally regarded as a security lock under present conditions; if this is operative from the one side only a very high number of differs are afforded – some 3,000 as opposed to 125 with a double-sided lock.

A striking plate of 8 to 10 inches (20 to 25 cm) in length is desirable to receive the bolt; the best types include a small box of strong metal behind the opening for the bolt to prevent the use of end pressure upon it to force it back into the lock case – this would be possible if there were purely a wood surround to the bolt end.

When these locks are used in doors where it is possible to attack the bolt by means of a hack-saw – in double non-rebated doors – better quality locks must be fitted which incorporate hardened steel rollers in the bolt. When reached by the hack-saw blade the rollers revolve freely giving no bite to the saw edge.

The numbers of levers can be progressively extended, at least one firm operates a series of locks with no fewer than ten levers providing almost limitless combinations for keys.

An alternative to the lever mortise lock is the cylinder mortise lock in which the lever mechanism is replaced by a cylinder with almost infinite differs. An appealing advantage of these lies in the small size of key that can be used. The question of 'master keys' will be dealt with later.

### *Combination or code locks*

The substantial forms of these are usually fitted to safes and strong-room doors. They have a number of discs each cut with a small notch. When the correct code is set on the dial, the notches are in alignment and the bolt can be withdrawn. The number of different combinations which are possible is, of course, dependent upon the number of discs in use. To give an example, if two discs only are utilized in conjunction with 100 different markings on the dial, the possible combinations are 100 multiplied by 100; these will increase proportionately as each additional disc is added.

These are keyless locks but the dial itself can be fitted with a key operation to prevent it being removed. The value of this type of lock is in the security of the combination number and it is essential to ensure that the person who is setting the mechanism should not be overlooked while he is doing so.

### *Time lock*

This is a special mechanism used in conjunction with safes, and strongrooms which operates independently of any key or a combination lock. It is set to go off at a specific hour and the door cannot be opened until this time is reached, even if any other locks have been unlocked.

### *Padlocks*

Padlocks come into a separate category. The locks themselves can be extremely good; a well-designed one should be made in one piece with the shackle close enough to the body of the lock to prevent an instrument being inserted to force it; the whole should be of hardened steel to prevent cutting. At least five and preferably more levers or pin tumblers should be contained in padlocks used for security purposes.

The main practical fault is one of sheer carelessness, in that the hasps and staples or the locking bars may be subject to cutting or be insufficiently fixed to the door or jamb. Like the padlock itself, the locking bar should be of hardened steel and with no external rivets. It should be bolted through the door to the inside and through a backing plate with the bolt ends burred over. Manufacturers sell padlocks and locking bars complete.

It is a good practice periodically to change the locations of padlocks in use. This will go towards defeating anyone who has an unauthorized copy of a key. A spare set of locks is recommended which can be incorporated in the rotation plan.

### *Locking bolts*

These are essentially bolts which can be locked into a closed position by means of a key, unfortunately that key is common to all the bolts the manufacturer produces in that range. They are of course only for internal locking, are usually required and supplied as pairs, and give a good degree of protection – provided that they are not accessible to a hack-saw blade because of an ill-fitting door or frame.

### *Master key 'suites' or 'systems'*

These have been designed to reduce to a minimum the number of individual keys that need to be carried by senior executives; for the same reason these will be of use to security staff. The best way to illustrate the function of such a system is to quote what will happen if it is installed in a works or office block. Locks from a planned grouping of locks will be fitted to all doors. Each of these locks will be different yet there will be one key which will open all of them – this will be the key of the managing director or senior executive on the site, the 'master' key.

In each subsidiary department there will again be one key which will open all the doors of that department but which will not fit the doors of any other department. These keys are 'sub-master' keys and will be held by the departmental heads. Finally, there will be one key which will only open one lock and this will be held by each individual who has responsibility for the enclosure behind that particular door.

This system will pin-point responsibility for keys and reduce the number that need to be in existence. Each person will only carry one key and this will give him access to as many – or as few – doors as he has the right to enter. None of the inferior keys in such a system can be converted into a master key and special care is taken by all manufacturers to ensure no extra keys can be obtained other than by properly authorized persons. Manufacturers keep a full record of the firms to whom such systems are sold to ensure that maximum security of keys is obtained.

### *Colour coding of key tabs*

It is possible, if it is thought necessary, to obtain keys with identifying coloured end-tabs. This can be of assistance in security offices where large numbers of keys are kept for a variety of departments. Each department will have a special identifying colour and a key would only be given to a member of that department presenting a similarly coloured numbered disc. Manufacturers will no doubt readily co-operate in supplying the needs for particular circumstances or this can be applied to a series of master key systems.

### *Conclusion*

Remember that the security of the most highly-priced lock is no better than that which is accorded to its key. If the key is misplaced by rank carelessness the lock can be valueless. Where security is of any importance the making of duplicate keys must be rigidly suppressed and the strictest records kept of the issue and return of keys.

### *Access control*

This is the term applied to the electrically-activated locking systems mentioned earlier in connection with computer security for which they are especially useful. There are numerous suppliers and a variety of options in the methods of operation; no matter how sophisticated the latter are, they will lack any attraction if it is possible to short out the wiring, or get at the workings of the control housing so as to bypass the means of validation and open the door. The feasibility of this should be raised with the vendors, some of whom may not have considered it.

The basic system of a card-reader into which a standard coded card is introduced, in the same way as a key into a lock, is not regarded as 'access control', neither is a control with a cypher reader only which responds to a pre-programmed 4 digit number, in the same way as a combination lock. Cards however can be encoded, by punched holes (obsolescent), magnetically, by coded copper foil, by infra-red, etc. to provide personal identification of the presenter via a central processing unit which can be adapted to record his entry, or exclude cards that have been cancelled. A further extension would be to restrict authenticity of the card to particular times or days, or, in a complex with several areas of restriction and multiple card-reader entry points, to selected areas only. Indeed, in one large unit with a high security risk, a person who wishes to enter a department for which he is not normally entitled, has to surrender his own card before receiving one which will allow him to do so. The whole transaction and movement is centrally computer-recorded.

The problem of lost cards, borrowed cards, or stolen cards can be blocked by cancellation once the fact is known but a safeguard can be built in by incorporating a cypher reader into the system so that the user has also to tap out his personal number, which should not be known to anyone else – he has also to be dissuaded from writing the number on his card so he will not forget it! Some cards themselves have been found a nuisance in the past but it is hoped that these problems have now been solved – they bent, cracked, or broke in use with all the inconvenience of cancellation and replacement that that entailed. Cards are in practice best fitted with a clip for lapel attachment because when put in the back pocket of jeans, it is only a matter of time before they become unusable. One manufacturer has recognized this difficulty by producing a card in the form of a key tab which will fit on to an ordinary key ring.

A widely-used alternative with most of the card facilities is provided by an electronic code-transmitting device which is carried by the user; this is sensed by a receiver as the user approaches the door and the lock is released. If an intruder or unauthorized person tries to enter at the same time an alarm will sound. Batteries have to be regularly recharged under this system and it is advisable to fit special warning devices to stop staff taking units home in error. An advantage lies in that it allows the user to have his hands free for carrying purposes. Using much the same principle, proximity systems have been introduced which function by having tuned resonant circuits embedded in a card with sensors installed in walls etc. adjacent to the door that react to the presence of the card.

Access control has not yet been fully exploited. It can be used with turnstiles; it can under certain circumstances replace the permanent presence of a guard or receptionist but when this is done, it must be remembered that this presupposes that the user is a person of integrity who does not merit surveillance. If necessary its security rating could be further enhanced by incorporating, say, a metal-detector doorway to prevent dangerous objects being taken in by an authorized user. There are certain drawbacks in the purely human failings of propping open doors, jamming card slots, wedging the tongues of locks, writing codes on walls, etc. with the exception of the last (which incidentally the authors have never encountered); all can be catered for by warning devices. This is a highly-competitive market where several quotations and different systems should be considered and evaluated – the cost justifies the time spent if for that purpose alone – and it must be anticipated that the protagonists of a particular system will be at pains to point out the defects of the others.

## *SAFES*

In a similar way to locks, many safes are not worthy of the name. This is particularly true of the older varieties where the backs, and with the oldest models even the sides, are riveted rather than welded together. In fact they are little more than strongboxes and can easily be ripped or cut open, especially if the back is readily accessible. Even the metal from which they have been manufactured is occasionally suspect and it has been known for a large safe to virtually disintegrate from a blow with a 14-pound (6.4 kg) hammer. Few of these old safes even live up to their fire resistant claims due to the deterioration of the fire resistant material they contain.

Modern safes really came into being with the First World War when the use of welding equipment became commonplace. The dangers of attack by explosives were recognized at an early stage and devices were introduced to combat this. However, only the most important safes were fitted with them and it is reasonable to assume that any safe manufactured before 1945 will not be explosive-resistant. The usual anti-explosive device consists of a spring-loaded steel plunger which is activated when an explosive charge is detonated in the safe door; this prevents movements of the safe bolts and the safe then requires factory attention to open it. Resistance to cutting by oxy-acetylene burners has also only been developed since about 1945.

The age of a safe can usually be established from the serial number put on by all the larger manufacturers. This may identify quality, date of manufacture, size and other details; it is frequently stamped into the metal of the outer side of the top hinge, other places are the inside front edge of the door, or the safe itself over the door, or on a metal strip rivetted to the door back. Difficulty may be encountered both in finding it and reading it because of repeated painting over. A useful guide is published by ABIS (Association of Burglary Insurance Surveyors) from their address: Aldermary House, Queen Street, London EC4. The J. W. Levy Group, Thavies Inn House, Holborn Circus, London EC1N 2HE produce a laminated card checklist as an *aide-mémoire* which they will supply. These figures are of importance to insurers in assessing what they are prepared to cover in such a safe, and they will be able to quote a figure over the telephone having been given this information.

It can be difficult to assess the value of a really old but substantial-looking safe. Tests can be made by striking the top with knuckles – if there is any hollow sound or feel the safe is virtually useless. The safe's key, too, may give some indication as to whether this was originally a superior type. The cutting on the key should be clean and sharp, there should be a minimum of seven levers, and if the key is 'double bitted' (cut at both sides) this is an indication of a good lock.

The ideal would be to progressively down-grade safes to storage of records and finally to scrap them, replacing them by more modern types. Regrettably, while new premises appear to merit new and strong safes, economic considerations are frequently quoted as reasons for not replacing them elsewhere, despite the fact that they can no longer fulfil their modern function.

If there is reluctance to replace a safe the alternative is to shield the metalwork from direct attack, so far as possible, leaving only the stronger part, the door, exposed. The best way to do this is to encase it in reinforced concrete – ordinary concrete could be peeled off and bricks are of little use. The same precaution can be taken in respect of any portable safe; to place any reliance in a safe which the criminals could take away, to open at leisure, is ludicrous. If setting in concrete is not practicable it should at least be possible to fix the safe's base into reinforced concrete by means of bolts.

Even a manufacturer of the most modern safes would be unwise to claim that his products are impregnable. If a thief has adequate time at his disposal and the most modern means – and it is not divulging a criminal trade secret to quote the thermic lance as a typical example – he will eventually cut his way into that safe. However, the longer the thief is delayed in attaining his objective the more likely he is to abandon the project and, if a well-designed safe is accorded further protection by well-devised alarm systems and periodic inspection, it should be adequately resistant to all the normal attacks.

There are few problems attached to the choice of the modern safe; production of these has been going on for many years and those firms that sold inferior products have long since become extinct. All manufacturers supply adequate literature describing their products with details of their capacity and resistance to types of attack. Insurance companies too, will advise on a type of safe commensurate with the risk that is being incurred.

### *Insurance requirements*

Special data has been compiled by insurance companies equating each make of safe with a cash figure representing the risk, against which they are prepared to insure it. Some of these amounts are surprisingly low to the layman. It is no use at all going ahead with a purchase which appears to fit all requirements, and then find the insurers are not prepared to cover the amount of money it is desired to lodge in it. Non-compliance with their wishes will simply result in subsequent claims being refused. It is obvious therefore that insurers' advice is all important and must be followed if they are involved.

### *Siting of safes*

The idea of safes being placed in out-of-the-way places where they may evade being found is obsolete (except for small wall safes – see below). Wherever possible safes should be put where they are visible to patrolling police officers, security staff, or even members of the public. Time and the opportunity to work unseen are the factors that the safe-breaker desires. The safe should be illuminated to facilitate inspection and the absence of the light will at once attract attention. A double-filament bulb is recommended. Observation points may be through peepholes in doors or walls and mirrors can be used to allow inspection round corners and where direct viewing is impossible.

### *Anchorage of safes*

Any safe which weighs less than a ton and is sited at ground floor level should be secured against being removed bodily. Above ground floor level suitable measures depend on the facilities available for removal – such as lifts and hoists. A further limiting factor would be what the fabric of the building will carry, which also might control the size of the safe.

Practically all the larger modern types have anchorage points so that the safe can be bedded firmly into reinforced concrete. Where this is not so, substantial rag-bolts can be sunk into the floor and grouted in, while corresponding holes are drilled in the base of the safe; it is then set over the threaded ends and mild steel locking plates, linking each pair of bolts, are fitted before the nuts are put on – the object of this is to spread the force and prevent the safe being levered over the nuts. Similar means can be used to attach the safe to walls by the sides or back instead of the base.

### *Combination safe locks*

The workings of these locks have been described earlier in this chapter; they are ideal where there are apt to be frequent changes of staff since the combination can be altered to suit the person then responsible for the security of the safe. Moreover, there is no possibility of the key being copied by a previous holder as could happen with normal locks. A further attribute lies in the difficulty of applying explosive substances, since there is no keyhole whereby they can be inserted in the door.

Having established a code for a combination lock the user should deposit a copy of it at some normally inaccessible point to obviate the gross inconvenience that might occur if he were to forget it or if the safe needed opening in his absence. The ideal place is at his bank. Any patrolling security officer, or person with security responsibilities, should occasionally glance at the side of the safe to ensure that no one has been so stupid as to write the code number there for convenience – this is not a flight of fancy, either!

Apart from the normally accepted free-standing kinds of safe there are smaller types which are used for specific purposes; the most important of these are described below.

### *Wall safes*

These are used for safe keeping of sums of money and small valuable personal belongings. They can be obtained in a variety of sizes from approximately that of one brick upwards. Their resistance against attack may be limited and this type of safe should be concealed in some out-of-the-way place. If they are to contain valuables of consequence they should be drill resistant, fitted with anti-explosive devices, and firmly fitted into the wall in a similar manner to which a safe is set into the floor. A substantial wall safe is of little use in an ordinary brick wall from which it can easily be prised. Smaller safes have key locks only but the larger ones may have either key or combination locks. The essential point is that of concealment.

### *Floor safes*

These are becoming increasingly common and are particularly valuable for guarding cash. Relatively speaking, they are much stronger than wall safes; they are easy to set in reinforced concrete; the removable tops are tapered and therefore cannot be knocked in; they are normally fitted with anti-explosive bolts; and they can easily be concealed beneath the floor tiles. In a variation money can be dropped into a floor safe by way of a chute, without a key being in the possession of the person making a deposit.

*Night safes and trap deposits*

These are simply means of enabling money to be put into a safe without opening it and without substantially reducing its security by the method employed. The form taken is usually that of having a rotary trap linked to the safe interior by a chute. Apart from use at banks, for the convenience of shopkeepers and businesspeople in general, such safes can be of value at petrol stations and the like in conjunction with normal safes. The main advantage is that substantial sums of money need not be kept on the premises under conditions of needless risk after normal working hours.

## *STRONGROOMS*

Erection of strongrooms is a job for the expert and should not be given to the ordinary building contractor. Current criminal practice is to attack strongrooms by way of the walls, floors, or ceilings – frequently from adjacent premises.

As with safes, strongrooms should ideally be sited where they can be regularly inspected, and there certainly should be some form of alarm system fitted to them. In new constructions, special consideration should be given to the siting of the strongroom at the drawing-board stage. Extremely resistant heavy doors and grilles are available but the other potential areas of attack, through doors, walls, and ceilings, must not be overlooked and these weaknesses countered by the use of strongly reinforced concrete or other methods.

# *APPENDICES*

# 1

# *Supervisors' security course*

The following items could be included in a curriculum:

1 *Current position in country.* Describe the fire losses experienced stressing that the published figures do not take account of consequential loss and firms forced out of business. Emphasize increases in crime by 500 per cent in the last 20 years coupled with a general lowering of moral standards.
2 *Current position in own firm.* Describe losses that have become known, how committed and how to prevent, and unaccountable losses which have occurred and how it is possible for them to have been concealed.
3 *Sources of loss–police–courts.* Define what constitutes stealing in law; refer to 'perks' with the probability of their escalating to become a serious drain on a company if not stopped; taking away and non-return of tools; clocking offences; how and when to obtain police assistance; court procedure.
4 *Firm's policy on prosecutions.* Give details of the policy and the procedures that are to be followed in the event of any contravention by (a) outsiders, (b) contractors and (c) own employees.
5 *Loyalty to firm–responsibilities of carrying authority.* The consequences of deliberately turning a blind eye to misconduct. The construction placed upon negative action. Emphasize that dishonest employees constitute a very small anti-social minority who will try to influence relationships to their own advantage.
6 *Conclusion.* Underline the value of loyalty, co-operation with

management generally and the security department where this exists, and any system of notification to that department to remove personal involvement.

# 2

# *Security policy checklist*

1 Where the firm is the complainant, what action will be taken against an employee whose guilt is admitted or proven beyond doubt in instances of dishonesty or attempted dishonesty? Where the sufferer is an employee and the offence takes place on the premises, does the same rule apply? Has a firm policy been agreed? If so:

(a) Is that policy clearly stated in works or company rules, or exhibited in such a way that no employee can claim lack of knowledge of it?
(b) Does the wording of the policy statement allow any misunderstanding or misinterpretation – as might happen by the use of 'may' instead of 'will'?
(c) Has the policy statement been strictly adhered to in the past, or has it been manipulated to convenience and could therefore be subject to challenge on grounds of bad precedents?
(d) Has a procedure been established for the dismissal of an offending employee? Is this in accordance with the Code of Practice on Disciplinary Rules and Procedures? Who is responsible for the checking and clearance of his personal belongings off the premises?

2 In the event of an employee's dishonesty:

(a) Who decides on prosecution or otherwise?

(b) If prosecution is not automatic, what is the reporting method to the decision maker? Can this be expedited? Does this allow for representations by the security head?
(c) What degree of discretion is allowed to the security head to instigate a prosecution by notifying the police without reference to higher management, and under what circumstances?

3 In the event of an offence against the firm by someone unknown what are the considerations that affect a decision to report the matter to the police for investigation? Who makes the decision? and what discretion is given to the chief security officer?
4 In the event of an offence of dishonesty by a non-employee legitimately on the premises (for example, a visiting driver or contractor), does the security head have absolute discretion in referring to the police, or is there an established procedure otherwise? (There are valid reasons why this should not be a decision made by the manager who has engaged a contractor.)
5 In the event of offences by persons unconnected with the firm, has the security head absolute discretion in police notification?
6 Is there a 'search clause' in the conditions of employment? If so:

(a) Is it regularly implemented, or has it lapsed through disuse?
(b) Is the selection for search sufficiently random to avoid criticism? Is searching done efficiently in a manner which will not cause offence? Are adequate records kept of persons or vehicles checked?
(c) Is provision made for search of females, or can failure to do so be justified to male employees who raise the fact?
(d) Have the staff doing the searching been fully instructed in their duties and, in particular, their action in face of refusal?

If not, has any attempt been made for the inclusion of this possibility in the conditions of employment? What are the circumstances in which an attempt for its acceptance would be likely to be successful?

7 Are sales of products or other goods on the premises permitted? If so, is the procedure and documentation sufficient to prevent collusion between employees, obviate the repeated use of a receipt to cover theft, and ensure payment is made? If sales are permitted of scrap or damaged products etc. are the precautions adequate to stop deliberate damage to create a cheap sale or make a totally unrealistically low charge to a friend?
8 Are employees allowed use of firm's facilities against repayment to make purchases from outside suppliers at discount rates? If so,

is the system sufficiently waterproof to stop the charges remaining against the firm? Similarly, is there an arrangement whereby work is done by the firm for employees against repayment – as in car servicing – and are the precautions in connection with this adequate?

9 Is there a clear ruling that no responsibility is accepted for the money, property or vehicles of employees whilst on the premises? Is this sufficiently publicized in works rules and by 'disclaimers' exhibited? Has safe storage been offered for the proceeds of collections etc.?

10 (a) What precautions are taken during recruitment of personnel for posts with a high loss/nuisance potential to the firm, for example drivers, computer operators, security officers, keyholders, etc.? Are references checked for sufficient period before a job offer is made? Is the application form so worded as to clarify the consequences of false statements?

(b) Is there, or has there been considered, a policy in connection with the working of notice by a person who has been dismissed, or has resigned from a position of special risk, or is known to be joining a competitor?

11 Is there a nominated communications source for the media to deal with questions relating to newsworthy items – accidents, industrial action, and losses?

12 Are there policies, or contingency plans, for emergencies which will affect the site as a whole, for example, bomb threats and evacuation, serious accident or major incident, and strike action coupled with picketing?

13 Have measures been taken to ensure the confidentiality of the contents of important documents? Is there an appropriate system of classification, safekeeping, handling, and destruction?

14 Are there adequate controls on purchasing, and awarding contracts to suppliers and for work to be done, to reduce the opportunity for corrupt practice?

15 Do the procedures for handling cash and cheques minimize chance of theft and misappropriation?

16 Are policies periodically checked for implementation and review?

# 3

# *Chief security officer job description*

It is suggested that the chief security officer's position and responsibility be on a par with other middle line management positions, as shown in the following diagram. Colleagues of similar status might include the stores manager, the maintenance manager, the work study manager, etc. Titles and relative status are obviously different in every organization.

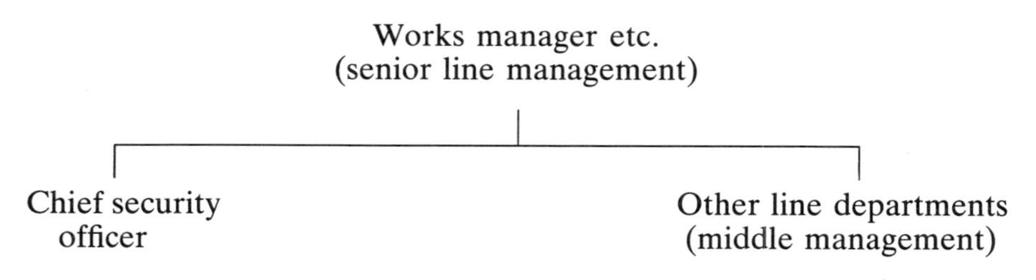


*WORK INVOLVED*

1 To organize the systematic supervision and patrolling of all factory boundary fences, storage areas, works and office buildings, etc., to ensure the safekeeping of company property.
2 To make recommendations to management on all matters where he/she considers the security of the company's plant, buildings,

materials, and other property can be improved. He/she will have similar responsibility in connection with the property of any company employee, but, where there is no legal liability on the company for such matters, this will be only in an advisory capacity to help the individual without legal obligation.

3 To be responsible for the recruitment and day-to-day administration of the security officers, their instruction in the various aspects of their duties and ensuring that they maintain a suitable standard of efficiency and deportment. He/she will also take steps to provide them with facilities to participate in periodic first aid and security training to keep their abilities in this field to a satisfactory level consistent with the requirements of the Factories Act.

4 To ensure that the security staff are fully conversant with the operation of all fire equipment in the works; that such equipment is fully and adequately maintained; that the satisfactory liaison is created and maintained at a high level with the local fire prevention officer and fire brigade; and that all fires occurring on the company property are fully investigated and reported upon.

5 To organize duties at all gatehouses, to ensure that the company's rules and regulations relating to the entry and exits of employees, contractors' employees, visitors, and vehicles belonging to company and other parties are observed. This will include the discretionary right to search persons and vehicles.

6 To cause such books to be kept in the main gatehouse as are necessary to ensure the permanent recording of commercial vehicle movement in and out of the works, with notation of purpose or load. He/she will also cause a day-to-day diary to be kept for reference purposes and such other records as he/she deems essential for the efficient functioning of the department.

7 To be responsible for the guarding of wages after receipt from the bank and during distribution to employees.

8 To enquire as necessary into and report upon any thefts within the works and use his/her discretion in connection with any of these matters he/she deems should be reported to the civil police.

9 To maintain the best possible co-operation at high level with the police and fire authorities of adjacent areas and be responsible for dealing with any enquiries from them.

10 To arrange the allocation of staff to carry out any miscellaneous duties which may be required in the company's interests; these will include:

   (a) Control and direction of all commercial and private traffic entering and leaving the works and office areas.

(b) Control of the parking of cars and motorcycles of the company's employees at the works.
(c) Supervision of the car parks reserved for visitors.
(d) Assisting the local police when necessary to control and expedite the exit of traffic from the works area into adjoining roads.

11 To prepare annual estimates of the expenditure to be incurred on the upkeep of the security staff, installations and equipment and submit these for inclusion in the annual budget.
12 To carry out such miscellaneous duties and enquiries as may be required of him in the company's interest.
13 To keep in touch with developments in mechanical and other aids to security; and in techniques by maintaining contacts with persons in parallel positions in other companies and professional associations.

*Requirements*

The position calls for a thorough training and experience in the administration of the law concerned with the prevention of crime and in the detection of offenders. The ability to train, organize, and supervise the duties of a security staff is also required.

*Initiative*

He/she will be expected to advise the company on matters within his/her experience and to perform his/her duties without immediate supervision and to make decisions. In his/her dealings with all levels of management, staff, and general workforce, he/she will be expected to maintain an amicable and co-operative relationship.

His/her normal hours of duty will be the same as those required of other members of management on Monday to Friday inclusive. He/she will have authority at his/her discretion to vary those hours in accordance with the needs of economy.

*Responsibility*

He/she will be directly responsible to the works manager for the proper performance of the duties described under 'work involved'.

*Alterations*

No alterations or additions to the duties of the chief security officer will be made without authority.

# 4

# *Security officer application form*

*CONFIDENTIAL*

*Complete in ink, block capitals please*
*No approach will be made to present employer without applicant's prior consent*

1 **Full name** Date of birth ..........................................
Nationality ..........................................
Married/Single ..........................................
Number of children ..................................

2 **Permanent address** ..........................................................................................
Owner-occupier YES/NO
Telephone number ...................................

3 **Next of kin:** Name ........................................ Relationship ..................................
Address .................................................................................................
.........................................................................................................

4 **Health**

Have you:

(a) Undergone any major operation ......................................................................
(b) Had any illness causing absence from work in last five years
.........................................................................................................
.........................................................................................................
(c) Any physical disability (e.g. impaired vision/hearing, hernia, slipped disc, etc.)
.........................................................................................................
.........................................................................................................

(d) Suffered from dermatitis YES/NO
asthma/bronchitis YES/NO
(If YES give details) ..........................................................................
..........................................................................................

N.B. Successful applicant will have to undergo a full medical examination.

5 **Prior employment:** Give details of present and previous employments, most recent first, include military service. Cover at least last ten years.

| Name and address of employer | Job | From–To | Reason for leaving |
|---|---|---|---|
| .................................... | ............. | ..................... | ...................................... |
| .................................... | ............. | ..................... | ...................................... |
| .................................... | ............. | ..................... | ...................................... |
| .................................... | ............. | ..................... | ...................................... |

6 **Experience and record:** N.B. Failure to disclose a conviction would mean immediate dismissal.

(a) Have you a current first-aid certificate? YES/NO
If 'No', list experience, if any ..........................................................

(b) Experience in combating fire? ..........................................................
..........................................................................................

(c) Have you a current driving licence? YES/NO

(d) What groups does your licence cover? ..................................................

(e) Have you been convicted of any offence involving dishonesty, if so, give offence, date and penalty.........................................................................
..........................................................................................
..........................................................................................

(f) List security experience and training ..................................................
..........................................................................................
..........................................................................................

7 **Hobbies**..................................................................................

8 **Any other relevant information** ..........................................................

I apply for employment as a security officer and vouch that the preceding details are correct. I understand that a misleading statement or unsatisfactory reference could lead to my subsequent dismissal.

I accept that my employment may involve regular shift and weekend working and, in emergency, duty change and/or overtime at short notice.

Signed..................................
Date ..................................

# 5

# Duty rosters

### *CONTINUOUS WORKING – THREE SHIFT, FOUR GANG 4 × 7 SYSTEM*

*Cycle starting Monday, 21 days working 7 days rest per month*

| | Day | 06.00 to 14.00 | 14.00 to 22.00 | 22.00 to 06.00 | Rest | Taking group 'A' only | | |
|---|---|---|---|---|---|---|---|---|
| | | | | | | Rest period | Rest hours | Hours worked |
| 1 | M | A | B | C | D | | | |
| 2 | T | A | B | C | D | | | |
| 3 | W | A | B | C | D | | | |
| 4 | T | A | B | C | D | Nil | Nil | 56 |
| 5 | F | A | B | B | C | | | |
| 6 | S | A | B | C | D | | | |
| 7 | Sun | A | B | C | D | | | |
| 8 | M | D | A | B | C | | | |
| 9 | T | D | A | B | C | | | |
| 10 | W | D | A | B | D | | | |
| 11 | T | D | A | B | C | Nil | Nil | 56 |
| 12 | F | D | A | B | C | | | |
| 13 | S | D | A | B | C | | | |
| 14 | Sun | D | A | B | C | | | |
| 15 | M | C | D | A | B | | | |
| 16 | T | C | D | A | B | | | |
| 17 | W | C | D | A | B | | | |
| 18 | T | C | D | A | B | Nil | Nil | 56 |
| 19 | F | C | D | A | B | | | |
| 20 | S | C | D | A | B | | | |
| 21 | Sun | C | D | A | B | | | |
| 22 | M | B | C | D | A | | | |
| 23 | T | B | C | D | A | 6 a.m. Mon | 168 | Nil |
| 24 | W | B | C | D | A | to | | |
| 25 | T | B | C | D | A | 6 a.m. Mon | No Pay | |
| 26 | F | B | C | D | A | (Pay Usually averaged per week | | |
| 27 | S | B | C | D | A | over duration of rota) | | |
| 28 | Sun | B | C | D | A | | | |

## *CONTINUOUS WORKING – THREE SHIFT, FOUR GANG CONVENTIONAL SEVEN-SHIFT CYCLE*

| | | Day | 06.00 to 14.00 | 14.00 to 22.00 | 22.00 to 06.00 | Rest | Rest period | Rest hours |
|---|---|---|---|---|---|---|---|---|
| | 1 | S | A | C | B | D | 2 p.m. Sat to | 72 |
| | 2 | M | A | C | B | D | 2 p.m. Tues | |
| | 3 | T | A | D | B | C | 10 p.m. Mon to | 72 |
| 1 | 4 | W | A | D | B | C | 10 p.m. Thurs | |
| | 5 | T | A | D | C | B | 6 a.m. Thurs to | |
| | 6 | F | A | D | C | B | 6 a.m. Sun | 72 |
| | 7 | Sat | A | D | C | B | | |
| | 8 | S | B | D | C | A | 2 p.m. Sat to | |
| | 9 | M | B | D | C | A | 2 p.m. Tues | 72 |
| | 10 | T | B | A | C | D | 10 p.m. Mon to | |
| 2 | 11 | W | B | A | C | D | 10 p.m. Thurs | 72 |
| | 12 | T | B | A | D | C | 6 a.m. Thurs to | |
| | 13 | F | B | A | D | C | 6 a.m. Sun | 72 |
| | 14 | Sat | B | A | D | C | | |
| | 15 | S | C | A | D | B | 2 p.m. Sat to | 72 |
| | 16 | M | C | A | D | B | 2 p.m. Tues | |
| | 17 | T | C | B | D | A | 10 p.m. Mon to | 72 |
| 3 | 18 | W | C | B | D | A | 10 p.m. Thurs | |
| | 19 | T | C | B | A | D | 6 a.m. Thurs | |
| | 20 | F | C | B | A | D | 6 a.m. Sun | 72 |
| | 21 | Sat | C | B | A | D | | |
| | 22 | S | D | B | A | C | 2 p.m. Sat to | 72 |
| | 23 | M | D | B | A | C | 2 p.m. Tues | |
| | 24 | T | D | C | A | B | 10 p.m. Mon to | 72 |
| 4 | 25 | W | D | C | A | B | 10 p.m. Thurs | |
| | 26 | T | D | C | B | A | 6 a.m. Thurs to | |
| | 27 | F | D | C | B | A | 6 a.m. Sun | 72 |
| | 28 | Sat | D | C | B | A | | |

## *CONTINUOUS WORKING – THREE SHIFT, FOUR GANG 3 × 2 × 2 'CONTINENTAL' SYSTEM*

| | | Day | 06.00 to 14.00 | 14.00 to 22.00 | 22.00 to 06.00 | Rest | Rest period | Rest hours |
|---|---|---|---|---|---|---|---|---|
| | 1 | S | A | B | C | D | 6 a.m. Sun to | |
| | 2 | M | A | B | C | D | 6 a.m. Wed | 72 |
| | 3 | T | A | B | C | D | | |
| 1 | 4 | W | D | A | B | C | 6 a.m. Wed to | |
| | 5 | T | D | A | B | C | 6 a.m. Fri | 48 |
| | 6 | F | C | D | A | B | 6 a.m. Fri to | |
| | 7 | Sat | C | D | A | B | 6 a.m. Sat | 48 |
| | 8 | S | B | C | D | A | 6 a.m. Sun to | |
| | 9 | M | B | C | D | A | 6 a.m. Wed | 72 |
| | 10 | T | B | C | D | A | | |
| 2 | 11 | W | A | B | C | D | 6 a.m. Wed to | 48 |
| | 12 | T | A | B | C | D | 6 a.m. Fri | |
| | 13 | F | D | A | B | C | 6 a.m. Fri to | |
| | 14 | Sat | D | A | B | C | 6 a.m. Sun | 48 |
| | 15 | S | C | D | A | B | 6 a.m. Sun to | |
| | 16 | M | C | D | A | B | 6 a.m. Wed | 72 |
| | 17 | T | C | D | A | B | | |
| 3 | 18 | W | B | C | D | A | 6 a.m. Wed to | 48 |
| | 19 | T | B | C | D | A | 6 a.m. Fri | |
| | 20 | F | A | B | C | D | 6 a.m. Fri to | 48 |
| | 21 | Sat | A | B | C | D | 6 a.m. Sun | |
| | 22 | S | D | A | B | C | 6 a.m. Sun to | |
| | 23 | M | D | A | B | C | 6 a.m. Wed | 72 |
| | 24 | T | D | A | B | C | | |
| 4 | 25 | W | C | D | A | B | 6 a.m. Wed to | 48 |
| | 26 | T | C | D | A | B | 6 a.m. Fri | |
| | 27 | F | B | C | D | A | 6 a.m. Fri to | 48 |
| | 28 | Sat | B | C | D | A | 6 a.m. Sun | |

## *TEN-MAN ROTA*

Example of 10-man rota designed to provide extra cover daily for the period 6 p.m. to 2 a.m., plus an additional man for wage payment on Thursdays and Fridays

| Week | 1 | 2 | 3 | 4 | 5 | 6 | 7 | 8 | 9 | 10 |
|---|---|---|---|---|---|---|---|---|---|---|
| Sunday | 7 a.m. | L | 10 p.m. | 2 p.m. | 7 a.m. | L | 10 p.m. | 2 p.m. | 6 p.m. | L |
| Monday | 7 a.m. | L | 10 p.m. | 2 p.m. | 7 a.m. | L | 10 p.m. | 2 p.m. | 6 p.m. | L |
| Tuesday | 7 a.m. | 10 p.m. | L | 2 p.m. | 7 a.m. | 10 p.m. | L | 2 p.m. | L | 6 p.m. |
| Wednesday | 7 a.m. | 10 p.m. | L | 2 p.m. | 7 a.m. | 10 p.m. | L | 2 p.m. | L | 6 p.m. |
| Thursday | 7 a.m. | 10 p.m. | 2 p.m. | L | 7 a.m. | 10 p.m. | 2 p.m. | L | 10 a.m. | 6 p.m. |
| Friday | 7 a.m. | 10 p.m. | 2 p.m. | L | 7 a.m. | 10 p.m. | 2 p.m. | L | 10 a.m. | 6 p.m. |
| Saturday | L | 10 p.m. | 2 p.m. | 7 a.m. | L | 10 p.m. | 2 p.m. | 6 p.m. | 7 a.m. | L |

NB: 2 p.m. denotes the period 2 p.m. to 10 p.m. 10 p.m.–10 p.m./7 a.m. 7 a.m.–7 a.m./2 p.m.
Average 40.8 hours per week.

## *SEVEN-MAN ROTA – 48-HOUR WEEK*

| Week | 1 | 2 | 3 | 4 | 5 | 6 | 7 |
|---|---|---|---|---|---|---|---|
| Sunday | 2 p.m. | 6 a.m. | 10 p.m. | 2 p.m. | 6 a.m. | L | 10 p.m. |
| Monday | 2 p.m. | 6 a.m. | 10 p.m. | 2 p.m. | 6 a.m. | 10 p.m. | L |
| Tuesday | L | 6 a.m. | 10 p.m. | 2 p.m. | 6 a.m. | 10 p.m. | 2 p.m. |
| Wednesday | 6 a.m. | L | 10 p.m. | 2 p.m. | 6 a.m. | 10 p.m. | 2 p.m. |
| Thursday | 6 a.m. | 10 p.m. | L | 2 p.m. | 6 a.m. | 10 p.m. | 2 p.m. |
| Friday | 6 a.m. | 10 p.m. | 2 p.m. | L | 6 a.m. | 10 p.m. | 2 p.m. |
| Saturday | 6 a.m. | 10 p.m. | 2 p.m. | 6 a.m. | L | 10 p.m. | 2 p.m. |

## *FOUR CREWS, THREE SHIFTS*

W = hours worked.

| Shifts | Week 1 | Week 2 | Week 3 | Week 4 |
|---|---|---|---|---|
| | S M T W T F S | S M T W T F S | S M T W T F S | S M T W T F S |
| Nights | *A A A A A A B* | *B B B B B C C* | *C C C C D D D* | *D D D A A A A* |
| Afternoons | *C C D D D D D* | *D A A A A A A* | *B B B B B B C* | *C C C C C D D* |
| Mornings | *B B B B C C C* | *C C C D D D D* | *D D A A A A A* | *A B B B B B B* |

| | W | W | W | W |
|---|---|---|---|---|
| Crews | *A* 48 | *A* 48 | *A* 40 | *A* 40 |
| | *B* 40 | *B* 40 | *B* 48 | *B* 48 |
| | *C* 40 | *C* 40 | *C* 40 | *C* 40 |
| | *D* 40 | *D* 40 | *D* 40 | *D* 40 |

| Duty | Week 9 | Week 10 | Week 11 | Week 12 |
|---|---|---|---|---|
| | S M T W T F S | S M T W T F S | S M T W T F S | S M T W T F S |
| Nights | *B B B B C C C* | *C C C D D D D* | *D D A A A A A* | *A B B B B B B* |
| Afternoons | *A A A A A A B* | *B B B B B C C* | *C C C C D D D* | *D D D A A A A* |
| Mornings | *C C D D D D D* | *D A A A A A A* | *B B B B B B C* | *C C C C C D D* |

| | W | W | W | W |
|---|---|---|---|---|
| Crews | *A* 48 | *A* 48 | *A* 40 | *A* 40 |
| | *B* 40 | *B* 40 | *B* 48 | *B* 48 |
| | *C* 40 | *C* 40 | *C* 40 | *C* 40 |
| | *D* 40 | *D* 40 | *D* 40 | *D* 40 |

| Duty | Week 17 | Week 18 | Week 19 | Week 20 |
|---|---|---|---|---|
| | S M T W T F S | S M T W T F S | S M T W T F S | S M T W T F S |
| Nights | *C C D D D D D* | *D A A A A A A* | *B B B B B B C* | *C C C C C D D* |
| Afternoons | *B B B B C C C* | *C C C D D D D* | *D D A A A A A* | *A B B B B B B* |
| Mornings | *A A A A A A B* | *B B B B B C C* | *C C C C D D D* | *D D D A A A A* |

| | W | W | W | |
|---|---|---|---|---|
| Crews | *A* 48 | *A* 48 | *A* 40 | *A* 40 |
| | *B* 40 | *B* 40 | *B* 48 | *B* 48 |
| | *C* 40 | *C* 40 | *C* 40 | *C* 40 |
| | *D* 40 | *D* 40 | *D* 40 | *D* 40 |

SEVEN-DAY WEEK Repeat at week 25

| Week 5 | Week 6 | Week 7 | Week 8 | |
|---|---|---|---|---|
| S M T W T F S | S M T W T F S | S M T W T F S | S M T W T F S | |
| *A A B B B B B* | *B C C C C C C* | *D D D D D D A* | *A A A A A B B* | |
| *D D D D A A A* | *A A A B B B B* | *B B C C C C C* | *C D D D D D D* | |
| *C C C C C C D* | *D D D D D A A* | *A A A A B B B* | *B B B C C C C* | |
| W | W | W | W | Hours |
| *A* 40 | *A* 40 | *A* 40 | *A* 40 | = 336÷8 = 42 |
| *B* 40 | *B* 40 | *B* 40 | *B* 40 | 336÷8 = 42 |
| *C* 48 | *C* 48 | *C* 40 | *C* 40 | 336÷8 = 42 |
| *D* 40 | *D* 40 | *D* 48 | *D* 48 | 336÷8 = 42 |

| Week 13 | Week 14 | Week 15 | Week 16 | |
|---|---|---|---|---|
| S M T W T F S | S M T W T F S | S M T W T F S | S M T W T F S | |
| *C C C C C C D* | *D D D D D A A* | *A A A A B B B* | *B B B C C C C* | |
| *A A B B B B B* | *B C C C C C C* | *D D D D D D A* | *A A A A A B B* | |
| *D D D D A A A* | *A A A B B B B* | *B B C C C C C* | *C D D D D D D* | |
| *W* | *W* | *W* | *W* | |
| *A* 40 | *A* 40 | *A* 40 | *A* 40 | = 336÷8 = 42 |
| *B* 40 | *B* 40 | *B* 40 | *B* 40 | 336÷8 = 42 |
| *C* 48 | *C* 48 | *C* 40 | *C* 40 | 336÷8 = 42 |
| *D* 40 | *D* 40 | *D* 48 | *D* 48 | 336÷8 = 42 |

| Week 21 | Week 22 | Week 23 | Week 24 | |
|---|---|---|---|---|
| S M T W T F S | S M T W T F S | S M T W T F S | S M T W T F S | |
| *D D D D A A A* | *A A A B B B B* | *B B C C C C C* | *C D D D D D D* | |
| *C C C C C C D* | *D D D D D A A* | *A A A A B B B* | *B B B C C C C* | |
| *A A B B B B B* | *B C C C C C C* | *D D D D D D A* | *A A A A A B B* | |
| *A* 40 | *A* 40 | *A* 40 | *A* 40 | = 336÷8 = 42 |
| *B* 40 | *B* 40 | *B* 40 | *B* 40 | 336÷8 = 42 |
| *C* 48 | *C* 48 | *C* 40 | *C* 40 | 336÷8 = 42 |
| *D* 40 | *D* 40 | *D* 48 | *D* 48 | 336÷8 = 42 |

# 6

# Code of Conduct

Those carrying out security duties, either in-company or for a contract security firm, can only do so satisfactorily if persons with whom they come into contact in the course of the duties respect their personal integrity, professional competence, and conscientious approach to the job. Failures in performance can have consequences, immediate or long term, more serious to employers, fellow workers, or contract clients than comparable actions by practically any other category of employee. Timidity and delay, or lack of knowledge of first aid could endanger an injured workman's life; lack of diligence in patrolling and powers of observation could result in gross fire or other damage to property; connivance in theft, or lack of action on information in connection with it, could lead to serious loss; indiscretion, tactlessness or unnecessary exercise of authority could jeopardize the entire industrial relations atmosphere in an organization and cause labour disputes of indeterminable consequences.

With these factors in mind, and the desirability of having a professional standards yardstick for reference, a more stringent code of conduct should be required from those engaged in security duties than other employees. Obviously this is reasonable since to do so will react to the benefit of all genuinely interested in their work and will make it easier to weed out the unreliable, untrustworthy undesirables.

Such a code has been formulated by the International Professional Security Association to be observed by its members and their employees. The form may be that of a disciplinary code but fundamental principles are expressed therein which merit recognition beyond the confines of the Association; with IPSA permission it is reproduced hereunder:

## *CODE OF CONDUCT*

Persons employed by members of the International Professional Security Association:

Article 1 Shall not neglect nor, without due and sufficient cause, omit to promptly and diligently discharge a required task whilst at work.

Article 2 Shall not leave a place of work without due permission or sufficient cause.

Article 3 Shall not knowingly make or sign any false verbal or written statement of whatever description.

Article 4 Shall not, without due and sufficient cause, destroy, mutilate, alter nor erase any document or record.

Article 5 Shall not, without authority divulge any matter which is confidential to the employer or his clients past or present.

Article 6 Shall not corruptly solicit or receive any bribe or other consideration from any person, or fail to account for the moneys or property received in connection with the employer's business.

Article 7 Shall not be uncivil to persons encountered in the course of work, or make unnecessary use of authority in connection with the discharge of the employer's business.

Article 8 Shall not act in a manner reasonably likely to bring discredit upon the employer, a client or fellow employee.

Article 9 Shall not feign or exaggerate any sickness or injury with a view to evading work.

Article 10 Shall not wear the employer's uniform or use his equipment without authority.

Article 11 Shall maintain proper standards of appearance and deportment whilst at work.

Article 12 Shall not work whilst under the influence of alcohol, or consume any alcohol whilst at work.

Article 13 Shall, on conviction for any criminal offence, notify the employer forthwith.

# 7

# *Standing orders for security staff*

1 Staff conditions of employment will apply to all security personnel.
2 They will be expected to maintain a standard of personal conduct and deportment in keeping with the nature of their duties.
3 Duties will be performed in the uniform provided unless wearing of plain or civilian clothing has been especially authorized. Medal ribbons should be worn. Shirt sleeve order in hot weather will only be adopted on authority of the security manager. Black footwear will be worn for all normal purposes. Wellington boots and protective clothing will be made available if necessary. Black or dark blue scarves may be used during inclement weather.
4 Uniforms remain the property of the company and must be returned on leaving its employment. Where two sets of uniform are held, the older will be used for night duty. They are to be kept in a clean and tidy condition to the credit of the wearer and the company. Uniform trousers may be worn during travelling to and from work, but jackets and caps are only to be used off company property when specifically directed. Any damage to a uniform which renders it unfit to wear must be reported immediately.
5 Instructions to the security staff will only be given by those directly responsible for their supervision.
6 All instructions on security measures that may be issued from

time to time are confidential and are not to be discussed with anyone outside the department. The security manager will exercise a discretion in this matter.

7 Members of the security staff are expected to keep themselves informed of all instructions affecting their duties and of any amendments that may be made. They will initial any written instructions given.

8 Security staff may be required to work reasonable hours of overtime as and when necessary in the company's interests. Holiday periods will be agreed annually and must be adhered to.

9 Security staff will be expected to make themselves proficient in first aid by passing the relevant primary examination of the Order of St John or the Red Cross at the earliest opportunity and by taking refresher courses when these are deemed necessary by management.

10 They will be fully responsible for the administration of first aid in the absence of the nursing staff. For this purpose they will familiarize themselves with the equipment used in the ambulance room, procedures for recording accidents and injuries, and the obtaining of outside specialized assistance.

11 Security staff will be expected to be conversant with the location and usage of all fire-fighting equipment on site and to undergo such training as may be required by management.

12 They will be expected to familiarize themselves with the usage of the weighbridge and those forms of deception that may be encountered in connection with it which might react to the company's detriment.

13 They will receive instruction in the automatic transmission of recorded teleprinter messages, the changing of teleprinter rolls, and the use of computer-ventilation temperature controls, or other mechanism that they may be asked from time to time to supervise in the company's interests.

14 In accordance with employees' conditions of employment they will be expected to carry out searches of company employees and their vehicles to detect the unauthorized removal of company property.

15 They will exercise the same discretionary right of search in respect of personnel and vehicles of contractors or others entering or working on the site.

16 They will be expected to make themselves proficient in controlling vehicular traffic in accordance with police practice.

17 They will familiarize themselves with the principles of dog handling and train with the dog(s) as and when required to become fully proficient.

18 As and when required they will attend external training courses to improve their proficiency.

19 The principal duties of security officers will be as follows:

(a) To protect the property of the company at all times against theft, be it committed by employees or by intruders.

(b) To ensure that all the regulations of the company affecting the security of its property and the property of employees are carried out.

(c) Where instances of theft or malicious damage have taken place or are taking place against company property, or the property of employees whilst on site, to detain any person:

(i) who is committing, or with reasonable cause is suspected to be in the act of committing, such offence; or

(ii) who with reasonable cause is suspected of having committed a known offence;

and to notify the security manager forthwith for further instructions.

(d) To ensure that any interference to the factory perimeter protection is located as soon as possible and reported upon.

(e) To protect the buildings of the company and their contents from damage by fire or water or from the effects of any adverse conditions, and prevent waste of materials.

(f) To pay attention to all water, steam, gas, and electrical installations to detect breakdowns and wastage; to take any immediate action that is necessary in the interest of safety and security.

(g) To see that fire-fighting equipment remains in designated locations and is not impeded to prevent immediate accessibility.

(h) To ensure that no unauthorized persons or vehicles gain access to company premises.

(i) To ensure that no employee, or employee of a contractor, or any vehicle leaves the premises in an irregular manner.

(j) To ensure that visitors of all kinds are courteously received, assisted, and directed in a manner which will reflect to the company's credit; to record, as directed, such details of visitors and vehicles as may be required.

(k) To record all vehicles visiting the works to collect or deliver materials; to issue gate passes for authorization by the departments concerned.

(l) To record the movement of all company goods vehicles and their loads; to collect the consignment sheets from them and ensure that these are returned to the appropriate departments.

(m) To implement the company's right of search of employees and contractors and their vehicles; to ensure that company property is not taken from the site without authority.

(n) To require the production of authorizations for the removal of company property including borrowed tools; to record and deal with those authorizations in accordance with practice.

(o) To accept all personal property found on the site; to maintain a register of lost and found property containing all relevant details of time, place, finder, etc.

(p) To accept the higher security-risk keys for safekeeping; to issue them to the authorized persons, keeping a key register for this purpose.

(q) To ensure that all payroll employees leaving the premises at irregular times are in possession of written passouts.

(r) To record all occurrences of security interest in a daily log book for the information of the security staff, management, and other interested persons.

(s) To regulate the movement of traffic; to organize and control the parking of vehicles and storage of bicycles whilst on company property.

(t) To check regularly the teleprinters during quiet periods and ensure that important messages are passed to the intended recipients as soon as possible; to transmit as required automatic telexes when this is necessary.

(u) To observe at intervals the temperature control on the computer room or any other recording instruments that it may be in the company's interest for them to observe during 'quiet' periods.

(v) To assist as weighbridge operators as and when required in emergency.

(w) To protect the company payroll during arrival and during paying out whilst on the firm's premises.

(x) To accept, act upon, or pass on any messages received from the company's drivers, other factories, clients, the police, etc., relating to the company's business; to do the same for emergency messages in respect of employees, during quiet periods.

(y) To attend withdrawals made for essential purposes from the general stores at times when it is not staffed; to receive and check the requisition for the materials obtained and ensure that it is passed to stores supervision immediately on reopening; to apply a like procedure to withdrawals from the engineering stores outside the hours during which it is manned.

(z) To check the weigh tickets handed in by drivers of loaded vehicles leaving the site and ensure that the figures recorded thereon do not reveal weights in excess of the permitted plated maximum; to take the necessary action of notification and reference back to the central dispatch depot when a contravention is found.

(aa) To warn relatives of fellow employees, on request by the staff of the ambulance centre, when, because of injury or serious illness, those employees are incapable of returning home at the time they would be normally expected.

(bb) To accept mail bags and packeted mail for outside factories and warehouses, dispatch them by the earliest possible company transport, and record accordingly; to accept incoming mail from company vehicles and other transport and ensure that it is transferred to the mailroom or dealt with as otherwise required.

(cc) To accept and record telephone calls of a precautionary nature from the security staff on 'one man duty' at other company factories; to endeavour to contact the expected caller after 10 minutes' delay in the making of a call, continue to do so until there has been a 30 minutes' delay, and then notify the police of the location concerned; to report subsequently in writing to the security manager.

(dd) To act as a reception for unexpected customers wishing to collect goods in emergency; to ascertain their requirements, contact the appropriate sales department, and ensure that the customer is sent to the correct warehouse(s) in accordance with sales instructions; to ensure that the recognized documentary procedure is complied with.

20 These instructions do not touch on all circumstances which may call for the attention of the security staff. Where a situation arises and no specific instructions have been issued which apply to it, the members of the security staff will be expected to use intelligence, imagination, and discretion to ensure that it is dealt with satisfactorily.

21 If breaches of industrial discipline are observed by the security staff, their action will be confined to reporting the matter to the security manager who will communicate the details to the appropriate authority. If immediate action is required, the matter

will first be reported to the supervisor of the department or section concerned.

22 Security officers should discuss any problems affecting their duties that they may have with the security manager and, when in doubt regarding their actions, similarly confer for directions.

23 Additions or amendments to these orders will only be made in writing and by the security manager.

24 On appointment security officers will be issued a personal copy of these orders and of any subsequent additions or amendments, which will be returned by them when they leave.

25 A copy of these orders and additions and amendments will be kept at the security department for ready reference when required by any member of the security staff.

# 8

# *Books and records needed in the security office*

These will, of course, depend upon the form of activity in the premises, which may require special data to be kept. However, the following are essential to any firm whose security function is well organized:

1. Standing orders for security staff.
2. Emergency plans with security staff instructions for:
   - (a) fire
   - (b) major disaster
   - (c) bomb threat and evacuation
   - (d) Miscellaneous.
3. Routine instruction file for security staff – that is, temporary instructions for dealing with day-to-day matters.
4. Copies of fire drill instructions and assembly points.
5. Daily occurrence book.
6. Telephone message book (and pads).
7. Names, addresses, and telephone numbers of senior and key personnel.
8. Key register for withdrawal and return against signature.
9. Vehicle register, both incoming and outgoing.
10. Visitors' register.
11. Lost and found property register.

12 Alarm test register.
13 Register of personnel passouts issued.
14 Register of tools out on loan.
15 Register of authorizing chits for scrap and purchased material taken out by employees.
16 Search register where applicable for:

(a) employees
(b) vehicles

17 Employee vehicle parking register, where applicable.
18 Internal organization chart.

*Forms to be available in gate office*

1 Vehicle passes. (NB: Type shown in Appendix 11 is dual purpose in so far as it can be used as a weighbridge ticket.)
2 Visitor's passes (see Appendix 11).
3 Blank passout and tool loan chits.
4 Blank cards for compiling employee car register (see p. 99).

*Miscellaneous books and forms*

1 Internal telephone directory.
2 Local telephone directory.
3 STD code book.
4 Large plan of works showing hydrants, fire points, etc.
5 Outline maps of premises if sufficiently large, for the guidance of visitors.
6 Supply of occurrence report forms.
7 Supply of fire report forms.
8 Supply of accident report forms.
9 Supply of first-aid treatment forms, if used.

*Miscellaneous equipment for security use*

1 Hand lamps.
2 Emergency warning lamps.
3 'No Parking' signs.
4 Emergency floodlighting.
5 Emergency breathing equipment.
6 Bad weather clothing and wellington boots.
7 Truncheons.
8 Supply of padlocks and chains.
9 Typewriter.

10 Adequate first-aid equipment.
11 Fire equipment.

*Additional material*

1 Signing-in book for contractors' employees.
2 Identity cards for contractors' employees.
3 Special information book – confidential for matters undesirable to be included in routine entries in occurrence book.

Many firms use preprinted report forms for such things as premises insecure, lights and appliances left switched on, parking offences, etc. It is debatable whether all these are necessary, and a small occurrence form or note will equally meet the purpose. However, special forms can be easily devised if wanted.

# 9

# *Lost property report form*

NAME AND ADDRESS OF COMPANY

SECURITY DEPARTMENT: Report of property lost or suspected stolen

Form serial number

| | |
|---|---|
| Full details of loser, or, if company property, of person reporting. | Full name ..........<br>Address ..........<br>Home phone number ..........<br>Work phone ext. .......... Dept. .......... |
| Full description of property, including make, colour, size, shape, and any distinguishing marks by which it could be identified. If money, how made up. | ..........<br>..........<br>..........<br>..........<br>..........<br>.......... |
| Date, time, and place when property last known to have been in loser's possession. | Date .......... Time .......... hours<br>Place .......... |
| Date, time, and place when loss discovered. | Date .......... Time .......... hours<br>Place .......... |
| If lost from the person, routes taken by loser since property last known to have been in possession. | ..........<br>..........<br>.......... |
| Has loser any further information which may assist recovery of the property? | YES/NO (Delete as required)<br>..........<br>.......... |
| Date, time, and to whom reported. | Date .......... Time .......... hours<br>To whom reported ..........<br>Log book page .......... refers. |

[*Action taken by the security department should be shown on the back of the form*]

# 10

# Found property report form

NAME AND ADDRESS OF COMPANY

SECURITY DEPARTMENT: Report of found property

Form serial number

---

Time, date, and place ..........................................................................

Where found ..........................................................................

Name, address, and department and phone extension of finder ..........................................................................

..........................................................................

Time and date of report ..........................................................................

Description of property ..........................................................................

..........................................................................

Initials of finder ..........................................................................

Report received by .................................... Log book page.......refers

DISPOSAL OF PROPERTY

Restored to loser — Received by ..................................................

Address..........................................................

Department .............. Date ................................

Finder informed of disposal and by whom ..........................................................................

Returned to finder

I, the undersigned, hereby acknowledge having received the above described property and undertake to indemnify (name of company) in respect of any claim which may be made against the company in connection with this property.

Signature............................................

Witness .................. Date ..................

---

[*Action taken by the security department should be shown on the back of the form*]

# 11

# *Vehicle and visitor's passes*

VEHICLE PASS

Registration No .............................. Serial No .......................................
Date ............................................ Time: In ......... Out ........................
Name of transport ........................... Supplier.........................................
Gross weight: .............tonnes........kgs Weigh office signature
Tare weight: .............tonnes........kgs .............................
LOAD/UNLOAD AT ..........................................................................

In CONTENTS Out

Signature.....................................

This form to be returned by the lorry driver to the gatehouse after being signed and weighed.

Please see that the tare weight is stated.

---

Serial No. .................. Time ................... Date ..............................................

VISITOR'S PASS

NB: THE HOLDER MUST CONFINE HIS VISIT TO THE PERSON(S) OR DEPARTMENT(S) SHOWN HEREUNDER.

THE PASS MUST BE HANDED IN AT THE GATE OFFICE WHEN LEAVING.

Name ............................................ Car number .......................................
Firm ............................................................................................................
Person to be seen ..........................................................................................
Reason...........................................................................................................
Issued by ....................................... Seen by ..........................................

# 12

# *Vehicle accident report form*

Nature of accident ........................................................................................................
Time ................. Date .................. Place ..........................................................
Weather conditions .........................................................................................................
Visibility ...........................................................................................................................
Road surface ......................................................................................................................
Vehicles involved (use additional sheet if necessary)

| | Vehicle 1 | Vehicle 2 |
|---|---|---|
| Reg. no. | ............................ | ................................. |
| Make | ............................ | ................................. |
| Type | ............................ | ................................. |
| Driver | ....................................... | ............................................. |
| Address | ....................................... | ............................................. |
| | ....................................... | ............................................. |
| | ....................................... | ............................................. |
| Owner | ....................................... | ............................................. |
| Address | ....................................... | ............................................. |
| | ....................................... | ............................................. |
| | ....................................... | ............................................. |
| Insurance company | ....................................... | ............................................. |
| Cert. no. | ....................................... | ............................................. |
| Driver's licence | Seen/Not seen<br>In/Not in order | Seen/Not seen<br>In/Not in order |
| Damage to vehicle | ....................................... | ............................................. |
| | ....................................... | ............................................. |

Damage to other property ....................................................................................
Persons injured ....................................................................................
Names, addresses; ....................................................................................
injury details ....................................................................................
....................................................................................
....................................................................................
Witnesses ....................................................................................
Names and addresses ....................................................................................
....................................................................................

[*Attach copies of any statements taken*]

Sketch plan of accident
(Show road widths, traffic signs and indicate
paths of vehicles by dotted lines.)

Synopsis of what happened

Reported by ............................................
Time ............... Date ................................

Signature of security officer
receiving report ..........................................................................

# 13

# Drug Abuse Checklist

| *Drug used* | *Physical symptoms* | *Look for* | *Effects* |
|---|---|---|---|
| Cannabis, marijuana, hashish (all hemp plant derivatives) | Euphoria and lassitude, lack of concentration, affects like alcohol, sleepiness, lack of muscular and mental co-ordination | Smell of burnt rope, small seeds, papers for rolling cigarettes. Stained fingers. Hashish – green/black compressed block | Least damaging; habit-forming but said not to induce dependence. Leads to harder drug usage. Reduces powers of perception. Hashish more likely to cause dependence |
| Glue and solvents (sniffing) | Drunken appearance, dreamy or blank look, may be violent, smells of glue or solvents | Tubes of glue, glue smears and smell on handkerchief, plastic or paper bags, aerosol cans. | Lung, brain, and liver damage, may die through suffocation, anaemia. |
| Amphetamines, Methedrine, cocaine, Drinamyl, Benzedrine ('uppers') | Hyperactivity, euphoria, poor appetite, aggression, dilated pupils, dry mouth; withdrawal symptoms – fatigue and depression | Pill, some double-scored on one side, varying colours, multi coloured capsules. Cocaine – white crystalline powder | Hallucinatory, liver and kidney damage, faculties impaired; cocaine causes dependence |
| Barbiturates, – depressants ('downers'). Tranquillisers – valium etc. | Drowsiness, slurred speech, physical and emotional instability – staggers. Withdrawal – twitchings, hand and facial | Pills and capsules, all types and colours | Addictive, can cause death, very dangerous with alcohol |

| *Drug used* | *Physical symptoms* | *Look for* | *Effects* |
|---|---|---|---|
| Mescaline, LSD, DMT, MDA, STP (hallucinogenics) | Hallucinations, odd behaviour, incoherence, cold extremities, enlarged iris. 'Flashback' days after taking | Powder, liquids, pills or capsules | Unpredictable, personality changes, psychotic behaviour. Possible permanent mental disorder |
| Morphine, heroin, codeine. (opiates) | Stupor, drowsiness, hallucinations, euphoria, needle marks, bruising and collapsed veins on arms, slurred speech | White, odourless, bitter crystalline powder, may be compressed into blocks or tablets. Needle or hypodermic syringe; rope string or belt for tourniquet. Burnt bottle tops or spoons | Addictive, overdose can kill. Hepatitis and other medical conditions |

# 14

# *Duties and records of police custody officer*

A custody officer is usually a police sergeant delegated specific duties under the Police and Criminal Evidence Act 1984 to deal with detained persons at police stations. Police stations which have been designated must have a custody officer there at all times. He is a person with very wide ranging responsibilities.

## *CUSTODY RECORDS*

As soon as a detained person arrives at the police station, a separate custody record sheet must be opened. This record sheet must be started as soon as practicable for each person who is brought to the station or who is arrested at the station, having attended there voluntarily. All information which has to be recorded under the Codes of Conduct of the Police and Criminal Evidence Act 1984 must be recorded on that custody sheet.

The custody officer is responsible for the accuracy and completeness of the custody record and for ensuring that the record or a copy of the record accompanies a detained person if he/she is transferred to another police station. The record shall show the time of and reason for transfer, and the time a person is released from custody.

When a person leaves police detention he/she or his/her legal representative shall be supplied on request with a copy of the custody record. This entitlement lasts for 12 months after his/her release.

The custody officer must inform a detained person of the following rights and of the fact that they need not be exercised immediately:

1 the right to have someone informed of his/her arrest;
2 the right to consult a solicitor;
3 the right to consult the Codes of Practice (kept at the police station).

The custody officer must give the detained person a written notice setting out the above three rights, the right to a copy of the custody sheet and the caution in the terms prescribed. The custody officer shall ask the person to sign the custody record to acknowledge receipt of this notice.

If the custody officer then authorizes a person's detention he must inform him/her of the grounds as soon as practicable, and in any case before that person is questioned about any offence. The person shall be asked to sign on the custody record, to signify if he/she wants legal advice at this point.

*Detained persons: special groups*

If the person does not understand English or appears to be deaf and the custody officer cannot communicate with him/her then the custody officer must as soon as practicable call an interpreter, and ask him to provide the information required above.

If the person is a juvenile, is mentally handicapped or is suffering from mental illness then the custody officer must as soon as practicable inform the appropriate adult of the grounds for his/her detention and his/her whereabouts, and ask the adult to come to the police station to see the person.

If the person is blind or seriously visually handicapped or is unable to read, the custody officer should ensure that his/her solicitor, relative, the appropriate adult or some other person likely to take an interest in him/her is available to help in checking any documentation. Where this code requires written consent or signification, then the person who is assisting may be asked to sign instead if the detained person so wishes.

The person can then be formally interviewed and a 'record of the interview' written down, or a 'statement after caution' taken and completed.

# 15

# *Codes of Practice for interviews (PACE Act 1984) – interview record form*

## *GENERAL RULES*

If an interview record is not made during the course of the interview it must be made as soon as practicable after its completion.

Written interview records must be timed and signed by the maker.

If an interview record is not completed in the course of the interview the reason must be recorded in the officer's pocket book.

Any refusal by a person to sign an interview record when asked to do so in accordance with the provisions of this code must itself be recorded.

A detained person may not be supplied with intoxicating liquor except on medical directions. No person who is unfit through drink or drugs to the extent that he is unable to appreciate the significance of questions put to him and his answers may be questioned about an alleged offence in that condition.

As far as practicable, interviews shall take place in interview rooms which must be adequately heated, lit and ventilated.

Persons being questioned or making statements shall not be required to stand.

Before the commencement of an interview each interviewing officer shall identify himself and any other officers present by name and rank to the person being interviewed.

Breaks from interviewing shall be made at recognized meal times. Short breaks for refreshment shall also be provided at intervals of approximately two hours, subject to the interviewing officer's discretion to delay a break if there are reasonable grounds for believing that it would:

(i) involve a risk of harm to persons or serious loss of, or damage to, property;
(ii) delay unnecessarily the person's release from custody; or
(iii) otherwise prejudice the outcome of the investigation.

If in the course of the interview a complaint is made by the person being questioned or on his behalf, concerning the provisions of this code then the interviewing officer shall: record it in the interview record.

*Records*

(a) An accurate record must be made of each interview with a person suspected of an offence, whether or not the interview takes place at a police station.
(b) If the interview takes place in the police station or other premises:
    (i) the record must state the place of the interview, the time it begins and ends, the time the record is made (if different), any breaks in the interview and the names of all those present; and must be made on the forms provided for this purpose or in the officer's pocket book or in accordance with the code of practice for the tape recording of police interviews with suspects;
    (ii) the record must be made during the course of the interview, unless in the investigating officer's view this would not be practicable or would interfere with the conduct of the interview, and must constitute either a verbatim record of what has been said or, failing this, an account of the interview which adequately and accurately summarizes it.

Where the appropriate adult or another third party is present at an interview and is still in the police station at the time that a written record of the interview is made, he shall be asked to read it (or any written statement taken down by a police officer) and sign it as correct or to indicate the respects in which he considers it inaccurate. If the person refuses to read or sign the record as accurate or to indicate the respects in which he considers it inaccurate, the senior officer present shall record on the record itself, in the presence of the person concerned, what has happened. If the interview is tape recorded the arrangements set out in the relevant code of practice apply.

*Where the parents or guardians of a person at risk are themselves suspected of involvement in the offence concerned, or are the victims of it, it may be desirable for the appropriate adult to be some other person.*

# INTERVIEW RECORD

Name of person interviewed (*surname*) Date of birth
(*forename(s)*)

Address

Tel. No.

Time and date record commenced concluded

Place of interview

Identity of note taker

Persons present at commencement

Other persons present at commencement (*include name and capacity*)

Important: An accurate record will be kept of all movements, breaks, facilities offered, changes in persons present and cautions given.
Only one side of the forms to be used.

| Time | Record |
|---|---|
| | Caution: I understand that I need not say anything unless I wish to do so and that what I say may be given in evidence.<br><br>Signed .................................................................. |
| | |

Signature of note taker ............................ Signature of witness(s) ............................

..........................................................

Signature of
person interviewed ................................. ..........................................................

INTERVIEW RECORD – continuation page no. ..........................................................

Full name of person interviewed ..............................................................................

| Time | Record |
|---|---|
| | |

Signature of note taker ............................ Signature of witness(s) .............................

..........................................................

Signature of
person interviewed ................................. ..........................................................

# 16

# *Codes of Practice for statements under caution (PACE Act 1984) – specimen statements*

All written statements after caution shall be taken in the following manner:

If a person says that he wants to make a statement he shall be told that it is intended to make a written record of what he says. *He shall always* be asked whether he wishes to write down himself what he wants to say. If he says that he cannot write, or would like someone to write it for him, a police officer may offer to write the statement for him.

Where the person wishes to write it himself, he shall be asked to write out and sign before writing what he wants to say, the following:

> I make this statement of my own free will. I understand that I need not say anything unless I wish to do so and that what I say may be given in evidence.

Any person writing his own statement shall be allowed to do so without any prompting except that a police officer may indicate to him which matters are material or question any ambiguity in the statement.

*Written by a police officer*

If a person says that he would like someone to write it for him, a police officer shall write the statement, but, before starting, he must ask him to sign, or make his mark, to the following:

> I, . . ., wish to make a statement. I want someone to write down what I say. I understand that I need not say anything unless I wish to do so and that what I say may be given in evidence.

Where a police officer writes the statement, he must take down the exact words spoken by the person making it and he must not edit or paraphrase it. Any questions that are necessary (for example, to make it more intelligible) and the answers given must be recorded contemporaneously on the statement form.

When the writing of a statement by a police officer is finished the person making it shall be asked to read it and to make any corrections, alterations or additions he wishes. When he has finished reading it he shall be asked to write and sign or make his mark on the following certificate at the end of the statement:

> I have read the above statement, and I have been able to correct, alter or add anything I wish. This statement is true. I have made it of my own free will.

If the person making the statement cannot read, or refuses to read it, or to write the above mentioned certificate at the end of it or to sign it, the senior police officer present shall read it over to him and ask him whether he would like to correct, alter or add anything and to put his signature or make his mark at the end. The police officer shall then certify on the statement itself what has occurred.

Security personnel should also follow these directions.

*Sample statements under caution*

*Accused person writes statement himself*

Name: John Smith Age: 36
Address: 8 Harold Street
Hull, Humberside

I make this statement of my own free will. I understand that I need not say anything unless I wish to do so and that what I say may be given in evidence.

[Signed] John Smith

[Body of statement as person writes it]

It is true I stole the purse from the lockers in the canteen. I was desperate for extra money. I have a great deal of debt, and spent the money settling a bill. I am very sorry.

I have read the above statement, and I have been able to correct, alter or add anything I wish. This statement is true. I have made it of my own free will.

[Signed] John Smith

[Caption]

This statement written by Mr John SMITH in my presence was commenced at 2.05 p.m. and ended at 2.20 p.m. on Thursday 1 October 1988 in the personnel office at William Brown Ltd, 62 Cavendish Street, Hull. Security officer Michael Oliver was present.

[Signed] William Wilson<br>Chief security officer

Michael Oliver<br>Security officer

NB: The only questions which are allowed to be put to the writer in connection with the main context of his statement, are in connection with matters which are material, i.e. 'What happened to the money in the purse?' or to question any ambiguity in the statement.

Any spelling mistakes or corrections should be initialled by the maker if altered, otherwise should be presented as written. No alterations by anyone except the writer are permissable.

*Accused person requests someone else write it for him*

Name: John Smith Age: 36<br>Address: 8 Harold Street<br>Hull, Humberside

(Written by chief security officer Wilson)

I, John Smith wish to make a statement. I want someone to write down what I say. I understand that I need not say anything unless I wish to do so and that whatever I say may be given in evidence.

[Signed by Smith] John Smith

[Dictated by Smith]

It is true I stole the purse from the lockers in the canteen. I was desperate for extra money. I have a great debt, and spent the money settling a bill. I am very sorry.

[At this point the person making the statement shall be asked to read it and to make any corrections, alterations or additions he wishes. He will then be asked to WRITE HIMSELF the following certificate.]

[Written by Smith]

I have read the above statement and I have been able to correct, alter or add anything I wish. This statement is true. I have made it of my own free will.

[Signed] John Smith

[Written by Chief Security Officer Wilson]

This statement was written by myself at the request of Mr John Smith. It was commenced at 2.05 p.m. and ended at 2.20 p.m. on Thursday 1 October 1988 in the personnel office at William Brown Ltd, 62 Cavendish Street, Hull. Mr Arthur Hemmings, the personnel officer was present and witnessed the making of this statement.

[Signed] William Wilson<br>Chief security officer

Arthur Hemmings<br>Personnel officer

NB: The exact words dictated by the maker of the statement must be written down. They must not be edited or paraphrased by the writer.

Any questions that are necessary and the answers given must be recorded contemporaneously on the statement form.

# 17

# *Cash security audit report form*

Factory ................................................ Date of survey ................................
Accountant/Cashier i/c .............................. Tel. No. ..........................................
Ext. No. ..........................................

1 *WAGE HANDLING*

(a) Collections from bank:

| | Day | Time (approx.) |
|---|---|---|
| (1) | .................... | ...................................... |
| (2) | .................... | ...................................... |
| (3) | .................... | ...................................... |

(b) Means of collection ..........................................................................................
..........................................................................................

(c) Wages cheque presentation to bank procedure ..........................................................................................
..........................................................................................
..........................................................................................

(d) Contract delivery point (carrier)
..........................................................................................

(e) Cash acceptance from carrier and checking procedure
..........................................................................................
..........................................................................................
..........................................................................................

(f) Carrier contract – comment on content
..........................................................................................
..........................................................................................

2 *CASH QUANTITIES*

(a) Normal cash deliveries:
1(a)(1) ..........................................................................................
1(a)(2) ..........................................................................................
1(a)(3) ..........................................................................................

(b) Abnormal amounts (current max.) – occasions:
(1) Spring Bank Holiday ..........................................................................................
(2) Summer Holiday ..........................................................................................
(3) Xmas week ..........................................................................................
(4) Profit sharing ..........................................................................................
(5)....................... (other)..........................................................................................

(c) Responsibility to inform carrier ....................................................................
by letter/telephone ....................................................................

3 *CASH-HOLDING FACILITIES*

(a) Safe (type): (1) ....................................................................
(2) ....................................................................
(3) ....................................................................
(b) Alarm if any ....................................................................
(c) Alarm keys held by ....................................................................
(d) Safe keys held or combination known by ....................................................................
(e) Keyholder holiday provisions ....................................................................
(f) Key duplicates held by and where ....................................................................
(g) Normal maximum 'after hours' amount held (wages, petty cash, stamps, etc.) ....................................................................
(h) Variations, if any, at abnormal amount occasions:
2(b)(1) ....................................................................
2(b)(2) ....................................................................
2(b)(3) ....................................................................
2(b)(4) ....................................................................
(i) Insurance rating of safe:
3(a)(1) ....................................................................
3(a)(2) ....................................................................
3(a)(3) ....................................................................
(j) Comment
....................................................................
....................................................................

4 *WAGES MAKE-UP*

(a) Office used ....................................................................
(b) Staff used ....................................................................
(c) Precautions during make-up ....................................................................
(d) Payslip issue:
(1) Day and time .......................... (2) Procedure ..........................................
....................................................................
....................................................................

(e) Packet contents checks ..............................................................................

..............................................................................................

(f) Comment

..............................................................................................

..............................................................................................

5 *PAYOUT*

| (a) | Day | Times (between) |
| --- | --- | --- |
| | ........................ | .................................... |
| | ........................ | .................................... |
| | ........................ | .................................... |

(b) Pay stations used ..............................................................................

..............................................................................................

(c) Staff used..............................................................................

..............................................................................................

(d) Identification/authorization of collector:

..............................................................................................

(1) Payroll..............................................................................

(2) Staff..............................................................................

(3) Bulk collection (if allowed) – precautions

..............................................................................................

..............................................................................................

..............................................................................................

(e) Box/packets handover procedure between issuing staff

..............................................................................................

..............................................................................................

..............................................................................................

(f) Precautions at pay station(s)

..............................................................................................

..............................................................................................

..............................................................................................

(g) Comment and recommendations

..............................................................................................

..............................................................................................

..............................................................................................

6 *RETENTION OF WAGES PENDING PAYOUT*

(a) In cashier's working hours ..............................................................................

(b) After working hours ..............................................................................

..............................................................................................

(c) Residual packets procedure (temporary)

..............................................................................................

..............................................................................................

(d) Uncollected wages – procedure

..........................................................................................................

..........................................................................................................

..........................................................................................................

(e) Payments during holiday periods

..........................................................................................................

..........................................................................................................

..........................................................................................................

(f) Procedure where payslips lost

..........................................................................................................

..........................................................................................................

..........................................................................................................

(g) Procedure where wrong collection alleged ..........................................................-

..........................................................................................................

..........................................................................................................

(h) Procedure for wage collection – other than by employee

..........................................................................................................

..........................................................................................................

..........................................................................................................

(i) Comments and recommendations

..........................................................................................................

..........................................................................................................

..........................................................................................................

7 *PETTY CASH*

(a) How brought in and frequency

..........................................................................................................

..........................................................................................................

..........................................................................................................

(b) Maximum holding

..........................................................................................................

(c) Amounts carried in (other than with wages): (1) Normal.......................................... (2) Abnormal ..........................................

(d) Mode of transport and precautions – if own staff used

..........................................................................................................

..........................................................................................................

..........................................................................................................

(e) Competent to authorize payment of expenses etc.

..........................................................................................................

(f) Manner of application for expenses

..........................................................................................................

..........................................................................................................

(g) Responsible for holding and recording petty cash

........................................................................................................

(h) Routine checking responsibility

........................................................................................................

(i) Check frequency

........................................................................................................

(j) Container used and location

........................................................................................................

(k) Key held by

........................................................................................................

(l) Key duplicates held by

........................................................................................................

(m) Postage procedure, recording, and checking

........................................................................................................

........................................................................................................

(n) Comment and recommendations

........................................................................................................

........................................................................................................

........................................................................................................

8 *OTHER MONIES ON SITE*

(a) Canteen proceeds:
- (1) Gross holding before transfer to bank ............................................
- (2) Temporary holding facilities ............................................
- (3) Means of transfer to bank ............................................
- (4) Responsibility for record checks ............................................
- (5) Frequency of checks ............................................

........................................................................................................

(b) Proceeds of savings or other schemes, vending machines, etc.:
- (1) Where held ............................................
- (2) Weekly amounts ............................................
- (3) Responsibility ............................................
- (4) Checking procedure ............................................

(c) 'Floats':

| | Purpose | Amount (max.) |
|---|---|---|
| (1) | .................................................... | .................................................... |
| (2) | .................................................... | .................................................... |
| (3) | .................................................... | .................................................... |

(d) Staff sales:
- (1) Procedure ............................................
- (2) Means of payment ............................................
- (3) Accounting process ............................................
- (4) Average weekly receipts ............................................

(e) Comment and recommendations

.................................................................................................

.................................................................................................

.................................................................................................

9 *CHEQUE HOLDING AND CASHING*

(a) Unsigned cheques held by:
  - (1) Ordinary.................................................................................
  - (2) Kalamazoo type ......................................................................

(b) Where held:
  - (1)..............................................................................................
  - (2)..............................................................................................

(c) Presigned cheques (bank and Giro):
  - (1) Purpose ...............................................................................
  - (2) By whom held.........................................................................
  - (3) Where held ............................................................................
  - (4) Maximum potential value .......................................................
  - (5) Insurer's comments ................................................................

  .................................................................................................

(d) Cheque-signing machine (if any):

(1) Where used..................................................................................

(2) Keys held by ................................................................................

(3) Where signature block stored ....................................................

(e) Employee cheque cashing:
  - (1) Maximum individual amount.....................................................
  - (2) Who authorizes individual's right ...........................................
  - (3) Time and date of encashing ....................................................
  - (4) Average weekly gross ..............................................................
  - (5) Source of cash..........................................................................

(f) Comments and recommendations

.................................................................................................

.................................................................................................

.................................................................................................

# 18

# *Schedule of potential weaknesses in internal cash handling*

1 No check of longstanding wage carrier's contract to ensure that its contents are still valid – for example, amounts are within specified limit, delivery points the same, and no variations made which might in any way provide legal liability loopholes.
2 Failure to inform carriers when excessive amounts are to be collected – thereby possibly putting them at insurance risk by inadequate manning, and exceeding your own contract maximum.
3 Safes purchased indiscriminately on vendor's assurances, without checking with insurers that they are acceptable to them for the amounts intended to be held.
4 Exceeding the insurer's specified maximums for retention in safes without notifying them and obtaining their sanction or conditions.
5 Safe keys made accessible to too many people over a period of time; keys or spare keys left in office drawers for convenience; combination lock number never altered.
6 Too tolerant an attitude taken to excluding fellow employees from the cashier's department.
7 Safes or strongrooms left open too often for convenience during normal working.
8 Cash left out on clerks' desks whilst they are absent from the office.

9 Bulk wage-packet collection by departmental nominees allowed without adequate checking by collector or signature(s) obtained.
10 Departmental transfer of packets between supervisory staff for issue to subordinates without any checking or acceptance signatures.
11 Neglect of precautions for safekeeping of wage packets held in departments pending issue to shift personnel, for example, unlocked drawers.
12 Unpaid wages packets or uncollected holiday pays retained for cashier's convenience too long before banking (month normally adequate with subsequent payment from petty cash); coupled with this, unnecessarily including them in boxes each week at pay stations instead of ruling that they be collected from cashier's office.
13 Petty cash: excessive amounts held; paying, recording, and routine checking all carried out by the same person; insufficient spot checks by senior person.
14 Expenses claims vouchers not initialled by departmental head to confirm validity before authorizing executive signature given.
15 Inadequate attention paid to security of presigned cheques – or appreciation of their potential risk value; failure to clarify insurers' reaction to their holding.
16 Monies accrued by catering contractors held in firm's custody pending banking without any agreement as to liability.
17 Lack of thought given to security when delivering accrued money to the bank – in contrast to the steps taken when withdrawing it.
18 Petty-cash box kept in filing cabinet with lock common to other cabinets, or with key number shown on lock which facilitates purchase of duplicate.
19 Absence of policy statements on wage-packet loss/theft liability or on safekeeping and responsibility for privately-collected monies, etc.

# 19

# *Office security checklist*

1 Has a person in a supervisory position been given specific responsibility for security measures and their enforcement in the offices?
2 Are the access points into the offices sufficiently supervised to prevent unchallenged entry by non-employees?
3 Can these access points be reduced to a minimum without offending against fire regulations or causing staff discontent?
4 Are receptionists/commissionaires/security staff/those occupying offices adjacent to unsupervised entrances, adequately briefed on dealing with visitors, invited, unexpected, or unwelcome? Have they lists and extension numbers whereby individuals can be readily contacted? Are they instructed on the information they may give to callers about the firm or employees without referring the enquiry to someone of higher authority?
5 Is there a routine for window cleaners, Telecom engineers, gas and electricity servicemen, etc. to report their arrival (and produce identification if necessary)?
6 Have telephonists been briefed and given written instructions on what to do if the following contingencies happen?

(a) obscene or nuisance calls;
(b) bomb threats;
(c) Fire alarm activation/fire outbreak.

Have they lists of external numbers whereby essential services, police, fire brigade, ambulance can be contacted without delay;

have they similar lists for senior members of the firm when they are at home?

7 Do arrangements for mail receipt preclude theft by outsiders before handling by employees (i.e. from letter boxes and the like); ensure quick sorting and circulation; safe keeping and recording of payments/coupons received; prevent confidential letters being left lying where they can be tampered with; cater adequately for letters to receive attention in the absence of the addressee?

8 Are cloakrooms sited where they cannot be accessed by outsiders without being observed; are disclaimers of employer responsibility for loss displayed; are warning notices concerning rings and watches posted over washbasins?

9 Have staff been warned not to leave cash and valuables in desks etc.; is there an arrangement for the safe lodging of football coupon or other collection proceeds? Are they briefed not to leave small valuable pieces of office equipment lying on desks when work is over, or keys to drawers in typewriter tops etc. where they can be easily found?

10 Are private secretaries, in particular, briefed on the importance of preserving the confidentiality of important documents; is a 'clean desk' policy advocated and enforced for all levels of employee; is the classification system for documents working; are the arrangements for handling, storage and destruction of confidential matter adequate; is all use of copying machines sufficiently controlled to prevent abuses?

11 Are the methods and timing of office cleaning satisfactory for security purposes; are there any petty thefts which might be attributable to cleaners?

12 Is the issue of keys restricted to a minimum; are they housed safely; is the recording sufficient to trace a missing key or responsibility for it?

13 Are the facilities for the holding of cash and valuables acceptable and checked out with insurers; are cheque books, travellers cheques, credit cards, agency cards, pre-signed cheques and cheque signing machines safely housed?

14 Are the fire emergency precautions fully understood with wardens appointed and notices posted; have drills been carried out to satisfaction; have local fire brigade prevention officers been consulted?

15 If burglar alarms/fire alarms are installed, are the staff liable to set them, or act upon their functioning, adequately briefed and with written instructions; are the police in possession of an up-

to-date list of keyholders; have all malfunctions been investigated to prevent recurrence?

16 Are the procedures for checking over and locking up the premises adequate for both fire and security purposes; are keyholders for premises containing high-value property warned to check call-out messages if at all suspicious; if watchmen/guards are employed, are they adequately equipped and instructed, are the checks that they do their job adequate; if a commercial security firm is used for spot visiting, is its performance monitored?

# 20

# *Shops and supermarkets: security checklist*

## STAFF

1 Are all references verified before employees are engaged?
2 Have conditions of employment been prepared?
3 Does every employee on engagement sign and receive a copy?
4 Do the conditions refer to:

(a) Bringing parcels or cases on to the premises.
(b) Handbags, baskets in the sales area.
(c) Not leaving sales area without permission.
(d) Searching of handbags etc.
(e) Leaving only by front entrance.

5 Purchasing conditions: Are these outlined in the conditions of employment? Do these deal with:

(a) Times of purchase.
(b) Dress when shopping.
(c) Use of wire baskets.
(d) Retention of till slips signed by checker.
(e) Sale of goods in short supply.
(f) Sale of goods reduced to clear.
(g) Amount of discount and how to claim.

## *KEYS TO FRONT DOOR*

6 Is keyholder registered with police? Is information up to date?
7 Who has responsibility for ensuring the premises are secure before being left?

## *FIRE DOORS*

8 Is any device fitted to activate an audible alarm if opened?

## *REAR OR DELIVERY DOORS*

9 Who holds the keys (a) during times premises are open, (b) when closed?
10 Are doors locked between deliveries?
11 Are delivery staff admitted to sales area?
12 Are they allowed to enter through front door?
13 If deliveries are made to sales area are they checked there together with any collections, for example stale bread, cakes, etc.?
14 Are composite delivery notes signed representing multiple deliveries?

## *DELIVERIES*

15

(a) How are specially valuable and attractive goods protected?
(b) If in a separate compound, who holds key?
(c) Does the goods-inward checker when tallying hold the delivery note at the same time?
(d) Are all over and under deliveries properly recorded?
(e) Are deliveries occasionally rechecked by a member of management?
(f) Is a goods inward book kept? If so, who makes the entries?

## *RUBBISH AND SALVAGE*

16 Who supervises the removal of rubbish and salvage?
17 Are bins examined from time to time to detect stolen goods?

*CASH TILLS*

18 Can customers see clearly the amount rung up on the till?
19 What pockets are there in the cashiers' overalls?
20 Have cashiers been told they must not serve relatives?
21 Do the cashiers:

- (a) Check their floats before accepting any more money?
- (b) Check the date recorder on their till?
- (c) Check that the total amount previously recorded on the till has not been changed?
- (d) Have a key to their tills?
- (e) Take money, handbags, baskets or cigarettes to their checkout point?
- (f) Ring up on the till each item separately?
- (g) Shut the drawer of the till after each completed sale?

*CASHING UP*

22 Is an independent count taken of each till at close of business and reconciled with the registered total?
23 Is a record kept of overs and unders and brought to notice of management?
24 Is the pen record in 22 and 23 signed by the till operator and the clerk concerned?
25 Are notes above requirements collected from time to time from the till and are records made and signed by the respective cashiers?
26 Where is any spare cash and takings, as in 25, kept until banked?
27 If in a safe, or other locked and satisfactory permanent container, where is key kept?
28 Are some of the day's cash takings taken to the bank during normal banking hours?
29 Is a bank safe used?

*SAFE*

30 Where is it located?
31 Can it be seen from outside the premises?
32 Would security of the safe be improved if a light was left burning?
33 Is it secured to the floor?
34 Who holds the key?
35 Where is the duplicate? In the bank?

## *SHOPLIFTERS AND THEFTS BY STAFF*

36 Has a policy for dealing with shoplifters and staff found stealing been decided and circulated?
37 Have 'Rules for the guidance of managers concerned with shoplifters' been circulated to them?
38 Are details of all employees found stealing company property and the result of any prosecutions posted in positions where they can be read by all employees?
39 Are records kept of all shoplifters whether prosecuted or not?

## *SECURITY STAFF (WHERE EMPLOYED)*

40 Are employees told of the responsibilities and authority of such staff and that they are liable to be asked to explain their possession of company property after leaving the premises?

# 21

# Driver application form

CONFIDENTIAL

*Complete in ink*
*Block capitals please*
*Applicant must provide driving licence at interview*
*No approach will be made to present employer without applicant's prior consent*

*1* **Full name** ............................................................ ............................................................

Date of birth ..........................................
Nationality..............................................
Married/single ..........................................
Number of children ..................................

2 **Permanent address** ..........................................................................................................

3 **Next of kin**
Name ............................................... Relationship............................................
Address ...........................................................................................................................

4 **Health**
(a) Details of illness or disablement in last five years
..........................................................................................................................
..........................................................................................................................

(b) Serious illness, injury or operation prior to last five years
..........................................................................................................................
..........................................................................................................................

(c) Is your hearing/vision impaired? YES/NO. If YES give details
..........................................................................................................................

5 **Prior employment**
Give details of present and previous employments, most recent first, include military service

| Name and address of employer | Job | From–to | Reason for leaving |
|---|---|---|---|
| .......................... | ...................... | .......................... | .............................. |
| .......................... | ...................... | .......................... | .............................. |
| .......................... | ...................... | .......................... | .............................. |
| .......................... | ...................... | .......................... | .............................. |
| .......................... | ...................... | .......................... | .............................. |

6 **Driving ability and record**
NOTE: disclosure of a conviction does *not* mean you will automatically be barred from employment; failure to disclose would, however, mean subsequent dismissal.

(a) Have you experience of customer delivery? ......................................................

(b) Are you experienced in keeping drivers' records? ..............................................

(c) List types of commercial vehicles driven ..........................................................
..........................................................
..............................................................................................................

(d) Give details of any driving convictions ............................................................
..............................................................................................................

(e) Have you been convicted of any offence involving dishonesty? If so, give offence, date and penalty ........................................................................................
..............................................................................................................
..............................................................................................................

(f) Has any load, part load or vehicle in your charge ever been stolen? If so, give details
..............................................................................................................

7 **Hobbies**.......................................................................................................

8 **Any other matters** .........................................................................................
..............................................................................................................

I apply for employment as a driver and vouch that the preceding details are correct.

I understand that a misleading statement or unsatisfactory reference could lead to my subsequent dismissal.

I accept that my employment may involve overnight stay away from my base.

Signed ................................................

Date...................................................

# 22

# *Weighbridge: general operating instructions*

1 Keep bridge surface clear of water, rubble, or other foreign matter.
2 Check and adjust if necessary the zero reading between weighings.
3 Check that the vehicle is correctly and evenly positioned on bridge, that the engine is switched off for safety purposes and any slight error it may cause, that the driver and passengers are out, and that nothing but the vehicle is impinging on the bridge.
4 Check that the vehicle is similarly positioned for both gross and tare weights if double weighing is required. Do not allow this if circumstances permit the trailer to be dropped and dealt with separately.
5 Ensure that the mechanism is at rest before making or taking reading. (NB: Wind effect can induce considerable fluctuation in readings and errors; this, or engine vibration if this is allowed to remain running, may stop a recording from being made where an equilibrium control is fitted.)
6 Check that any printout is legible. If not, make it so.
7 Tell the driver verbally what the weights are so that he can contest them if he so desires. Give him a written or printout copy of the figures.
8 In taring, cast a glance at the tare shown on the vehicle if this is visible, and compare.

9 Make any book entries at once and in sequence.
10 Report any suggestions of inducements to falsifying readings before the driver concerned has left the premises.
11 Familiarize yourself with the ways whereby you can be misled by a dishonest user, and do not hesitate to report suspicions. If equipment seen on the vehicle when taring is not on when grossing, do not weigh pending clarification. (NB: Battens that have been used to facilitate handling and are no longer required by carriers may have been thrown off at the disposal point, thereby showing more material delivered than the actual.)
12 If the vehicle is overloaded, tell the driver not to leave the premises and inform the department responsible for loading.

# 23

# *Fire instructions*

THE PERSON DISCOVERING A FIRE SHOULD:

Give the alarm by telephoning the company's telephone operator stating location of fire.

Use the fire extinguishers and, if necessary, the fire hose reels provided (see plan). Only use a hose reel in event of a major fire.

Smother burning clothing and similar small fires with coats, curtains, carpets, etc.

Switch off current before using the fire extinguishers or fire hose reels on electrical apparatus.

EXIT AND MEANS OF ESCAPE:

Employees should make themselves acquainted with all the means of exit from the premises. Staircases, landings, ordinary and emergency exits should be kept clear and unobstructed.

ON RECEIVING FIRE WARNING:

Close all doors and windows.

Act on the instructions of the floor fire stewards who will be responsible for the safety of the staff during the period of the fire. Instructions should be carried out immediately and without question.

DURING A FIRE:

Do not evacuate unless ordered by a fire steward. If evacuation is ordered:

Do not use lifts.
Maintain silence.
Do not rush.
Do not attempt to pass others.
*Do not panic*.

YOUR FLOOR FIRE STEWARDS ARE ................................

# 24

# *Fire report form*

To............................... Company secretary    Date ...................................................
Copies to .............................................
................................. Factory manager    .................................................

*A* Location and extent of fire ..........................................................................................
..........................................................................................................................
..........................................................................................................................

*B* Fire found at ........... hours on ........... Date by ...................................................

*C* Area affected last seen in order ........... hours on .......... date by .............................

*D* Public fire brigade
Called at ............... hours by ............
Appliances attending (1)............. arrived.............hours left ...................... hours
(2)............. arrived.............hours left ...................... hours
(3)............. arrived.............hours left ...................... hours
Officer in charge............. Station............ Tel No ..............................................
Cause to which outbreak attributed (if given) ........................................................
..........................................................................................................................
..........................................................................................................................

*E* Factory personnel notified

| | | |
|---|---|---|
| ................................. | Chief engineer | at ............................. hours |
| ................................. | Production supervisor | at ............................. hours |
| ................................. | | at ............................. hours |
| ................................. | | at ............................. hours |
| ................................. | Safety officer | at ............................. hours |
| ................................. | Chief security officer | at ............................. hours |
| * ................................. | Fire assessor | at...... hours ................. date |

*This is responsibility of senior engineer/production supervisor

*F* Brief description of outbreak and details of any injuries

*Person reporting*

*G* INVESTIGATING OFFICER'S REPORT
Appraisal of damage and likely effect on production

*H* Comments and recommendations

(Other sheets attached as required)    Investigating officer's
Signature ................................................

# 25

# *Legislation – fire prevention and control*

*Fire Services Act 1947*. Establishment of fire authorities. Responsibility for advice on prevention, segregation, means of escape. Authorizes breaking into premises without permission for fire-fighting purposes. Penalties for malicious/false calls, and deliberate obstruction of people or appliances.
*Theatres Act 1843 and 1968*. Theatres – licensing, safety of staff and public.
*Explosives Acts 1875 and 1923*. Controls on storage of explosives, safety cartridges, fireworks, etc.; registration and licensing.
*Cinematograph Acts 1909 and 1952*. Requires licensing, adequate exits, equipment and staff, controls on smoking, and standards of wiring, etc.
*Motor Vehicle Regulations 1929*. Storage and licensing requirements for motor fuel. Offences in disposal of spirit into sewers and drains.
*Public Health Acts 1936 and 1961*. Powers in respect of schools, theatres, clubs, etc., to insist on adequate means of escape, passageways, ingress and egress, etc.
*Thermal Insulation (Industrial Buildings) Act, 1957 and Regs 1958*. Insulation value and standard of fire resistance in buildings where heating is provided.
*Factories Act 1961*. Omnibus application to factories (see Chapter 6).
*Consumer Protection Act 1961*. Standards for gas, electric and oil heaters.

*Offices Shops and Railways Premises Act, 1963*. Omnibus application to shops etc. (see Chapter 6).
*Licensing Act, 1964*. Affords means of objection to issue of Justices Licences if fire authority requirements not met.
*Building Regulations 1972*. Part E – structural fire precautions, tests for fire resistance, etc.
*Fire Precautions Act 1971*. Now the main governing legislation on fire prevention, supplemented by a variety of orders and regulations.
*Fire Precautions Act 1971 (Modifications) Regulations 1976* (SI 1976 No. 2007).
*Fire Precautions (Factories, Offices, Shops, and Railway Premises) Order 1976* (SI 1976 No. 2009).
*Fire Certificates (Special Premises) Regulations 1976* (SI 1976 No. 2003).
*Fire Precautions (Non-Certificated Factory, Office, Shop and Railway Premises) Regulations 1976* (SI 1976 No. 2010).
*Health and Safety at Work Act 1974*. Has something of an omnibus application.
*Petroleum Acts and Regulations*. Empowers conditions for safekeeping of petroleum spirit during storage and in transit. Also storage and conveyance of combustible gases.
*Local Authority Acts*. Individual items – making safe of derelict fuel tanks, access for brigades, control of places for music and dancing, etc.

# 26

# *Industrial accident report form*

| Date and Time | | Clock Number | Surname | Place of accident | Location of injuries |
|---|---|---|---|---|---|
| of Report | of Injury | | | | |
| | | | | | |

Stated cause
.................................................................................................................................

| For use of supervision Occupation | Shift | From To | Action after accident – delete as necessary off work/light work/own job |
|---|---|---|---|

What was the accident and how did it happen? ...........................................................
.................................................................................................................................
.................................................................................................................................
.................................................................................................................................
.................................................................................................................................

If due to machinery state–
1 Part of m/c causing injury
2 In motion by mechanical power. YES/NO
3 Type of lifting m/c or crane.

Witnesses 1 Clock number
2 Clock number
Exact location

Action taken to prevent recurrence
.................................................................................................................................
.................................................................................................................................
.................................................................................................................................
.................................................................................................................................
.................................................................................................................................
.................................................................................................................................

Signature Supervisor Date

Management comment and action ...........................................................................
.................................................................................................................................
.................................................................................................................................
.................................................................................................................................

Signature Date

# 27

# *Bomb threat: checklist for telephone operators*

Instructions to the telephonist should include:

1 Let the caller finish his message without interruption.
2 Get the message exactly – bearing in mind the points shown on the pro forma.
3 If it is possible to tie the supervisor or another operator into the conversation, do so.
4 Ensure senior management or a predesignated person are told exactly the contents of the call as soon as possible.
5 If the caller is apparently prepared to carry on a conversation, encourage him to do so and try to get answers to the following:

   (a) Where has the bomb been put?
   (b) What time will it go off?
   (c) Why has it been done?
   (d) When and how was it done?

In general, if the caller is prepared to continue, try to get him to talk about possible grievances as they affect the firm and anything which bears upon the truthfulness of the message and the identity of the caller.

It is essential that senior management should be told as soon as possible so that there is no delay in implementing policies and procedures.

Signal your supervisor and conform to prearranged drill for nuisance calls: tick through applicable words below, insert where necessary.

FOR GUIDANCE OF TELEPHONE OPERATOR
BOMB THREAT CHECKLIST (n.b. 99 per cent are hoaxes)

Signal your supervisor and conform to prearranged drill for nuisance calls: tick through applicable word below, insert where necessary.

Time .................................. Date ..........................................................................

Origin: STD Coin box Internal

Caller: Male Female Adult Juvenile

| VOICE | SPEECH | LANGUAGE | ACCENT | MANNER | BACKGROUND NOISES |
|---|---|---|---|---|---|
| Loud | Fast | Obscene | Local | Calm | Factory |
| Soft | Slow | Coarse | Regional | Rational | Road traffic |
| Rough | Distinct | Normal | Foreign | Irrational | Music |
| Educated | Blurred | Educated | | Coherent | Office |
| High pitch | Stutter | | | Incoherent | Party atmosphere |
| Deep | | | | Deliberate | Quiet |
| Disguised | | | | Hysterical | Voices |
| | | | | Aggrieved | Other |
| | | | | Humorous | |
| | | | | Drunken | |

Text of conversation:

# 28

# *Bomb threat: checklist of general precautions*

1 Inaugurate a registration system to immediately identify bona fide employee cars – if not already in existence.
2 Control entrance of visitors, suppliers and contractors to the site. Visitor and vehicular passes can be used to ensure a record and that the person and/or vehicle leaves.
3 Do not allow visitors to enter upon any pretext without prior confirmation that they are expected/welcome. Arrange collection from reception point in cases of doubt.
4 Review physical protection of buildings, i.e. adequacy of fencing, external lighting, doors, ground-floor windows, fire escapes, alarms, etc.
5 Ensure that the standard of housekeeping around buildings is such that unfamiliar objects will at once become noticeable.
6 Restrict parking outside particularly important facilities, i.e. computer installations, gas terminals and sub-power stations.
7 Establish a central control point which will not be evacuated, unless there is a substantiated positive danger to it. This should have means of communication with outside authorities should the firm's switchboard be obliged to close down and it should be easily accessible to incoming police and fire services.
8 In the event of a bomb warning being given, small rough parcels and plastic-type shopping bags left in odd corners or near entrances should be at once suspect, likewise unfamiliar cars parked haphazardly, especially those with Irish registration plates.

# 29

# *IPSA Ethical Code of Conduct*

This Ethical Code of Conduct is prescribed by the International Professional Security Association in the absence of legislative control over persons providing a service or otherwise employed in the security industry. It consists of those basic requirements deemed essential to ensure that members involved in the field of industrial, commercial or domestic security maintain proper standards.

Members of whatever category must uphold the aims, objects and purposes of the Association. Failure to do so or adhere to the principles set out below shall lead to consideration of suitability for membership.

Members shall, where appropriate to the category of membership:

Article 1 Maintain proper records concerning contracts leading to the provision of services or equipment. Full details of that service or equipment shall be kept together with information related to any person employed.

Article 2 Provide an administrative office where such records, together with all professional and business documents, certificates, correspondence, files and the like, necessary to the proper conduct of business transactions, shall be kept.

Article 3 Prepare annual accounts, certified by an accountant or solicitor, with complete details of expenditure and income.

Article 4 Be prepared to submit business records and accounts for examination by the International Council or suitably-qualified nominee.

Article 5 Ensure the confidentiality of information received whilst tendering for or carrying out any contract to supply a service or equipment. This applies not only at the time but subsequently and agreements entered into later shall not override this obligation.

Article 6 Act at all times, in a manner consistent with a client's interests, to protect and enhance the image and reputation of the International Professional Security Association.

*Relationship with the police service*

Article 7 Not enter into any commitment assuming the powers and authority of the civil police.

Article 8 Endeavour to maintain good relations and liaison with members of the police service and co-operation, in particular with those involved in the prevention and detection of crime and community relations.

Article 9 Ensure the local crime prevention officers are supplied with details of the nature of any company formed and security service offered.

Article 10 Act at all times in a competent and professional manner which does not contravene any respect of the criminal or civil law.

*Insurance cover*

Article 11 Take out public and employer liability insurance cover at a level commensurate with the nature of the business undertaken and number of persons employed.

Article 12 Use only insurance approved vehicles for the carriage of cash and other valuables.

*Vehicles and other equipment*

Article 13 Ensure that vehicles and other equipment are suitable for the use intended and operated in accordance with the law.

Article 14 When using equipment in connection with the undertaking or supplying equipment to a client ensure that it conforms to approved standards.

*Persons employed*

Article 15 Employ only persons of competence and integrity preferably on a full-time basis. Where employment is on occasional or part-time basis pre-employment checks must be thorough, comprehensive and of no lesser standard than procedures adopted in respect of full-time employees.

Article 16 Supervise all employees in accordance with health and safety legislation and to ensure proper conduct of the undertaking.

Article 17 Not employ any person on a part-time basis who is in receipt of unemployment benefit contemporaneously.

Article 18 Not require any employee to work abnormally long hours to the detriment of health or efficiency. Only in exceptional circumstances should the working day exceed twelve hours.

Article 19 Provide training of such duration and scope as is compatible with the efficient discharge of the task involved and safety of the employee. That training shall include instruction on at least the basic principles of security work, powers and authority at law, first aid and fire fighting preferably given at periods divorced from normal work hours.

Article 20 Carry out full pre-employment enquiries to ensure that only suitably qualified persons are engaged. To this end an application form, approved by the International Professional Security Association, must be used which requires the support of not less than three referees. The Code of Conduct for persons employed in industrial, commercial and domestic security must be reproduced on the application form and acknowledged by the prospective employee as a condition of employment.

Article 21 When making unsolicited telephone calls to consumers in order to promote the sale of security systems, members should observe the following guidelines:

(a) Callers should identify themselves and their

company and make clear the purpose of the call at the start of the call. They should ask whether the timing of the call is convenient. Calls should not be made after 9 p.m.

(b) Callers should not ask consumers for details of their security arrangements and should not play on consumers' fear of intruders, or otherwise mislead, exaggerate or use partial truths. They should answer any questions honestly and fully.

# *Index*